Advances in Industrial Control

Other titles published in this series:

Digital Controller Implementation and Fragility
Robert S.H. Istepanian and James F. Whidborne (Eds.)

Optimisation of Industrial Processes at Supervisory Level
Doris Sáez, Aldo Cipriano and Andrzej W. Ordys

Robust Control of Diesel Ship Propulsion
Nikolaos Xiros

Hydraulic Servo-systems
Mohieddine Jelali and Andreas Kroll

Model-based Fault Diagnosis in Dynamic Systems Using Identification Techniques
Silvio Simani, Cesare Fantuzzi and Ron J. Patton

Strategies for Feedback Linearisation
Freddy Garces, Victor M. Becerra, Chandrasekhar Kambhampati and Kevin Warwick

Robust Autonomous Guidance
Alberto Isidori, Lorenzo Marconi and Andrea Serrani

Dynamic Modelling of Gas Turbines
Gennady G. Kulikov and Haydn A. Thompson (Eds.)

Control of Fuel Cell Power Systems
Jay T. Pukrushpan, Anna G. Stefanopoulou and Huei Peng

Fuzzy Logic, Identification and Predictive Control
Jairo Espinosa, Joos Vandewalle and Vincent Wertz

Optimal Real-time Control of Sewer Networks
Magdalene Marinaki and Markos Papageorgiou

Process Modelling for Control
Benoît Codrons

Computational Intelligence in Time Series Forecasting
Ajoy K. Palit and Dobrivoje Popovic

Modelling and Control of Mini-Flying Machines
Pedro Castillo, Rogelio Lozano and Alejandro Dzul

Ship Motion Control
Tristan Perez

Hard Disk Drive Servo Systems (2nd Ed.)
Ben M. Chen, Tong H. Lee, Kemao Peng and Venkatakrishnan Venkataramanan

Measurement, Control, and Communication Using IEEE 1588
John C. Eidson

Piezoelectric Transducers for Vibration Control and Damping
S.O. Reza Moheimani and Andrew J. Fleming

Manufacturing Systems Control Design
Stjepan Bogdan, Frank L. Lewis, Zdenko Kovačić and José Mireles Jr.

Windup in Control
Peter Hippe

Nonlinear H_2/H_∞ Constrained Feedback Control
Murad Abu-Khalaf, Jie Huang and Frank L. Lewis

Practical Grey-box Process Identification
Torsten Bohlin

Control of Traffic Systems in Buildings
Sandor Markon, Hajime Kita, Hiroshi Kise and Thomas Bartz-Beielstein

Wind Turbine Control Systems
Fernando D. Bianchi, Hernán De Battista and Ricardo J. Mantz

Advanced Fuzzy Logic Technologies in Industrial Applications
Ying Bai, Hanqi Zhuang and Dali Wang (Eds.)

Practical PID Control
Antonio Visioli

(continued after Index)

Arun K. Samantaray • Belkacem Ould Bouamama

Model-based Process Supervision

A Bond Graph Approach

Springer

Arun K. Samantaray, PhD
Department of Mechanical Engineering
Indian Institute of Technology
721302 Kharagpur
India

Belkacem Ould Bouamama, PhD
LAGIS-UMR CNRS 8146
Ecole Polytechnique de Lille (Polytech'Lille)
USTL, Cité Scientifique
59655 Villeneuve d'Ascq Cedex
France

ISBN 978-1-84800-158-9 e-ISBN 978-1-84800-159-6

DOI 10.1007/978-1-84800-159-6

Advances in Industrial Control series ISSN 1430-9491

British Library Cataloguing in Publication Data
Samantaray, Arun K.
Model-based process supervision : a bond graph approach. -
(Advances in industrial control)
1. Process control - Mathematical models 2. Bond graphs
I. Title II. Bouamama, Belkacem Ould
670.4'27'015115
ISBN-13: 9781848001589

Library of Congress Control Number: 2007941548

Cover design: eStudio Calamar S.L., Girona, Spain

Printed on acid-free paper

9 8 7 6 5 4 3 2 1

springer.com

Professor Emeritus O.P. Malik
Department of Electrical and Computer Engineering
University of Calgary
2500, University Drive, NW
Calgary
Alberta
T2N 1N4
Canada

Professor K.-F. Man
Electronic Engineering Department
City University of Hong Kong
Tat Chee Avenue
Kowloon
Hong Kong

Professor G. Olsson
Department of Industrial Electrical Engineering and Automation
Lund Institute of Technology
Box 118
S-221 00 Lund
Sweden

Professor A. Ray
Pennsylvania State University
Department of Mechanical Engineering
0329 Reber Building
University Park
PA 16802
USA

Professor D.E. Seborg
Chemical Engineering
3335 Engineering II
University of California Santa Barbara
Santa Barbara
CA 93106
USA

Doctor K.K. Tan
Department of Electrical Engineering
National University of Singapore
4 Engineering Drive 3
Singapore 117576

Professor Ikuo Yamamoto
The University of Kitakyushu
Department of Mechanical Systems and Environmental Engineering
Faculty of Environmental Engineering
1-1, Hibikino,Wakamatsu-ku, Kitakyushu, Fukuoka, 808-0135
Japan

To my wife Trupti, mother Urmilla, and daughters
Atishna (Kunmun) and Anouska (Chunmun).
-A.K.S.

To my wife Souheir, mother Fatma Doulmoul,
and children Fouzi, Nassim and Sirine.
- B.O.B.

Series Editors' Foreword

The series *Advances in Industrial Control* aims to report and encourage technology transfer in control engineering. The rapid development of control technology has an impact on all areas of the control discipline. New theory, new controllers, actuators, sensors, new industrial processes, computer methods, new applications, new philosophies..., new challenges. Much of this development work resides in industrial reports, feasibility study papers and the reports of advanced collaborative projects. The series offers an opportunity for researchers to present an extended exposition of such new work in all aspects of industrial control for wider and rapid dissemination.

All evolving engineering disciplines first create a body of fundamental knowledge and then move on to new problem areas. Control engineering has now reached this level of maturity and is tackling new theoretical and applications areas. The field of nonlinear systems is receiving much research attention as are the problems of industrial supervisory control. The twin drivers of research into supervisory control are the use of new technology (computer networks and distributed sensor networks, for example) and the search for theoretical techniques to describe and solve supervisory control application problems.

In this field, the *Advances in Industrial Control* series has recently published an optimisation approach to supervisory control of in-building traffic systems by Sandor Markon and colleagues (*Control of Traffic Systems in Buildings*). A monograph dealing with the second type of research impetus was that by Emilia Villani and colleagues entitled *Modelling and Analysis of Hybrid Supervisory Systems.* This particular entry to the series dealt with process industry applications, and is the first book in the series to use hybrid system techniques. Continuing with this trend of work on supervisory systems, we are pleased to welcome *Model-based Process Supervision: A Bond Graph Approach* by Arun Samantaray and Belkacem Ould Bouamama into the *Advances in Industrial Control* monograph series.

It is really surprising to learn that bond graph techniques have been developing over a forty-five year history since they were invented by Henry Paynter of MIT in 1959. Their main advantages are their generality of application to a very wide range of systems and their rigour and consistency. Thus, the bond graph rigorously describes the physical processes in a system, shows the cause and effect relationships active, and allows the further analysis of properties of the process being modelled. Although the pictorial representation of the bond graph may look rather abstract, software is now available for the production, analysis and transformation of bond graph information.

In this monograph on model-based process supervision, the authors have written a truly encyclopaedic account of how bond graphs can be used to provide the framework and techniques for process fault monitoring and analysis. It is particular noteworthy that Chapters 1–5 allow the reader to learn about the core bond graph, control, and fault detection concepts and some of these chapters even include problem sections so that the reader can test the skills they have learned. The remainder of the book describes a sequence of highly relevant advanced topics, and applications. A particular strength of the book is the many examples used to illustrate the techniques and the real-world case study material presented.

We believe the systematic and thorough development of bond graph, process modelling and fault detection methods presented by the authors will meet the requirements of a very wide range of readers. The early chapters of the book can be used in taught courses or for self-study whilst the later chapters may be more appropriate for the process control expert. Industrial engineers and control academics alike will be interested in the application and case study material presented. This is a very welcome addition to the *Advances in Industrial Control* monograph series that extends the available range of contributions on supervisory control methods for industry.

Industrial Control Centre
Glasgow
Scotland, UK
2007

M.J. Grimble
M.A. Johnson

Preface

Today's process engineering systems are of ever-increasing complexity. Modern automated industrial processes use complex control laws in order to improve the final production quality, the safety and the efficiency. Moreover, process automation includes process supervision to assist the human operator in the management of alarming situations, and to increase reliability, availability and safety of the process. The tasks of an integrated supervision platform are many fold: plant-wide monitoring, database and knowledgebase maintenance, fault detection, fault diagnosis and control, which has to include further some degree of fault tolerance. Therefore, design of new process engineering systems has to consider the complete system at all stages of the design.

Modern process engineering systems are, inherently, multidisciplinary. These processes are composed of interacting subsystems or parts from different engineering disciplines requiring an integration of process engineering, chemical engineering, mechanical engineering, electrical engineering and control engineering. The interactions between different subsystems of a process are often very complex. Therefore, a concurrent design approach is needed for analysis of such systems. Moreover, design of proper control laws almost always requires a well developed process model.

Various books have been written so far to address the above-mentioned issues in modern process supervision. Most of these books use abstract mathematical models instead of a structured representation suitable for multidisciplinary systems. However, in our opinion, the choice of a proper modeling method, which is suitable for the analysis of the multidisciplinary system's behavior, control synthesis and at the same time, clearly exposes the interplay between various subsystems, is very important because it can greatly reduce the development time. This is the reason for using bond graph models in this book.

Bond graphs were introduced as early as 1959 by Professor Henry Paynter of the Massachusetts Institute of Technology, Cambridge, U.S.A. Since then, application of bond graphs to various domains has seen rapid growth. Bond graph (BG) modeling has also been successfully applied to model various process engineering systems. Furthermore, BG based techniques have been developed for the analysis of structural control properties, sensor placements, fault diagnosis, system inversion, input-output decoupling, system identification, parameter estimation, model order reduction, robustness studies and actuator sizing; to name a few. These recent developments in the field of systems and control engineering have been integrated in this book to develop reliable supervision systems.

One aim of this book is to present the in-depth state of the art of applications of bond graph modeling to fault diagnosis and control of process engineering systems while giving adequate importance to the broader spectrum of engineering applications. This book arose from our individual experiences in research and in teaching spanning more than two decades. During our professional lives we have had the opportunities to interact personally with many leading personalities who have made significant contributions to the fields covered in this book. Consequently, these interactions have shaped our research directions and motivated us to write this book.

From our teaching experiences as well as those of our colleagues, we understand that bond graph based teachings of physical system modeling and model based control are well appreciated by student communities because the methodologies are graphical, follow a step-by-step approach to model building and more importantly, retain close relationship to the physical system so that mathematical complexities do not obscure the ability to analyze and reason intuitively. This book is written for the student and research community who are interested in studying bond graph model based control and supervision.

Finally we would like to express our thanks to the following persons. Professor Amalendu Mukherjee of Indian Institute of Technology-Kharagpur and Professor Genevieve Dauphin-Tanguy of Ecole Centralle de Lille introduced bond graph modeling, respectively, to the first and the second authors. Their constant encouragement and support are the pillars which have sustained and fostered our research activities. Prof. Marcel Staroswiecki introduced us to the field of process supervision and his initiatives were instrumental in our various collaborative works. Our research students, Dr. Kamel Medjaher and Dr. Sanjoy K. Ghoshal, have carried out most of the work presented in this book. Their works have served as the basic skeleton on which we have structured this book.

In no lesser terms, we acknowledge the support we have received from Professors Ranjit Karmakar, Ranjan Bhattacharyya, Surjya Kanta Pal, Kingshook Bhattacharyya and Goutam Chakraborty of IIT-Kharagpur. We also appreciate the encouragement we received from Professors Rochdi Merzouki and Abdel Aitouche of University of Lille-1.

Last but not the least, we are greatly indebted to our families and friends for their unwavering tolerance and patience with us. Without their sacrifice and support, we would never have completed this work.

Kharagpur, Lille,
February 2008

Arun Kumar Samantaray
Belkacem Ould Bouamama

Contents

Abbreviations xix

1 Introduction to Process Supervision 1
1.1 Process Supervision 1
1.1.1 Basic Diagnosis Tasks 3
1.1.2 Fault, Failure and Safety 4
1.2 Diagnostic System 7
1.2.1 Specification of Diagnostic Systems 7
1.2.2 Classification of Diagnostic Systems 8
1.3 Organization of the Book 11

2 Bond Graph Modeling in Process Engineering 13
2.1 The Bond Graph Methodology 13
2.1.1 Introduction 13
2.1.2 Concepts and Definitions 13
2.1.3 Why Use Bond Graphs? 18
2.2 Generalized Variables in Bond Graph Models 19
2.2.1 Power Variables 19
2.2.2 Energy Variables 20
2.2.3 Word Bond Graph and Block Diagram 21
2.3 Pseudo Bond Graph 22
2.3.1 Why Pseudo Bond Graph? 22
2.3.2 Pseudo Power Variables 24
2.3.3 Pseudo Energy Variables 25
2.4 Basic Bond Graph Elements 26
2.4.1 One Port Passive Elements 26
2.4.2 Active Elements 37
2.4.3 Junctions 38
2.4.4 Transformers and Gyrators 41
2.4.5 Information Bonds 43
2.5 Causality 43
2.5.1 Introduction 43
2.5.2 Sequential Causality Assignment Procedure (SCAP) 45

2.5.3 Bicausal Bond Graphs ... 47
2.5.4 State-space Equations ... 48
2.5.5 Model Structure Knowledge ... 50
2.6 Single Energy Bond Graph ... 52
2.6.1 Bond Graphs for Mechanical Systems ... 52
2.6.2 Bond Graphs for Thermal Processes ... 52
2.7 Formal Generation of Dynamic Models ... 59
2.7.1 Bond Graph Software ... 59
2.7.2 Application ... 59
2.8 Coupled Energy Bond Graph ... 62
2.8.1 Representation ... 62
2.8.2 Thermofluid Sources ... 63
2.8.3 Thermofluid Multiport R ... 63
2.8.4 Thermofluid Multiport C ... 66
2.8.5 Application: Bond Graph Model of a Thermofluid Process ... 68

3 Model-based Control ... 81
3.1 Introduction ... 81
3.2 Classical Model-based Control ... 84
3.2.1 Conversion of Bond Graph Models to Signal Flow Graph Models ... 84
3.2.2 Transfer Function from State-space Models ... 91
3.2.3 Conversion of Bond Graph Models to Block Diagram Models ... 93
3.2.4 Example I: Physical Model-based Control ... 93
3.2.5 Example II: Physical Model-based System Design ... 95
3.3 Causal Paths ... 100
3.3.1 Transfer Functions from Bond Graph Models ... 101
3.3.2 Delay and Attenuation Dynamics ... 103
3.4 Augmented Controller and Observer Design ... 104
3.4.1 Pole Placement ... 104
3.4.2 Example: Active Flow-induced Vibration Isolation ... 107
3.4.3 Pole Placement Architecture in Bond Graph Models ... 109
3.4.4 Discrete-time Augmented Controller and Observer ... 111
3.4.5 Current Estimator ... 112
3.5 Structural Analysis of Control Properties ... 113
3.5.1 Structural Rank ... 113
3.5.2 Structural Controllability ... 114
3.5.3 Structural Observability ... 116
3.5.4 Example I: Two Spools in a Cylinder ... 118
3.5.5 Example II: A Hybrid Two-tank System ... 121
3.5.6 Example III: A Biomechanics Problem ... 124
3.5.7 Infinite Zeroes and Relative Degree ... 128
3.5.8 Zero Dynamics ... 133

4 Bond Graph Model-based Qualitative FDI 141
4.1 Model Order Reduction 141
4.2 FDI Using Bond Graphs and Qualitative Reasoning 154
4.2.1 Determination of Initial Fault Set 155
4.2.2 Fault Disambiguation 158
4.3 Qualitative Analysis Using Tree Graphs 159
4.4 Qualitative FDI Using Temporal Causal Graphs 163
4.4.1 Fault Hypothesis Generation 164
4.4.2 Fault Hypothesis Validation 166
4.5 Hybrid Diagnosis with Temporal Causal Graphs 169
4.6 Remarks on Model Linearization 170

5 Bond Graph Model-based Quantitative FDI 177
5.1 Introduction 177
5.2 Classical Quantitative FDI and Residual Generation 180
5.2.1 Observer-based Methods 181
5.2.2 Observer-based Residuals 183
5.2.3 Unknown Input Observers 185
5.2.4 Parity Space Residuals 191
5.3 Analytical Redundancy Relations and Fault Signature 195
5.3.1 Residual and Decision Procedure 195
5.3.2 The Fault Signature Matrix 196
5.4 Structured Approach to ARR Derivation 198
5.4.1 Behavior Model 198
5.4.2 Constraints and Variables 201
5.4.3 Derivation of ARRs 202
5.5 ARR Generation from Bond Graph Models 204
5.5.1 Constraints and Variables 204
5.5.2 Algorithm for Generation of ARRs 207
5.5.3 Example 209
5.6 Causality Inversion Approach for ARR Derivation 214
5.6.1 Example I: A Mechanical System 215
5.6.2 Example II: A Two-tank System 217
5.7 An FDI Application 218
5.7.1 Residual Evaluation and Fault Signature Matrix 218
5.7.2 Single Fault Hypothesis and Fault Isolation 220
5.7.3 Simulation Results 221

6 Application to a Steam Generator Process 229
6.1 Introduction 229
6.1.1 Process Description 229
6.1.2 Nomenclature 231
6.1.3 Word Bond Graph Model of the Process 233
6.2 Bond Graph Models of Steam Generator's Components 234
6.2.1 Bond Graph Model of the Storage Tank 234

6.2.2 Bond Graph Model of the Supply System 235
6.2.3 Bond Graph Model of the Boiler 236
6.2.4 Bond Graph Model of the Steam Expansion System 238
6.2.5 Bond Graph Model of the Condenser 239
6.2.6 Bond Graph Model of the Condensate Discharge Valves 243
6.3 Model Validation 244
6.4 Design of the Supervision System 248
6.4.1 Determination of Hardware Redundancies 249
6.4.2 Derivation of ARRs 250
6.4.3 Practical Fault Signature Matrix and Residual Sensitivity . . . 253
6.4.4 Effect of Hybrid Components 254
6.4.5 Selection of Decision Procedure 256
6.5 Online Implementation 257
6.5.1 Data Acquisition and Toolbox Integration 257
6.5.2 Native Interface 261
6.6 Experimental Validation of Fault Scenarios 262
6.6.1 Process Faults 262
6.6.2 Sensor Faults 265
6.6.3 Actuator Faults 266
6.6.4 Controller Faults 267
6.7 Reconfiguration 268

7 Diagnostic and Bicausal Bond Graphs for FDI 271
7.1 Diagnostic Bond Graph 271
7.1.1 Derivation of ARR 274
7.1.2 Example of a Non-resolvable System 276
7.1.3 Fault Signature Matrix from Causal Paths 280
7.2 Simulation and Real Time Implementation of the Residuals 281
7.2.1 Integrated System Simulation: Coupling the Models 282
7.2.2 Simulation Results 285
7.3 The Initial Conditions Problem 289
7.3.1 Order of Extra Derivatives 292
7.3.2 Fault Scenario Simulation 294
7.4 Matching Problems in Classical Bond Graph Modeling 294
7.4.1 Notion of Bicausality 298
7.4.2 Algorithm for ARR Generation and Construction of FSM . . 300
7.5 Example I: A Two-tank Process 300
7.5.1 Sensor Placement by Using Bicausal Bond Graphs 300
7.5.2 Residual Generation: Symbolic Method 304
7.5.3 Residual Evaluation and Fault Scenario Simulation 305
7.6 Example II: A Servo-valve Controlled Motor Transmission System . 306
7.6.1 System Description and Bond Graph Model 306
7.6.2 ARRs and FSM 308
7.6.3 Validation Through Simulation 310
7.7 The Fault Isolation Problem 311

8 Actuator and Sensor Placement for Reconfiguration ... 315
8.1 Introduction ... 315
8.1.1 Minimal Sensor and Actuator Placement ... 315
8.1.2 Sensor Placement for FDI and FTC ... 316
8.2 External Model ... 316
8.2.1 External Model in a Bond Graph Sense ... 317
8.2.2 Services ... 317
8.2.3 User Selected Operating Mode (USOM) ... 318
8.2.4 Operating Mode Management ... 319
8.3 Application to a Smart Pneumatic Valve ... 320
8.3.1 Description of the System ... 321
8.3.2 Bond Graph Model of the Smart Actuator ... 322
8.3.3 Missions and Versions ... 325
8.3.4 Operating Mode Management of the Smart Actuator ... 325
8.3.5 Monitoring of the Smart Actuator ... 328
8.4 Reconfiguration of a Thermo-fluid System ... 329
8.4.1 Minimal Sensor and Actuator Placement ... 329
8.4.2 Determination of Direct and Deduced Redundancies ... 332
8.4.3 Analytical Redundancy Relations and FSM ... 333
8.4.4 Sensor and Actuator Loss ... 335
8.4.5 Automaton Representation of Equipment Availability ... 336
8.4.6 Operating Modes of the Thermo-fluid System ... 338
8.5 Application to a Steam Generator Process ... 339
8.5.1 Operating Modes of the Steam Generator Process ... 340
8.5.2 Experimental Results ... 342

9 Isolation of Structurally Non-isolatable Faults ... 347
9.1 Introduction ... 347
9.2 Residuals and Robustness ... 348
9.3 Localization of Fault Subspace ... 350
9.4 Methodology for Single Fault Isolation ... 352
9.4.1 Parameter Estimation ... 352
9.4.2 Parallel Simulation of Bank of Fault Models ... 353
9.5 Application to a Controlled Two-tank System ... 355
9.5.1 ARRs and FSM ... 356
9.5.2 Parameter Estimation ... 359
9.5.3 Improvement of Isolability Using Bank of Fault Models ... 361
9.5.4 Validation Through Simulation ... 363
9.5.5 Qualitative Trend Analysis ... 365

10 Multiple Fault Isolation Through Parameter Estimation ... 373
10.1 Introduction ... 373
10.1.1 Adaptive Thresholds for Robust Diagnosis ... 374
10.1.2 Localization of Fault Subspace ... 379
10.2 Fault Isolation by Parameter Estimation ... 380

10.3 Example I: A Linear Two-tank System 383
10.3.1 Output Error Minimization 384
10.3.2 Optimization of Least Squares of ARRs 387
10.3.3 Optimization by Using Diagnostic Bond Graph 391
10.4 Example II: A Refrigerator Subsystem 393
10.4.1 Bond Graph Model and the ARRs 395
10.4.2 Fault Isolation Through Parameter Estimation 397
10.5 Example III: A Non-linear Two-tank System 402
10.5.1 The System and Its Bond Graph Model 402
10.5.2 Residual Generation and Fault Detection 404
10.5.3 Fault Isolation Through Parameter Estimation 405
10.6 Optimization by Using Residual Sensitivity 409
10.6.1 Gauss-Newton Optimization 411
10.6.2 Example 411
10.7 Sensitivity Bond Graphs 414
10.7.1 Diagnostic Sensitivity Bond Graphs 415
10.7.2 Example of the Use of Sensitivity Bond Graphs for FDI 417

11 Fault Tolerant Control 423
11.1 Introduction 423
11.2 Classical System Inversion Algorithms 425
11.2.1 Linear Time-Invariant (LTI) System Inversion 426
11.2.2 Implicit Inversion of Strictly Proper Systems 427
11.2.3 Examples of System Inversion 428
11.2.4 Example of Input Reconstruction 429
11.2.5 Example of Bond Graph Model Based Implicit System Inversion 431
11.2.6 Bond Graph Model Based Explicit System Inversion 432
11.2.7 Example of Bond Graph Model Based Explicit System Inversion 434
11.3 Parameter Estimation 435
11.4 Benchmark Problem: Active FTC of a Two-tank System 437
11.4.1 Fault Quantification with Single Fault Hypothesis 437
11.4.2 Fault Quantification with Multiple Fault Hypotheses 440
11.4.3 Fault Accommodation Through Fault Tolerant Control 442
11.4.4 System Inversion 443
11.4.5 Actuator Sizing 443
11.5 Passive FTC: Robust Overwhelming Control 447
11.5.1 Overwhelming Controller Design 447
11.5.2 Example: A Robust Level Controller 450

References 453

Index 467

Abbreviations

AEM	Abnormal Event Management
ANN	Artificial Neural Network
ARR	Analytical Redundancy Relation
BBG	Bicausal Bond Graph
BG	Bond Graph
CETF	Coupling Element for Thermo-Fluids
CUSUM	CUmulative SUM
CV	Control Volume
DB	DataBase
DBG	Diagnostic Bond Graph
DOF	Degrees of Freedom
DSS	Decision Support System
EKF	Extended Kalman Filter
FDI	Fault Detection and Isolation
FDT	Fuzzy Decision Tree
FSM	Fault Signature Matrix
FTC	Fault Tolerant Control
HILS	Hardware-in-the-Loop Simulation
I/O	Input-Output
KB	KnowledgeBase
LFT	Linear Fractional Transformation
LTI	Linear Time Invariant
MIMO	Multi-Input Multi-Output
MORA	Model Order Reduction Algorithm
MPC	Model Predictive Control
MSPC	Multi-variate Statistical Process Control
NLS	Non-linear Least Squares
ODE	Ordinary Differential Equation
OM	Operating Mode
PCA	Principal Component Analysis
PDE	Partial Differential Equation

PID	Proportional, Integral and Derivative
PMIGW	Pulsed-Metal Inert Gas Welding
QR	Qualitative Reasoning
QTA	Qualitative Trend Analysis
SCAP	Sequential Causality Assignment Procedure
SDBG	Sensitivity Diagnostic Bond Graph
SFG	Signal Flow Graph
SISO	Single-Input Single-Output
TB	ToolBox
TCG	Temporal Causal Graph
UIO	Unknown Input Observer
USOM	User Selected Operating Mode
XML	eXtended Markup Language

Chapter 1
Introduction to Process Supervision

1.1 Process Supervision

Modern process engineering plants use complex equipment, control laws, and myriads of operation sequences and instrumentation. In complex and safety critical systems, such as chemical plants, nuclear power plants and airplanes, total failure of a component can be extremely hazardous; the Chernobyl and Bhopal tragedies are indicators of the extent of such damage. Even partial failures or malfunctions of process components or instrumentation increase the operating costs of the plant. Gross failures, such as accidents, are definitely more serious. These failures could be due to faulty design, faulty operation and human errors, to which sabotage and acts of terrorism may be added these days.

Therefore, modern process engineering requires prompt detection and classification of process anomalies, which would minimize the overall repair time by assisting field and plant technicians in the diagnosis of system degradation. Moreover, it is essential to detect incipient faults and to locate the deteriorating or the deteriorated components as early as possible. In other words, some form of prediction of each device's condition and likelihood of its failure is necessary to undertake timely and appropriate maintenance work. The health (condition) of the system must be monitored at each and every instant and the purpose of the diagnostic system is to detect quickly any fault which could seriously degrade the performance of the system. A specialized field called condition monitoring and diagnosis is devoted to determine the life span of operating devices and helps in scheduling their maintenance. However, some devices tend to fail abruptly without showing any symptoms or warnings. Moreover, complex control laws often compensate for and conceal the faults. That is why real time process supervision assumes a central role in timely fault detection and isolation of complex systems. Real-time detection and diagnosis of faults while the plant is still operating in a controllable region can help avoid abnormal event progression and reduce productivity loss.

Automated Fault Detection and Isolation (FDI) procedures implemented in the supervision platform are meant to ensure safe operation of industrial processes, because faults in a process will often cause an undesired sequence of events and the consequences could be damaging to the plant, the personnel and the environment. Over the years, automated production systems have aided operators in controlling the process in order to improve the final production quality, the safety and the efficiency of industrial units. Recently, another challenge has appeared in the form of automation of the supervision aimed at providing assistance to the human operator in the management of alarming situations to increase reliability, availability and safety of the process. Therefore, the design of complex dynamic systems has to take into account additional functionality such as maintenance, management and supervision facilities.

Thus, FDI algorithms are central components of an Abnormal Event Management (AEM) system, which deals with the timely detection, diagnosis and correction of abnormal conditions in a process. Figure 1.1 depicts the components of a general fault diagnosis framework where a controlled system and the different sources of failures in it are indicated.

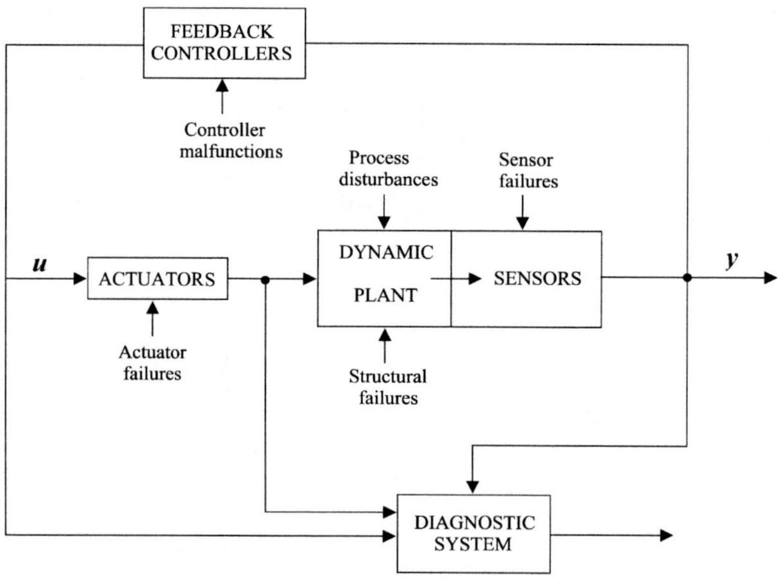

Fig. 1.1 A general diagnostic framework

Process supervision is performed by means of a set of tools and methods, which ensure safe process operation in the normal situation as well as in the presence of failures or undesired disturbances. The primary activities concerned with process supervision are Fault Detection and Isolation (FDI level), diagnosis (determining the root cause of the fault) and the decision making to accommodate the fault, whenever possible (fault accommodation level). Fault accommodation is done through reconfiguration, *i.e.* by using alternate standby devices called hardware redundancies, and/or Fault Tolerant Control (FTC), *i.e.* by changing the control laws and the associated control problem to maintain the desired output. The presence of a fault is detected at the monitoring level, which determines whether the process is in normal operation or not. Other tools associated with diagnosis and further high level tasks are executed only after detection of an abnormal process state.

1.1.1 Basic Diagnosis Tasks

The diagnosis problem can be divided into four distinct phases:

- Fault detection/monitoring: this concerns the activities to determine if the process dynamics has deviated beyond an acceptable limit from its normal operation model. If an unacceptable process behavior is detected then an alarm state is declared.
- Fault isolation: if there is an alarm then the objective of this stage is to locate one or more faulty component(s), called fault candidate(s). This stage further concerns filtering of the fault candidates such that they are kept to a minimum and also filtering out false alarms (due to process and measurement uncertainties). One or more decision procedures are used in parallel (multiple decision procedures, whose output are matched), series (a multi-stage procedure, where output of one decision procedure is filtered by another and so on), or hybrid (a network of parallel and series decision procedures).
- Fault quantification/identification: the aim of this stage is to determine the severity of the fault and its type. If the fault is due to a parametric change and it is not too severe, then the approximate values of the parameter has to be estimated. If the fault is due to a complete failure (hard failure), which changes the system morphology or its associated model structure, then the new system model has to be identified. There are several system identification and parameter estimation techniques available in control theory, which are readily used for fault quantification. If the results of fault quantification indicate that a component is so severely damaged that it cannot be further used even with acceptable degraded functionality, then the faulty component is removed from a database of available equipment maintained by the AEM. Note that as and when a previously faulty component is repaired or replaced with a new one, and it is ready for use, it has to be manually added to the equipment availability database.
- Fault accommodation: the decision regarding whether a fault can be accommodated or not is taken at this stage. If a fault cannot be accommodated, then the AEM system must determine the further sequence of events (*e.g.* to enter a controlled shut-down phase); if a fault can be accommodated, then the AEM system must determine how and to what extent it can be done. The degree of fault severity is used at this stage to determine whether to use process reconfiguration, FTC or manual supervision. Furthermore, development of control laws for FTC

requires an explicit model and approximate values of model parameters with limited uncertainty, both of which have to be available from the fault identification stage.

1.1.2 Fault, Failure and Safety

Fault is defined as a departure from an acceptable range of an observed variable or a calculated parameter associated with a process [265]. The underlying cause of a fault is called the basic event or the root cause. Unstructured uncertainties, *e.g.* process noise and measurement noise, are mainly faults that are not modeled *a priori* and are outside the scope of fault diagnosis. Process noise refers to the mismatch between the actual process and the predictions from model equations, whereas measurement noise refers to high frequency additive components in the sensor measurements. Usually, unstructured uncertainties are filtered out (*e.g.* Kalman filter) or decoupled (*e.g.* robust control) before they are used in a diagnosis system, or they are accommodated within the FDI and alarm filtering stages.

Different malfunctions can be classified into three types as follows:

- Gross parameter changes or parametric faults arise from disturbances to the process due to exogenous (independent) variables, whose dynamics are not provided with that of the process. An example of such a malfunction is a change in the heat transfer coefficient due to fouling in a heat exchanger. Here, the parameter for heat transfer coefficient is an exogenous variable. Another example of such a malfunction is a blockage in a pipeline resulting in modification of the overall discharge coefficient. This type of fault is also termed a multiplicative fault, because in the differential equation of the system, functions of system parameters are multiplied with the state variables.
- Structural changes refer to hard failures in equipment, which may change the model structure, *i.e.* disturb the information flow between various variables. The diagnostic system would require remodeling, *i.e.* the model equations have to be restructured in order to describe the present faulty state of the process. A failure of a controller, a broken transmission system, a stuck valve or a broken pipe may be categorized under this class of failure. Note that a model for a hybrid process is usually an ensemble of several models, each corresponding to a particular structure of the process. The system transits from one model structure to another depending on the evolution of its variables; these transitions and the corresponding conditions are generally represented in the form of a state-flow-graph. On the contrary, structural changes due to a fault results in new process dynamics, which may be known (*i.e.* a model exists for this kind of fault) or unknown (*i.e.* no corresponding model is immediately available for the new process structure).
- Malfunctioning of sensors and actuators may result due to a constant bias, intermittent disturbance, saturation, or an out of the range failure. Some feedback

signals, which are essential for the control of a plant, are provided by the measurement instruments. A failure in one of those instruments could cause the instability of the plant and seriously degrade the plant performance unless the failure is detected promptly and remedial actions are taken in time. These faults are also referred to as additive faults. In general, additive faults refer to undesirable change of inputs and outputs, and undesirable (often unknown) inputs, *e.g.* environmental effects.

A component's malfunction is called a fault when it is possible to take appropriate measures to recover from it without replacing the component, *i.e.* the fault can be accommodated through fault tolerant control. When the malfunctioning of the component is too severe, or there is a structural change to the system, we usually refer to it as a failure. A failure can be accommodated only through system reconfiguration, *i.e.* replacing the malfunctioning device by other devices rendering similar function if appropriate redundant hardware is/are available. Classification of different malfunctions as faults and failures depends on the specific process under consideration and its instrumentation architecture.

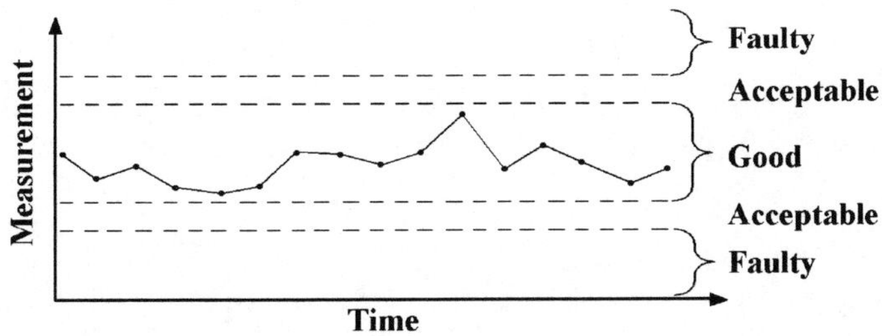

Fig. 1.2 One-dimensional limits on process variables

Because a fault is defined as departure from the acceptable behavior, it is necessary to fix the threshold for acceptable behavior. This threshold may put directly on process measurements as shown in Figure 1.2, where points marked by '·' are process measurements acquired at fixed or variable intervals. Such a scheme would require separate thresholds on separate measurements and it becomes difficult to establish correlations between them. This is why a multi-dimensional threshold is more suitable.

A schematic representation of a cross-section of a multi-dimensional threshold and the various operating regimes of the process are shown in Figure 1.3, in which y_1 and y_2 are two independent measurements. The dotted lines in Figure 1.3 show the corresponding one-dimensional thresholds. During normal process behavior, the outputs (shown by ·'s) stay in zone-I. Faults and failures shift the outputs outside zone-I. If the outputs are in zone-II, then the fault can be accommodated to bring the process behavior inside zone-I. However, if the process behavior enters zone-III, then the fault cannot be accommodated. Zone-IV represents the unsafe process oper-

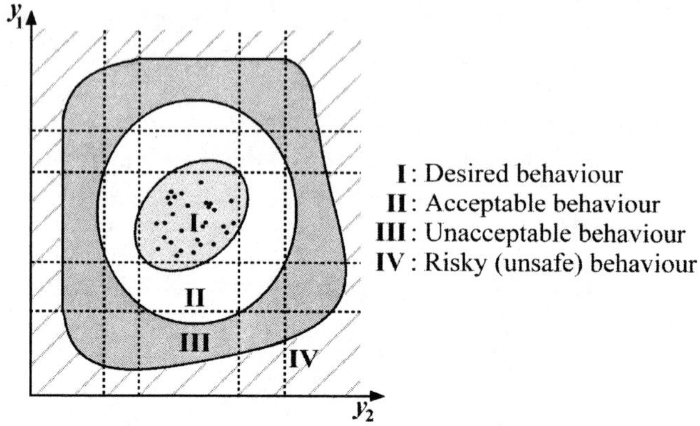

Fig. 1.3 Threshold zones in a cross-section of the space spanned by the process variables

ation. The aim of the safety system is to continuously monitor the process behavior such that it does not drift into zone-IV from zone-III.

Controller actions often keep the output within desired ranges even when one or more faults are present in a process; *i.e.* controllers can mask some fault effects.

Therefore, the phase-space defining various zones of process behavior must include the controller outputs and process inputs. In such a multi-dimensional phase-space, cross-sections in two-dimension can be examined to determine the process state *vis-a-vis* various operating zones. Often, various variables in this phase-space (inputs and outputs) are correlated among themselves. Therefore, the dimension of the phase-space can be reduced by selection of a set of basis vectors called principal components. Principal Component Analysis (PCA) is a statistical approach to FDI, in which deviation of various principal components (a projection or map of inputs and outputs) are analyzed vis-a-vis various fixed or time varying multi-dimensional thresholds (zones in Figure 1.3).

Furthermore, occurrence of faults may be classified into four different types [16, 107]:

- Abrupt, *i.e.* all of a sudden a normally operating system starts behaving abnormally
- Progressive, *i.e.* a component's behavior is gradually drifting away from its normal behavior
- Intermittent, *i.e.* a component behaves abnormally for a small duration of time and this abnormality is repeated at uncorrelated time intervals
- Incipient, *i.e.* a component's behavior is normal, but is continuously in the borderline between zones I and II in Figure 1.3

1.2 Diagnostic System

1.2.1 Specification of Diagnostic Systems

A common set of requirements or standards that a diagnostic system should possess [16] may be identified as follows:

- Quick detection and diagnosis: the diagnostic system should respond quickly in detecting and diagnosing process malfunctions. However, designing a system for quick response attenuates noise sensitivity and hence some sort of trade off is usually made to achieve tolerable performance.
- Isolability: the diagnostic system should be able to differentiate between various failures. Note that fault isolability depends on availability of various redundant information. Thus, there is a trade off between desired degree of isolability and the cost incurred in providing too much instrumentation. Furthermore, if an abnormality is detected, but cannot be isolated or classified, then the diagnostic system must be able to classify the new fault and identify this type of fault on further occurrence.
- Robustness: the diagnostic system should be robust to various disturbances, noises and uncertainties. Robustness needs often compromise with process performance.
- Adaptability: process operating conditions rarely remain the same over the passage of time due to normal wear and tear, environmental degradations, changing climatic conditions, and periodic maintenance work, *etc.* Therefore, the diagnostic system should be able to adapt to newer situations and operating conditions; *i.e.* in other words the diagnostic algorithm should not be hard-coded and it should be flexible.
- Operator assistance: a diagnostic system should not only provide the list of faults and their place of occurrence, but it should also assist the operator in explaining the origin of the fault, *i.e.* the root cause, and the sequence of the fault propagation from the root cause to the presently detected malfunction. The diagnostic system must be able to do a qualitative reasoning, *i.e.* hierarchically organize different cause and effect relationships, and convey them to the operator in simple language. This means the diagnostic system should clearly state why certain fault hypothesis has been proposed and why other fault hypothesis are not proposed [265]. A quantitative structured FDI approach is good at selection of appropriate fault hypothesis while rejecting some others, but the operators must be trained to interpret the results obtained from such structural analysis.
- Resource intensiveness: a diagnostic system should have low modeling requirements and low computational and storage requirements. Furthermore, it should be easy to implement and be adaptable to various situations.

As of now, there is no single method which satisfies all of the above-mentioned desirable characteristics. Therefore, recent research focuses its attention on development of hybrid diagnostic systems, where a diagnostic system comprises of several complimentary tools and the root cause analysis is done by using a voting scheme. European project CHEM [1] was one such attempt, under which several toolboxes were developed by using different diagnostic approaches and they were then used simultaneously to diagnose reliably faults in several industrial test cases.

1.2.2 Classification of Diagnostic Systems

The first level in FDI is the fault detection step, wherein the objective is to decide whether the system is in normal operating condition or not and, if not, to generate the appropriate alarms. *A priori* knowledge representation forms the core of this step, which receives the process measurements as inputs and produces a set of alarms as outputs.

The second level in FDI deals with alarm interpretation, *i.e.* deciding the actual fault and what its characteristics are (occurrence time, fault size, consequences, *etc.*). The input to this step is a set of alarms while the output is the isolated fault with its characteristic. Fault characterization often relies on explicit availability of fault models as *a priori* knowledge to the diagnostic system. Fault characterization and quantification is required to determine the process state and to determine whether the fault can be accommodated such that the process is able to carry on its mission.

Diagnostic search strategy is usually a function of the knowledge representation scheme, which in turn is influenced by the kind of *a priori* knowledge available. The basic *a priori* knowledge that is needed for fault diagnosis is the set of failures and the relationship between the observations or symptoms and the failures, *i.e.* the knowledge representation forms a set of causal or inferential relationships. A diagnostic system may have them explicitly or it may be inferred from some source of domain knowledge. The *a priori* domain knowledge may be developed from a fundamental understanding of the process using first-principles knowledge. Such knowledge is referred to as deep, causal or model-based knowledge. On the other hand, it may be gleaned from past experience with the process. This knowledge is referred to as shallow, compiled, evidential or process history-based knowledge. The model-based *a priori* knowledge can be broadly classified as qualitative or quantitative. The model is usually developed based on some fundamental understanding of the physics of the process. In quantitative models this understanding is expressed in terms of mathematical functional relationships between the inputs and outputs of the system. On the other hand, in qualitative models, these relationships are expressed in terms of qualitative functions centered around different units in a process [266]. In contrast to the model-based approaches where *a priori* knowledge about the model (either quantitative or qualitative) of the process is assumed, in process history based methods only the availability of large amount of historical process data is assumed. There are different ways in which this data can be transformed and presented as *a priori* knowledge to a diagnostic system [76, 139, 267]. This is known as the feature extraction process from the process history data, and is done to facilitate later diagnosis. This extraction process can be quantitative or qualita-

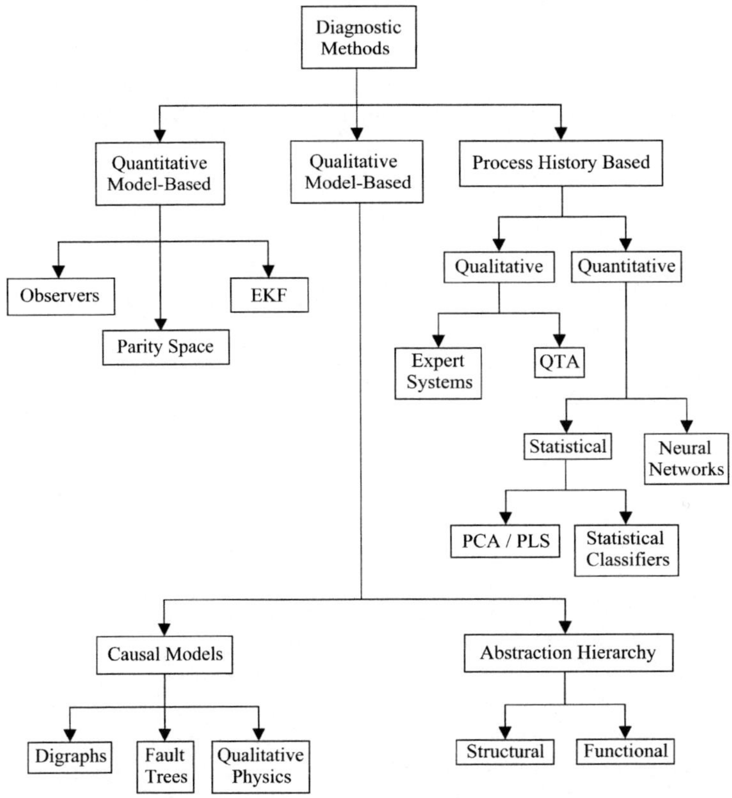

Fig. 1.4 Classification of diagnostic algorithms

tive feature extraction. In quantitative feature extraction one can perform either a statistical or non-statistical feature extraction.

The classification of diagnostic systems is shown in Figure 1.4. There is always an overlap between the various approaches. For instance, it is clear that all models need data for estimating some of the parameters in the model and all the methods based on process data need to extract some form of a model to perform fault diagnosis. Different approaches for the FDI procedures have been developed, depending on the kind of knowledge used to describe the process model. Each system or process has its own complexities, which need to be described by following a specific methodology. Venkatasubramaniam *et al.* [265–267] have classified various fault detection and diagnosis methods into quantitative model based methods, qualitative model and search based methods and process history based methods. Further sub-classifications and detailed reviews on each approach are given therein. Differentiation of quantitative, qualitative and process history methods, in the opinion of the authors of [265–267] and authors of this book, provides a classification in terms of the manner in which these methods approach the problem of fault diagnosis.

The quantitative approach relies on advance information processing techniques such as state and parameter estimation and adaptive filtering. The qualitative approach makes use of causal analysis (cause and effect/antecedents and consequences relationships), which links individual component malfunctions expressed in qualitative form with deviations in measured values. This approach can be used when precise numerical models are not available, especially in the design stage. We broadly classify FDI methods into two groups, namely, model based and non-model based.

Model based FDI methods require a mathematical model representing the behavior of the system. This mathematical model can be statistical or analytical. In this book, mainly analytical model based FDI methods are considered.

Modeling is an important and difficult step because of the complexity in representing the behavior of the monitored system along with its control equipment. Note that modeling errors seriously degrade performance of the FDI system. A model for FDI should neither be too complex (*i.e.* include minor or secondary dynamics) nor be too simple (*i.e.* exclude some essential dynamics). Therefore, there needs to be some trade-off to create a reduced order process model, which would reliably replicate essential process dynamics under given operating conditions.

Bond Graph modeling [20, 39, 46, 59, 127, 128, 172, 173, 199, 256, 257], which is a unified multi-energy domain representation and is also suitable for developing models of process engineering systems [258] involving various energetic couplings and control equipment, is used in this book. Note that bond graph modeling has attracted attention in the recent past for its application in various qualitative and quantitative FDI approaches [65, 72, 135, 150, 171, 189–192, 226, 233, 250, 284].

An important property of bond graph models relates to power conservation in a so-called junction structure, which neither consumes nor produces power. Therefore, when a modeler constructs a model from one viewpoint, *e.g.* dynamics (effort balance) or kinematics (flow balance), the other set is automatically satisfied by the constraints in a bond graph. This is why bond graph modeling is so useful in creating models of physical systems involving passive interactions, especially when different energy domains are interfaced and parts of the model can be modeled easily in some energy domains using effort constraints while other parts belonging to different energy domains may be modeled using flow constraints.

Online fault detection requires sensors, which capture the change in states of the system. This means that faults can be detected only when they occur in the observable subspace of the system's model. Moreover, to accommodate those faults, that subspace must be controllable. Therefore, controllability and observability properties assume central roles in the design of supervision systems. Controllability and observability properties of a system are elegantly derived from bond graph models.

1.3 Organization of the Book

This book presents the latest developments made in the field of bond graph modeling and its applications to process supervision. Fault detection and isolation of both linear and non-linear systems are considered.

There have been various recent developments in the field of control engineering by using various graphical models. This book utilizes those developments in the field of control engineering to develop FDI and fault accommodation algorithms. It is shown that the use of bond graphs helps to integrate many of the recently developed advances made in the field of control engineering into development of complex supervision systems. Special emphasis is given in this book to physical model based control and supervision.

This book follows a pedagogic approach to introduce various bond graph model based process supervision methods. The next four chapters are intended to develop a basic understanding of the subject. They may be part of an undergraduate course and therefore we have provided some unsolved problems at the end of these chapters. The rest of the chapters are on advanced topics and applications, which are more suitable for postgraduate level courses and as reference material for research.

The concept of bond graph modeling and the notion of causality are introduced in the second chapter. Emphasis is given to modeling of thermo-fluid and other process engineering systems.

The third chapter is devoted to introduce certain concepts in classical and modern control. We specifically introduce the principles of physical model based control and demonstrate the advantages offered by it during the design stage in development of complex control systems. Structural analysis and the notions of bicausality in bond graph models are introduced as tools for this purpose.

Fault detection by using qualitative reasoning methods is introduced in the fourth chapter. Therein, the importance of model order reduction is discussed followed by a few model order reduction schemes. The reduced order bond graph models are then used for fault detection and diagnosis through various qualitative analysis methods.

The fifth chapter forms the core of this book. In this chapter, the concepts of quantitative model based fault detection and isolation are introduced, which include observer based, parity relation based and frequency domain approaches. Thereafter, the analytical redundancy based fault detection and structural fault isolation are introduced. The analytical redundancy relations are first derived by following a classical approach and then the algorithm for their derivation from bond graph models is presented. Some examples are given to illustrate various steps involved in deriving the redundancy relations and to show how they can be used obtain residuals (fault indicators) for fault detection and isolation.

An industrial example is considered in the sixth chapter. The theory developed in the fifth chapter is used in this chapter to develop an actual supervision system. Practical results from an actual process are shown to illustrate the applicability of the developed approach. Moreover, the implementation issues (communication protocols, data storage, handling of multiple diagnosis tools, *etc.*) are also discussed.

The methodologies developed in the fifth chapter suffer from some inadequacies, which are particularly exposed when supervision of not so well instrumented processes is considered. Therefore, the concepts of the so-called diagnostic bond graphs and bicausalled bond graphs are introduced in the seventh chapter. These two developments allow for direct use of bond graph models in online process supervision because they solve the fault detection and isolation problem in the numerical domain. Moreover, the problem of sensor placement for fault isolation is solved in the seventh chapter with the aid of bicausalled bond graphs.

In the eighth chapter we deal with fault accommodation through reconfiguration. Management of process resources and selection of proper operating mode is discussed in this chapter. Operating modes are constructed by qualitatively synthesizing various functionalities offered by various healthy and redundant devices. Functional models, represented by external models, are integrated with bond graph models to add a quantitative/deterministic layer to operating mode management. Moreover, redundant hardware is determined through structural analysis and their availability in good health is determined from fault detection algorithms.

Sometimes it may not be possible to differentiate between two or more fault effects by using a purely structural approach. In this case, one way of fault disambiguation from a set of fault candidates is to use a bank of models and to simulate them with some fault hypotheses so that their outputs can be compared with the process outputs to isolate the actual fault. Such a way of diagnosis is developed in the ninth chapter by using bond graph models.

All the chapters leading up to the tenth chapter consider single fault hypothesis, *i.e.* how to detect and isolate a single fault occurring in a process. In a more general framework, more than one fault may occur in a process. While it is often possible to detect those faults (if the fault effects do not cancel each other), it is generally impossible to decouple the fault effects. One way of isolating multiple process faults is to estimate the process parameters by using the process response. This is formally called identification and parameter estimation based fault diagnosis. We introduce the basic parameter estimation techniques in the tenth chapter and show that the use of analytical redundancy relations for parameter estimation in a least-square-optimization problem is a better approach due to its speed and simplicity of implementation. Moreover, residual sensitivities may be obtained by representing the diagnostic bond graph of the process in its corresponding sensitivity bond graph form and it is shown to aid in quicker convergence of the optimization problem.

In the eleventh and final chapter, passive and active fault tolerant control problems are discussed. Bicausalled bond graphs are used in this chapter for parameter estimation and system inversion. Bond graph model based implicit and explicit system inversion schemes are used to construct the input sequences, which are further tested against actuator capacities before being used for active fault tolerant control of the plant. As a passive approach, a bond graph model based robust overwhelming controller is designed in this chapter to control process variables in the presence of parametric uncertainties.

Chapter 2
Bond Graph Modeling in Process Engineering

2.1 The Bond Graph Methodology

2.1.1 Introduction

Because of the multidisciplinary nature of most industrial processes (mechanical, thermal, electrical, ...), a unified modeling method is needed for analysis and model synthesis. The bond graph tool is well suited for this purpose. This methodology allows integrated modeling (independent of the physical nature of the studied system), precisely due to its graphical nature of display of the power exchange in a system, which include storage, dissipation and transformation [218]. Furthermore, the bond graph modeling methodology allows for the generation of not only a behavioral model [18, 27, 217], but also it can be used for structural and causal analysis which are essential to design control and monitoring systems [81, 85, 86, 122, 262]. A bond graph model can be refined by graphically adding more elements like thermal losses or inertia and storage effects, without having to start all over again. More significantly, from the simulation point of view, the causal properties of the bond graph language enable the modeler to resolve the algorithmic level of modeling (*e.g.* singularity, invertibility, *etc.*) in the formulation stage, even before the detailed equations have been derived. Furthermore, the structural and causal properties provided by this graphical representation can be used (as developed in this book) for design of supervision systems. Therefore, bond graph modeling may be considered as an integrated computer aided design tool in the field of system engineering.

2.1.2 Concepts and Definitions

Before introducing bond graph methodology, let us define the notions of multiport, bond and port. The multiport represents any component (DC motor, pump, tank, ...) designed to accomplish some specific missions or objectives. When the process is complex, the multiport is a subsystem where the inputs and outputs are a pair of variables associated with certain number of points at which the subsystem could be connected to other subsystems to form the process. Consider a feed water supply installation as shown in Figure 2.1a. The system is composed of several subsystems (electric motor, pump, storage tank, ...). To achieve its global function (supply water to the user), each subsystem (multiport) is designed to achieve a specific subfunction. When two subsystems are physically connected together, a pair of complementary variables (called *power variables*) are constrained to be equal for the two subsystems. When the electric motor is connected to the hydraulic pump, the torque and angular speed at the output of the motor must be identical to the torque and speed at the pump inlet. The power exchanged between two connected subsystems can be expressed as a product of the two power variables and it is represented as shown in Figure 2.1b by connections called power bonds or, simply, bonds. Two subsystems could as well exchange information (controller, computer, ...), which is represented by an information bond or signal as given in Figure 2.1c. In the system shown in Figure 2.1a, the valve is controlled by an external signal u_c, which is represented by an information bond (modulation) in the submodel used to represent the valve dynamics.

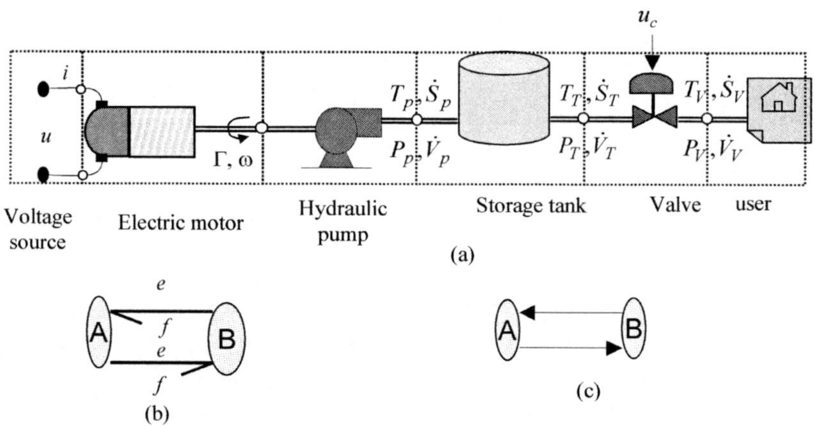

Fig. 2.1 Engineering multiport connection. **a** Feed water supply process. **b** Power exchange representation. **c** Information exchange representation

In bond graph theory, elementary multiport elements are represented by alphanumerical symbols and represent nodes in a bond graph. In Figure 2.2a, five multiports elements are represented by 1, R, I, Se and GY symbols. Those symbols represent, respectively, a junction (where energy conservation is a constraint), resistance, inertia, source of effort and gyrator elements. *Ports* (see Figure 2.2b) of each multiport are represented by line segments, which are used to connect multiport element to other multiports. For example, in Figure 2.2b, 1 is a four port element, R is a one port element and GY is a two port element. The *bonds* are formed when two ports are connected between a pair of elements. Connecting all elements, we obtain a bond graph model (Figure 2.2c) of the electric motor considered in Figure 2.1a.

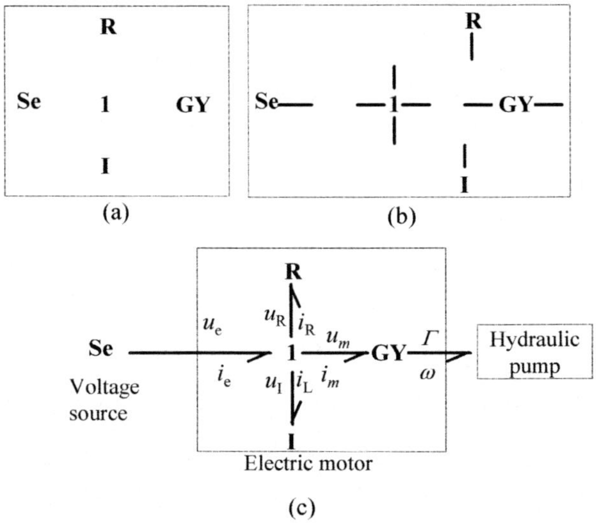

Fig. 2.2 Concepts and definitions. **a** Multiport elements. **b** ports and elements. **c** Bond graph

In this graphical model, Se represents a source of effort (voltage), and R and I are the resistance and the inductance in the electrical part, respectively. The transformation of electrical energy into mechanical energy is represented by GY element. The dynamical constraint, represented by 1, is a junction meaning that equal magnitude of current in equal sense passes through R and I (in serial connection).

Definition 2.1. A bond graph is a collection of multiport elements connected together. In the general sense, it is a linear graph whose nodes are multiport elements and whose branches are bonds. The key feature of bond graph modeling is the representation (by a bond) of exchanged power as the product of generalized efforts (e) and generalized flows (f) with elements acting between these variables and junction structures (algebraic constraints) to reproduce the global model as interconnected subsystems. The power variables are the generalized flow and the generalized effort variables as developed below.

In bond graph theory, bonds (or *power bonds*) represent power (consequently energy) exchange between connected multiports. This power is the product of two power variables: a potential variable (*e.g.* pressure, electrical potential, temperature, chemical potential, force, *etc.*) called generalized effort (e) and a current variable (*e.g.* volume flow, current, entropy flow, velocity, molar flow, *etc.*) referred to as generalized flow (f). The positive sense of power flow is represented by the half-arrow on the bond. The energy is exchanged according to the sense represented by a power direction (half arrow head) only when the two power variables have the same sign, *i.e.* either both are positive or both are negative; otherwise, the energy transfer is in the opposite direction to the assigned power direction. Choice of coordinate system is usually decided by the concerned engineering domain (*e.g.* Cartesian, cylindrical or spherical coordinate systems in mechanics, assumed direction of current flow in electrical networks, *etc.*). In some engineering domains it is very difficult to choose appropriate coordinate system (*e.g.* chemical engineering, thermodynamics). Further difficulty arises when subsystems belonging to different energy domains are to be coupled together. This is why, for multi-disciplinary model synthesis, power direction is taken as a generalized way of choice of coordinate system in bond graph theory.

The causality (cause and effect relationship) is the most important concept in bond graph theory. Indeed, the determination of causes and effects in the system is directly deduced from the graphical representation. In a bond graph model it is denoted by a cross-stroke at one end of a bond. The cause and effect relationship has physical meaning in most cases, but in general it represents which unknown variables can be calculated from which known variables. When causality is viewed and assigned from the point of view of eliminating unknown variables to obtain a mathematical model, then it is termed computational causality. This is equivalent to organizing left and right hand side variables in each equation and sequencing many such equations (with appropriate initialization) during development of any computer code.

Finally, as shown in Figure 2.3, the concise bond graph notation gives us four pieces of information: the *existence of physical link* between two systems by the bond, the *type of power* (electric, mechanical,...) by the power variables, the *power direction* by the half arrow and the *causality* by the stroke.

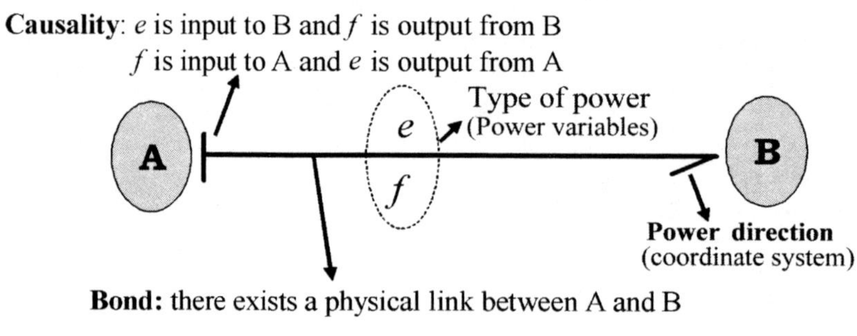

Fig. 2.3 Information portrayal in a bond graph representation

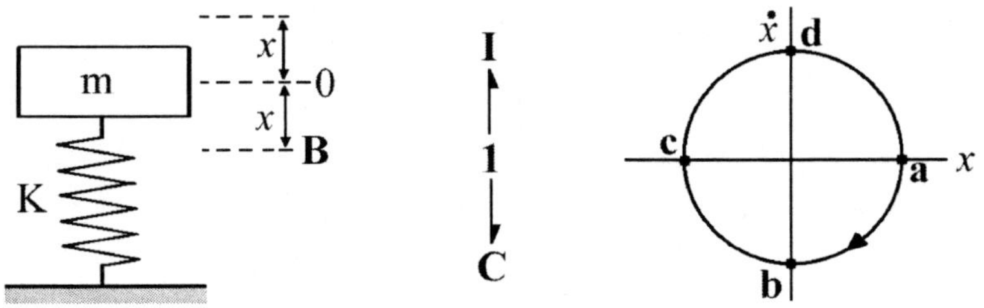

Fig. 2.4 Concept of power direction

Consider a simple spring-mass system as shown in Figure 2.4 to explore the concept of power direction. Assume that the spring-mass system is given an initial condition by pulling the mass upwards and then it is left to perform free vibrations. The bond graph model of the system is given by an I and a C-element connected to a 1-junction, which means that the velocity of the mass is the same as the rate of deformation of the spring. On the bond graph model, power directions show power flow towards the I and C elements. Then one may ask where does the power come from such that both the mass and the spring get positive power?

Consider the velocity *vs* displacement diagram given in the right side of Figure 2.4. The initial point is marked 'a', at which there is some initial displacement and no velocity. This represents the initial state, when the spring has maximum stored potential energy and the mass has zero kinetic energy. In bond graph terms, the spring now has a positive effort (tension). As the system is left from rest, it moves downwards, *i.e.* the flow variable in the spring is negative whereas the effort is still positive (the spring remains in extended state until it moves below the equilibrium position). Therefore, the product of effort and flow variables in the C-element is negative, *i.e.* the spring is loosing potential energy. At the same time, the mass point is moving downwards (negative flow) and the force acting on it is directed downwards by the pull of the spring (negative effort). Thus, the product of the two power variables corresponding to the I-element is positive, *i.e.* the mass is gaining kinetic energy. As the mass point moves further and passes the static equilibrium point 'b' as shown in Figure 2.4, then the spring starts being compressed (flow variable in the spring is negative) while the force acting on the spring is downwards (due to inertial effect). Therefore, after point 'b', the product of the two power variables in the C-element becomes positive, *i.e.* the spring starts to accumulate potential energy. At the same state, the mass is moving downwards (negative flow) whereas the force acting on it is directed upwards; thereby making the product of power variables negative, *i.e.* the I-element starts loosing the stored kinetic energy.

Power direction assigned to a bond is only a reference. Power transfer is in the indicated direction only when both effort and flow variables have the same sign; otherwise, it is in the opposite direction.

2.1.3 Why Use Bond Graphs?

The invention of bond graphs was driven by the need for a common language to model systems involving different energetic domains. Bond graphs were invented by Henry Paynter of MIT to replace intensive, but inefficient, use of block diagrams for servo-controls and simulation problems [199]. It has subsequently grown into a fully developed field. More recent references are [20, 30, 31, 46, 59, 127, 128, 172, 173, 257, 258]. A bond graph model can be introduced in a graphical form and simulated using special software [22, 35, 47, 68, 100, 174, 260]. This tool is widely used in the industry for mechatronics [100, 215, 262] and integrated system design [176, 177].

Four levels of modeling can be represented by a bond graph model:

1. *The technological level.* This level represents the architecture of the system by the assembling different sub-system models in iconic form, which correspond to different plant items (heat exchanger, boiler, pipe,...). The technological level can be represented by the so-called *word bond graph*. This means that the system is represented as interconnected subsystems. Unlike block diagrams (where the inputs and outputs are information variables), the interconnection in word bond graph is realized by the power variables (pair of efforts and flows).
2. *The physical level.* At this level the modeling uses an energetic description of the physical phenomena. One uses the basic concepts of physics such as dissipation of energy, transformation, accumulation, sources and so on by considering the relevant physical phenomena (inertia, friction, and compressibility,...), in a discretised form, to introduce the system dynamics through a graphical description. Here, the bond graph is used as a universal language for all the domains of physics.
3. *Mathematical level.* The mathematical model is represented by the mathematical equations (algebraic and differential equations) which describe the system behavior. The mathematical level from a bond graph is obtained by writing the constitutive equations of the components and constraints.
4. *Algorithmic level.* The algorithmic level is connected directly with information processing. This level indicates how the mathematical models are calculated. The algorithmic problem is solved in the bond graph methodology by systematic assignment of causalities, which are shown by the perpendicular lines at the ends of the bonds.

In conclusion, the advantages of bond graph modeling are that:

- It is a unique language for all physical domains.
- It clearly shows the cause and effect relations in the model.
- It allows further possible development and evolution of the model.
- It is also a tool for analyzing the system's structural properties.

2.2 Generalized Variables in Bond Graph Models

2.2.1 Power Variables

Power interactions are always present when two multiports are passively connected. In bond graph language, the various power variables are classified in a universal scheme so as to describe all types of multiports in a common language. Power variables are generally referred to as *effort* and *flow*; additional qualifications pertaining to the domain under consideration may also be included in the model. Table 2.1 gives effort and flow variables for some of the physical domains. The power P_u exchanged at the port is the product effort and flow:

$$P_u(t) = e(t)f(t) \tag{2.1}$$

This type of bond graph is then called true bond graph.

Table 2.1 Power variables in a true bond graph

Domain	Effort $e(t)$	Flow $f(t)$
Electrical	Voltage u [V]	Current i [A]
Mechanical rotation	Torque Γ [N.m]	Angular velocity ω [Rad/s]
Mechanical translation	Force F [N]	Velocity v [m/s]
Hydraulics	Pressure [Pa]	Volume flow rate $\dot{V}$ [m^3/s]
Thermodynamics	Temperature [K]	Entropy flow [J/(K.s)]
Chemical transformation	Chemical potential μ [J/mole]	Molar flow rate $\dot{n}$ [mole/s]
Chemical kinetics	Chemical affinity A [J/mole]	Rate of reaction $\dot{\xi}$ [mole/s]
Economics	Unit price P_u [$/unit]	Order flow rate f_c [units/period]

In the chemical systems, two kinds of power variables can be distinguished: those modeling the chemical transformation of the reaction, and those modeling the kinetic phenomena of the reaction. Regarding transformation phenomena of the reaction, the effort is the chemical potential μ and the flow is the molar flow $\dot{n}$ leading to a true bond graph. Concerning the kinetic phenomena of the reactions, the chemical affinity A (J/mole) is used as effort variable and the rate of reaction $\dot{\xi}$ (mole/s) is used as the flow variable.

Some historical antecedents apply analogy approaches to economic modeling using the equivalent Kirchoff laws. Recently, some researchers have applied bond graph modeling (usually applied to physical systems) to represent economic and sociological system dynamics [32–34]. In these works, as in the pseudo bond graph, the power is used as a flow variable.

Economic bond graph does not rely only on physical analogy but attempts to use a unified approach to integrated modeling. In economic systems, the effort variable is the *unit price* (\$/unit) denoted P_u. The flow variable is the *order flow rate,* denoted by f_c, which is of the same magnitude and has the same unit as the flow of commodity (units/period) [32]. The positive flow is in the same direction as the cash flow CF, which is given by

$$CF = P_u.f_c. \tag{2.2}$$

2.2.2 Energy Variables

Two additional physical quantities are used in bond graph modeling. They are called energy variables and are important for dynamic system representation and are associated with state variables. Two kinds of energy variables are used: *generalized momentum, p(t),* and *generalized displacement, q(t).* They are obtained by integration of the power variables with respect to time:

$$\begin{aligned} p(t) &\equiv \int_{-\infty}^{t} e(\tau)d\tau = p_0 + \int_{t_0}^{t} e(\tau)d\tau, \\ q(t) &\equiv \int_{-\infty}^{t} f(\tau)d\tau = q_0 + \int_{t_0}^{t} f(\tau)d\tau. \end{aligned} \tag{2.3}$$

Energy variables for some modeling domains are given in Table 2.2.

Table 2.2 Energy variables in a true bond graph

Domain	Impulse $p(t)$	Displacement $q(t)$
Electrical	Flux linkage λ [Vs]	Charge q [C] or [As]
Mechanical rotation	Angular momentum p_ω [N.m.s]	Angle ω [Rad]
Mechanical *translation*	Momentum p [N.s]	Displacement x [m]
Hydraulics	Pressure momentum p_p [Pa.s]	Volume V [m^3]
Thermodynamics	-	Entropy S [J/K]
Chemistry	-	Molar mass N [mole]
Economy	Economic momentum p_e	Accumulation of orders q_e

In chemical and thermal domains the time integral of efforts (temperature or chemical potential) do not have any direct physical meaning. This is why the generalized momentum *p(t)* is not defined in Table 2.2 for these domains.

Let us explain why those variables are termed "energy variables". Equation 2.3 can be rewritten in differential form as follows:

$$\frac{dp(t)}{dt} = e(t) \Rightarrow e(t)dt = dp(t),$$
$$\frac{dq(t)}{dt} = f(t) \Rightarrow f(t)dt = dq(t). \tag{2.4}$$

Energy $E(t)$ is computed by integration of the power:

$$E(t) = \int_{-\infty}^{t} P_u(\tau)d\tau = \int_0^t e(\tau)f(\tau)d\tau + E_0. \tag{2.5}$$

Based on Equation 2.4 (*e(t)dt=dp(t)* and *f(t)dt=dq(t)*), it is easy to express E as a function of energy variables q or p:

$$E(t) = \int_0^t e(\tau)\frac{dq(\tau)}{d\tau}d\tau + E_0 \text{ or } E(t) = \int_0^t f(\tau)\frac{dp(\tau)}{d\tau}d\tau + E_0. \tag{2.6}$$

As it will be shown later, the variables *e(t)* and *f(t)* can be expressed as functions of q and p : *e(t)=e(q)* and *f(t)=f(p)*. Finally energy E (Equation 2.6) can be rewritten as function of energy variables q or p:

$$E(q) = \int_{q_0}^{q} e(q)dt + E_{(q=q_0)}, \tag{2.7}$$

$$E(p) = \int_{p_0}^{p} f(p)dt + E_{(p=p_0)}. \tag{2.8}$$

For example, let us consider potential energy E_p stored by a spring. The power flow into the spring (port) leads to storage of energy E. If no energy is lost during the process then energy E is

$$E_p(t) = \int_0^t e(\tau)f(\tau)d\tau + E_0, \tag{2.9}$$

where effort e and flow f are respectively the force and the velocity. If the force is proportional to the displacement x ($F = kx$), where $x = q = \int f(\tau)d\tau$ is the energy variable, then the stored energy is

$$E_p(t) = \int_{q_0}^{q_1} e(q)dq = \int_{x_0}^{x_1} kxdx = k\left(x_1^2 - x_0^2\right). \tag{2.10}$$

2.2.3 Word Bond Graph and Block Diagram

Any process can be considered to be composed of interconnected subsystems [204, 226]. Engineers are more familiar with block diagram representation, where the input and output are both signals. Every block represents a functional relation (linear, non-linear, ...) between its inputs and outputs. A signal may not be real; it may be some abstraction made by the user. Essentially, a signal represents the causal signal to calculate some variables on the left hand side of an equation from the variables on the right hand side of that same equation. This representation neither requires nor ensures that the relations embedded in the block comply with the first principles of physics. The block diagram is, therefore, a computational structure and it does not reflect the physical structure of a system[145]. The block diagram of the system described in Figure 2.1 is given in Figure 2.5a, where the connection(s) between two subsystems represent only a signal. The equations of submodels are causal, which means that for the system in Figure 2.5a, the variable u must be known to compute the variable Γ.

Contrary to the block diagram representation, a word bond graph (Figure 2.5b) represents the physical structure of the system in which the inputs and outputs are the power variables. There is no need to decide whether an interface variable is an input or output, *i.e.* the interconnections are acausal.

The half arrow indicates the power direction. The computational structure can be superimposed on a bond graph by adding a perpendicular stroke to each bond at one of its ends. Then the calculation (simulation) structure can be deduced as it is shown in the causal word bond graph (Figure 2.5c). Depending on the stroke position, different causalities can be derived. Causality assignment rules are developed later.

The used input and output power variables depend on the assumed modeling hypothesis. If the thermal effect is considered in the process then the word bond graph would be as given in Figure 2.6. The additional power variables can be represented by the pair of power variables, temperature-entropy flow $(T-\dot{S})$ or temperature-enthalpy flow $(T-\dot{H})$.

Definition 2.2. A word bond graph is a bond graph where the vertices represent subsystems or technological components denoted by a word and the inputs and outputs are power variables.

2.3 Pseudo Bond Graph

2.3.1 Why Pseudo Bond Graph?

True bond graphs, as shown in Table 2.2, have the physical power as product of two conjugate effort and flow variables. More precisely, it is the instantaneous power exchanged through a bond between two ports. Such bond graphs are well suited for mechanical, hydraulic and electrical systems.

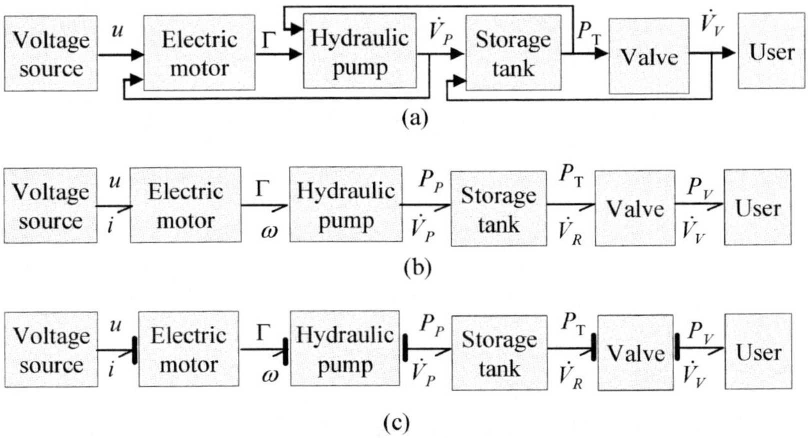

Fig. 2.5 Block diagram (**a**), and non causal (**b**) and causal (**c**) word bond graphs of the system in Fig. 2.1

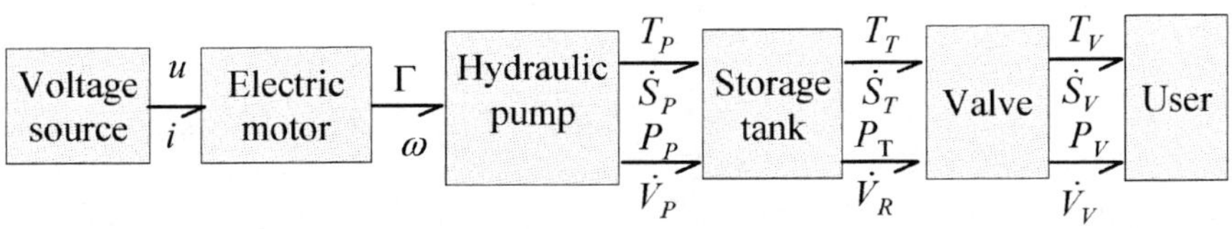

Fig. 2.6 Word bond graph including thermal effect

In process engineering systems, each plant item is associated with a set of process variables. For instance, in thermo-fluid process, the circulating fluid is characterized by a set of variables, namely pressure P and temperature T (or specific enthalpy h) for the efforts, volume flow $\dot{V}$ (or mass flow $\dot{m}$ for compressible fluid) and the entropy flow $\dot{S}$ for the flows. Other variables can be added such as the quality of steam X, internal energy flow $\dot{U}$ or enthalpy flow $\dot{H}$.

Furthermore, true bond graphs in thermal and chemical engineering introduce complex variables like entropy and chemical potential, which do not obey simple conservation laws. This is why, it is convenient to choose a *pseudo bond graph* (introduced by Karnopp [123]) where the product of effort and flow variables does not have the physical dimension of a power. However, in such a representation, the theory and the systematic derivation of equation from bond graph, causal and structural properties for control and monitoring analysis, *etc.* still remain applicable. Pseudo bond graph modeling is used in chemical, thermodynamic and thermal engineering. Several such applications are given in [258]. The selection of power variables in such process plants including thermo-fluid, biological and chemical phenomena is not trivial and depends on the type of the process.

2.3.2 Pseudo Power Variables

In complex processes, several phenomena (chemical, thermal, hydraulic) are coupled [82, 106, 123, 125, 126]. Thermofluid systems involve additional complexity in the modeling task, since the mass that flows through the process carries the internal energy which is stored in it, and which is thus transported from one location to another in a non-dissipative fashion. If one has to include chemical reactions in addition to thermal phenomena then three energy domains are involved in a general process engineering system. Power variables of process engineering models are thus presented in vectorial form as

$$E = \left[e_h\ e_t\ e_c\right]^T, \text{ and } F = \left[f_h\ f_t\ f_c\right]^T, \tag{2.11}$$

where e_t, e_h and e_c represent respectively the thermal effort (the specific enthalpy or the temperature), the hydraulic effort (the pressure), and the chemical effort (the chemical potential, chemical affinity or the concentration); and f_t, f_h and f_c represent respectively the thermal (or entropy) flow (by conduction $\dot{Q}$ or by convection $\dot{H}$, *i.e.* enthalpy flow), hydraulic flow (mass flow $\dot{m}$ or volume flow $\dot{V}$) and chemical flow (mole flow $\dot{n}$).

Remark 2.1. In this book, the term conduction will be used to describe heat transfer between the fluid (or the wall) and the wall and the term convection represents internal energy transported by the fluid, *i.e.* a forced convection phenomena.

Table 2.3 shows the pseudo power variables used in modeling of process engineering systems. More details and motivations for the specific selection of power variables is given in [258].

Table 2.3 Power variables in process engineering

Domain	Pseudo bond graph power variables	
	Effort (e)	Flow (f)
Chemical		
transformation	C [mole/m^3]	$\dot{n}$[mole/s]
kinetics	C [mole/m^3]	$\dot{n}$[mole/s]
Hydraulics	P [Pa]	$\dot{m}$ [kg/s]
Thermal		
conduction	T (K)	$\dot{Q}$ [J/s]
convection	T or h [J/kg]	$\dot{H}$ [J/s]

In the hydraulic domain, the pair $(\dot{m}, P)$ is preferred because $\dot{m}$ acts as a common variable for both incompressible and compressible fluid flow modeling, and if thermofluidics is considered, then the energies being coupled, the enthalpy flow $\dot{H}$ is calculated from mass flow $\dot{m}$. In the convection of thermal energy (from a Lagrangian point of view), the pair enthalpy flow-temperature(or specific enthalpy) $(\dot{H}, h \text{ (or } T))$ are used as power variables. In the case of thermal conduction (from the Eulerian point of view [15]), the pair $(T, \dot{Q})$ is used.

In chemical systems the concentration C is used as the effort variable instead of the chemical potential and the molar flow $\dot{n}$ is used as the flow variable. Other research works use the concentration of the substance as effort variable instead of the chemical potential and the derivative of the concentration as the flow variable. This approach is extensively used in [145] because it allows one to manipulate variables more intuitively (therefore easy to simulate) and the power variables (*e.g.* concentration) are measurable quantities.

2.3.2.1 Other Variables

In addition to power variables e and f, other variables can be expressed as functions of state variables or power variables. For example, the saturated steam pressure in a boiler may be expressed by using steam tables:

$$\begin{gathered} h_l(P) = \Sigma_{i=0}^{k} \alpha_i P^i, \quad h_v(P) = \Sigma_{i=0}^{l} \beta_i P^i, \\ \frac{1}{\rho_l(P)} = v_l(P) = \Sigma_{i=0}^{m} \gamma_i P^i, \quad \frac{1}{\rho_v(P)} = v_v(P) = \Sigma_{i=0}^{n} \delta_i P^i, \\ T(P) = \Sigma_{i=0}^{o} \varepsilon_i P^i, \quad L_V(T) = \Sigma_{i=0}^{p} \lambda_i P^i, \end{gathered} \tag{2.12}$$

where h, v, T and L are respectively the specific enthalpy, specific volume, temperature and latent heat; α, β, γ, δ, ε and λ are constants, and the subscripts l and v, respectively, represent steam and liquid. These coefficients in Equation 2.12 are determined experimentally by using least square techniques [258]. The temperature can be calculated using Equation 2.67 if the fluid is under saturated. In this regime c_P varies weakly with temperature and the density is considered constant.

- *Mixture quality X*

The mass of 1 kg of mixture with the quality X contains $(1 - X)$ kg of liquid and X kg of vapor. For $X = 0$, we are in the liquid state and with $X = 1$ we have only saturated dry vapor. X is a function of specific enthalpy h_M and specific volume v_M of the steam-water mixture: $X = f(v_M, h_M)$.

2.3.3 Pseudo Energy Variables

The energy variables (or state variables) in the pseudo bond graph of process engineering systems are time integral of mass flow $(m = \int \dot{m}dt)$, time integral of enthalpy flow $(H = \int \dot{H}dt)$, time integral of mole flow $(N = \int \dot{n}dt)$ and time integral of thermal flow $(Q = \int \dot{Q}dt)$. Consequently the energy variables (as summarized in Table 2.4) are:

- Mass m stored by any accumulator
- Total enthalpy H (or internal energy) stored in any heated accumulator
- Number of moles N accumulated in a reactor
- Thermal energy Q stored by any metallic body

The time integrals of temperature or chemical potential have no physical meaning. This is why impulse variables do not exist for thermal or chemical processes. To be more precise, impulse variables are associated with inertial motion. However, thermal processes are irreversible and the kind of chemical kinetics considered in this book is also irreversible, *i.e.* in such processes, there will be no inertial effects. Note that thermal equivalent of inertia is undefined, as it would violate the laws of thermodynamics.

Table 2.4 Energy variables in process engineering

Domain	Pseudo bond graph energy variables	
	Impulse (p)	Displacement (q)
Chemical	-	Number of moles N [mole]
Hydraulic	Pressure momentum p_p	Mass m, [kg]
Thermal	-	
conduction	-	Thermal energy Q, [J]
convection	-	Internal Energy U, [J]

2.4 Basic Bond Graph Elements

In bond graph language, two active elements (namely sources) (Se and Sf), three generalized passive elements (I, Cand R), two junctions (0 and 1) and two transducers (TF and GY) are used to model any energetic process. When the exchanged power is negligible, or is drawn from some unmodeled external source (*e.g.* the tank circuit in an amplifier), then it is represented by an information bond having full arrows to show the direction of information imposition like in a block diagram. The full arrow on the information bond (also called activated bond) can represent the signal transmitted by a sensor, integrator, sum member, *etc.*

There are nine basic multiport elements, grouped into four categories according to their energy characteristics. These elements and their definitions are summarized in Table 2.5.

2.4.1 One Port Passive Elements

Passive bond graph elements transform received power either into dissipated power (R-element), store as potential energy (C-element) or as kinetic energy (I-element). They are also called one port elements because of the fact that, in the simplest case, they can be connected to another system through only one port. In such elements, by convention, the half arrow is oriented towards the port as given in Figure 2.7.

Depending on how the power is transformed by the port, passive elements are classified into resistive (R), storage (C) and inertial (I) elements.

Table 2.5 Definitions of the basic bond graph elements

		Symbol	Constitutive equation	Name
Sources		Se:e $\overset{e}{\underset{f}{\rightharpoonup}}$	$\begin{cases} e(t) \text{ given by the source} \\ f(t) \text{ arbitrary} \end{cases}$	Source of effort
Sources		Sf:f $\overset{e}{\underset{f}{\rightharpoonup}}$	$\begin{cases} f(t) \text{ given by the source} \\ e(t) \text{ arbitrary} \end{cases}$	Source of flow
Passive elements	Dissipator	$\overset{e}{\underset{f}{\rightharpoonup}}$ R	$\Phi_R(e,f)=0$	Resistance
Passive elements	Energy stores	$\overset{e}{\underset{f}{\rightharpoonup}}$ C	$\Phi_C(e,q)=0$	Capacitance
Passive elements	Energy stores	$\overset{e}{\underset{f}{\rightharpoonup}}$ I	$\Phi_I(f,p)=0$	Inertance
Junctions	Transducers	$\overset{e_1}{\underset{f_1}{\rightharpoonup}}$ TF :m $\overset{e_2}{\underset{f_2}{\rightharpoonup}}$	$\begin{cases} e_1 = me_2 \\ f_2 = mf_1 \end{cases}$	Transformer
Junctions	Transducers	$\overset{e_1}{\underset{f_1}{\rightharpoonup}}$ GY :r $\overset{e_2}{\underset{f_2}{\rightharpoonup}}$	$\begin{cases} e_1 = rf_2 \\ e_2 = rf_1 \end{cases}$	Gyrator
Junctions	Junctions	$\overset{e_1}{\underset{f_1}{\rightharpoonup}}$ 0 $\overset{e_2}{\underset{f_2}{\rightharpoonup}}$, with f_3, e_3 bond entering 0 from above	$\begin{cases} e_1 = e_2 = e_3 \\ f_1 - f_2 + f_3 = 0 \end{cases}$	Zero junction : common effort junction
Junctions	Junctions	$\overset{e_1}{\underset{f_1}{\rightharpoonup}}$ 1 $\overset{e_2}{\underset{f_2}{\rightharpoonup}}$, with f_3, e_3 bond entering 1 from above	$\begin{cases} f_1 = f_2 = f_3 \\ e_1 - e_2 + e_3 = 0 \end{cases}$	One junction : common flow junction
Sensors	Sensors	$\overset{e}{\underset{f=0}{\rightarrow}}$ De:e	$\begin{cases} e = e(t) \\ f = 0 \end{cases}$	Sensors (Detectors)
Sensors	Sensors	$\overset{e=0}{\underset{f}{\rightarrow}}$ Df:f	$\begin{cases} f = f(t) \\ e = 0 \end{cases}$	Sensors (Detectors)

Fig. 2.7 Representation of 1-port passive elements

2.4.1.1 Resistive Elements: R-element

Definition 2.3. A bond graph element is of type R-element (resistor or dissipater) if it defines an algebraic constitutive equation between effort and flow. This element models all energy dissipation type phenomena.

Figure 2.8 shows the bond graph model and sketches of resistors in several energy domains. As technological systems representing an R-element, one can cite electrical resistors, mechanical frictions, hydraulic resistors and chemical resistances (membrane in a fuel cell, for example). The type of the phenomenon, represented by an R-element, is usually identified by the associated power variables or subscripts (some nomenclature) given to the associated parameter used in the element's constitutive equation: R_e, R_m, R_h or R_c.

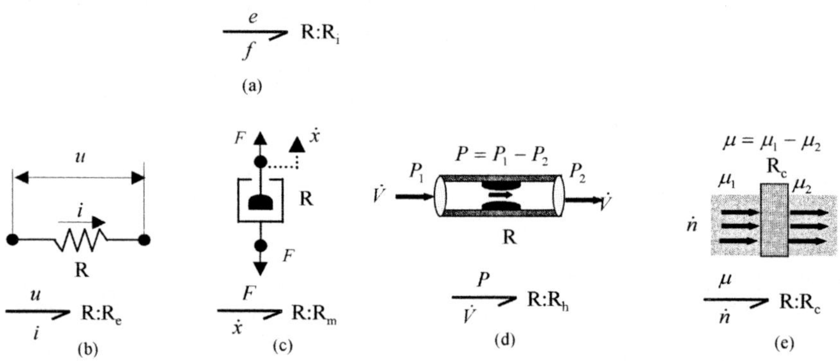

Fig. 2.8 R element. **a** Bond graph. **b** Electrical resistance. **c** Mechanical damper. **d** Hydraulic resistor. **e** Chemical membrane

The constitutive equation of R-element is a static function:

$$\Phi_R(e, f) = 0. \tag{2.13}$$

This relation can be linear or non-linear. Examples are:

- Ohm's law in the electrical domain: $u - R_e i = 0$
- Mechanical friction: $F - R_m \dot{x} = 0$,
- Bernoulli's law in hydraulics: $\Delta P - R_h \dot{V}^2 = 0$
- Chemical diffusion law between two substances of different potential through a membrane: $\Delta\mu - R_C \dot{n} = 0$

The R-element can be modulated by an external signal. For example consider a valve controlled by "On-Off" controller as shown in Figure 2.9.

The control signal $u_c = \{0, 1\}$ is modeled by an information bond (Figure 2.9b). The volume flow rate $\dot{V}$ is then computed as follows:

$$\begin{cases} \dot{V} = \sqrt{\Delta P / R_h}, \text{ if } u_c = 1; \\ \dot{V} = 0, \text{ if } u_c = 0. \end{cases} \tag{2.14}$$

Concerning the process plant items associated with dissipation processes, the driving force which leads to energy dissipation or material transportation is the effort

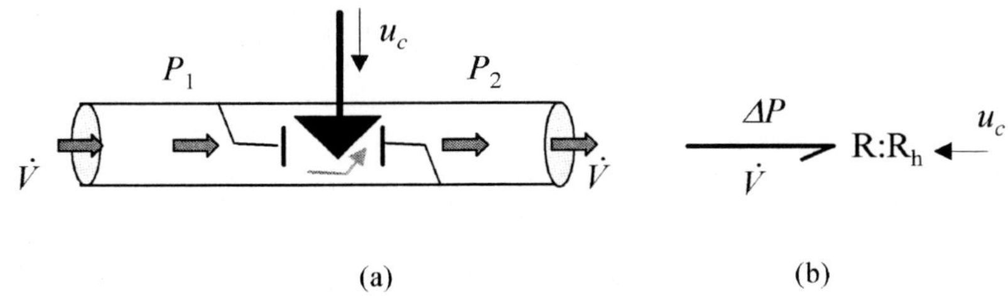

Fig. 2.9 Modulated R element. **a** Schematic form of a controlled valve with variable restriction. **b** Bond graph model

gradient (difference between the upstream and downstream efforts: pressure drop, chemical potential difference, temperature difference). From a functional point of view, resistive processes receive power and dissipate it. They are valves, restrictions, friction, evaporation, resistors, *etc.*

The resistance can be negative if it supplies power. For example, the dry friction phenomena, flow induced excitation phenomena, torque-speed characteristic of motors or pressure difference-volume flow characteristic for a pump can be modeled as R -elements [188] with negative values for parameter R.

It should be noted that the R-element usually withdraws power from the system and transforms it mostly into useless heat, illustrating the fact that the dissipation phenomenon is irreversible.

- *Economic R-element*

The R-element can be used for modeling economic systems characterized by information exchange. The typical R-type economic component is a functional relationship deduced from the supply and demand curves in market places.

A specific example of this kind of constitutive relation for supply and demand curves is

$$f_c = \left(\frac{P_u}{P_{u0}}\right)^E, \tag{2.15}$$

where E is a constant defined as sensitivity or elasticity in economics literature.

2.4.1.2 R and RS Elements in Conduction Heat Transfer

To introduce the R-element in thermal domain, let us consider bond graph modeling of conduction heat transfer widely used in heat exchangers, thermal losses through walls of furnaces, *etc.* In Figure 2.10, heat transfer through a wall of thermal resistance R_t is represented. For a wall with thermal conductivity of the material λ, thickness ℓ and area A, R_t is:

$$R_t = \frac{\ell}{\lambda A}. \tag{2.16}$$

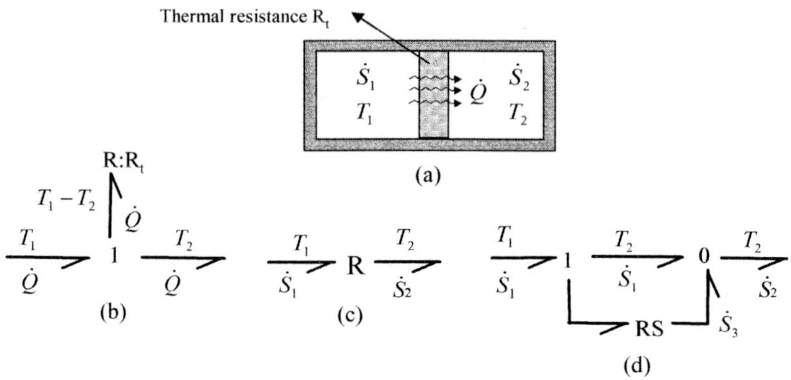

Fig. 2.10 **a** Thermal conduction. **b** Pseudo-bond graph model. **c, d** True bond graph models

By analogy with electrical systems, the heat flow $\dot{Q}$ (quantity of heat transmitted through the wall per unit of time), is a power resulting from the product of a voltage T and a current $\dot{S}$:

$$\dot{Q} = T\dot{S}. \tag{2.17}$$

The unit of thermal flow is expressed like any other power as J/s or Watt, the temperature T has its own unit, called Kelvin (K) and the entropy flow $\dot{S}$ has unit $J/(K.s)$. This equation is analogous to the equation of the electric power: $P = u.i$. As electric current i is the flow of an electric charge, the thermal current is the entropy flow, and the heat is the flow of power.

Modeling of thermal conduction (Figure 2.10a) can be realized using both a true bond graph form with the pair entropy flow-temperature $(\dot{S}, T)$ as power variables or in the pseudo bond graph form with heat flow - temperature $(\dot{Q}, T)$ as power variables. In practice, the latter is more convenient, because it leads to a simple constitutive equation, also called Fourier's law of conduction:

$$\dot{Q} = \frac{1}{R_t}(T_1 - T_2) \Rightarrow \Phi_R(\Delta T, \dot{Q}) = 0. \tag{2.18}$$

The elements "1" (respectively "0") in the bond graph models (Figure 2.10b,d) are junctions used to connect physical components (bond graph elements) having common flows (respectively common efforts). Those junction elements are developed later.

Equation 2.18 uses pseudo bond graph variables. It represents a constitutive static equation relating effort (ΔT) and flow $(\dot{Q})$, analogous to the electric one linking difference of potential with current. But the use of entropy flow (true bond graph) introduces complex non-linear equations. Indeed, considering that the heat flow from chamber 1 is instantaneously equal to the heat flow received by the chamber 2 (*i.e.* there is no intermediate storage giving rise to some time constant), we can write:

$$T_1\dot{S}_1 = T_2\dot{S}_2 = \dot{Q}, \tag{2.19}$$

which means the temperatures (effort variables) in the two chambers are different as well as the entropy flows (flow variables $\dot{S}_1 = \dot{Q}/T_1$ and $\dot{S}_2 = \dot{Q}/T_2$,) are unequal. Therefore, neither of the two junction structures available in bond graph theory (0 and 1) can be used. In fact, heat transfer and radiation phenomena are represented by an R-field element [172] in true bond graphs (Figure 2.10b). The entropy lost during this heat transfer is then:

$$\dot{S}_2 - \dot{S}_1 = \frac{1}{R_t}\frac{(T_1 - T_2)^2}{T_1 T_2} > 0, \tag{2.20}$$

where $\Delta\dot{S} = \dot{S}_2 - \dot{S}_1$ is positive since there is an entropy generation and it means that the entropy of the universe increases (as always with any R-element) and it is an irreversible phenomena.

Equation 2.20 is an R-constitutive relation relating the entropy flow and the effort (temperature). This resistance is a two-port element represented by a bond graph model in Figure 2.10c. If the resistance element represents any technological component (electronic circuit, for example) then the entropy of the system should increase when any thermal flow circulates through this element. Since this transducer does not store energy, the principle of power conservation must be respected ($T_1\dot{S}_1 = T_2\dot{S}_2$). The generated entropy $\dot{S}_3$ deduced from "0" junction can be explicitly shown in the bond graph model (see Figure 2.10d). To take into account the generation of entropy, the symbol "RS" [166, 258] is introduced. The letter "S "exposes the source character of the element.

Another complexity of the relation in Equation 2.20 is due to the computational problem, which is a causality issue and is discussed later in the section 2.5.

With multiport RS, irreversibility and energy conservation of multiport R are summarized as follows: power in any thermal bond must always be positive because, if it is otherwise, then it signifies reversible phenomena. In other words, this means that the power directions assigned to thermal bonds connected to R-elements show actual direction of power flow, which is unlike the notion of power direction in a general bond graph.

2.4.1.3 Storage C-element

Definition 2.4. A bond graph element is called C-element (storage) if it is defined by a dynamic constitutive equation relating displacement (time integral of flow) and effort. This element represents any system that transforms the received power into potential energy without loss.

In Figure 2.11 the bond graph symbol and some physical examples of related storage phenomena are given. Typical examples are electrical capacitor, mechanical spring, storage reservoir and torsion bars *etc.* The constitutive law for C-element always relates the effort e to the time integral of flow (displacement q):

$$\Phi_C(e,q) = 0\,, \text{ or } \Phi_C\left(e, \int f(\tau)d\tau\right) = 0. \tag{2.21}$$

Remark 2.2. The constitutive equation associated with bond graph elements are called *behavioral equations. The* physical laws expressing how the energy is transformed are mathematically described by the behavior model. In a bond graph model, they describe the physical phenomena which are represented by lumped-parameter bond graph elements (R, C and I). These equations are called "constitutive laws":

$$F_B = \{F_C\} \cup \{F_I\} \cup \{F_R\}\,. \tag{2.22}$$

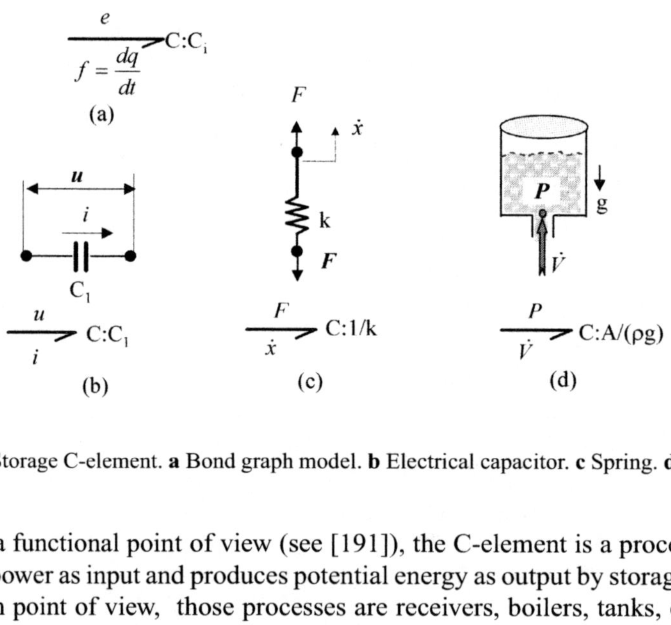

Fig. 2.11 Storage C-element. **a** Bond graph model. **b** Electrical capacitor. **c** Spring. **d** Storage tank

From a functional point of view (see [191]), the C-element is a processor which receives power as input and produces potential energy as output by storage. From the plant item point of view, those processes are receivers, boilers, tanks, condensers, *etc.*, which are classified as storage processes.

Thus if the flow f is the input to a C-element, it is first integrated to find q and then e is an output computed from q. The linear form of such constitutive law is

$$e = \frac{1}{C}\int f dt = \frac{q}{C}. \tag{2.23}$$

In the linear case, the constitutive law is $u = \frac{1}{C}\int i dt = \frac{q}{C}$ in electrical systems, $F = \frac{1}{C}\int \dot{x} dt = Kx$ in mechanical systems, and $P = \frac{1}{C}\int \dot{V} dt = \frac{V}{C} = \frac{\rho g}{A} V$ in hydraulic storage tanks having uniform cross-section.

For mechanical systems, it is common to use the *spring constant* K, rather than the *compliance* $C \equiv 1/K$, which is analogous to the electrical capacitance. For a spring, the potential energy of a linear compliant element (Equation 2.10) is given by

$$\Delta E = \int_{x_0}^{x_1} Kx.dx = \frac{K}{2}\left(x_1^2 - x_0^2\right), \tag{2.24}$$

or

$$\Delta E = \int_{t_0}^{t_1} e.f.dt = \int_{t_0}^{t_1} e.\frac{1}{K}\frac{de}{dt}.dt = \int_{e(t_0)}^{e(t_1)} \frac{e}{K}de = \frac{1}{2K}\left(e^2(t_1) - e^2(t_0)\right). \tag{2.25}$$

In a hydraulic system (gravity tank), the capacitance C is equal to $A/(\rho g)$. Indeed the pressure of the fluid of the density ρ (kg/m^3) at the bottom of the cylindric tank of cross-section A can be expressed as a function of the volume V, as $P(V) = \frac{\rho g}{A} V$. Based on Equation 2.7, the potential energy stored by a hydraulic capacitor (tank or accumulator) can be evaluated as follows:

$$E_p = \int_{V_0}^{V_1} P(V)dV = \frac{\rho g}{2A}\left(V_1^2 - V_0^2\right). \tag{2.26}$$

C-element does not dissipate energy but stores it and discharges it on demand by the rest of the system.

The stored energy $E(q)$ may be interpreted graphically as shown in Figure 2.12. The area under the curve represents the energy E. If q (volume, entropy, charge,...) goes from 0 to q_0 then energy is being stored; if q starts returning to 0, stored energy is being released (*e.g.* discharge of a capacitor). During this process, no energy is lost.

- *Thermal C-element*

To define C-element in thermal domain, consider a body (of global thermal capacity C_t) heated by an external heat flow $\dot{Q}$. The schematic is given in Figure 2.13a.

The temperature of the body (if the mechanical work is neglected) depends on the stored heat:

$$Q = \int_0^t \dot{Q} dt + Q_0. \tag{2.27}$$

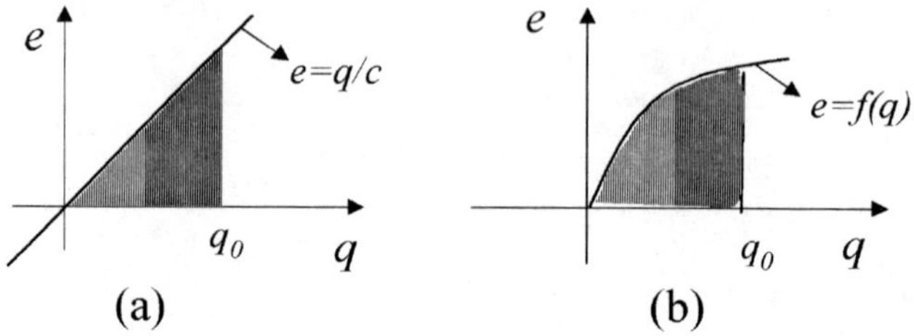

Fig. 2.12 Energy storage in a C-element. **a** Linear case. **b** Non-linear case

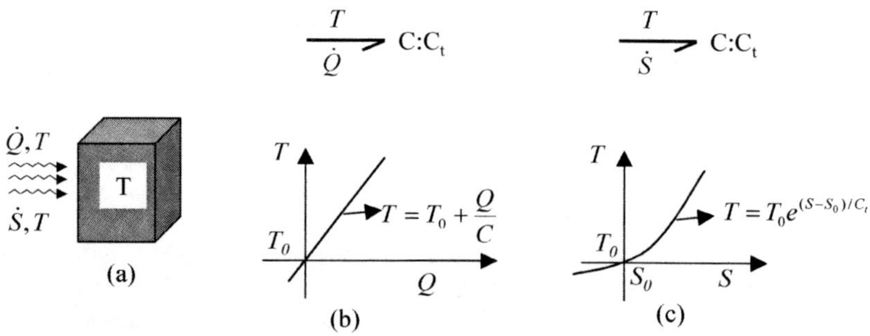

Fig. 2.13 Thermal C-element. **a** Bond graph model. (b,c) Relation between effort and charge in pseudo and true bond graphs

If the thermal capacity C_t ($C_t = mc_t$), where m is the global mass of the body and c_t is the specific heat at constant volume) is assumed to be constant, then the temperature is

$$T = \frac{Q}{C_t} + \frac{Q_0}{C_t} = \frac{Q}{C_t} + T_0. \tag{2.28}$$

This equation is a constitutive law of a pseudo bond graph relating the generalized displacement (Q) and the effort variable (T) in Figure 2.13b.

By using true bond graph power variables ($T - \dot{S}$), we obtain

$$S - S_0 = \int_0^t dS = \int_0^Q \frac{dQ}{T} = \int_0^Q \frac{dQ}{T_0 + \frac{Q}{C_t}} = C_t \left[\ln(T_0 + \frac{Q}{C_t}) - \ln T_0 \right], \tag{2.29}$$

which gives

$$T = T_0 e^{\frac{S-S_0}{C_t}}. \tag{2.30}$$

This equation is a constitutive law of a true bond graph relating the generalized displacement (S) and the effort variable (T). The corresponding curve is given in Figure 2.13c.

Contrary to dissipative element, a storage element cannot be controlled by an information bond (external signal). Indeed this would violate the principle of energy conservation[127]. Modulated C-element must be replaced by a two-port C. This element is developed below.

- *Economic C-element*

The accumulation of orders can be given as integral of order flow:

$$q(t) \doteq \int_0^t f_c(t)dt + q(0). \tag{2.31}$$

Note that the negative value of q is interpreted as commodity for which no orders have been received, that is to say a negative q represents inventory. The compliance effect is interpreted as an accumulation of unfilled orders or as a response to the build-up of inventory. The effort variable to the market price P_u is associated with the accumulation of unfilled orders or with build-up of inventory.

Based on bilinear models of economic dynamics given in [33], the constitutive relation can be written as

$$q = \ln\left(\frac{P_u}{P_{u0}}\right)^{\gamma}. \tag{2.32}$$

The price P_u increases exponentially with the number of unfilled orders (positive values of q) and diminishes with the build-up of inventory (negative values of q).

2.4.1.4 Inertia I-element

Definition 2.5. A bond graph element is called I element (storage) if it is defined by a dynamic constitutive equation relating momentum (time integral of effort) and flow. This element models any system that transforms the received power into kinetic energy without loss.

I-element (inertia) transforms the received power into kinetic energy for mechanical systems and magnetic field energy for electrical systems. The bond graph symbol and some physical examples are shown in Figure 2.14. The I-element is used to model inductance phenomenon in electrical systems, mass or inertia effects in mechanical or hydraulic process. The constitutive law is a dynamic relation relating integral of effort (momentum) and flow. *An inertia element does not dissipate energy but stores it.* A well known example from mechanics is the flywheel with its large moment of inertia. When it is brought back to its original state, *e.g.* when the flywheel is slowed down, it returns the entire energy to the system. The usual relation given for I-element is between the time derivative of the flow and the effort, or in other words, between flow and the integral of effort.

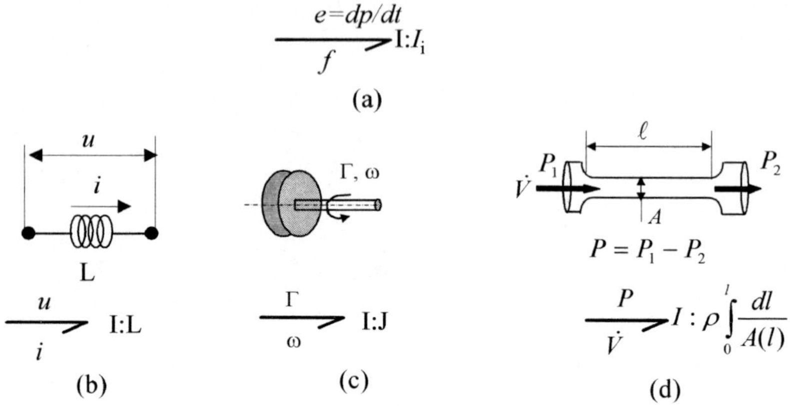

Fig. 2.14 Inertia or I-element. **a** Bond graph model. **b** Electrical example. **c** Mechanical example. **d** Hydraulic example

$$\Phi_I(f,p) = 0 \text{ or } \Phi_I\left(f, \int e(\tau)d\tau\right) = 0. \tag{2.33}$$

Examples: $\int u dt = Li$ or $\Phi = Li$ in electricity, $\int \Gamma dt = J\omega$ or $p_\omega = J\omega$ in mechanical rotation and $\int F dt = m\frac{dx}{dt}$ or $p = m\frac{dx}{dt}$ in linear motion. The inductance, mass or moment of inertia is the inertia parameter.

In hydraulics, the inertia due to the mass of the fluid of density ρ flowing in a pipe of length l and cross-sectional area A can be derived (Figure 2.14d) by assuming that the driving force is due to the difference of pressure ΔP. Newton's law for the mass of the fluid $(m = \rho A\ell)$ circulating in the pipe yields

$$\rho A\ell \frac{d(\dot{V}/A)}{dt} = A(P_1 - P_2) \Rightarrow \dot{V} = \frac{A}{\rho\ell}\int \Delta P dt = \frac{1}{I}p_h, \tag{2.34}$$

where p_h is the pressure momentum.

The parameter of inertia of the fluid is then $I = \frac{\rho\ell}{A}$ for a volume of constant area A. If A is not constant, $I = \rho\int_0^l \frac{d\ell}{A(\ell)}$.

If we consider $\dot{x}$ as the linear velocity of a fluid control volume of mass m then Equation 2.36 can be written as

$$p_h = \frac{\rho\ell}{A}\dot{V} = \frac{m}{A^2}\dot{V} = \frac{m}{A^2}A\dot{x} = \frac{1}{A}m\dot{x}. \tag{2.35}$$

From Equation 2.35, the pressure momentum can be considered as momentum per unit surface area:

$$p_h = \frac{p}{A}. \tag{2.36}$$

Based on the equation $(p = m\dot{x})$, the kinetic energy stored in a mechanical inertial element of mass m can be obtained:

$$E_c = \int_{p_0}^{p_1} f(p)dp = \int_{p_0}^{p_1} \frac{p}{m} dp = \frac{1}{2m}\left(p_1^2 - p_0^2\right). \tag{2.37}$$

Note that both C and I elements are associated with energy storage.

- *Economic inertia*

In economics, inertia is associated with the acceleration of production, $\frac{df_c}{dt}$. This build-up is associated with the commitment of cash to the purchase of production equipment with capital investment. The constitutive relation can be deduced from the investment inertia:

$$\beta = I.\ln(f_c), \tag{2.38}$$

which is equivalent to the equation

$$I.\frac{df_c}{dt} = P_u f_c. \tag{2.39}$$

The rate of change of production is proportional to the cash flow $P_u f_c$. I represents the capital cost of increasing production by one unit.

2.4.2 Active Elements

Definition 2.6. Sources are called active elements in bond graph notation. They supply power to the process. Their type is indicated either by a subscript "e" or "f" depending on whether source imposes an effort or a flow on the system.

Depending on the type of power variable the source provides, bond graph modeling has two source elements: effort source (Se) and flow source (Sf). Figure 2.15a shows the bond graph representation of such elements. Since sources provide power, the bond is conventionally oriented towards the system to which the power is imposed.

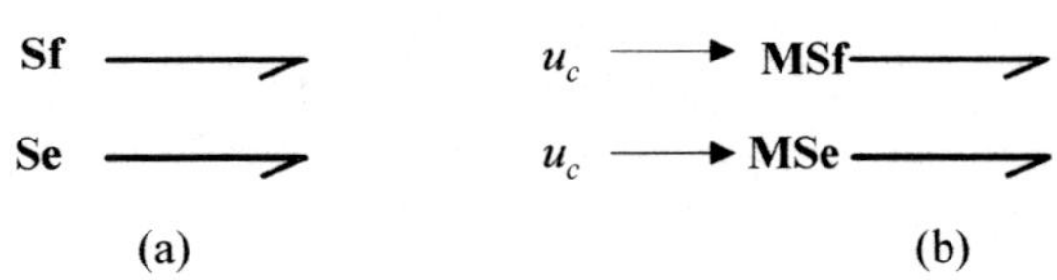

Fig. 2.15 Active elements. **a** Flow and effort sources. **b** Modulated flow and effort sources

2.4.2.1 Effort Source

It imposes an effort, which can be a function of time, but independent of the required flow. Examples are the electric voltage generator, pressure pump, temperature source, gravity, *etc.*

2.4.2.2 Flow Source

It provides a flow independent of the required effort. Examples are electric current generator, imposed velocity in mechanics, and hydrostatic pump driven with constant rotation frequency, where the volume flow is practically independent of the opposed pressure. If such a pump is equipped with an ideal pressure regulator then it becomes an effort source with the effort (pressure) independent of the flow.

2.4.2.3 Modulated Sources

Many effort and flow sources are controlled (by a signal). They are named modulated sources (MSe or MSf) and are highlighted by a letter *M* "modulated" preceding the source symbols as given in Figure 2.15b. For example, a pump can be controlled by an external control signal to regulate its speed.

2.4.3 Junctions

To express other constraints of the global system being modeled, bond graph elements (R, C, I, ...) are interconnected by "0" junctions when interconnected ports have a common effort and by "1" junction if their flow is the same. Formerly, these junctions have been called *P* (parallel)- and *S* (series)-junctions by author Jean Thoma. This is better for generating computer code but has been abandoned for international standardization because in fact the series and parallel aspects of the junctions are more obvious in the electrical than in the mechanical systems. Sometimes called three ports, the junctions allow to connect the elements (in parallel and series connections). Bond graph representation of "0" and "1" junctions are shown in Figure 2.16.

Remark 2.3. The equation deduced from junction are called *structural equations*: they represent a set of conservation laws (of mass, energy, *etc.*) and/or equilibrium equations.

In all junctions we have power conservation, which can be expressed as

$$\sum_{i=1}^{n} e_i f_i = 0, \tag{2.40}$$

Fig. 2.16 "1" and "0" junctions in bond graphs

where n is the number of ports and $\sum$ is the algebraic sum. The sign is taken (+) when the power direction is towards the junction and (−) when it is outwards from the junction. For the bond graph model of Figure 2.16, two power directions are into the junction (the bonds 1 and 4) and two power direction are out of the junction (the bonds 2 and 3); thereby giving the conservation law:

$$e_1 f_1 + e_4 f_4 - e_2 f_2 - e_3 f_3 = 0. \tag{2.41}$$

- *0-junction (common effort junction)*: it associates elements with the same effort, which means parallel circuit in electricity and oil hydraulics, and series circuit in mechanics. This corresponds to the Kirchoff's current law in electricity. The constitutive equation and representation of such junction are described by Figure 2.16a. The efforts on all bonds are identical, and the algebraic sum of flows is equal to zero. The constitutive equations (from Figure 2.16) may be written as

$$\begin{cases} f_1 + f_4 - f_2 - f_3 = 0, \\ e_1 = e_2 = e_3 = e_4 = 0. \end{cases} \tag{2.42}$$

In mechanical engineering, "0" junction represents a geometric compatibility for a situation involving a single force and many velocities which algebraically sums to zero. It represents for electrical systems a Kirchhoff's law for a node where conductors join. In hydraulic systems, it represents a conservation of volume/mass flow rate at a point where several pipes join. Those physical interpretations are illustrated in Figure 2.17.

- *1-junction (common flow junction)*: it associates elements with the same flow, which means series circuit in electricity and oil hydraulics but parallel circuit in mechanics.. The flows in all bonds are identical, and algebraic sum of efforts is equal to zero. The constitutive equations (from Figure 2.16) are written as

$$\begin{cases} e_1 + e_4 - e_2 - e_3 = 0, \\ f_1 = f_2 = f_3 = f_4 = 0. \end{cases} \tag{2.43}$$

Examples of 1-junctions are given in Figure 2.18. "1" junction represents in mechanical engineering a dynamic equilibrium of forces associated with a common

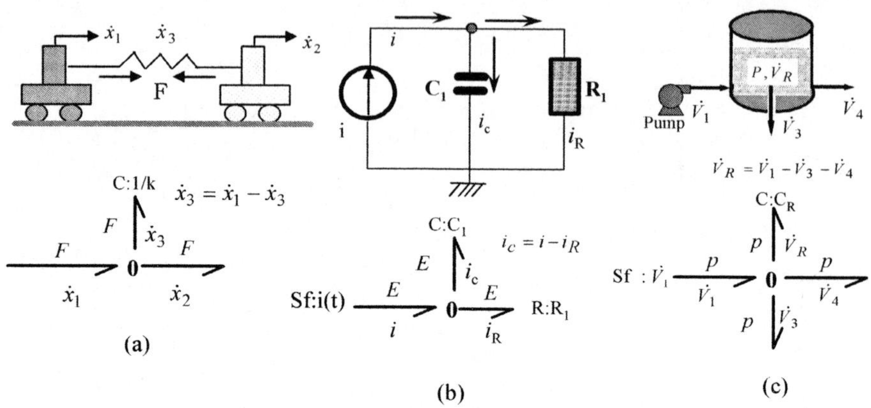

Fig. 2.17 0-junction in **a** mechanical, **b** electrical and **c** hydraulic systems

velocity (Figure 2.18a). If an inertia is involved, Newton's law for the mass element can be deduced. In electrical circuit (Figure 2.18b), 1-junction represents circuits in series connection: Kirchoff's law for voltage can be written. When hydraulic components are in serial connection, the flow is the same and the pressure drops around the circuit are summed up. Corresponding to Figure 2.18c, the constitutive equation is

$$Pe - \Delta P_1 - \Delta P_2 - P_{atm} = 0, \tag{2.44}$$

where $\Delta P_1 = Pe - P_1$, $\Delta P_2 = P_1 - P_{atm}$ and atmospheric pressure $P_{atm} = 0$.

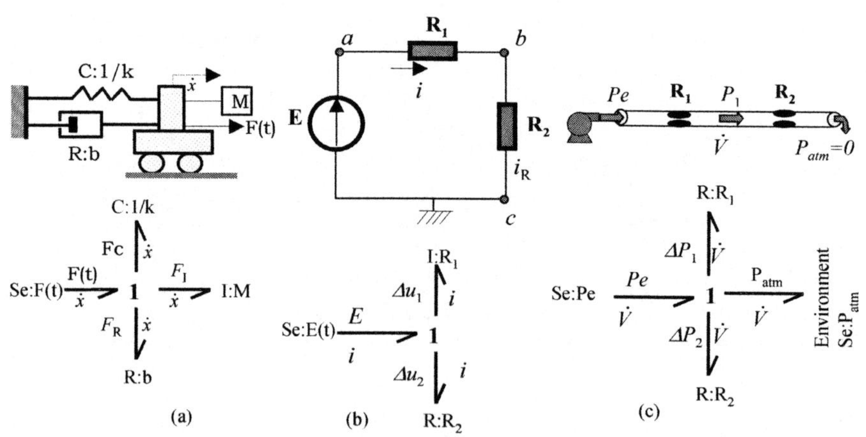

Fig. 2.18 1-junction in **a** mechanical, **b** electrical and **c** hydraulic systems

2.4.4 Transformers and Gyrators

- *Transformer element - TF*

Definition 2.7. A transformer element, denoted by TF, is a conservative two-port bond graph element which scales like power variables, sometimes transforming energy from one domain into another. Its constitutive equations (Equation 2.45) algebraically link inlet and outlet efforts and inlet and outlet flows. The input effort is proportional to the output effort and the output effort is proportional to the input effort.

This energy conservative element models electric transformers, gear reducers in rotary mechanics or simply levers in rectilinear mechanics.

The constitutive relations of a transformer element are between input and output efforts, and between input and output flows:

$$\begin{cases} e_1 = m e_2, \\ f_2 = m f_1, \end{cases} \tag{2.45}$$

where m is called the modulus of the transformer. Obviously, $e_1 f_1 = m e_2 \times f_2 / m = e_2 f_2$. Since TF element does not store energy, we have power direction of one bond pointing towards the element and the other one is oriented away from the element. If the modulus m of the transformer is not constant, but depends on time or any other parameter, then it is called an MTF (*modulated transformer*) element.

Examples for transformers are pumps and cylinders in fluid power engineering, and levers and gear boxes in mechanical engineering. (Figure 2.19). The constitutive equations for the piston-cylinder (Figure 2.19b) mechanism of cross-section A is

$$\begin{cases} P = \dfrac{F}{A}, \\ \dot{x} = \dfrac{\dot{V}}{A}. \end{cases} \tag{2.46}$$

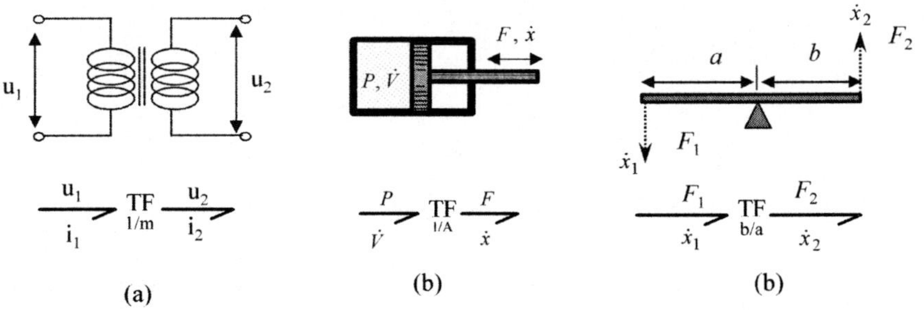

Fig. 2.19 Examples of TF element. **a** Electrical transformer. **b** Hydraulic piston. **c** Mechanical lever

Transformers are of two different kinds:

1. Impedance transformers, where the input and output variables are of the same energy domain or class.
2. Class transformers, where the input and output variables belong to different energy domains or classes, *e.g.* hydraulic cylinders and pumps connecting hydraulic variables with mechanical variables. Hydraulic power is transduced into mechanical power in the piston.

- *Gyrator element - GY*

Definition 2.8. A Gyrator element, denoted by GY, is a conservative two-port bond graph element which scales dissimilar power variables, generally transforming energy from one domain into another. The input effort is proportional to the output flow and the output effort is proportional to the input flow (Equation 2.47).

Gyrators are also called transducers, and they allow one to transform energy from one domain into another (gyroscope, electric motor). An example is the electric motor that transforms electric power to mechanical rotary power (Figure 2.20a). The constant of proportionality r is called *gyrator ratio* or *modulus*. The gyrator may be adjustable, and its conversion ratio can be modulated by a signal applied through an activated bond. This is represented by the symbol MGY(*modulated gyrator*). An example is the electric motor with variable field strength (Figure 2.20b).

Unlike the TF-element, the constitutive equation for GY is "crossed", *i.e.* between dissimilar power variables:

$$\begin{cases} e_1 = r f_2, \\ e_2 = r f_1. \end{cases} \tag{2.47}$$

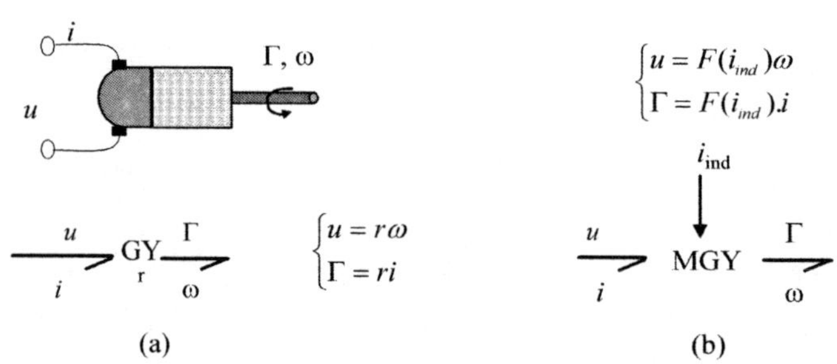

Fig. 2.20 **a** Gyrator element. **b** Modulated gyrator element

Note that the modulated signal (activated bond) means that no power is associated with the change in m and r (in both TF and GY elements), thus the power $e_1 f_1$ is equal to $e_2 f_2$. This is contrary to other passive elements, *e.g.* C-element where the stored energy changes when the information signal changes. One example of the latter case is a capacitor with movable plates.

2.4.5 Information Bonds

In measurement and control process, the energy transferred by the signal is negligible compared to those exchanged between physical components. This signal is represented by an information bond, which corresponds to block diagrams arrows. It is shown as a full arrow on the bond and can represent the signal transmitted by a sensor, integrator, sum member, *etc.*

Figure 2.21 shows the representations of the effort sensor De and the flow sensor Df. The causality assignment decides the type of the sensor. The concept of causality is discussed in the next section.

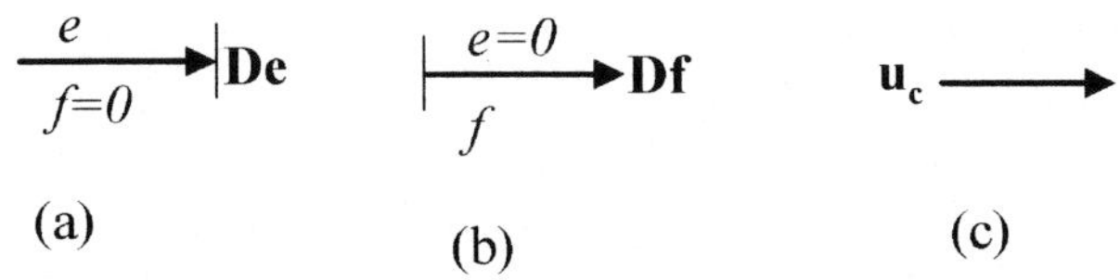

Fig. 2.21 Information bond. **a** Effort detector. **b** Flow detector. **c** Transmitted or control signal

This means that only one variable (effort or flow) is used in a signal bond. The effort sensor for example provide to the user or to operator only information about the effort measurement. There is no reaction from the sensor to the system. This is why the flow is considered equal to zero (Figure 2.21a). The bond is denoted as activated. *A bond is called activated, if one of its conjugate power variable is set to zero.*

2.5 Causality

2.5.1 Introduction

The energetic coupling between two interconnected physical components is represented by a bond as developed before. Such a model is acausal. Each component is described by its constraint (called constitutive equation). The physical component must behave according to its constraint defined in the terms of the conjugate power variables. But if we need to simulate the physical phenomena (the model), we have to decide in which order the variables (efforts and flows) will be computed. Thus we will introduce the block diagram simulation which is causal. Consequently we need to make a series of cause and effect decisions, alternatively called causal assignment. This concept of *causality* is central to any computational model development. To make the bond graph model causal, the founders of this theory introduced a perpendicular stroke on each bond. This single mark on a bond, called a *causal stroke*,indicates how e and f are simultaneously determined on a bond.

Let us consider interaction between two sub-systems A and B (hydraulic pump and a tank) as shown in Figure 2.22. The pump may deliver as output a flow (volume flow $\dot{V}_P$) or an effort (effort P_P). The conjugate power variables for the tank are pressure (P_P) and volume rate ($\dot{V}_R$). Computation of the models can be performed by two causal bond graphs and their corresponding block diagrams: P_P is known for B (Figure 2.22a) or $\dot{V}_R$ is known for B (Figure 2.22b).

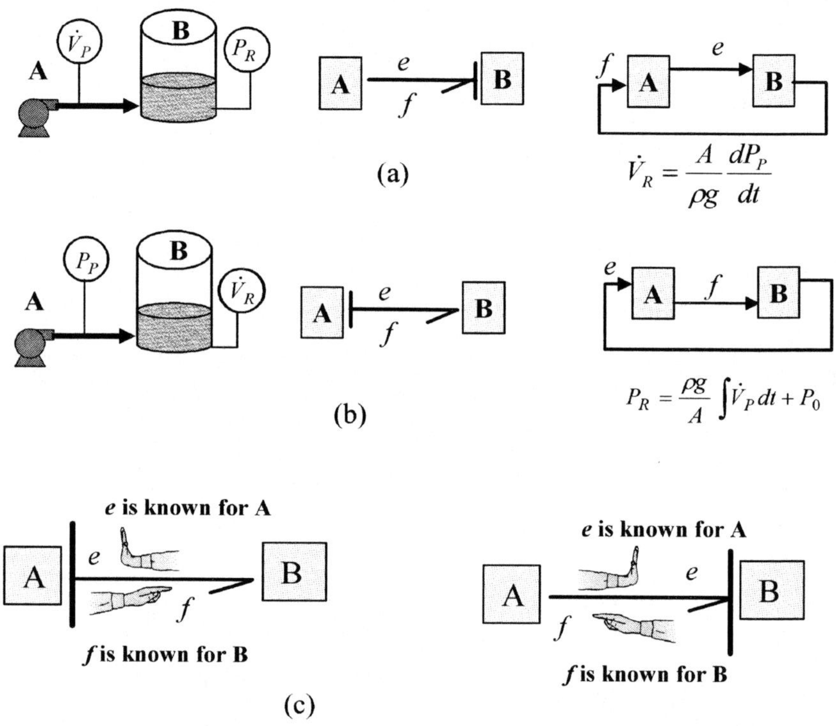

Fig. 2.22 Causality assignment rule

Based on the causality concept, the imposition of causes and effects in the system is directly portrayed on the graphical representation and shows at the same time the transition to the block diagram. The simulation algorithm is then indicated by the position of the causal stroke in the bond graph which becomes *causal bond graph*. The rule is as follows:

Remark 2.4. The causal stroke is placed at the end of the bond which receives the effort information. The other end receives the flow information. In physical systems, effort and flow information are always counter oriented. However, we will show later that in non-physical systems, such as when we use available information for diagnosis, they can be co-oriented.

The causality convention is easily remembered through the graphical representation given in Figure 2.22c.

2.5.2 Sequential Causality Assignment Procedure (SCAP)

In order to predict how the final equations will turn out and to avoid inconsistency, the procedure given below must be followed in strict order:

1. The sources impose always one causality: imposed effort by effort sources and imposed flow by flow sources. Then choose any Se or Sf and assign its required causality. Immediately extend the causal implications, using all 0,1,TF and GY restrictions that apply, as shown in Table 2.6.
2. Choose any C- or I-element and assign integral causality (well adapted to numerical calculations) as shown in Table 2.6. Again extend the causal implications of this action, using 0, 1, TF and GY restrictions.
3. Choose any R-element that is unassigned and give it an arbitrary causality. In linear R-elements, the causality is in principle indifferent, but indicates whether resistance or conductance need to be entered as parameter. In non-linear R-elements, equations are more comfortable in one direction according to the equation form. Extend the causal implications, using 0, 1, TF and GY restrictions.

Causal forms of different bond graph elements, their corresponding causal equations and block diagrams, and the causality assignment rules are given in Table 2.6.

As an example consider the causal bond graphs of the thermal conduction phenomenon described in Figure 2.10 by true and pseudo bond graphs (Equations 2.18 and 2.20). Figure 2.23 represents causal pseudo bond graph models and corresponding block diagrams for thermal resistor R. The 1-junction is used because the thermal flow, as a function of the difference between two thermal efforts (temperatures), is the same.

In Figure 2.23a, T_3 is known for R because the causal stroke is placed near R-element. T_1 and T_2 are imposed by systems $\sum 1$ and $\sum 2$. Based on discussed causality rules (Table 2.6), constitutive causal equations for the junction and R-element can be written as follows:

$$\begin{aligned}&\textbf{1 junction}\text{, for efforts: } \Delta T = T_1 - T_2\text{; for flows: } \dot{Q}_3 = \dot{Q}_1, \dot{Q}_3 = \dot{Q}_2.\\&\textbf{R element, } \dot{Q}_3 = \frac{\Delta T}{R_t} \text{ (Fourier's law in linear case).}\end{aligned} \tag{2.48}$$

From the bond graph model given in Figure 2.23b, we obtain

$$\begin{aligned}&\textbf{1 junction}\text{, for efforts: } T_1 = T_2 + T_3\text{; for flows : } \dot{Q}_3 = \dot{Q}_1, \dot{Q}_2 = \dot{Q}_1.\\&\textbf{R element, } T_3 = R_t\dot{Q}_1.\end{aligned} \tag{2.49}$$

Table 2.6 Causality assignment for bond graph elements

Element	Bond graph	Causal equation	Block diagrams	Rule
Effort source	Se:e ⟶|	e known	f ← System Se:e → System	Output of Se (of Sf) is an effort (flow) input for the system. *Rule: the causality is compulsory*
Flow source	Sf:f |⟶	f known	e ← System Sf:f → System	
0 Junction	**e₁**/**f₁** ⟶| **0** ⟶| e₃/f₃; e₂/f₂ (from top), f₄/e₄ (from bottom)	$e_2 = e_1$ $e_3 = e_1$ $e_4 = e_1$ $f_1 = -f_2 + f_3 - f_4$	f_2 → (-), f_3 → (+), f_4 → (-) → Σ → f_1 e_1 → [$e_2 = e_1$; $e_3 = e_1$; $e_4 = e_1$] → e_2, e_3, e_4	Only one effort (here e_1) is input. *Rule: only one bond can have causal stroke near a 0-junction*
1 Junction	**e₁**/**f₁** |⟶ **1** |⟶ e₃/f₃; e₂/f₂ (from top), f₄/e₄ (from bottom)	$f_2 = f_1$ $f_3 = f_1$, $f_4 = f_1$ $e_1 = -e_2 + e_3 - e_4$	e_2 → (-), e_3 → (+), e_4 → (-) → Σ → e_1 f_1 → [$f_2 = f_1$; $f_3 = f_1$; $f_4 = f_1$] → f_2, f_3, f_4	Only one flow (here f_1) is input. *Rule: only one bond can have causal stroke away from a 1-junction*
TF	e_1/f_1 |⟶ **TF** :m |⟶ e_2/f_2 e_1/f_1 ⟶| **TF** :m ⟶| e_2/f_2	$\begin{cases} e_1 = m e_2 \\ f_2 = m f_1, \end{cases}$ $\begin{cases} e_2 = \frac{1}{m} e_1 \\ f_1 = \frac{1}{m} f_2, \end{cases}$	e_2, f_1 → $\begin{bmatrix} m & 0 \\ 0 & m \end{bmatrix}$ → e_1, f_2 e_1, f_2 → $\begin{bmatrix} \frac{1}{m} & 0 \\ 0 & \frac{1}{m} \end{bmatrix}$ → e_2, f_1	Only one effort and one flow are inputs *Rule: one causal stroke near TF*
GY	e_1/f_1 |⟶ **GY** :r ⟶| e_2/f_2 e_1/f_1 ⟶| **GY** :r |⟶ e_2/f_2	$\begin{cases} e_1 = r f_2 \\ e_2 = r f_1, \end{cases}$ $\begin{cases} f_2 = \frac{1}{r} e_1 \\ f_1 = \frac{1}{r} e_2, \end{cases}$	f_2, f_1 → $\begin{bmatrix} r & 0 \\ 0 & r \end{bmatrix}$ → e_1, e_2 e_1, e_2 → $\begin{bmatrix} \frac{1}{r} & 0 \\ 0 & \frac{1}{r} \end{bmatrix}$ → f_2, f_1	Two efforts or two flows are inputs *Rule: two or no causal stroke near GY*
C	|⟶ **C:C₁** (e/f) ⟶| **C:C₁** (e)	$e = \Phi_C(\int f dt) = \Phi_C(q)$ $f = \frac{d}{dt}(\Phi_C^{-1}(e))$	f → [$\Phi_C(\int f dt)$] → e e → [$\frac{d}{dt}(\Phi_C^{-1}(e))$] → f	Integral causality: effort is an output Derivative causality: flow is an output
I	⟶| **I:I₁** (e/f) |⟶ **I:I₁** (e/f)	$f = \Phi_I(\int e dt) = \Phi_I(p)$ $e = \frac{d}{dt}(\Phi_I^{-1}(f))$	e → [$\Phi_I(\int e dt)$] → f f → [$\frac{d}{dt}(\Phi_I^{-1}(f))$] → e	Integral causality: flow is an output Derivative causality: effort is an output
R	|⟶ **R:R₁** (e/f) ⟶| **R:R₁** (e/f)	$e = \Phi_R(f)$ $f = \Phi^{-1}{}_R(e)$	f → [$\Phi_R(f)$] → e e → [$\Phi^{-1}{}_R(e)$] → f	Resistance causality: output is an effort Conductance causality: flow is an output

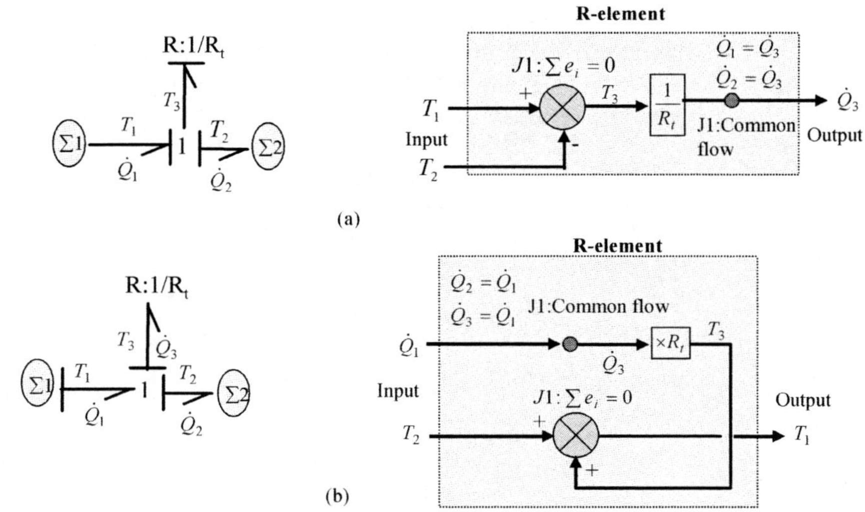

Fig. 2.23 Causal pseudo-bond graph and block diagram models of thermal conduction

Note that $1/R_t$ is the heat transfer coefficient.

2.5.3 Bicausal Bond Graphs

As we have seen, causality imposes a certain propagation throughout the bond graph. It implies that if effort acts in one sense, flow acts in the reverse sense, which is true for all physical systems. However, in system inversion and fault diagnosis, other rules are necessary[119]. One generalization of the causality selection proposed in [80, 83, 177] is the bicausal bond graph. There one divides the causality stroke into two. One half, on the same side as the half arrow indicating power direction, indicates the direction of flow, the other indicates the direction of effort, as shown in Figure 2.24.

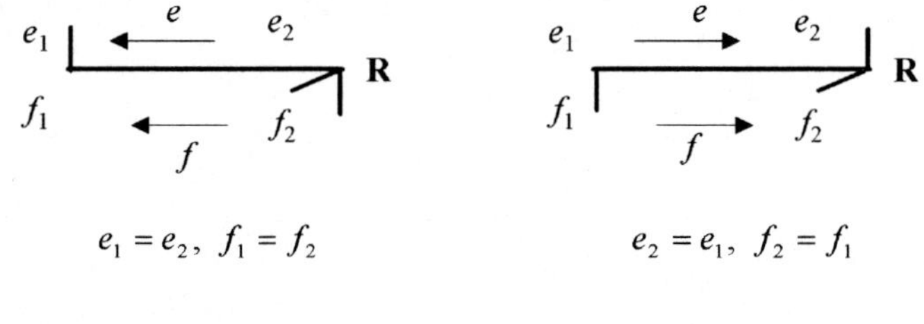

Fig. 2.24 Bicausal bond graph

In the first case shown in Figure 2.24, the causality of R indicates that both effort and flow are available to a block representing the constitutive relation of the R-element. This allows one to calculate the resistance R, *i.e.* it helps in parameter estimation. More details on bicausality notation and its use are given in Chapter 7.

2.5.4 State-space Equations

In control system theory, the state equation form is well suited for control analysis and synthesis. They can be systematically deduced from a bond graph in linear or non-linear form:

$$\begin{aligned}&\text{Linear form:}\begin{cases}\dot{x}=\mathbf{A}x+\mathbf{B}u\\ y=\mathbf{C}x+\mathbf{D}u\end{cases},\\ &\text{Non-linear form:}\begin{cases}\dot{x}=F(x,u)\\ y=G(x,u)\end{cases},\end{aligned}\tag{2.50}$$

where **A**,**B**, **C** and **D** are matrices with appropriate dimensions, u is the input vector (sources Se and Sf), y is the vector of measured variables (effort and flow detectors) and x is the state vector. The state vector x is composed of the variables p and q, *i.e.* the energy variables of storage elements (C- and I-elements) in integral causality:

$$x=\begin{bmatrix}p_I\\ q_C\end{bmatrix}.\tag{2.51}$$

2.5.4.1 Properties of State Variables

- The state vector does not appear on the bond graph, but only its derivatives appear:

$$\dot{x}=\begin{bmatrix}e_I\\ f_C\end{bmatrix}=\begin{bmatrix}\dot{p}_I\\ \dot{q}_C\end{bmatrix}.\tag{2.52}$$

- The dimension of the state vector is equal to the number of C- and I-elements in integral causality.
- If among the n C and I-elements, n_l are in derivative causality, then the order of the model is $n-n_l$.

The state-space equations can be deduced from a bond graph model in the following steps:

- Write structural laws associated with junction structure (0, 1, TF, GY).
- Write constitutive equations of sources and each passive element (R, C, I).
- Combine these different laws to obtain state equation through sequential ordering and substitutions.

2.5.4.2 Example: State-space Equation of an Electrical System.

Consider a simple electrical system example given by its schema and causal bond graph model in Figure 2.25a and 2.25b, respectively.

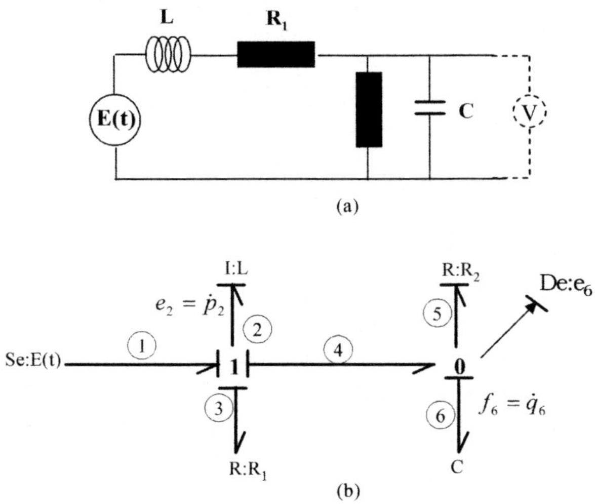

Fig. 2.25 **a** An example electrical system. **b** Its bond graph model

The state-space equation will be derived in the form: $\dot{x}(t) = \mathbf{A}x(t) + \mathbf{B}u$ and the outputs will be written in the form $y = \mathbf{C}x + \mathbf{D}u$.

Where the state variable x, the control input u and the measurement y are respectively the energy variables in dynamic bond graph element (C and I), the source (Se) and the effort detector (De):

$$\begin{aligned} x &= \begin{bmatrix} p_2 \\ q_6 \end{bmatrix} = \begin{bmatrix} \int e_2 dt \\ \int f_6 dt \end{bmatrix}, \\ u &= Se = E(t), \\ y &= De = e_6. \end{aligned} \tag{2.53}$$

The structural equations are the constitutive equations of "0" and "1" junctions (energy or power balance equations):

$$\begin{aligned} &\text{1 junction:} \begin{cases} f_1 = f_2, \;\; f_3 = f_2, \;\; f_4 = f_2, \\ e_2 = e_1 - e_3 - e_4. \end{cases} \\ &\text{0 junction:} \begin{cases} e_4 = e_6, \;\; e_5 = e_6, \\ f_6 = f_4 - f_5. \end{cases} \end{aligned} \tag{2.54}$$

The following equations are the constitutive behavioral equations of bond graph I, C and R elements:

$$\begin{cases} f_2 = \dfrac{1}{L}\int e_2 dt = \dfrac{p_2}{L}, \\ e_6 = \dfrac{1}{C}\int f_6 dt = \dfrac{q_6}{C}, \\ e_3 = R_1 f_3, \\ f_5 = \dfrac{1}{R_2} e_5. \end{cases} \tag{2.55}$$

Combining Equations 2.54 and 2.55, we obtain the state-space equation:

$$\begin{bmatrix} \dot{p}_2 \\ \dot{q}_6 \end{bmatrix} = \begin{bmatrix} -\frac{R_1}{L} & -\frac{1}{C} \\ \frac{1}{L} & -\frac{1}{R_2 C} \end{bmatrix} \begin{bmatrix} p_2 \\ q_6 \end{bmatrix} + \begin{bmatrix} 1 \\ 0 \end{bmatrix} E(t), \tag{2.56}$$

$$y = e_6 = \begin{bmatrix} 0 & \frac{1}{C} \end{bmatrix} \begin{bmatrix} p_2 \\ q_6 \end{bmatrix}. \tag{2.57}$$

More details of sequential ordering to obtain state equations are given in [172]. Note that for some specific forms, state equations cannot be obtained as simply through sequential ordering and substitutions. These situations arise due to algebraic loops (discussed in Chapter 7) and differential causality. A causal loop [172] is another situation where it becomes impossible to eliminate unknown variables.

2.5.5 Model Structure Knowledge

The bond graph is an advantageous modeling tool because it exposes both the structure and the behavior of the studied system. In order to illustrate this property let us consider a simple electrical example given by its schema and causal bond graph models (Figure 2.26). The integral causality (Figure 2.26b) is recommended for engineering simulations. Here the dynamics evolves from given initial conditions. The derivative causality (Figure 2.26c) is more used in FDI because initial conditions are treated as unknown variables in process supervision. In both the models, the current and the voltage are measured respectively by the flow sensor (Df) and effort sensor (De). The sensors, assumed to be ideal, are connected to the junction structure by means of signal bonds indicating that there is no power transfer between the system and sensors.

The bond graph model can be represented in a vectorial way (Figure 2.27) showing the junction structure and the different fields it is composed of.

The key vectors associated with the representation are: X : the state vector (generalized momentum "p" on "I-element" and generalized displacement "q" on "C-element") divided into x_i and x_d; the subvectors respectively associated with the components in integral and derivative causality $\dot{X}$: the time derivative of the state vector divided into $\dot{x}_i$ and $\dot{x}_d$, Z: the complementary state vector ("f" on "I" and "e" on "C") divided into Z_i and Z_d, U: source input vector, Y: sensor output vector, D_{in}: input vector to "R" field, and D_{out}: output vector from "R" field. This representation leads to the junction structure matrix **S** such that

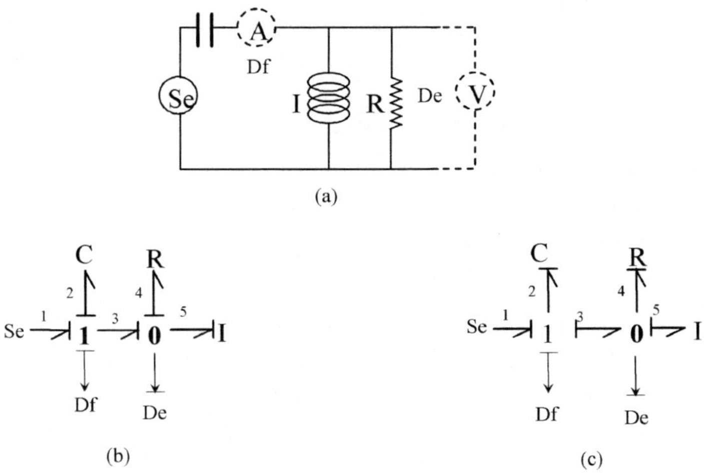

Fig. 2.26 An electrical system (**a**) and its bond graph model in (**b**) integral and (**c**) derivative causality

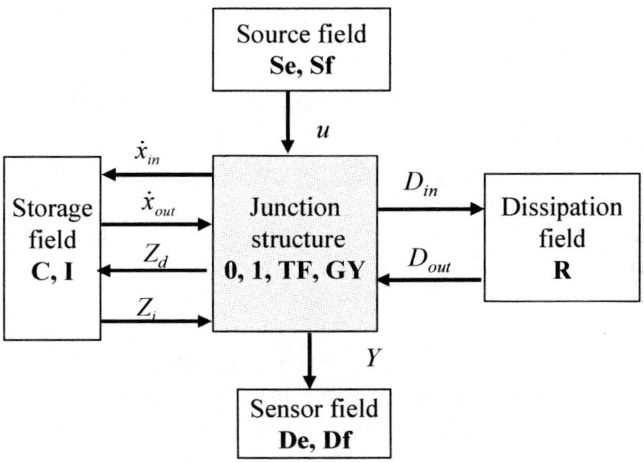

Fig. 2.27 Partitioned representation of model structure

$$\begin{bmatrix} \dot{x}_i \\ Z_d \\ D_{in} \\ Y \end{bmatrix} = \begin{bmatrix} S_{11} & S_{12} & S_{13} & S_{14} \\ S_{21} & S_{22} & S_{23} & S_{24} \\ S_{31} & S_{32} & S_{33} & S_{34} \\ S_{41} & S_{42} & S_{43} & S_{44} \end{bmatrix} \begin{bmatrix} Z_i \\ \dot{x}_d \\ D_{out} \\ U \end{bmatrix} \tag{2.58}$$

The matrix **S** has $(n_C + n_S)$ rows, where n_C and n_S are respectively the numbers of components (I,R or C) and sensors in the systems (De and Df). This matrix is composed of $0, -1, +1, m, r$ or $1/m$ and $1/r$, where m and r are the moduli of transformer TF and gyrator GY elements. More details on the model structure and its implications can be found in [127].

2.6 Single Energy Bond Graph

2.6.1 Bond Graphs for Mechanical Systems

It is easy to construct a bond graph model for electrical and hydraulic systems. Electrical and hydraulic systems can be represented as networks and then it becomes possible to apply the method of gradual uncover, method of point potential or a mix of the two methods (see [172]) to arrive at compact bond graph representations. For mechanical systems, method of flow map and method of effort map are developed in [172]. A chapter titled 'art of creating system bond graphs' in [172] details various steps to arrive at beautiful and compact bond graph structures for mechanical and electrical systems. With practice, it becomes possible to produce the bond graph model directly by looking at the system morphology, skipping all intermediate steps. For complex systems, one can follow the steps shown in Figure 2.28:

1. Fix a reference axis for velocities.
2. Consider all different velocities (absolute velocities for mass and inertia) and relative velocities for others.
3. For each distinct velocity, establish a 1-junction, attach to the 1-junction corresponding bond graph elements.
4. Express the relationships between velocities. Add 0-junction (used to represent those relationships) for each relationship between 1-junctions.
5. Place sources.
6. Link all junctions taking into account the power direction.
7. Eliminate any zero velocity 1-junctions and their bonds.
8. Simplify bond graph by condensing 0- and 1-junctions having two bonds: for example: $1 \rightharpoondown 0 \rightharpoondown 1$ is replaced by $1 \rightharpoondown 1$ and then simply by 1.

The final bond graph obtained in the last step could as well be derived in a much simplified manner. One may note that the velocity of the mass point and the applied load must be the same; moreover the velocity of the mass point is equal to the rate of extension of the spring and the damper. Although these rates (velocities and strain rates) are qualitatively different, they have equal magnitude and hence they should be connected to the same 1-junction.

2.6.2 Bond Graphs for Thermal Processes

Bond graph modeling methodology for thermal processes is considered in this section. Let us consider as an example a thermal process given in Figure 2.29a, where

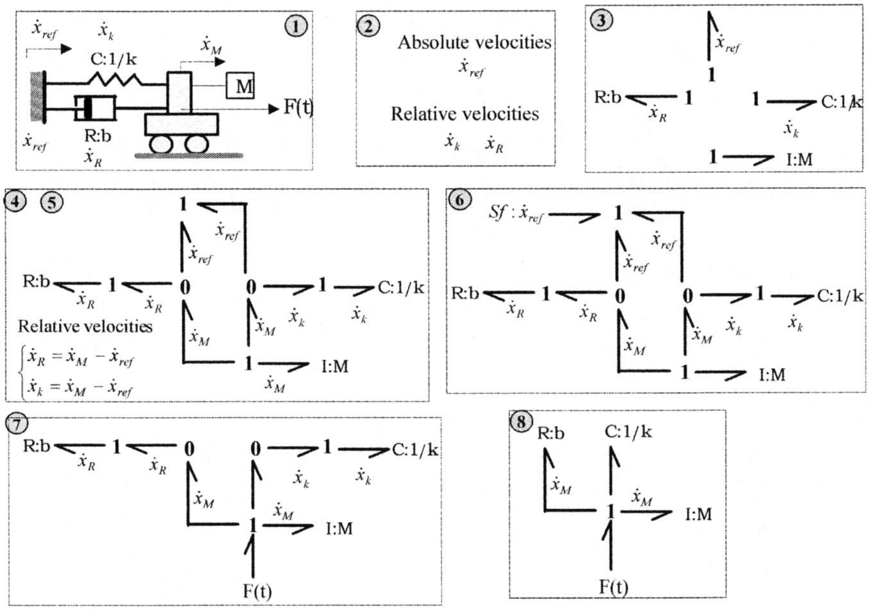

Fig. 2.28 Steps for construction of bond graph models of mechanical translation

only the thermal phenomenon is considered. The coupling with hydraulic energy is developed in the Section 2.8.

The process is heating of a thermal bath (3) by exchanging heat through the metallic walls (2) of temperature T_p with a warm fluid (1) kept at a given temperature T_f. Heat is also leaving through the free surface to the environment supposed to be at a temperature T_a. So there is a thermal resistor between the bath (3) and the external environment at temperature T_a. This thermal resistor may be linear (conduction only) or non-linear (conduction and radiation). The temperature of the bath noted T_r is measured by a temperature sensor TI.

Figure 2.29b gives the analogous electrical schema, based on the $(T, \dot{Q})$ analogy, that is a pseudo bond graph. So voltage u takes the place of temperature T and current i the place of heat flow $\dot{Q}$.

A step-by-step approach to construct the bond graph model of this system is shown in Figure 2.30.

2.6.2.1 Word Bond Graph

The word bond graph of the studied process is given Figure 2.31. The used conjugated variables are temperature T and thermal flow $\dot{Q}$.

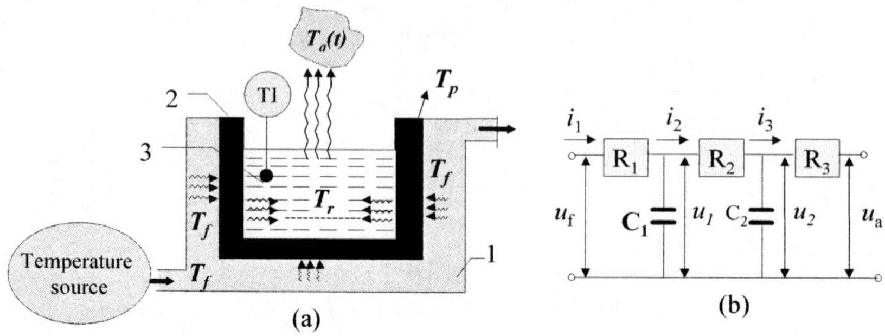

Fig. 2.29 Thermal system and its electrical equivalence

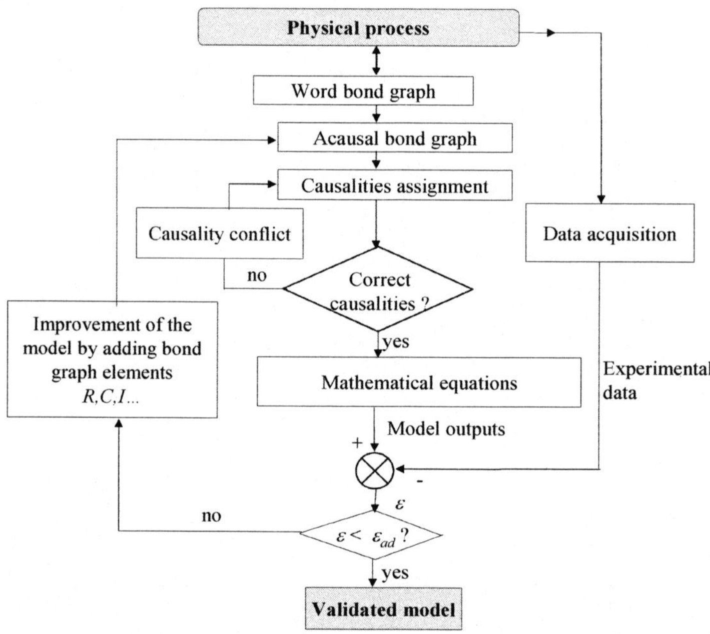

Fig. 2.30 Different steps in bond graph modeling and validation

2.6.2.2 Bond Graph Model

This step represents a physical level of modeling. Based on an assumed modeling hypothesis, the physical phenomena can be graphically represented. At first, suppose that the thermal capacity of the wall, C_m, is negligible, while the thermal capacity of the bath, C_r, is taken into account. To make the bond graph easy to read, all bonds are numbered. The bond graph model is given Figure 2.32a.

The bond graph model is constructed from an energetic consideration of the process. The temperature T_f is imposed as an external effort source noted $Se : T_f$.

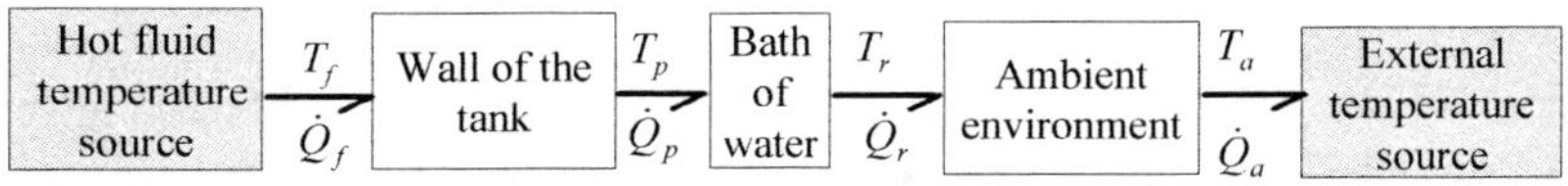

Fig. 2.31 Word bond graph of the thermal process

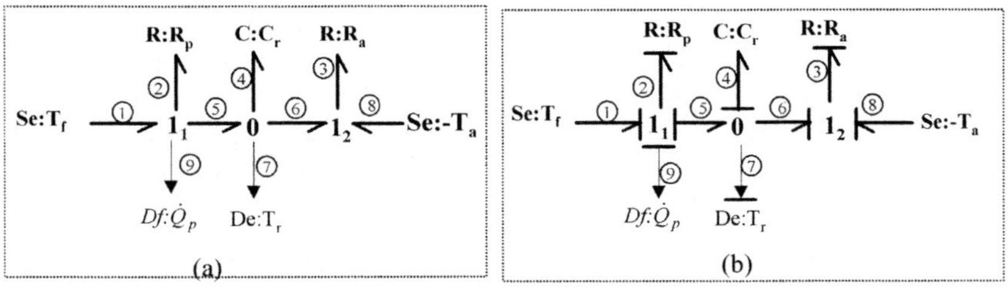

Fig. 2.32 Acausal (**a**) and causal (**b**) bond graph of the thermal system where the thermal capacity of the wall is neglected

There is a thermal transfer by conduction (modeled by R_p) from the hot fluid T_f through the wall to the bath. Part of the transferred heat is accumulated in the thermal capacity of the water $C_r$and the other part is dissipated through R_a to the environment at the temperature T_a (modeled as a negative source as indicated by the orientation of the bond). All these effects are coupled by the parallel and series junctions (0- and 1-junctions). The 0-junction expresses the common temperature points. The 1-junction represents the common thermal flows. The bond graph model is then constructed as given above. The bond graph can be realized using a pedagogical procedure as follows:

1. Fix reference axis for thermal flows (generally from high to low temperature).
2. Place energy sources (temperature source $Se : T_a$ and thermal flow source $Sf : \dot{Q}$).
3. A 0-junction is associated with each different temperature (this corresponds to nodes of different point-potentials in electricity). If a energy storage occurs in this junction, attach C-element ($C : C_r$, here) to this junction.
4. Place a 1-junction between two zero junctions or between 0-junction and thermal effort sources, attach R-elements to model dissipation and heat transfer.
5. Place sensors ($De : T_r$) and $Df : \dot{Q}_p$).
6. Link all junctions taking into account the power directions.

2.6.2.3 Assignment of the Causalities

At this stage, the computational structure of the model which is associated with block diagram structure is deduced. The assignment of causalities is often guided by the necessity to resolve numerical problems: for simulation, preferred integral causality for dynamic elements is imposed. For model based fault detection and isolation purposes derivative causality is imposed. In our case, where the model is used for simulation, let us impose integral causalities for the bond graph model considered in Figure 2.32a. We start from the compulsory causalities: they are effort sources ($Se : T_f$ and $Se : T_a$). Then assign integral causality to $C : C_r$. In the next step, causality propagation through junction rules (*e.g.* only one causal stroke near a 0-junction) automatically forces causalities in the remaining elements $R : R_p$ and $R : R_a$. The causal bond graph model in then obtained (Figure 2.32b).

Sometimes, to avoid derivative causality, spurious elements are added to the model. These elements are often given some physical meaning. Such problems usually arise from our modeling assumptions, *i.e.* from over complexifications or over simplifications. As an example, if the thermal resistance R_p is neglected, we would have derivative causality on a $C : C_r$ element. One could avoid this by a flow source instead of an effort source, or by adding a thermal resistance of some value and keep the effort source as it is. These problems require ingenuity and experience. The most common structure used to break differential causalities are coupling capacitors [258] in hydraulic and thermo-fluid systems, and soft and hard pad structures [172] in mechanical systems.

2.6.2.4 Writing Mathematical Equations

The mathematical structure of the model is deduced systematically from a causal bond graph. According to the assigned causalities, we have:

- Structural equations (conservation equations from junctions)

$$\begin{array}{l} \text{Junction } 1_1 : \; e_2 = T_f - e_5, \; f_1 = f_5 = f_9 = \dot{Q}_p = f_2 \\ \text{Junction } 0 : \; \dot{Q}_4 = f_4 = f_5 - f_6, \, e_5 = e_7 = e_6 = T_r = e_4, \\ \text{Junction } 1_2 : \; e_3 = e_6 - T_a, \; f_8 = f_6 = f_3. \end{array} \tag{2.59}$$

- Behavioral equations (constitutive equations for Bond graph elements)

$$\begin{array}{l} \text{R:R}_\text{p} \text{ element: } f_2 = \dfrac{1}{R_p} e_2 = \dfrac{T_f - e_5}{R_p}. \\ \text{C:Cr element: } e_4 = \dfrac{1}{C_r} \int_0^t f_4 dt \; + e_4(0) = \dfrac{Q_4}{C_r}. \\ \text{R: R}_a \text{ element: } \; f_3 = \dfrac{1}{R_a} e_3 = \dfrac{1}{R_a} (e_6 - T_a) , \end{array} \tag{2.60}$$

where C_r is the global thermal capacity represented as a parameter for C-element in the bond graph model. It depends on the specific heat capacity of the water in the bath at constant volume (c_v) and the total water mass, m_r:

$$C_r = m_r c_v. \tag{2.61}$$

Parameters R_p and R_a are thermal resistances of the hot fluid-water bath interface and the reservoir-environment interface, respectively. Note that the thermal resis-

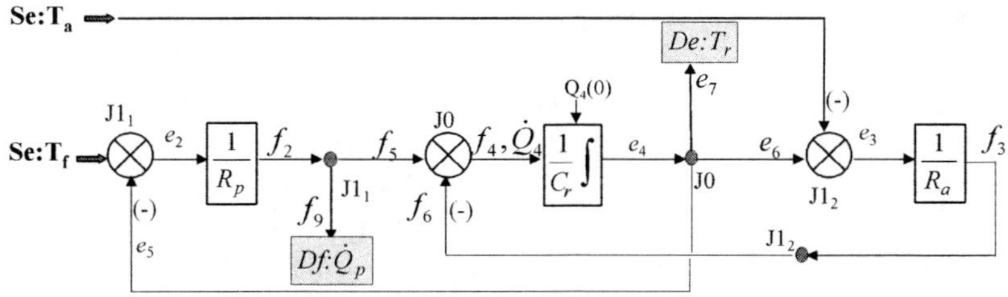

Fig. 2.33 Block diagram for simulation

tances R are equal to the reciprocal of the heat transfer coefficient K_c or conductance ($R = 1/K_c$).

2.6.2.5 Simulation

- *Block diagram*

As developed before, a block diagram representation (Figure 2.33) can be easily arrived at from Equations 2.59 and 2.60.

- *State equations*

Since we have neglected hydraulic energy and assumed the mass of the fluid constant, we have only one C-element in integral causality and therefore there is only one state variable. The state variable x is the accumulated thermal energy in the tank (generalized displacement), the inputs u are thermal effort sources and the measured variables y are given by the detectors:

$$x = Q_4 \equiv Q_r,\ u = \left[T_f\ T_a\right]^T, y = \left[\dot{Q}_p\ T_r\right]^T. \tag{2.62}$$

Similar to the electrical system example given in Figure 2.25, the following state equation is obtained:

$$\dot{Q}_r(t) = -\left(\frac{1}{R_pC_r} + \frac{1}{R_aC_r}\right) Q_r(t) + \left(\frac{1}{R_p} + \frac{1}{R_a}\right) \begin{bmatrix} T_f(t) \\ T_a(t) \end{bmatrix},$$

$$y = \begin{bmatrix} \dot{Q}_p(t) \\ T_r(t) \end{bmatrix} = \begin{bmatrix} -\dfrac{1}{R_pC_r} \\ \dfrac{1}{C_r} \end{bmatrix} Q_r(t) + \begin{bmatrix} \dfrac{1}{R_p} & 0 \\ 0 & 0 \end{bmatrix} \begin{bmatrix} T_f(t) \\ T_a(t) \end{bmatrix}. \tag{2.63}$$

2.6.2.6 Refinement of the Model

Suppose that the error between the model and the real process is bigger than the admissible error, which is usually about 5%. It means that some physical phenomenon has been neglected. There is no need to write the equations from the beginning, but we just have to add a bond graph element (or set of elements forming a structure) corresponding to the newly accounted physical phenomenon. As an example, we can include the thermal capacity C_m (neglected in the first version) of the wall of the bath 3. The revised bond graph model is then given in Figure 2.34. The thermal capacity $C : C_m$ is shown within the doted lines.

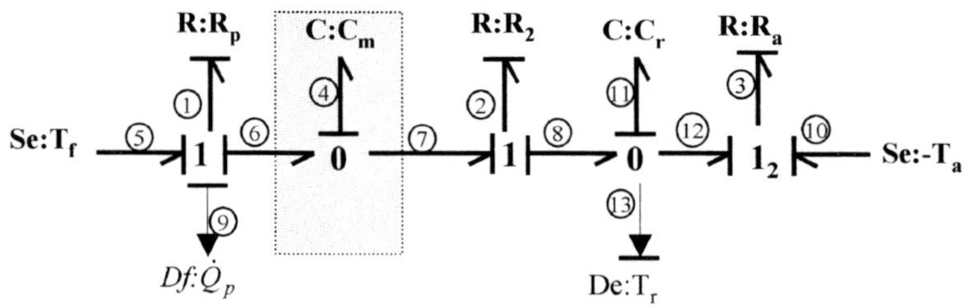

Fig. 2.34 Refined bond graph model which includes the thermal capacity of the wall

In this case, there is conduction heat transfer, modeled by R_p, from the hot fluid T_f to the wall of the heated bath. One part of this energy is accumulated by the thermal capacity of the wall, modeled by C_m. The other part is transmitted by R_2 into the bath.

Now the system has two state variables, which come from energy storage in the wall of the heat bath (Q_m) and from the heat bath itself (Q_r). The state equations are easily deduced from the constitutive equations of two C-elements and three R-elements:

$$\dot{x}(t) = \begin{bmatrix} \dot{Q}_m \\ \dot{Q}_r \end{bmatrix} = \begin{bmatrix} -\left(\frac{1}{R_1 C_m} + \frac{1}{R_2 C_m}\right) & \frac{1}{R_2 C_r} \\ \frac{1}{R_2 C_m} & -\left(\frac{1}{R_2 C_r} + \frac{1}{R_a C_r}\right) \end{bmatrix} \begin{bmatrix} Q_m \\ Q_r \end{bmatrix} \tag{2.64}$$

$$+ \begin{bmatrix} \frac{1}{R_1} & 0 \\ 0 & -\frac{1}{R_a} \end{bmatrix} \begin{bmatrix} T_f \\ T_a \end{bmatrix},$$

$$y(t) = \begin{bmatrix} \dot{Q}_p(t) \\ T_r(t) \end{bmatrix} = \begin{bmatrix} -\frac{1}{R_1 C_m} \\ \frac{1}{C_r} \end{bmatrix} \begin{bmatrix} Q_m \\ Q_r \end{bmatrix} + \begin{bmatrix} \frac{1}{R_1} & 0 \\ 0 & 0 \end{bmatrix} \begin{bmatrix} T_f \\ T_a \end{bmatrix}.$$

2.7 Formal Generation of Dynamic Models

2.7.1 Bond Graph Software

Bond graph models are supported by many commercial software: the model can be graphically introduced in the software and the software automatically generate the dynamic models. The model can be automatically transformed into a simulation or control or supervision program. Author Samantaray presents a review of various commercially available bond graph software in [220]. Moreover, bond graph models can be described in Modelica language as connection of submodels [68] and the constitutive equations can be introduced by the user. The 20-sim software [260] developed at the University of Twente (Netherlands) is based on the well known block oriented TUTSIM simulation program. Another software developed by A. Mukherjee and A.K. Samantaray from Indian Institute of Technology in Kharagpur, allows powerful of use bond graph as integrated computer aided design tool [174] for large and complex systems. Indeed, this software, called Symbols, is an object oriented hierarchical modeling, simulation and control analysis software. It allows users to create models using bond graph, block-diagram and equation models. Recent modules [191] have been developed in order to use this software not only for modeling but for control and monitoring [96, 97, 161, 189, 190, 225]. Thanks to a developed generic plant item database[188] which consists of a set of predefined process, controller and sensor classes, the designer can easily build the dynamic and functional models of most thermofluid processes from the Process and Instrumentation Diagram (P&I D) of the plant, and automatically generate mathematical models. The model can also be used for design of supervision systems and associated algorithm generation. Several applications in process engineering were developed using this software in the framework of European project CHEM [1] at Ecole Universitaire Polytechnique de Lille.

2.7.2 Application

For illustration, let us consider briefly how the dynamic modeling of the studied thermal system (Figure 2.29a) can be automated. In the next chapters it will be shown how the obtained model can be used for control and diagnostic design.

By using the few bond graph elements (R, C, I, TF, GY, Se, Sf and Junctions) displayed in a list, the model is graphically entered (Figure 2.35a). The power directions and causalities are automatically assigned and they may be altered later by the user. The dynamic model (Figure 2.35b) is then generated. In this model, the variable f4 represents the thermal flow in the bond number 4 and corresponds to the time derivative of the generalized displacement Q4. R2 and R3 are the parameter values of the resistances R_p and R_a respectively. K4=1/C4 is the capacitance of C_r.

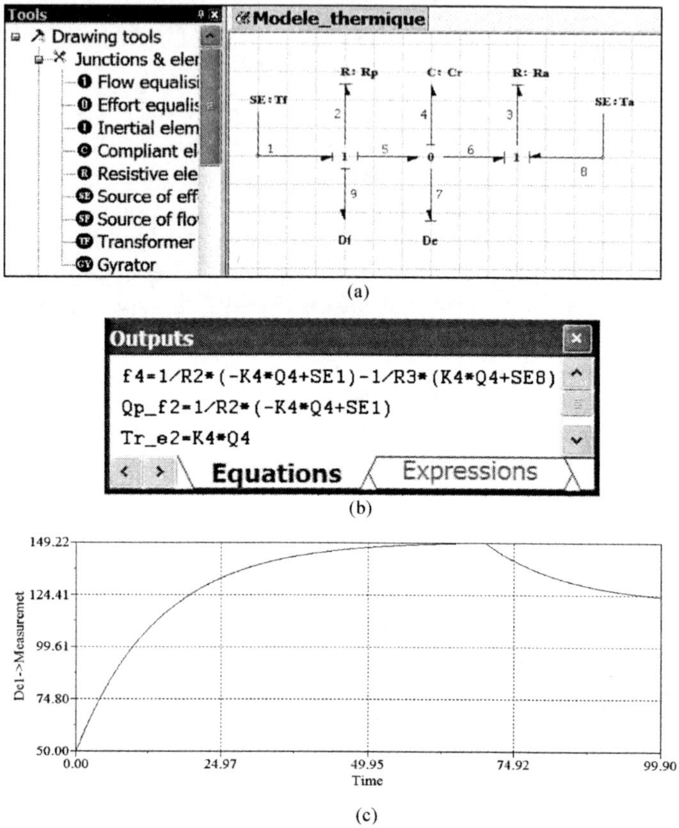

Fig. 2.35 Automated modeling. **a** Graphical User Interface in Symbols-2000. **b** Dynamic model in symbolic format. **c** Simulation result

SE1 and SE8 are efforts sources at the bonds 1 and 8 and represent, respectively, $Se : T_a$ and $Se : T_f$. Tr_e2 and Qp_f2 are the measurement vector. The generated model can be rewritten as follows:

$$\dot{Q}_4 = \left(\frac{K4}{R2} + \frac{K4}{R3}\right) Q_4 + \left(\frac{1}{R2} - \frac{1}{R3}\right) \begin{bmatrix} SE1 \\ SE8 \end{bmatrix},$$
$$y = \begin{bmatrix} Q_{p}\_f2 \\ T_r\_e2 \end{bmatrix} = \begin{bmatrix} -\dfrac{K4}{R2} \\ \dfrac{1}{C4} \end{bmatrix} Q4 + \begin{bmatrix} \dfrac{1}{R2} & 0 \\ 0 & 0 \end{bmatrix} \begin{bmatrix} SE1 \\ SE8 \end{bmatrix}. \tag{2.65}$$

which is identical to Equation 2.63.

The numerical simulation results for specific parameter values is given in Figure 2.35c. The given simulation scenario represents a step input of T_f followed by a fall in the ambient temperature. The used bond graph software has its own

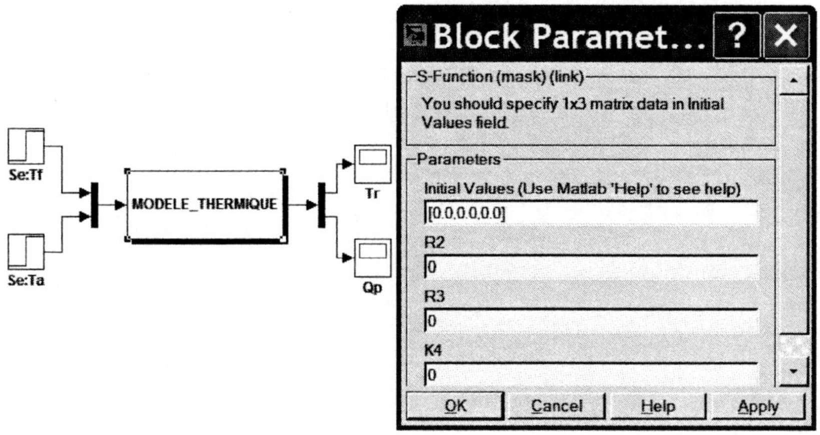

Fig. 2.36 Simulink® S-functions to construct block diagram from a bond graph model and its block parameters

simulation interface. Moreover, the bond graph model can be converted into a MATLAB®-Simulink® S-function [238] as shown in Figure 2.36. The obtained Simulink® model can be simulated by entering numerical values of bond graph parameters (R2, R3 and K4).

If the modeling hypothesis is to be changed, then the bond graph model can be modified graphically and the corresponding state equations, manually obtained in Equation 2.64, are automatically generated (Figure 2.37).

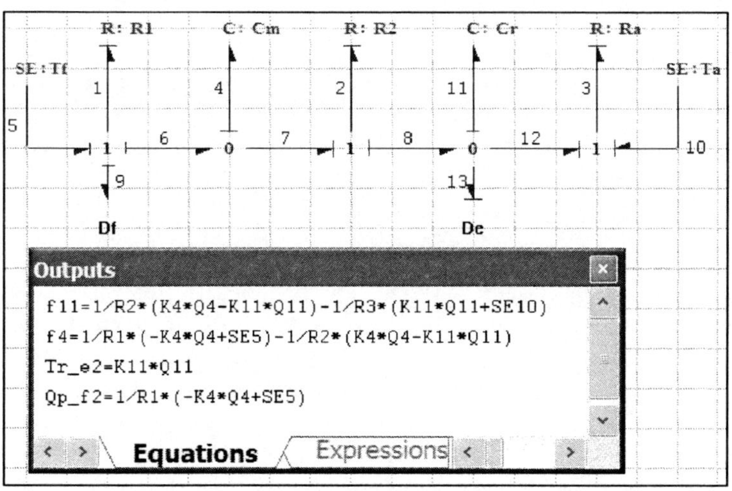

Fig. 2.37 Graphical modification of bond graph model (thermal capacity of the wall is included)

2.8 Coupled Energy Bond Graph

2.8.1 Representation

In thermofluid process, thermal and hydraulic energies are coupled [232]. Their coupling can be represented by a small ring around the bond (Figure 2.38a) or by a multi-bond [18, 29, 43] as shown in Figure 2.38b. Other authors [127] use a different symbolism shown in Figure 2.38c, whose corresponding decoupled version is shown in Figure 2.38d.

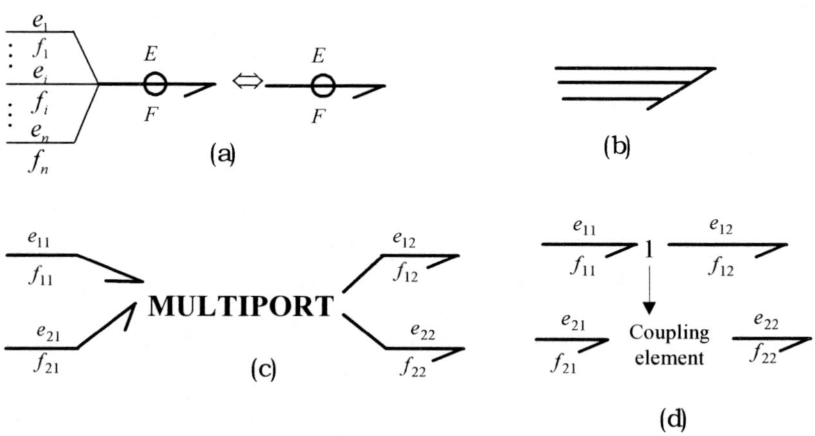

Fig. 2.38 Representation of coupled bond graphs

To distinguish energy domains, hydraulic power is represented in full line and thermal power is given in dotted line. The coupling is modeled by some fictive element (usually R-element). This representation is well suited for analysis.

A general expression for convected energy is

$$\dot{H} = \dot{m}\left(u + \frac{P}{\rho} + \frac{\rho v^2}{2}\right) = \dot{m}\left(h + \frac{\rho v^2}{2}\right). \tag{2.66}$$

Taking into account that in practice the kinetic energy term $\frac{\rho v^2}{2}$ is negligible for sufficiently low flow velocities v and that the fluid is under saturated, the convected energy $\dot{H}$ is calculated from the mass flow $\dot{m}$ and the specific thermal capacity c_p :

$$\dot{H} = \dot{m}.h = \dot{m}c_pT. \tag{2.67}$$

Finally, the thermofluid power variables (efforts and flows of a pseudo-bond graph) are

$$e = \left[e_h \; e_t\right]^T = \left[P \; T\right]^T, \; f = \left[f_c \; f_t\right]^T = \left[\dot{H} \; \dot{m}\right]^T. \tag{2.68}$$

2.8.2 Thermofluid Sources

The thermofluid source consists of hydraulic and thermal conjugate flows, transferred by the convection of the fluid. A thermofluid sources is characterized by the coupling of hydraulic and thermal energies, as defined in Equation(2.67). In bond graph form, it can be represented as given in Figure 2.39.

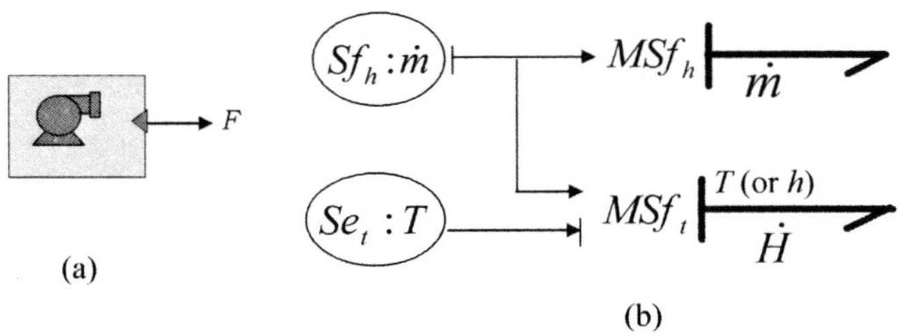

Fig. 2.39 Thermofluid source. **a** Pump. **b** Its bond graph model

This representation is important from the physical and system analysis points of view. In fact, the real actuator gives hydraulic flow $Sf_h : \dot{m}$. The enthalpy flow $\dot{H}$ source is considered as fictive actuator since it is a consequence of hydraulic flow variation (see Equation 2.67). The temperature source $Se_t : T$ can be considered as a parameter or variable (if it is provided by an external source for instance). The source can be modulated by any external control signal, which regulates the flow rate.

2.8.3 Thermofluid Multiport R

In thermodynamic processes, the dissipation phenomenon is modeled by a two-port R-element. The main phenomena modeled by such multiport R-elements are evaporation, condensation, convection through a pipe, valve, nozzle, *etc.* Concerning the thermofluid plant items associated with dissipation processes, the driving force which leads to energy dissipation or material transportation [38, 47] is the difference between the upstream and downstream efforts (pressure drop, chemical potential difference, temperature difference).

Usually the causality for this type of plant item in thermofluid processes is the following: the input is the driving force (effort) and the output is the flow. But, in practice, the causality assignment for this component will be arbitrary. It depends on the inlet and outlet signals imposed by the connected processes at upstream and downstream sides; *i.e.* the element could be in conductance causality or in resistance causality. These cases appear when connecting several multiport R-elements (pipes) in a serial connection.

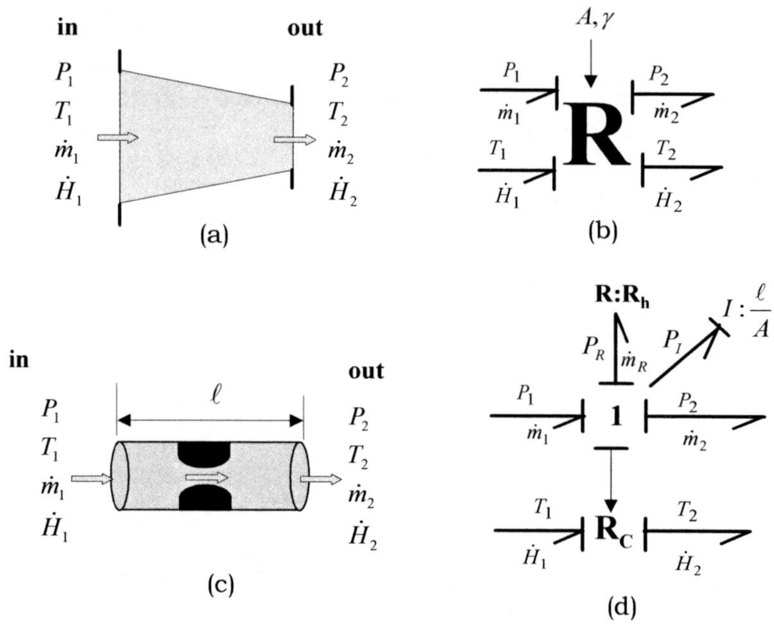

Fig. 2.40 R multiport. **a** Nozzle. **b** Bond graph of nozzle. **c** Pipe with orifice. **d** Bond graph of pipe with orifice

The flows leaving the R-field, because of continuity ($F_{in} = F_{out}$), are calculated as follows:

$$F = \begin{bmatrix} \dot{m} \\ \dot{H} \end{bmatrix} = \begin{bmatrix} \Phi_{RH}(T_{in}, T_{out}, P_{in}, P_{out}) \\ \Phi_{RT}(T_{in}, T_{out}, P_{in}, P_{out}) \end{bmatrix}, \tag{2.69}$$

where the constitutive equations given by functions Φ_{RH} and Φ_{RT} are usually nonlinear and depend mainly on the thermodynamic state of the fluid and on the modeling hypothesis. According to conductance causality imposed to R-element (Figure 2.40b), the outputs of the two port restrictor are thermal flow $\dot{H}$ and hydraulic flow $\dot{m}$. The suffixes "*in*" and "*out*" represent the physical input and output sides of the modeled system.

In the following, we consider two types of technological components modeled by R-multiport: a nozzle and a pipe with a restriction.

2.8.3.1 Nozzle

The general form of bond graph model (Figure 2.40a) is useful when a restrictor can choke due to supersonic flow at its throat. Then the mass flow rate depends not only on the pressure difference, but also the pressure ratio P_r :

$$P_r = \frac{P_{out}}{P_{in}}. \tag{2.70}$$

It is then possible to predict, with acceptable accuracy, the mass flow rate by using the upstream and downstream pressures [127]:

$$\begin{aligned} \dot{m} &= A\frac{P_{in}}{\sqrt{T_{in}}}(P_r)^{\frac{1}{\gamma}}\sqrt{\frac{2\gamma}{R_G(\gamma-1)}\left[1-(P_r)^{\frac{\gamma-1}{\gamma}}\right]}\ , \text{ if } P_r \geq b; \\ \dot{m} &= A\frac{P_{in}}{\sqrt{T_{in}}}\sqrt{\frac{\gamma}{R_G}\left(\frac{2}{\gamma+1}\right)^{\frac{\gamma-1}{\gamma-1}}}\ , \text{ if } P_r \leq b; \end{aligned} \tag{2.71}$$

where

$$b = \left(\frac{2}{\gamma+1}\right)^{\frac{\gamma-1}{\gamma-1}}, \tag{2.72}$$

R_G is the gas constant, A is the exit area, $\gamma = c_p/c_v$, c_p is the specific heat at constant pressure and c_v is the specific heat at constant volume.

2.8.3.2 Thermofluid Transport Phenomena in a Pipe

The drawback of the aforementioned modeling approach is that in the opposite causality, the effort cannot be calculated from such a non-linear equation (Equation 2.71). In industrial process engineering, we generally assume low velocity or subsonic flow. Thus it is sufficient to assume that the mass flow rate $\dot{m}$ is a function of the pressure difference. We assume first that the flow-field properties are constant in a long pipe and the accumulation of mass, momentum and energy in such a component are of negligible order. The flow is regular and the fluid is Newtonian. The bond graph model, with these simplifying assumptions, is given in Figure 2.40b, where the I-element (developed later) is included to take fluid momentum into account. Readers may look at [173] for detailed models of hydraulic transport phenomena with flow saturation characteristics, which are derived from first principles.

The energy coupling (Equation 2.67) is modeled by the fictive thermal resistor R_c modulated (using information bond) by the hydraulic flow variable $\dot{m}$.

According to the conductive causality assigned to R-element, the commonly used form of the constitutive relation is given by the Bernoulli's law, *i.e.*

$$\begin{bmatrix} \dot{m} \\ \dot{H} \end{bmatrix} = \begin{bmatrix} \frac{1}{R_H}.sign\left(P_{in} - P_{out}\right)\sqrt{|P_{in} - P_{out}|} \\ \frac{1}{R_H}.sign(P_{in} - P_{out})c_p\sqrt{|P_{in} - P_{out}|}.T_{in} \end{bmatrix}, \tag{2.73}$$

where

$$R_H = \frac{D^4\pi}{128\ell\nu} \tag{2.74}$$

is the coefficient of hydraulic losses, ρ is the fluid density, ν is the dynamic viscosity of the fluid, D and ℓ are respectively the diameter and the length of the pipe, and c_p is the thermal capacity of the fluid at constant pressure. If the fluid state is saturated, the temperature T_{in} is calculated as a function of the pressure P_{in} by using thermodynamic tables.

- *Hydraulic momentum*

We assume now that the accumulation of mass is neglected, *i.e.* incompressible fluid flow or compressible fluid flow under steady state, but consider the momentum of the transported fluid in the model. The inertia of the fluid is modeled by I-element as shown in the bond graph model in Figure 2.40d. Assuming a linear hydraulic resistance, the constitutive equations of the model are given by

$$\begin{cases} \text{junction } 1 : P_I = P_1 - P_2 - P_R, \\ \text{R element} : P_R = R_h\dot{m}, \\ \text{I element} : \dot{m} = \frac{1}{I}\int P_I dt = \frac{p}{I}. \end{cases} \tag{2.75}$$

The pressure momentum p has been discussed before (see Equations 2.35 and 2.36).

Based on Equation 2.75, the dynamics of the fluid in the pipe can be expressed by a second order differential equation:

$$\tau\left(\frac{d\dot{m}}{dt}\right) + \dot{m} = \frac{(P_1 - P_2)}{R_h}, \tag{2.76}$$

where $\tau = \dfrac{I}{\rho R_h} = \dfrac{\ell}{A\rho R_h}$ is a time constant.

Then the mass flow response to a step input given through pressure difference ΔP_e becomes

$$\dot{m}(t) = \frac{1}{R_h}\Delta P_e\left(1 - e^{-\frac{t}{\tau}}\right) - \dot{m}(0). \tag{2.77}$$

For $t \to \infty$, the steady state mass flow rate is $\dot{m}(t) = \dfrac{1}{R_h}\Delta P_e$ in the linear case and $\dot{m}(t) = \dfrac{1}{R_H}\sqrt{\Delta P_e}$ in the non-linear case.

2.8.4 Thermofluid Multiport C

In thermofluid storage, the hydraulic and thermal energies are stored together. In integral causality, the flows are imposed for the multiport C-element or thermofluid energy storage (Figure 2.41). The two 0-junctions model the conservation of mass and energy. F_{in} and F_{out} are respectively the inlet and outlet flows; the subscripts T and H, respectively, correspond to thermal and hydraulic ports:

$$f_{in} = \left[\sum f_{Hin} \sum f_{Tin}\right]^T, \quad f_{out} = \left[\sum f_{Hout} \sum f_{Tout}\right]^T.$$

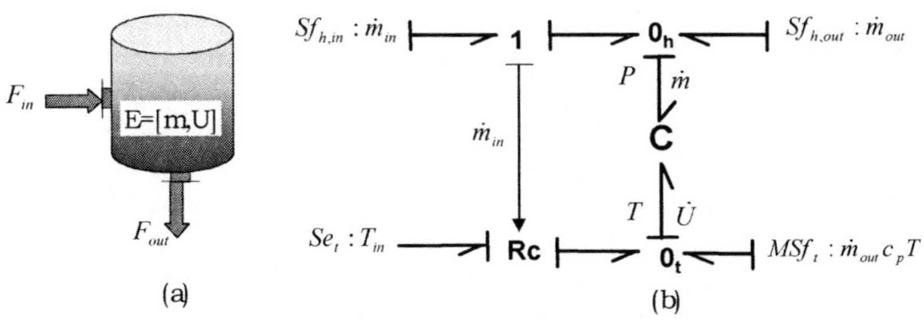

Fig. 2.41 C multiport. **a** Thermofluid storage tank. **b** Its bond graph model

The outputs of the accumulator are calculated by the constitutive equation of the multiport C. These constitutive relations link the system states (integral of flow variables, *i.e.* stored mass m and total enthalpy H) to the effort variables (temperature T and pressure P) as follows:

$$e = \begin{bmatrix} P \\ T \end{bmatrix} = \begin{bmatrix} \Phi_{CH}\left(\int (f_{in}, f_{out}) dt\right) \\ \Phi_{CT}\left(\int (f_{in}, f_{out}) dt\right) \end{bmatrix} = \begin{bmatrix} \Psi_{ch}(m, H) \\ \Psi_{ct}(m, H) \end{bmatrix}, \tag{2.78}$$

where Φ_{CH} and Φ_{CT} are strongly non-linear relations and depend on the thermodynamic state of the fluid (saturated, under saturated, two phase, ...).

As an example, in the case of a mixture of water and steam at saturated, but not superheated, regime, the temperature and pressure are not independent. The pressure P, temperature T and the mixture ratio X are determined for the 2-port C-element by resolving the following mixture equation:

$$\begin{cases} h_M = \dfrac{H}{m} = h_v(P)X + h_l(P)(1 - X), \\[2ex] v_M = \dfrac{V_b}{m} = \dfrac{X}{\rho_v(P)} + \dfrac{1 - X}{\rho_l(P)}, \end{cases} \tag{2.79}$$

where v_M and h_M represent, respectively, the specific volume and the specific enthalpy of the mixture; and X is the steam quality, *i.e.* the ratio of the mass of steam and the total mass of the mixture.

Typical thermofluid storage elements are accumulators of saturated steam which are used in the industry as boiler or steam generator. Taking into account the fact that this component is among the commonly used processes, many works have been devoted to steam generator modeling using non-linear ordinary differential equations (ODEs) [9, 140, 185], partial differential equations (PDEs) solved through finite element approximation [175, 200] and lumped parameter models based on bond graph representations [188].

2.8.5 Application: Bond Graph Model of a Thermofluid Process

2.8.5.1 Bond Graph Model

Here we consider a heated tank as an example of a thermofluid process. The main aim of the system is to provide a continuous water flow to a consumer. The process, as given in Figure 2.42, consists of a water tank which is heated by a modulated thermal source, $\dot{Q}_e$, to maintain a constant fluid temperature. The schematic representation of the process and its bond graph model are given in Figures 2.42 and 2.43. To explain the details in the bond graph model, bonds have been numbered.

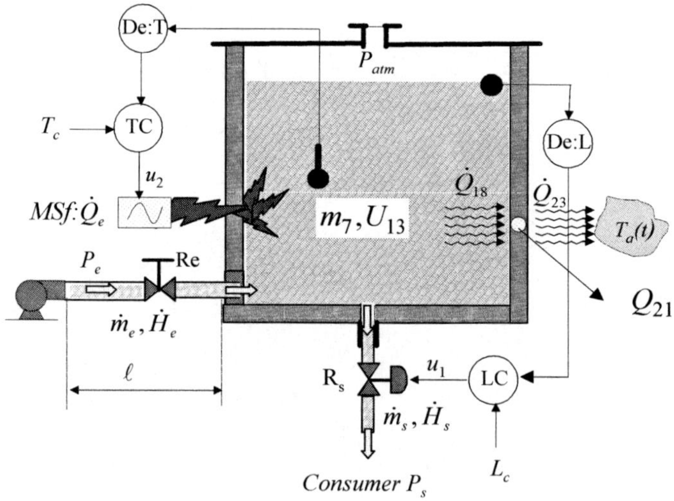

Fig. 2.42 Hot water supply unit as an example thermofluid process

The fluid is assumed to be under saturated, *i.e.* there is no gaseous phase. The mixture is ideal and the temperature in the reservoir and the body are uniform. The system is not insulated, *i.e.* there are thermal losses to the environment, which is at a given temperature T_a.

We are interested in the mass flow rate of the fluid being delivered to the end user as well as its temperature. Two sensors have been installed in the process: level sensor *De:L* and temperature sensor *De:T*. There are two actuators, namely a pump (or a large reservoir) generating a constant pressure $P_e(t)$ and a heater delivering thermal power $\dot{Q}_e(t)$. The ambient pressure and temperature are P_s and T_a, respectively. The feed water (of density ρ and temperature T_e) supply is made up of the pump and a pipe of length ℓ and is assumed to be well insulated. The inertia of the fluid in the inlet pipe is taken into account and is represented by I:ℓ/A in the bond graph model given in Figure 2.43.

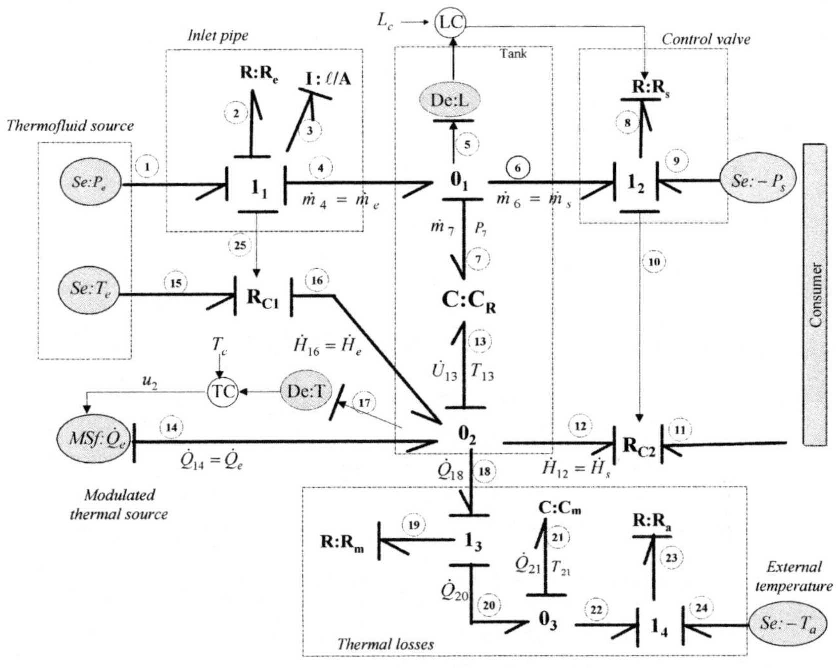

Fig. 2.43 Bond graph model of the hot water supply system

The pump is represented as an effort source. It can, of course, be modeled by a more complex bond graph by taking mechanical and hydraulic energies into account. Such models can be consulted in [188].

The water leaving the tank with mass flow $\dot{m}_s$ through the controlled valve Rs is flowing to an external system (consumer for example) which imposes a back pressure P_s. The valve state depends on the Boolean information signal u_1 (0 or 1) provided by an "On-Off" level controller, LC as follows:

$$u_1 = \begin{cases} 1 \text{ if } L < L_{\min}, \\ 0 \text{ if } L \geq L_{\max}. \end{cases}$$

In the bond graph model, the valve is represented by element $R{:}R_s$ (coefficient of hydraulic losses) and it is modulated by the control signal u_1. The dynamics of the valve, *i.e.* its actuation dynamics, is neglected. A detailed bond graph model of such dynamics is given in [189]. $MSf : \dot{Q}_e$ is the thermal power delivered by the heater, which is modeled as a flow source modulated by a control signal u_2 provided by the temperature controller, TC. The temperature controller follows similar On-Off control law as the level controller.

The accumulation of energy in the tank is modeled by a two port C element, C:C_R. This element is associated with storage of two coupled energies, namely hydraulic and thermal. The storage of thermal energy by the body of the reservoir is modeled by simple C-element, C:C_m. Different heat transfer phenomena are represented by R-elements. The parameters R_m and R_a are thermal resistance representing, respectively, heat transfer by conduction from the fluid in the reservoir to the wall and from the wall to the environment. The pair $(T - \dot{Q})$ is used as the power variables to model such thermal losses.

2.8.5.2 Dynamic Equations

- **Junction 1_1** : common flow (same mass flow through the pipe) or the continuity equation and the pressure drop across the valve:

$$\left\{\begin{array}{l} e_3 = e_1 - e_2 - e_4 \Rightarrow P_3 = P_e - P_2 - P_4, \\ \left\{\begin{array}{l} f_1 = f_3 \\ f_2 = f_3 \\ f_4 = f_3 \\ f_{25} = f_3 \end{array}\right. \Leftrightarrow \left\{\begin{array}{l} \dot{m}_1 = \dot{m}_3 \\ \dot{m}_2 = \dot{m}_3 \\ \dot{m}_4 = \dot{m}_3 \\ \dot{m}_{25} = \dot{m}_3 \end{array}\right. \end{array}\right. \tag{2.80}$$

- **R:R$_\mathbf{e}$ element**: hydraulic losses according to Bernoulli's law:

$$e_2 = \Phi_{\mathrm{Re}}(f_2) = R_e.f_3^2 \Rightarrow \Delta P = P_2 = R_e \dot{m}_3^2. \tag{2.81}$$

Looking at the causality, the mass flow rate, $\dot{m}_3$, is imposed to R element and it is calculated from the constitutive equation of the I element (bond number 3).

- **I : $\ell/$A element**: inertial phenomena due to mass of the fluid

$$f_3 = \frac{1}{I}\int e_3 dt + f_3(0) \Rightarrow \dot{m}_3 = \frac{1}{I} p_3 + \dot{m}_3(0). \tag{2.82}$$

Recall that p_3 is a generalized impulse which represents the pressure momentum per unit surface.

- **Junction 0_1**: mass conservation law: common effort (pressure in the bottom of the tank is uniform) and a conservation of mass:

$$\begin{cases} f_7 = f_4 - f_6 \Rightarrow \dot{m}_7 = \dot{m}_4 - \dot{m}_6 = \dot{m}_e - \dot{m}_s \\ \begin{cases} e_4 = e_7 \\ e_5 = e_7 \\ e_6 = e_7 \end{cases} \Rightarrow \begin{cases} P_4 = P_7 \\ P_5 = P_7 \\ P_6 = P_7 \end{cases} \end{cases} \tag{2.83}$$

- **Junction 0_2**: energy conservation law: common effort (temperature is uniform in the tank) and a conservation of energy:

$$\begin{cases} f_{13} = f_{16} + f_{14} - f_{12} - f_{18} \Rightarrow \dot{U}_{13} = \dot{H}_{16} + \dot{Q}_e(u_2) - \dot{H}_{12} - \dot{Q}_{18} \\ \begin{cases} e_{12} = e_{13} \\ e_{14} = e_{13} \\ e_{16} = e_{13} \\ e_{17} = e_{13} \\ e_{18} = e_{13} \end{cases} \Rightarrow \begin{cases} T_{12} = T_{13} \\ T_{14} = T_{13} \\ T_{16} = T_{13} \\ T_{17} = T_{13} \\ T_{18} = T_{13} \end{cases} \end{cases} \tag{2.84}$$

The information bond (number 17) represents an effort detector and receives temperature T_{13} information, but the reaction flow is zero.

- **C:C$_R$ Multiport**: the multiport C-element represents two coupled capacities, namely thermal C_{Rt} and hydraulic C_{Rh}. These two capacities are associated with constitutive equations relating efforts (P or T) and generalized displacements (time integral of flow): mass ($\int \dot{m}dt = m$) in the hydraulic domain and internal energy ($\int \dot{U}dt = U$) in the thermal domain.

  Hydraulic capacity :

$$\begin{aligned} \mathrm{C{:}C_{Rh}} &\Rightarrow e_7 = \Phi_{Ch}(q_7, q_{13}) = \Phi_{Ch}\left(\int f_7 dt, \int f_{13} dt\right) \\ &\Rightarrow P_7 = \frac{1}{C_h} \int f_7 dt = \frac{m_7}{C_h} + P_7(0). \end{aligned} \tag{2.85}$$

where $C_h = \frac{A}{g}$ is the hydraulic capacity parameter, and A and g are, respectively, the cross section of the cylindrical reservoir and the gravity.

The level measured by the detector *De:L* can be expressed as a function of the state variable for mass m_7 as

$$L = \frac{P_7}{\rho g} = \frac{m_7}{A\rho}. \tag{2.86}$$

Thermal capacity:

$$\begin{aligned} \mathrm{C{:}C_{Rt}} &\Rightarrow e_{13} = \Phi_{Ct}(q_7, q_{13}) = \Phi_{Ct}\left(\int f_7 dt, \int f_{13} dt\right) \\ &\Rightarrow T_{13} = \frac{1}{C_t(m_7)} \int \dot{U}_{13} dt = \frac{U_{13}}{c_V m_7} + T_{13}(0). \end{aligned} \tag{2.87}$$

The mean temperature in the reservoir, measured by the detector *De:T*, is

$$T_{13} = \frac{U_{13}}{c_V m_7}. \quad (2.88)$$

The internal energy, U_{13}, is calculated by time integration of the thermal flow to C:C_{Rt} (Equation 2.84).

Remark 2.5. The global thermal capacity $C_t(m_7) = \frac{1}{c_V m_7}$ depends on the specific heat capacity c_V and the hydraulic state variable m_7. The hydraulic capacity C_h is constant in the under saturated regime. In saturated cases, the effort variables (P and T) are dependent (as in a boiler, for example). Such models are found in [9, 169, 188].

- **Junction 1_2**

$$\begin{cases} e_8 = e_6 + e_9 \Rightarrow P_8 = P_6 - P_s, \\ f_6 = f_9 = f_{10} = f_8 \Rightarrow \dot{m}_6 = \dot{m}_9 = \dot{m}_{10} = \dot{m}_8. \end{cases} \quad (2.89)$$

- **Controlled valve R:$\mathbf{R_s}$**

$$\begin{aligned} f_8 &= \frac{1}{R_s(u_1)}.sign(e_8).\sqrt{|e_8|} = \frac{u_1}{R_s} sign(e_8)\sqrt{|e_8|} \\ &\Rightarrow \dot{m}_8 = \dot{m}_s = \frac{u_1}{R_s} sign(P_8).\sqrt{|P_8|}. \end{aligned} \quad (2.90)$$

- **R:$\mathbf{R_m}$ and R:$\mathbf{R_a}$ elements**

$$\begin{aligned} f_{19} &= \frac{1}{R_m}(e_{18} - e_{20}) \Rightarrow \dot{Q}_{19} = \frac{1}{R_m}(T_{18} - T_{20}), \\ f_{23} &= \frac{1}{R_a}(e_{22} + e_{24}) \Rightarrow \dot{Q}_{23} = \frac{1}{R_a}(T_{22} - T_a). \end{aligned} \quad (2.91)$$

The thermal resistances are generally non-linear functions calculated from experiments or empirical formula (see [258]).

- **Junctions 1_3 and 1_4** : energy conservation laws for heat transfer from the fluid to the metal (1_3) and from the metal to the environment (1_4) :

$$\begin{aligned} &J1_3 : \begin{cases} e_{19} = e_{18} - e_{20} \Rightarrow T_{19} = T_{18} - T_{20}, \\ f_{18} = f_{19}, f_{20} = f_{19} \Rightarrow \dot{Q}_{18} = \dot{Q}_{19}, \dot{Q}_{20} = \dot{Q}_{19}. \end{cases} \\ &J1_4 : \begin{cases} e_{23} = e_{22} + e_{24} \Rightarrow T_{23} = T_{22} - T_a, \\ f_{22} = f_{23}, f_{24} = f_{23} \Rightarrow \dot{Q}_{22} = \dot{Q}_{23}, \dot{Q}_{24} = \dot{Q}_{23}. \end{cases} \end{aligned} \quad (2.92)$$

- **C : $\mathbf{C}_m$ element**: storage of thermal energy in the metallic wall:

$$e_{21} = \frac{1}{C_m}\int f_{21}dt + e_{21}(0) \Rightarrow T_{21} = \frac{Q_{21}}{C_m} + T_{21}(0). \tag{2.93}$$

The thermal capacity of the metal is

$$C_m = V_m \rho_m c_m,$$

where V_m is the volume of the metallic body, ρ_m is the density and c_m the specific heat of the metal.

- **Junction $\mathbf{0}_3$**

$$\begin{cases} f_{21} = f_{20} - f_{22} \Rightarrow \dot{Q}_{21} = \dot{Q}_{20} - \dot{Q}_{22}, \\ e_{20} = e_{21}, e_{22} = e_{21} \Rightarrow T_{20} = T_{21}, T_{22} = T_{21}. \end{cases} \tag{2.94}$$

- **Coupling elements $\mathbf{R}_{c1}$ and $\mathbf{R}_{c2}$**

$$\begin{cases} \mathrm{R_{C1}}: f_{16} = f_{15} = f_{25} c_p e_{15} \Rightarrow \dot{H}_{16} = \dot{H}_e = \dot{m}_3 c_p T_e. \\ \mathrm{R_{C2}}: f_{11} = f_{12} = f_{10} c_p e_{12} \Rightarrow \dot{H}_{11} = \dot{H}_s = \dot{m}_8 c_p T_{13}. \end{cases} \tag{2.95}$$

2.8.5.3 Simulation of the Global Model

- *State equations*

The state variables, x, of the global model are the energy variables associated with storage elements, *i.e.* I and C elements in integral causality. They are:

- The momentum of the fluid in the inlet pipe, p_3 (from element $I : l/A$)
- The stored mass in the reservoir, m_7 (C_h element)
- The stored internal energy in the reservoir, U_{13}((C_t element)
- The accumulated thermal energy in the metallic body of the reservoir, Q_{21} (C:C_m element):

$$x = \begin{bmatrix} p_3 \; m_7 \; U_{13} \; Q_{21} \end{bmatrix}^T. \tag{2.96}$$

The input vector u consists of sources:

$$u = \begin{bmatrix} P_e \; \dot{Q}_e \; T_a \; P_s \end{bmatrix}^T. \tag{2.97}$$

The measured variables (level L and temperature T) are expressed in terms of the state variables:

$$y = \begin{bmatrix} L \\ T \end{bmatrix} = \begin{bmatrix} \dfrac{m_7}{A\rho} \\ \\ \dfrac{U_{13}}{c_V m_7} \end{bmatrix}. \tag{2.98}$$

Thus, the state equation under non-linear form (because of the coupling of the two energies) $\dot{x}(t) = f(x,u)$ can be written after minor transformations as

$$\begin{cases} \dot{p}_3 = P_e - R_e\left(\frac{p_3}{I}\right)^2 - \frac{m_7}{C_h}, \\ \dot{m}_7 = \frac{p_3}{I} - \frac{u_1}{R_s}\text{sign}\left(\frac{m_7}{C_h} - P_s\right)\sqrt{\left|\frac{m_7}{C_h} - P_s\right|}, \\ \dot{U}_{13} = \left(\frac{p_3}{I}\right)c_P T_e - \frac{U_{13}}{c_V m_7}\left(\frac{1}{R_m} + c_P\frac{u_1}{R_s}sign\left(\frac{m_7}{C_h} - P_s\right)\sqrt{\left|\frac{m_7}{C_h} - P_s\right|}\right) \\ \quad + \frac{Q_{21}}{R_m C_m} + \dot{Q}_e(u_2), \\ \dot{Q}_{21} = \frac{U_{13}}{m_7 c_V R_m} - \frac{Q_{21}}{C_m}\left(\frac{1}{R_m} + \frac{1}{R_a}\right) + \frac{T_a}{R_a}. \end{cases} \tag{2.99}$$

In Equation 2.84, the ratio Q/C (in the thermal domain) represents temperature, which is analogous to the electricity where the ratio of charge and capacitance gives voltage. On the other hand, the temperature (voltage) of the fluid in the reservoir is the ratio of internal energy (analogous to charge) to a variable capacity which depends on mass or the hydraulic charge. The model under state space format can be easily represented in block diagram form as shown in Figure 2.44.

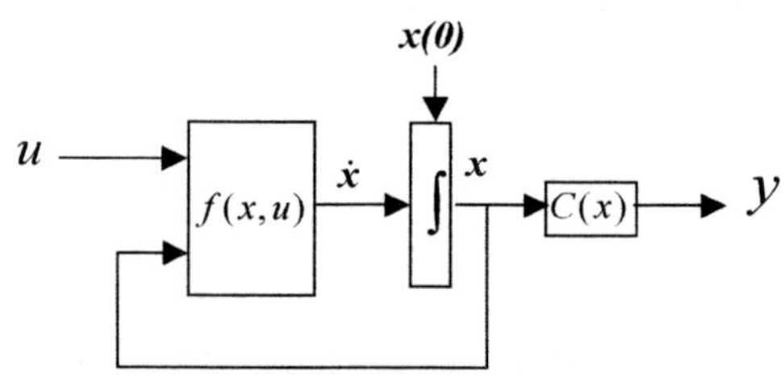

Fig. 2.44 Block diagram representation of state equations

2.8.5.4 Block Diagram

The block $f(x,u)$ represents the behavioral equations. It may be produced by synthesizing various terms of the equation or by following a more efficient approach of analyzing causalities in the bond graph model given in Figure 2.43. This block diagram (Figure 2.45) can then be simulated by using MATLAB®-Simulink® interface.

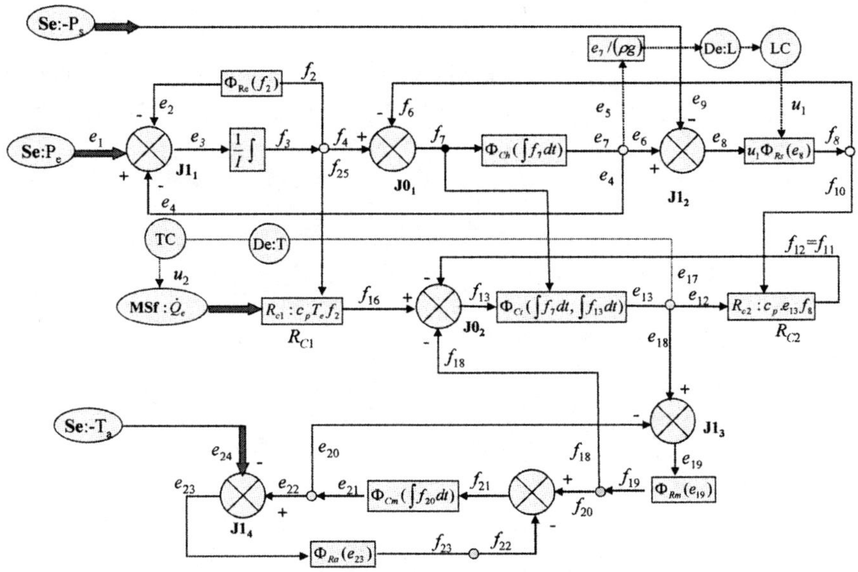

Fig. 2.45 Complete block diagram of the hot water supply system for simulation through MATLAB®-Simulink®

Problems

Modeling Problems

2.1. Consider the hydraulic process depicted in Figure 2.46. The process consists of three tanks: T_1, T_2 and T_3. The water flows (Q_{12}) through the valve V_3 from T_1 to T_2. Tank T_1 is filled by a pump P_1 whose flow output is controlled by a controller, LC, in order to keep the water level h_1 in T_1 constant. Tank T_3 is supplied water through valves V_1 and V_2 giving mass flow rates Q_1 and Q_2, respectively. Tank T_3 is further used as the sump (source of supply water) of the pump.

The flow Q_p provided by the pump depends on the control signal u and the pressure at the bottom of tank T_3. For simplicity, the constitutive equation of the pump is assumed to be linear:

$$Q_p = kuP_3 \tag{2.100}$$

where k is a coefficient (modulus of a modulated gyrator element, MGY, in a bond graph).

R_1, R_2 and R_3 are hydraulic resistances of the valves. A_i $(i = 1,2,3)$ are the cross sectional areas of cylindrical tanks (m^2), g is the gravity (m/s^2) and ρ is the density of water (kg/m^3).

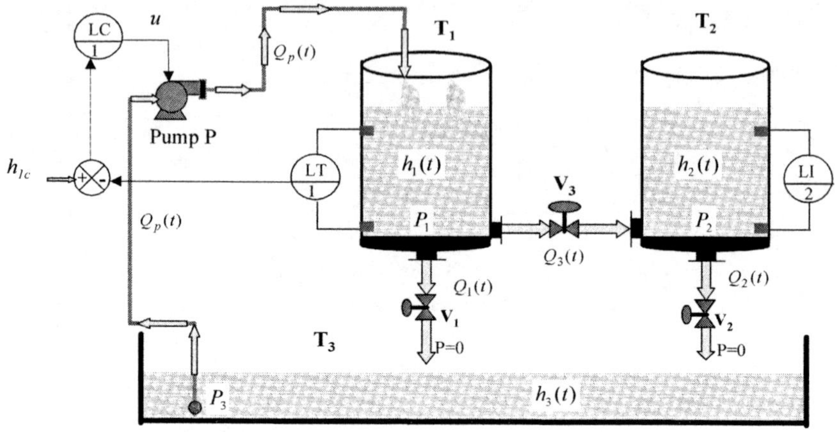

Fig. 2.46 A three-tank controlled hydraulic system

The objective is to study the dynamics of water levels h_1 and h_2, which are indicated by sensors LI_1 and LI_2.

a. Draw word bond graph of the system.
b. Construct a causal bond graph model.
c. Derive state equations.
d. Draw block diagram model corresponding to the bond graph model.

2.2. Consider the two mechanical systems in Figure 2.47. In the schematic representation of the mechanical system in Figure 2.47a, the lever can be considered elastic (with a bending stiffness k) or rigid.

a. Derive bond graph models considering elastic and rigid rod.
b. Draw block diagram representations of both bond graph models.
c. Discuss the causality problem.

The second system represents a vehicle suspension system by considering a so-called quarter-car model. The anti-leak solid rubber tyre has viscoelastic nature and it is usually approximated by a finite chain of mass-spring-damper combinations. Here we make further simplifications and assume that the viscoelastic phenomena can be described by damper b_g, springs k_{g1} and k_{g2} and mass m_g. M, b and k are the mass of the quarter vehicle, damping coefficient and stiffness of the suspension system. By considering only vertical dynamics (heave motion of the vehicle) and that the road imposes velocity $\dot{x}_r$ as a flow source.

a. Construct a bond graph model of the system.
b. Draw the corresponding block diagram representation from the bond graph model.

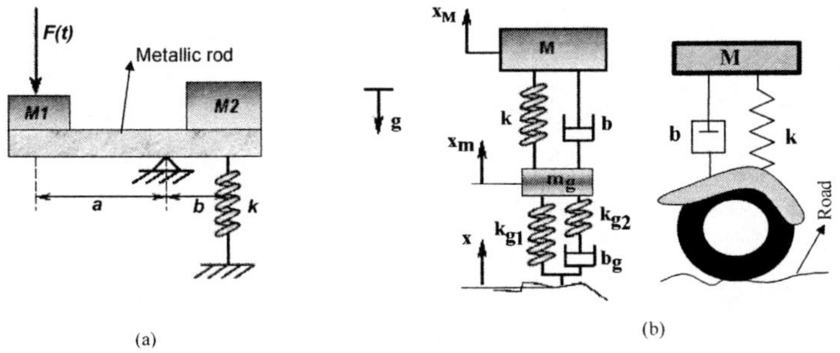

Fig. 2.47 Mechanical systems. **a** Lever. **b** Vehicle suspension system

2.3. The equivalent circuit of an electrical device is shown in Figure 2.48a.

a. Draw its bond graph model.
b. Convert the bond graph model to the corresponding block diagram form.

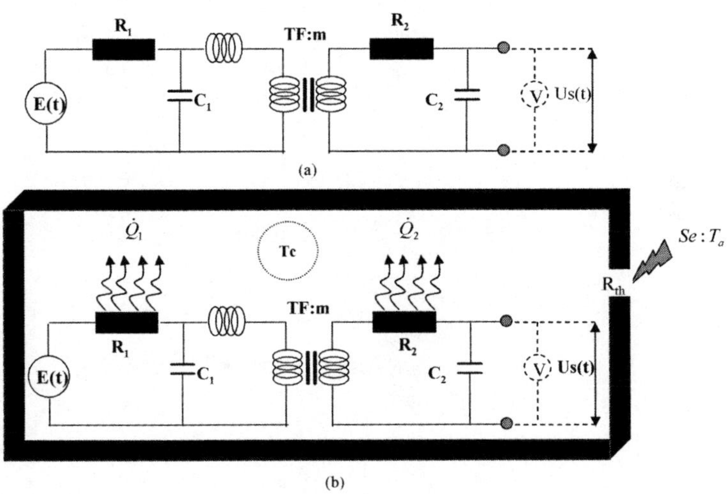

Fig. 2.48 Electrical system. **a** Isothermal network model. **b** Thermal effects on the electric circuit

Consider the case when the heat generated by the resistances are not immediately dissipated to the environment. Resistances R_1 and R_2 generate thermal power $\dot{Q}_1$ and $\dot{Q}_2$ as shown in Figure 2.48b, depending upon the current flowing through the resistors (i^2R is the dissipated thermal power). Assume T_a to be the fixed environment temperature and R_{th} to be the overall thermal resistance between the inside of the box and the external environment. Further assume that the box is thin and the temperature T_c inside the global system influences the values of resistances, such as due to thermal expansion (the reverse concept is used in Wheatstone bridge based strain gages). The mass of air inside the box is m_a. Considering only conduction heat transfer and suitably assuming nomenclature for other parameters (specific heats, densities *etc.*):

a. Develop the bond graph model of the system.
b. Draw the corresponding block diagram.
c. Discuss how the bond graph model would change if the volume inside the box is vacuum (or at a very low pressure) and thus radiation (as is usual in such cases) is the dominant mode of heat transfer. In this case, the energy stored by the mass of air inside the box is neglected, whereas heat energies stored by the resistors themselves (of masses m_1 and m_2, respectively) and the metallic box (of mass m_b) must be considered in the model. The usual form of Stefan-Boltzmann law with suitable view factors and gray body emissivities should be used as the constitutive relation of non-linear R-elements modeling the heat transfer phenomena. Note that the resistance offered by other parts of the circuit (*e.g.* the cell) are already considered in the equivalent circuit.

2.4. The schematic of a DC motor driven rotor shaft with a flywheel and bearing damping is shown in Figure 2.49. The field of the DC motor is externally energized such that the rotor's spinning speed is proportional to the current. By assuming that the bearing is rigid and neglecting the bending modes (whirling motion of the cantilever part of the rotor):

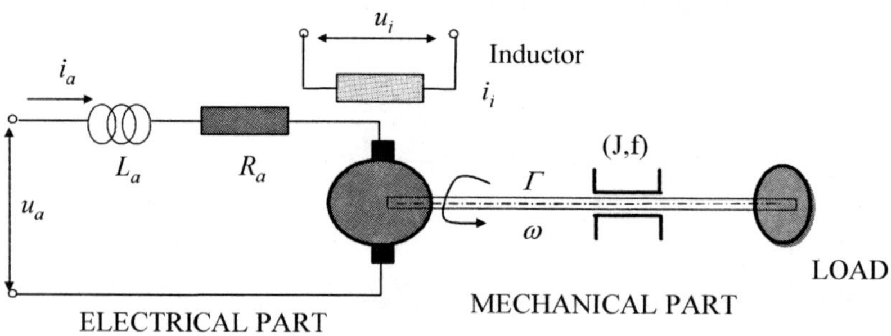

Fig. 2.49 An electromechanical system

a. Draw the word bond graph of the system.
b. Develop the bond graph model of the complete electromechanical system.
c. Draw the corresponding block diagram model.
d. Derive the state equations of the system and find the expression for the generated back emf.

N.B. For detailed models of different types of electric drives, rotor dynamics and related stability and control issues, see [172].

Project Type Simulation Problems

2.5. The liquid-spring given in Figure 2.50 has piston area A_p (excluding holes made through it) and the piston rod area is A_r. Height of the piston is t. The total mass of the liquid inside the spring is m, which is a known value. Determine the pressure on two sides of the liquid-spring supporting the given load, W, in equilibrium. What is the equilibrium position of piston (from top or bottom)? Find the stiffness of the spring for small displacement by assuming quasi-static conditions (no resistance to flow).

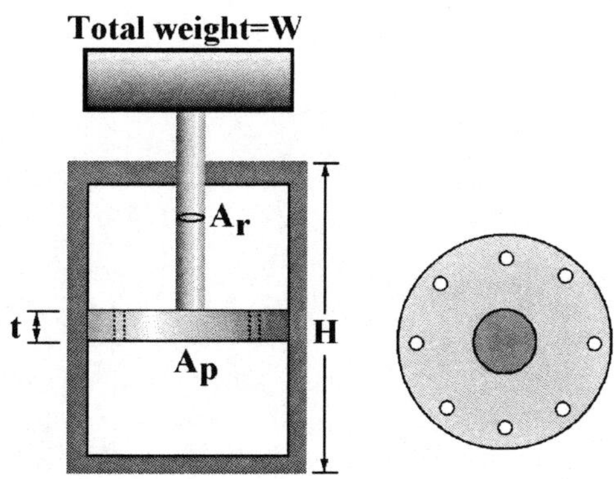

Fig. 2.50 A liquid spring

When the spring has to be used as a shock isolation system [221], holes are small and the flow through them is governed by normal flow through pipe (Bernoulli) laws. Assuming laminar flow and a total discharge coefficient of nC_d for n holes and isothermal operation, draw a bond graph model for the shock absorption system. Create a Simulink® block diagram model and simulate it for 4 s with given data and initial momentum given as 10000 kg.m/s. Assume the initial piston position at the middle of the spring and neglect wall friction. Plot the displacement of the load, pressure in the two chambers and the mass flow rate through the orifices.

Data: weight $W = 1000g$, maximum liquid column length $H - t = 0.5$m, gravity $g = 9.81$ m/s^2, piston area $A_p = 0.005$ m^2, piston rod area $A_r = 0.002$ m^2, total discharge coefficient $C_d = 1.5 \times 10^{-4}$ kg$^{1/2}$.m$^{1/2}$, compressibility of silicone fluid $\beta = 7.5 \times 10^8$ Pa and free-space (uncompressed) density of silicone fluid $\rho_0 = 970$ kg/m^3.

2.6. Consider that an air-filled chamber, as shown in Figure 2.51, is loaded by a weight to setup free vibrations of the system. Derive the constitutive relations for variation of pressure, temperature and chemical potential in terms of the state variables (volume, entropy and mass) from first principles by assuming ideal gas law. Draw the model of thermodynamic storage C-field with pressure-volume flow rate, temperature-entropy flow rate and enthalpy-mass flow rate ports [15, 173].

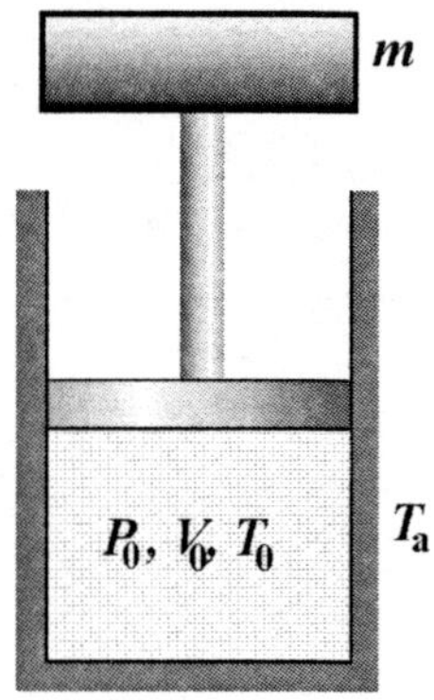

Fig. 2.51 A gas spring

a. Create a block diagram model of the entire system from a true bond graph model and by using Simulink®, simulate the system for 10 s by assuming the following data. Plot the pressure-volume diagram to show thermo-dynamic damping.
Data: initial pressure is 10^5 kg/m^2, initial temperature is 300 K, initial volume is 0.1 m^3, load mass is 1500 kg, internal radius of cylinder is 0.25 m, external radius of cylinder is 0.3 m, length of the cylinder is 0.5 m, ambient temperature is 300 K, the gas is air and the cylinder material is steel (finding proper value of heat transfer coefficient of the material as well as that for a cylinder of variable length is part of the project).

b. Change the cylinder material to silver (high thermal conductivity) and reduce wall thickness to 0.001 m to simulate approximate isothermal behavior; and increase wall thickness to 10^{100} m and select the wall material as lead (low thermal conductivity) to simulate approximate adiabatic behavior. Show that ideal isothermal and adiabatic behaviors (both are reversible) do not damp the free vibrations and thus prove that damping is due to irreversible thermodynamics. Plot the obtained isothermal and adiabatic pressure-volume diagrams in a single graph for comparison.

Chapter 3
Model-based Control

3.1 Introduction

A control system refers to a set of devices used to manipulate the output of a system by altering its inputs, called commands. There are two types of control systems: manual and automatic. We will devote our discussion to the latter in this chapter.

The desired or prescribed system output is called the reference. A controller manipulates the inputs to a system to obtain the desired effect on one or more outputs of the system, which are specified as different references. As an example, consider the speed control of an electric automobile in cruise mode. In this case, the reference behavior is specified as a constant vehicle speed. The input variable is the armature current of the drive motor, which regulates the driving torque. As we have discussed before, the electric current and mechanical torque are represented by a GY element in bond graphs, which also relates the motor speed to the back emf. The motor speed changes depending on the roughness and elevation (upward or downward slope) of the road as well as other external factors such as tail wind, tyre pressure, ambient temperature, *etc.* This means that during motion the vehicle will not be able to move at constant speed without changing its torque output, which consequently means without altering armature current of the motor. For this purpose, one needs to take note of the vehicle's velocity and modify the armature current accordingly. Such a system requires simultaneous use of sensors, controllers and actuators, and is called a closed-loop feedback control system. The controller monitors the vehicle's speed and appropriately adjusts the armature current to maintain the desired speed. The feedback compensates for disturbances to the system, such as changes in slope of the ground, road roughness, wind speed, *etc.*

A control system needs to be robust, *i.e.* it should be responsive to commands whereas it should not be affected by minor changes in system parameters and external disturbances. A robust controller should achieve its objective (reference tracking) even if the mathematical model used to design it is little imperfect. This is because no real physical system would truly behave according to its behavioral model (the set of differential equations). Moreover, controller design is usually based on reduced order system models; the simplification of mathematical models is necessary for arriving at simpler and implementable control laws as well as for fast computation (in computer or microprocessor based digital control). Furthermore, even when a perfect mathematical model of a system is used to design its controller, the numerical parameter values cannot be determined with absolute precision, *i.e.* there will always be some measurement or estimation error. Therefore, a robust controller will have to behave correctly even when the connected physical system's parameter values are marginally different from their corresponding nominal parameter values used in the controller synthesis. An adaptive control strategy uses on-line system identification and estimation of the process parameters, or modification of controller gains, so that strong robustness properties are obtained.

The advantages of closed-loop controllers over manual and open-loop controllers may be summarized as follows:

1. Disturbance rejection
2. Robustness with respect to process and measurement uncertainties, low sensitivity to process parameter changes
3. Stability of the processes
4. Better trajectory or reference behavior tracking

Some control systems implement simultaneous closed-loop and open-loop (feedforward) control, where the open-loop control helps improve the reference tracking performance.

One commonly used closed-loop controller is the PID (proportional-integral and derivative) controller. If the PID controller parameters, *i.e.* the proportional, integral and derivative gains, are improperly chosen, then the controlled process may become unstable or show steady/unsteady flutter behavior. Therefore, one needs to adjust optimally the controller gains, either statically or dynamically, to obtain the desired system response. Such adjustments are called tuning of the control loop. Improperly tuned controllers are inefficient, inaccurate, and often dangerous. Some controllers implement automatic tuning, *e.g.* a self-tuning PID controller. The tuning specifications may change from system to system depending upon its response time; while guarantying stability of the system without oscillations. Some process may require that there will be no overshoot above the setpoint while some other process may require minimizing the time or energy spent in reaching the setpoint. In process engineering systems, which are generally non-linear, PID controller gains need to be adjusted all along the start up and shut down process as well as when the setpoint changes.

One of the offline tuning method of a PID control loop involves experimentally cataloguing step response of the system at different operating conditions with different gain combinations and using a database to store this information, which would be later used during online process operation. Two of the well known online methods are Ziegler-Nichols method and Cohen-Coon method.

Sometimes feedback systems are constructed as subsystems and then they can be combined in many ways. This way of design offers flexibility of choice as well as the flexibility of switching between controllers. One example is the series connection of two controllers, referred to as cascaded control, where one control loop tries to maintain a setpoint, but then its output becomes a setpoint to another controller instead of becoming an actuating signal to an actuator. The second controller may have another setpoint and then it may act on the actuator. This way, usually two or more dynamically related process variables (*e.g.* temperature and pressure) are simultaneously controlled.

In process engineering, one has to control many process outputs and most often some of them are strongly coupled. The class of control strategies followed in this case is called a multivariable control. Model predictive control, or MPC, is one of the multivariable control methods. Many chemical process plants, power plants and oil refineries use MPC. One of the requirements of designing an MPC system is the availability of a well developed process model, which may be either analytical or empirical (*e.g.* by system identification, statistical analysis or training of a neural network model). These models are used to predict the process behavior and output with respect to the changes in the input process parameters. As an example, the weld bead geometry and strength of the welded joint may be the output parameters while pulse frequency, pulse voltage, background voltage, wire feed rate, table feed rate, and inert gas pressure, *etc.* may be the input parameters in an automated pulsed metal inert gas (PMIG) welding system. In chemical batch production processes, input parameters are generally setpoints of controllers actuating various valves (valve positioners) and heating and stirring elements, while the output variables are generally the process constraints, maximum productivity, product quality, safe operation limits.. Obviously, MPC has to be developed from optimality principles. An MPC system uses the process model and measurements from different sensors to precalculate the required changes in the setpoints such that operational constraints are satisfied.

To summarize, an MPC algorithm is an iterative finite horizon optimization of a plant model. At every sampling time, a numerical optimization algorithm computes the commands for a very short time span. The first step of this calculated command is then implemented and the output is sampled to be used again in the optimization algorithm. This way, the prediction horizon keeps on shifting forward. Therefore, MPC is also called receding horizon control. Because of the fact that MPC algorithms are executed at every sampling step, they must be very fast; *i.e.* they should be able to converge before the next sampling step. Creating such software for real time implementation is a specialized task. Honeywell, AspenTech, and Emerson are some of the well-known MPC software vendors; they also customize their software for the specific process.

One of the basic objectives of a control system is to ensure stability of the closed-loop behavior at all times. In linear systems, this may be achieved through direct feedback control or pole placement. However, non-linear systems pose great degree of difficulty in controller design because they may have multiple equilibriums (stable/unstable nodes, foci and saddle points), they may exhibit limit cycle, bifurcation and chaos behavior, their response may contain several sub and super harmonics of the input frequency and finally, they do not respect the principle of superposition. Stability of non-linear systems is often defined by Lyapunov's stability criterion, which tests the asymptotic stability of the output while disregarding the internal system dynamics. However, every process has some constraints on its internal variables.

It is sometimes possible to linearize piece-wise a non-linear system's model and then apply linear control techniques, such as gain scheduling control and adaptive control. Some control systems introduce auxiliary non-linear feedback in such a way that the system can be treated as linear for purposes of controller design, *e.g.* feedback linearization control. However, in many cases, specific control strategies are developed for specific non-linear systems. Such control strategies usually try to achieve stability in the Lyapunov sense, *e.g.* back-stepping control and sliding mode control.

From the preceding discussions, we understand that a well-developed model is a prerequisite for development of its control system. In this chapter, we will illustrate some of the physical model based control strategies. Moreover, we will emphasize the structural properties of systems which can aid in the preliminary design of control strategies and to foresee some of the problems in implementing the control algorithm. These theories establish certain thumb-rules to design quickly the required instrumentation architecture for specific control systems.

3.2 Classical Model-based Control

3.2.1 Conversion of Bond Graph Models to Signal Flow Graph Models

Bond graphs for linear systems can be easily converted to signal flow graphs (SFG) and transfer functions between inputs and outputs can be obtained for further analysis with linear control theory. The procedure for such conversion is laid out here.

3.2.1.1 Nodes of SFG (Junction Laws)

- A signal flow graph is constituted of nodes connected by a set of directed lines called branches or edges. Each branch has an associated gain and each node signifies an algebraic sum. Unlike bond graphs, SFGs do not have any power direction (the coordinate system is represented by gains $+1$ and -1), but causality is portrayed by the direction of the edges. Therefore, an SFG is a linear causal graph. The branches in an SFG simply specify traversal path and are associated with only one signal. Each bond in a bond graph is thus represented by two nodes in an SFG (effort and flow node). Activated or signal bonds are represented by only one node.

- All bonds connected to a 1-junction have the same flow and are thus represented by a single flow node. Similarly, all bonds connected to a 0-junction are represented by a single effort node. The nomenclature for a flow node representing flows in bonds of a 1-junction is $f_{i,j,k}$, where $i, j, k,...$ are the bond numbers of bonds connected to that junction. A common effort node for bonds at a 0-junction is similarly represented by $e_{i,j,k,...}$.
- All bonds connected to a 1-junction contribute separate effort nodes (*viz*, e_i, e_j,... where i, j,.. are bond numbers) leaving apart those which have been already represented in SFG. Similarly, All bonds connected to a 0-junction contribute separate flow nodes (*viz*, f_i, f_j... where $i, j, ...$ are bond numbers) except those which have been already represented in SFG.

3.2.1.2 Branches and Gains (Constitutive Relations)

- For an I-element in integral causality, the equation is $f = m^{-1} \int edt + f_0$. Because of the fact that initial conditions cannot be considered in the frequency domain, the term f_0 is neglected. Taking the Laplace transform, $f(s) = e(s)/(ms)$ or $f(s)/e(s) = 1/(ms)$. Thus the gain associated with an integrally causalled I-element is $\frac{1}{ms}$ while the branch is directed from the effort to the flow node (cause to effect). Similarly, for a differentially causalled I-element, the gain is ms and the branch is directed from the flow to the effort node.
- Cause and effect relation for an integrally causalled C-element is given by $e = K \int fdt + e_0$. Taking the Laplace transform of both sides, $e(s) = Ke(s)/s$ or $e(s)/f(s) = K/s$. Thus the gain associated with an integrally causalled C-element is K/s (or $\frac{1}{Cs}$) while the branch is directed from the flow to the effort node (cause to effect). For a differentially causalled C-element, the gain is s/K and the branch is directed from the effort to the flow node.
- The relationship between the cause and the effect for R-elements is not described by differentiation or integration. Thus the gain term doesn't contain the Laplace variable s. The gain for the R-element in resistive causality is $e(s)/f(s) = R$ whereas, for conductive causality, it is $f(s)/e(s) = 1/R$.
- Similar relationships can be established for two-ports TF and GY. The bond graph elements in different causalities and their corresponding signal flow graph representations are shown in Table 3.1.

3.2.1.3 Receptors (Junction Algebra)

- The 1-junction is a flow equalizing and effort sum junction. The strong bond for the 1-junction is the bond that has causality away from the junction. This strong bond thus provides information of flow to the junction. The weak variables of the 1-junction are efforts. The effort equation for the 1-junction is written for the weak variables (efforts), where the effort in the strong bond is expressed as

Table 3.1 Signal flow graph of bond graph elements

Bond graph	Signal flow graph
⇀| I:m	$e(s) \xrightarrow{1/ms} f(s)$
|⇀ I:m	$f(s) \xrightarrow{ms} e(s)$
|⇀ C:k	$f(s) \xrightarrow{k/s} e(s)$
⇀| C:k	$e(s) \xrightarrow{s/k} f(s)$
|⇀ R:r	$f(s) \xrightarrow{r} e(s)$
⇀| R:r	$e(s) \xrightarrow{1/r} f(s)$
$\vert\!\rightharpoonup_1$ $\overset{\mu}{\ddot{\mathrm{TF}}}$ $\vert\!\rightharpoonup_2$	$e_1(s) \xleftarrow{\mu} e_2(s)$ $f_1(s) \xrightarrow{\mu} f_2(s)$
$\rightharpoonup\!\vert_1$ $\overset{\mu}{\ddot{\mathrm{TF}}}$ $\rightharpoonup\!\vert_2$	$e_1(s) \xrightarrow{1/\mu} e_2(s)$ $f_1(s) \xleftarrow{1/\mu} f_2(s)$
$\rightharpoonup\!\vert_1$ $\overset{\mu}{\ddot{\mathrm{GY}}}$ $\vert\!\rightharpoonup_2$	$e_1(s) \xrightarrow{1/\mu} f_2(s)$ $e_2(s) \xrightarrow{1/\mu} f_1(s)$
$\vert\!\rightharpoonup_1$ $\overset{\mu}{\ddot{\mathrm{GY}}}$ $\rightharpoonup\!\vert_2$	$f_2(s) \xrightarrow{\mu} e_1(s)$ $f_1(s) \xrightarrow{\mu} e_2(s)$

signed sum of efforts in other bonds. This weak effort variable of the strong bond is called the receptor of the junction. For the 1-junction shown in Figure 3.1 the junction algebra equation is $e_2 = e_1 - e_3$. The signal flow graph representation is shown to the right of Figure 3.1.

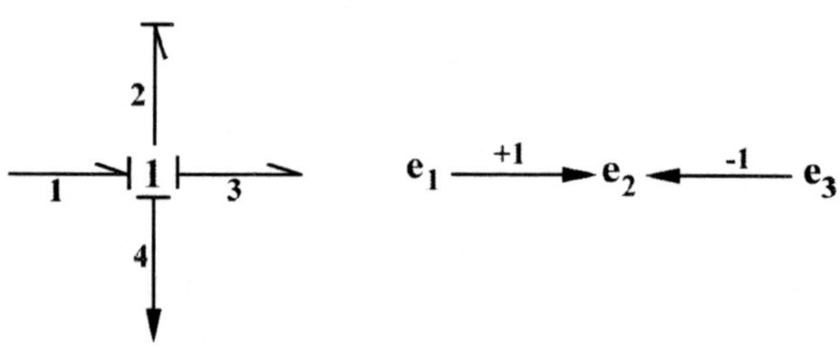

Fig. 3.1 SFG representation of 1-junction

Bond number 2 is the strong bond and the weak variable e_2 is the receptor node. Signals from other nodes for weak variables are added to this node. The gain for all branches in SFG corresponding to bonds that are power directed in opposite direction as compared to the strong bond (*i.e.* counter-oriented) is 1. Otherwise, *i.e.* for all co-oriented bonds (bonds having same power orientation as the strong bond as seen from the junction), the gain is -1.

- The receptor for a 0-junction is the weak flow variable of the strong bond (the bond that decides effort of the 0-junction, *i.e.* causalled near the junction). For the 0-junction shown in Figure 3.2, the junction algebra can be represented at the receptor node f_2 as shown to the right of Figure 3.2.

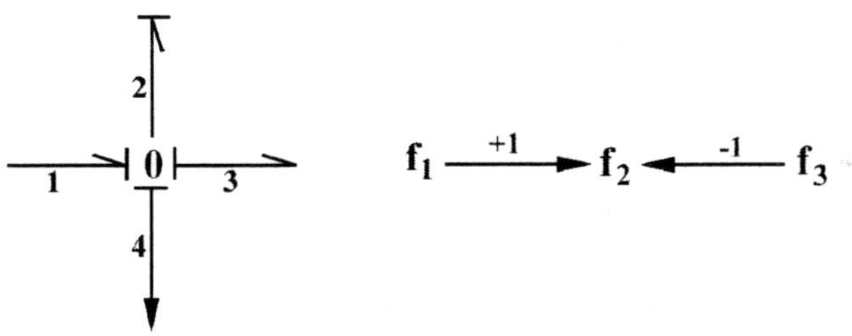

Fig. 3.2 SFG representation of 0-junction

3.2.1.4 Example

Let us consider an electromechanical system whose schematic diagram and bond graph model are shown in Figure 3.3. In the first step, the left-most 1-junction and elements connected to it in the bond graph model are considered. Node e_1 represents the input source. The part SFG is shown in Figure 3.4.

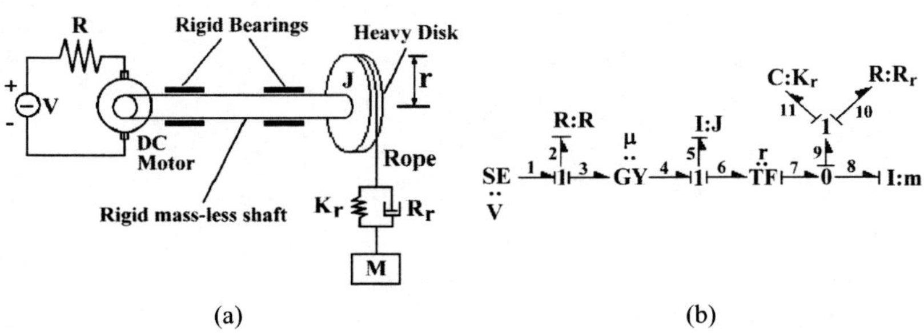

Fig. 3.3 A DC motor driven rotor

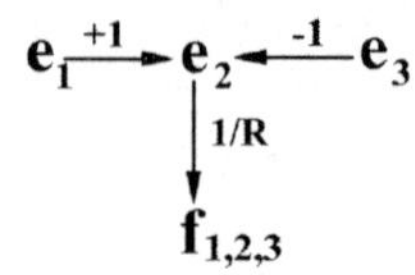

Fig. 3.4 First step in SFG construction

The next junction considered is the 1-junction for disk's angular velocity connected to bonds 4, 5 and 6. The resulting part SFG is shown in Figure 3.5.

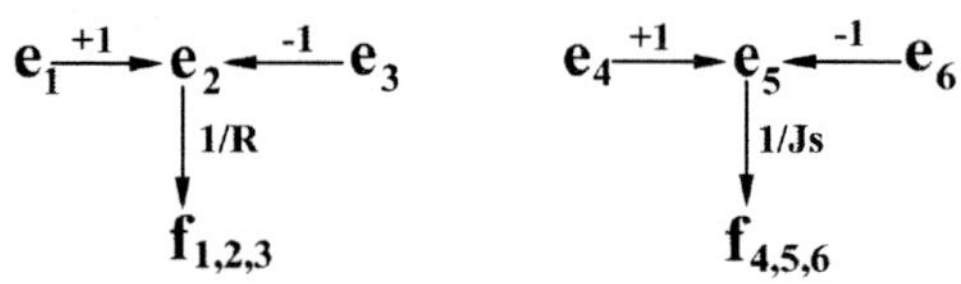

Fig. 3.5 Second step in SFG construction

The gyrator between the two 1-junctions is then represented in SFG (Figure 3.6).

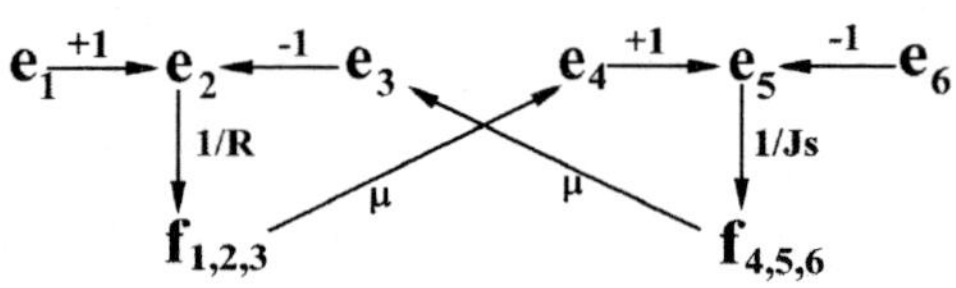

Fig. 3.6 Third step in SFG construction

In Figure 3.7, the next junction in the bond graph model (0-junction connected to bonds 7, 8 and 9) is represented in the SFG. The transformer representing rotational to linear velocity transform is then represented in the SFG in Figure 3.8. Thereafter, the last 1-junction for rope dynamics (connected to bonds 9,10 and 11) is modeled in SFG as shown in Figure 3.9.

Now the system has been completely represented as a signal flow graph. All available effort and flow nodes have been added. Signal from any flow node can be

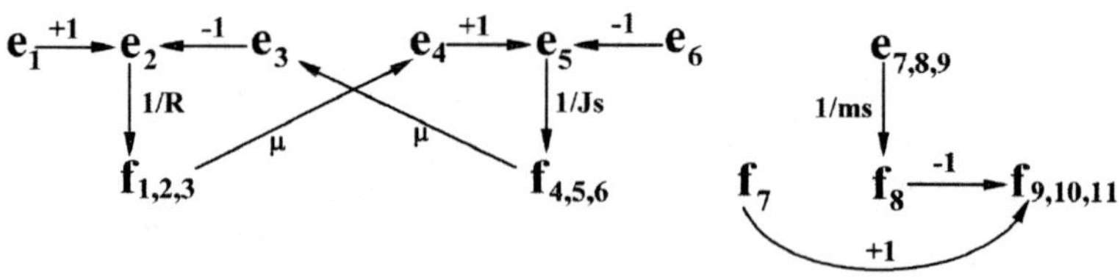

Fig. 3.7 Fourth step in SFG construction

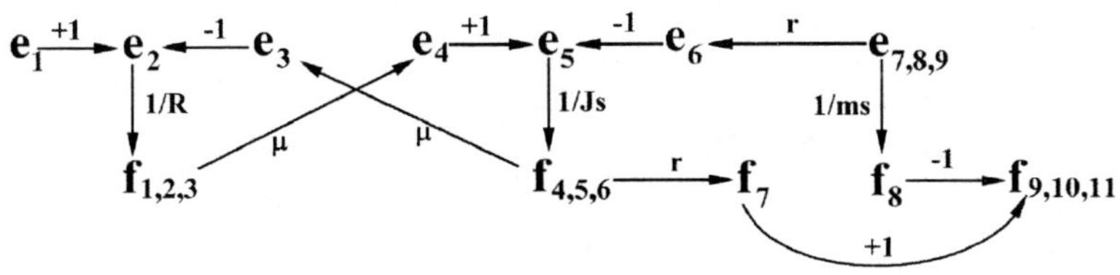

Fig. 3.8 Fifth step in SFG construction

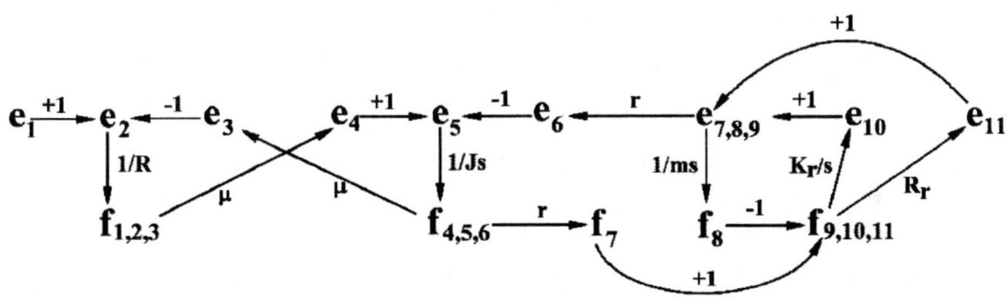

Fig. 3.9 Signal flow graph model of the example system

integrated (gain $= 1/s$) to represent a displacement node as shown in Figure 3.10. Similarly, effort signal may be integrated to get the momentum node, flow signal may be differentiated (gain $= s$) to get the acceleration node, *etc.*

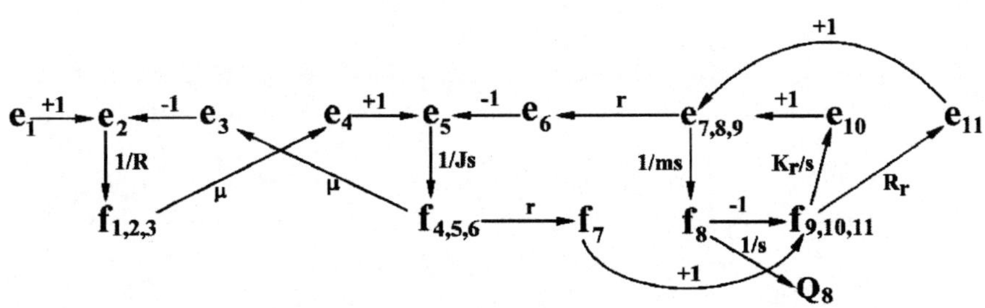

Fig. 3.10 Signal flow graph with modified output port

Let us now try to derive transfer function between the voltage source (node e_1) and the displacement of the mass (node Q_8). Lumping the loops in a way very similar to that used with block-diagrams can reduce the above SFG. The transfer function from input to output can be derived using Mason's gain rule as follows:

$$G(s) = \Sigma_i P_i \Delta_i / \Delta, \tag{3.1}$$

where P_i = gain of the i-th forward path, the graph determinant

$$\begin{aligned}\Delta = 1 &- \Sigma\text{all individual loop gains}\\ &+\Sigma\text{all possible gain products of two non-touching loops}\\ &-\Sigma\text{all possible gain products of three non-touching loops}\\ &+\ldots\end{aligned}$$

and Δ_i = the Δ for the part of the SFG which does not touch i-th forward path.

The path from node e_1 to node f_9 is in the forward path, which then follows two paths: one *via* e_{10} and the other *via* e_{11} to the terminating node Q_8. Thus the number of forward paths is 2. The number of loops in the SFG is five: (1) $e_2 \to f_{1,2,3} \to e_4 \to e_5 \to f_{4,5,6} \to e_3 \to e_2$, (2) $e_5 \to f_{4,5,6} \to f_7 \to f_9 \to e_{11} \to e_{7,8,9} \to e_6 \to e_5$, (3) $e_5 \to f_{4,5,6} \to f_7 \to f_9 \to e_{10} \to e_{7,8,9} \to e_6 \to e_5$, (4) $f_8 \to f_9 \to e_{11} \to e_{7,8,9} \to f_8$, and (5) $f_8 \to f_9 \to e_{10} \to e_{7,8,9} \to f_8$, having corresponding loop gains L_1 to L_5. These loop gains are

$$\begin{aligned}L_1 &= -\mu^2/(RJs),\\ L_2 &= -r^2 R_r/(Js),\\ L_3 &= -r^2 K_r/\left(Js^2\right),\\ L_4 &= -R_r/(ms),\\ L_5 &= -K_r/\left(ms^2\right).\end{aligned} \tag{3.2}$$

The forward path gains are

$$\begin{aligned}P_1 &= \mu r R_r/(RJms^3),\\ P_2 &= \mu r K_r/(RJms^4).\end{aligned} \tag{3.3}$$

Both forward paths touch all loops and hence $\Delta_1 = \Delta_2 = 1$. Gain products of two non-touching loops are L_1L_4 and L_1L_5. There are no three or more non-touching loops.

Thus the numerator of the transfer function is $(\mu r R_r s + \mu r K_r)/(RJms^4)$.

The denominator is $1 + \mu^2/(RJs) + r^2R_r/(Js) + r^2K_r/\left(Js^2\right) + R_r/(ms)$ $+K_r/\left(ms^2\right) + \mu^2/(RJs) \times R_r/(ms) + \mu^2/(RJs) \times K_r/\left(ms^2\right)$.

Multiplying both numerator and denominator by $RJms^4$, the transfer function is obtained as

$$\frac{Q_8(s)}{e_1(s)} = \frac{\mu\, r\, R_r\, s + \mu\, r\, K_r}{\begin{array}{c}RJms^4 + (\mu^2\, m + r^2R_rR\, m + R_rR\, J)s^3\\ + (r^2K_rR\, m + K_rR\, J + \mu^2R_r)s^2 + \mu^2K_rs\end{array}}. \tag{3.4}$$

Once the transfer function is obtained with symbolic coefficients, one may apply Routh's criteria to obtain the stability conditions for both the open and closed loop systems. Symbolic algebra can be performed manually or by using software like Reduce®, Mathematica®, *etc.* Assigning parameter values reduces the transfer function to a ratio of two polynomials in s with numeric coefficients and usual control theoretical approaches can be used both in frequency domain (Bode, Nyquist plots *etc.*) and time domain (through inverse Laplace transform). Root loci analysis for stability analysis may be conducted. The transfer function can be converted to digital domain by taking z-transforms and then digital control system analysis can be performed.

3.2.2 Transfer Function from State-space Models

The state-space model for a linear system is easily obtained from the state equations derived from a bond graph model. The state-space description of a dynamic system is given in the form

$$\begin{aligned} \dot{x} &= \mathbf{A}x + \mathbf{B}u, \\ y &= \mathbf{C}x + \mathbf{D}u, \end{aligned} \tag{3.5}$$

where x is the vector of states (Ps and Qs), n is the number of states, $\mathbf{A}$ is $n \times n$ square matrix, $\mathbf{B}$ is $n \times m$ matrix, m is the number of sources, u is the vector of sources (Ses and Sfs), y is the vector of outputs, l is the number of such outputs, $\mathbf{C}$ is $l \times n$ matrix and $\mathbf{D}$ is $l \times m$ matrix. Taking Laplace transform of Equation 3.5,

$$\begin{aligned} \mathbf{I}x(s) &= \mathbf{A}sx(s) + \mathbf{B}u(s) \text{ or } (s\mathbf{I} - \mathbf{A})x(s) = \mathbf{B}u(s), \\ y(s) &= \mathbf{C}x(s) + \mathbf{D}u(s), \text{ or } y(s) = \mathbf{C}\,(s\mathbf{I} - \mathbf{A})^{-1}\,\mathbf{B}u(s) + \mathbf{D}u(s), \end{aligned} \tag{3.6}$$

where $\mathbf{I}$ is $n \times n$ identity matrix.

The transfer function matrix $\mathbf{G}(s)$ between input vector(s) $u(s)$ and output vector(s) $y(s)$, satisfying $y(s) = \mathbf{G}(s)u(s)$, is given by

$$\begin{aligned} \mathbf{G}(s) &= \mathbf{C}\,(s\mathbf{I} - \mathbf{A})^{-1}\,\mathbf{B} + \mathbf{D} \\ &= \frac{\mathbf{C}\ \text{Adjoint}\,(s\mathbf{I} - \mathbf{A})\ \mathbf{B} + \mathbf{D}}{|s\mathbf{I} - \mathbf{A}|} \\ &= \frac{\mathbf{N}(s)}{D(s)}, \end{aligned} \tag{3.7}$$

where $\mathbf{N}(s)$ is matrix of the numerator polynomials and the denominator polynomial $D(s) = |s\mathbf{I} - \mathbf{A}|$. The poles of the system are, thus, the eigenvalues of matrix $\mathbf{A}$.

Symbolically deriving transfer functions of higher order systems is resource and time intensive. However, if numeric parameter values are assigned, the transfer function matrix can be easily obtained from numeric **A**, **B**, **C** and **D** matrices. Let us consider the system and its bond graph model used in the earlier example (Figure 3.3).

The state equations for the model are

$$\begin{aligned}
\dot{P}_8 &= R_r(rP_5/J - P_8/m) + K_r Q_{11}, \\
\dot{P}_5 &= \mu/R(SE_1 - \mu P_5/J) - r(R_r(rP_5/J - P_8/m) + K_r Q_{11}), \\
\dot{Q}_{11} &= rP_5/J - P_8/m,
\end{aligned} \tag{3.8}$$

where the state vector $x = \begin{bmatrix} P_8 \; P_5 \; Q_{11} \end{bmatrix}^T$.

Let us define only one observed output as the velocity of the hung-mass, or the variable f_8. From equations, $f_8 = P_8/m$. Thus, the matrices in the state space quadruple are

$$\mathbf{A} = \begin{bmatrix} -R_r/m & R_r r/J & Kr \\ R_r r/m & -\mu^2/RJ - r^2 R_r/J & -rK_r \\ -1/m & r/J & 0 \end{bmatrix}, \quad \mathbf{B} = \begin{bmatrix} 0 \\ \mu/R \\ 0 \end{bmatrix},$$
$$\mathbf{C} = \begin{bmatrix} 1/m \; 0 \; 0 \end{bmatrix} \text{ and } \mathbf{D} = 0. \tag{3.9}$$

Thus,

$$s\mathbf{I} - \mathbf{A} = \begin{bmatrix} s + R_r/m & -R_r r/J & -Kr \\ -R_r r/m & s + \mu^2/RJ + r^2 R_r/J & rK_r \\ 1/m & -r/J & s \end{bmatrix} \tag{3.10}$$

and the characteristic polynomial is

$$\begin{aligned}
|s\mathbf{I} - \mathbf{A}| &= (s + R_r/m) \times (s \times (s + \mu^2/(RJ) + r^2 R_r/J) \\
&\quad + r^2 K_r/J) - R_r r/J \times (s \times R_r r/m + rK_r/m) \\
&\quad - K_r * (R_r r^2/(mJ) - (s + \mu^2/(RJ) + r^2 R_r/J)/m), \\
D(s) &= s^3 + \left(\frac{\mu^2}{RJ} + \frac{r^2 R_r}{J} + \frac{R_r}{m}\right) s^2 + \left(\frac{K_r}{m} + \frac{r^2 K_r}{J} + \frac{\mu^2 R_r}{mRJ}\right) s \\
&\quad + \left(\frac{K_r \mu^2}{mRJ}\right).
\end{aligned} \tag{3.11}$$

The term in the first row and second column of the adjoint matrix of $s\mathbf{I} - \mathbf{A}$ is $K_r/J + (R_r r/J)s$. Other terms of the matrix are not relevant since both matrices **C** and **B** are sparse. So the numerator polynomial is $(1/m)(K_r/J + (R_r r/J)s)(\mu/R)$. After multiplying both numerator and denominator by mRJ, the transfer function obtained is

$$\frac{y(s)}{u(s)} = \frac{\mu\, r\, R_r\, s + \mu\, r\, K_r}{\begin{array}{l} RJms^3 + (\mu^2\, m + r^2 R_r R\, m + R_r R\, J)s^2 \\ + (r^2 K_r R\, m + K_r R\, J + \mu^2 R_r)s + \mu^2 K_r \end{array}}. \tag{3.12}$$

This transfer function, from the input excitation to the velocity of the hung mass, can be integrated (by multiplying with $1/s$) to obtain the transfer function up to the displacement of the mass. The integrated transfer function is the same as that obtained through signal flow graph method. However, state-space models can be used for advanced control theoretical analysis, whereas transfer function models from SFG can be used for frequency domain analysis only. Note that initial conditions cannot be included in the time domain analysis by using transfer function models. Transfer function models can be realized in many possible state-space forms.

3.2.3 Conversion of Bond Graph Models to Block Diagram Models

Bond graph models may be directly converted into corresponding block diagram form, as has been discussed in the previous chapter. For linear systems, the bond graph model may as well be converted into a block diagram form by using a signal flow graph representation as an intermediate step. Each node of an SFG having more than one input represents a sum block in a block diagram. The branch gains in an SFG are represented as blocks in the block diagram form. Thus, conversion to a block diagram form is straightforward and can be easily algorithmized.

However, SFG representation is limited to linear systems only. A block diagram representation is more general in the sense that it allows representation of non-linear constitutive relations. However, these two forms of representation are at a mathematical level and they have no relation to the structural or physical level of the system, as portrayed clearly in a bond graph model.

3.2.4 Example I: Physical Model-based Control

Consider an articulated electrical vehicle shown in Figure 3.11. We assume pure rolling (without slip) and further neglect the rolling friction with the ground. The velocity of the front wagon is measured and we want to control it through an actuator (a current controlled electric motor) at the rear wagon. The objective is to determine a suitable feedback gain.

The torque produced at the driving wheel is proportional to the armature current, where the constant of proportionality is μ_m. This is modeled as a bond graph GY element. Furthermore, the driving force times wheel radius (r) is the torque, which is modeled as a bond graph TF element. The consecutive GY and TF elements reduce to an equivalent GY element with modulus $\mu = \mu_m/r$, as given in the bond graph model of the system (Figure 3.12).

To determine the feedback gain, we consider the open-loop system and represent it as an SFG in Figure 3.13. This SFG can be further reduced to a form shown in Figure 3.14 where the input and output nodes are clearly mentioned.

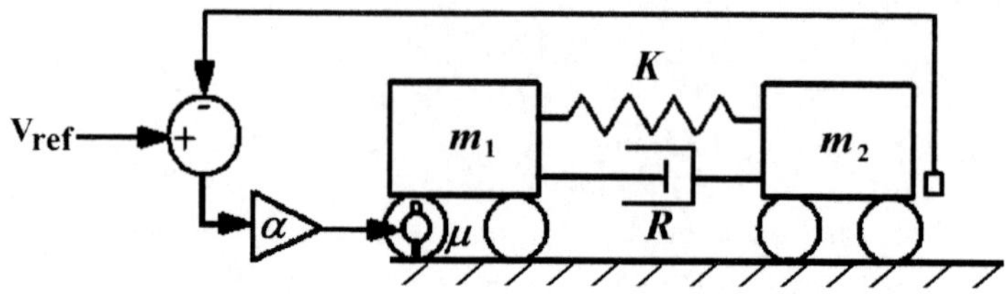

Fig. 3.11 An articulated vehicle

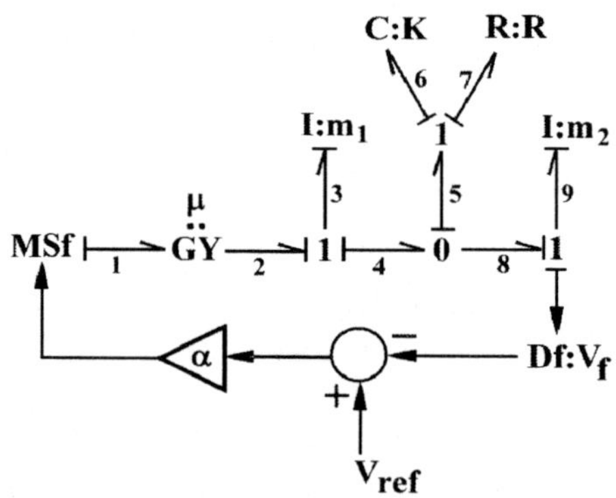

Fig. 3.12 Bond graph model of articulated vehicle

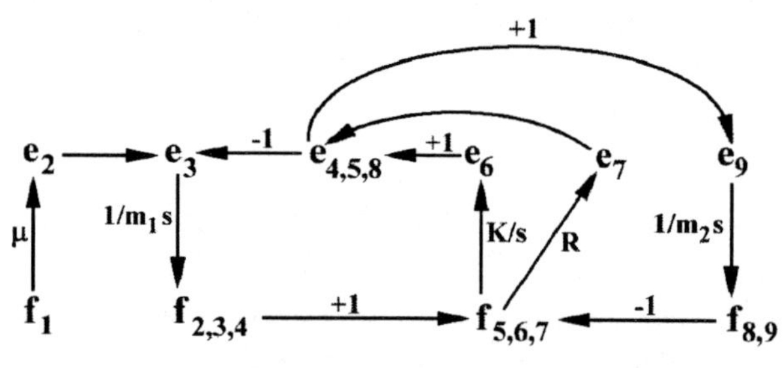

Fig. 3.13 Signal flow graph model of articulated vehicle

Considering the two loops in Figure 3.14 and no non-touching loops, the input-output transfer function is derived as

$$G(s) = \frac{y(s)}{u(s)} = \frac{\mu R s + \mu K}{m_1 m_2 s^3 + (m_1 + m_2) R s^2 + (m_1 + m_2) K s}. \tag{3.13}$$

Now one may construct the Routh array for the characteristic polynomial

$$\Phi(s) = m_1 m_2 s^3 + (m_1 + m_2) R s^2 + ((m_1 + m_2) K + \alpha \mu R) s + \alpha \mu K \tag{3.14}$$

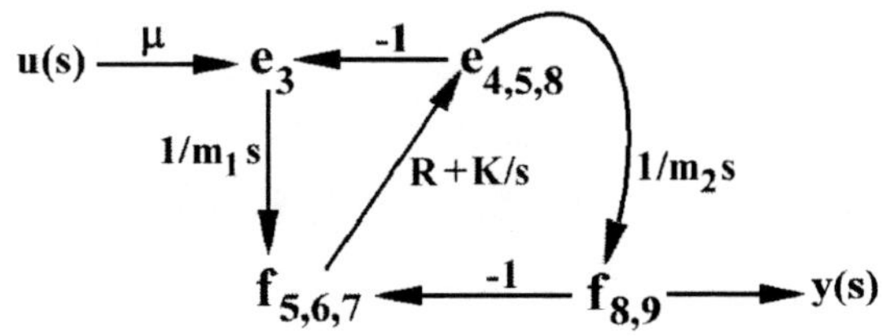

Fig. 3.14 Simplified SFG model of articulated vehicle

as

$$\begin{array}{lll} s^3: m_1 m_2 & (m_1+m_2)K+\alpha\mu R & 0 \\ s^2: (m_1+m_2)R & \alpha\mu K & 0 \\ s^1: (m_1+m_2)K+\alpha\mu R-\dfrac{m_1 m_2 \alpha\mu K}{(m_1+m_2)R} & 0 & \\ s^0: \quad \alpha\mu K & & \end{array}. \tag{3.15}$$

From the second row, we find that one of the necessary conditions to stabilize this system is $R > 0$. From the fourth row, we observe that $\alpha > 0$ is required and finally considering the third row, the overall stability domain is given by

$$R > 0 \text{ and } 0 < \alpha < \frac{KR(m_1+m_2)^2}{m_1 m_2 \mu K - \mu(m_1+m_2)R^2}. \tag{3.16}$$

With a representative set of parameter values chosen as $m_1 = 10$ kg, $m_2 = 1$ kg, $K = 1000$ N/m, $R = 5$ Ns/m, $\mu_m = 0.2$ and $r = 0.1$ m ($\mu = 2$), the stable gain range is found to be $0 < \alpha < 31.105$. The step responses of the system for various gain values in the stable regime are plotted in Figure 3.15. Note that because of the gain appearing in the forward path, the closed loop transfer function is $G_f(s) = \alpha G(s)/(1+\alpha G(s))$. The step response is the plot of inverse Laplace transform of $G_f(s)/s$, because the Laplace transform of unit step is $1/s$.

One may observe from these step responses that when the feedback gain is low, the settling time is large whereas when the feedback gain is on the higher side, there is considerable overshoot. As a trade off, a gain value of 10 seems proper for the considered system.

3.2.5 Example II: Physical Model-based System Design

Consider the schematic diagram of a vigilance system antenna as shown in Figure 3.16. The antenna is very light and is easily excited by air flow. Air flow across the bluff-body creates lift forces in the same direction of motion of the body. Such a phenomenon is called a flow induced vibration and it is also a concern in design of aircraft structures, tall buildings, *etc*. The flutter due to airflow can be considered as a damper with negative damping, which is anchored to the inertial frame. Such a damper is also referred to as a sky-hook damper. In Figure 3.16, M_e, K_e and R_e are,

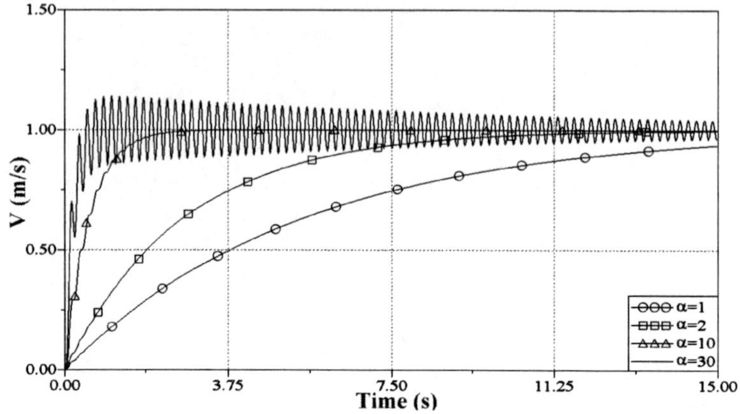

Fig. 3.15 Response of the articulated vehicle with different feedback gains

respectively, the equivalent mass, stiffness and damping of the slender equipment with a tip load. M_f, K_f and R_f are the foundation mass, stiffness and damping, respectively. The sky hook damper is R_s and scaled down values of all the parameters are mentioned in the figure.

We will consider the dynamics of the system in the vertical direction. The objective is to design proper foundation damping such that flutter instability can be avoided.

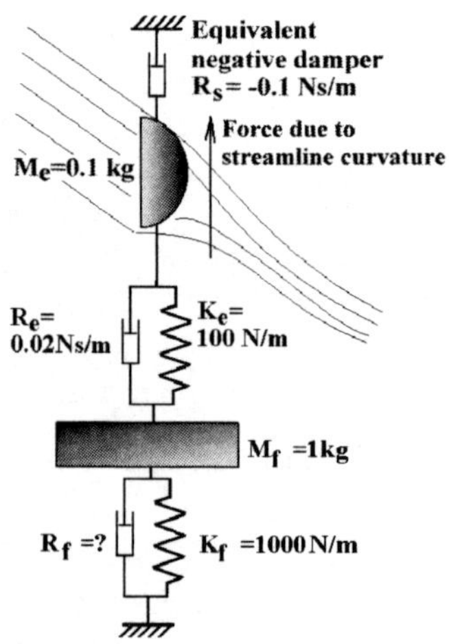

Fig. 3.16 Schematic diagram of a vigilance system antenna

We recognize that the foundation damper applies a force (effort) on the mass M_f, which is proportional to the velocity of the mass M_f and in the opposite direction to the velocity. Therefore, we can convert the above problem to a control problem as shown in Figure 3.17 and try to find a proper gain value R_f (a proportional feedback gain) for stability of the system. Such a way of looking at a system design problem from a control theoretic perspective is called a *driving point impedance* design (in fact, M_f, K_f and R_f, together, constitute the driving point impedance).

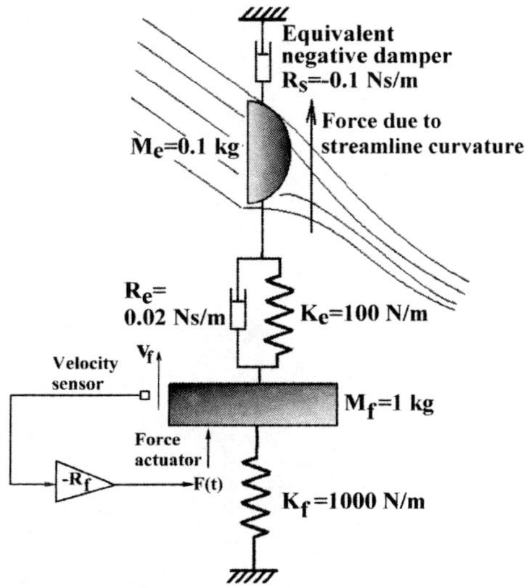

Fig. 3.17 Representation of foundation damping from an alternative view-point

We will deal with this problem numerically and start from the state space equations obtained from the bond graph model of the system shown in Figure 3.18.

The equations of motion of the open-loop system (after removing the feedback) derived from the above bond graph model are

$$
\begin{aligned}
\dot{P}_8 &= -K_e Q_6 - R_e\left(-P_1/M_e + P_8/M_f\right) - K_f Q_9 + F, \\
\dot{P}1 &= -R_s P_1/M_e + K_e Q_6 + R_e\left(-P_1/M_e + P_8/M_f\right), \\
\dot{Q}_9 &= P_8/M_f, \\
\dot{Q}_6 &= -P_1/M_e + P_8/M_f, \\
V_f &= P_8/M_f.
\end{aligned}
\tag{3.17}
$$

By substituting numerical values of parameters in Equation 3.17, the state space form with state vector $x = \left[P_8\ P_1\ Q_9\ Q_6\right]^T$ and output $y = V_f$ can be written in the form of the following matrices:

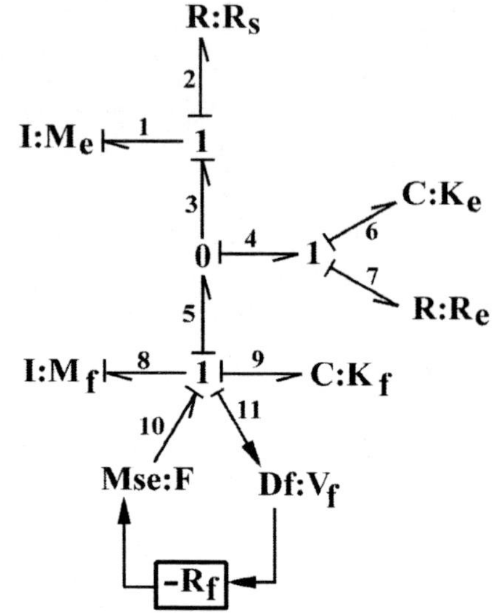

Fig. 3.18 Bond graph model of the vigilance system antenna

$$\mathbf{A} = \begin{bmatrix} -0.02 & 0.2 & -1000 & -100 \\ 0.02 & 0.8 & 0 & 100 \\ 1 & 0 & 0 & 0 \\ 1 & -10 & 0 & 0 \end{bmatrix}, \mathbf{B} = \begin{bmatrix} 1 \\ 0 \\ 0 \\ 0 \end{bmatrix},$$
$$\mathbf{C} = \begin{bmatrix} 1\ 0\ 0\ 0 \end{bmatrix}, \text{ and } \mathbf{D} = 0. \tag{3.18}$$

These state space matrices will be used later in this chapter for different ways of system stabilization. Note that the eigenvalues of matrix **A** are $0.131651 \pm 37j$ and $0.25834 \pm 27j$; which means that all the eigenvalues of the open loop system have positive real parts and hence the open loop system is unstable. We first derive the transfer function of this SISO (single input single output) system:

$$G(s) = \frac{s^3 - 0.8s^2 + 1000s}{s^4 - 0.78s^3 + 2099.98s^2 - 900s + 1000000}. \tag{3.19}$$

The poles of the above transfer function are $0.131651 \pm 37j$ and $0.25834 \pm 27j$ (same as the eigenvalues of matrix **A** if there is no pole-zero cancellation) and the zeroes are 0.0 and $0.4 \pm 31.6j$. Routh stability analysis of the above transfer function shows that the feedback system will be stable for the gain range (value of R_f) 1.226 to 124.97. However, we will draw a root loci plot of the transfer function in Equation 3.19 for further examination, as given in Figure 3.19.

As the gain is increased from 0 (poles marked with ×) to ∞ (zeroes marked with o), the eigenvalues move along the loci and all roots become stable when the gain exceeds 1.226. However, as the gain is increased further to 124.97 and beyond, one of the roots becomes unstable (as shown in the zoomed root loci in Figure 3.20).

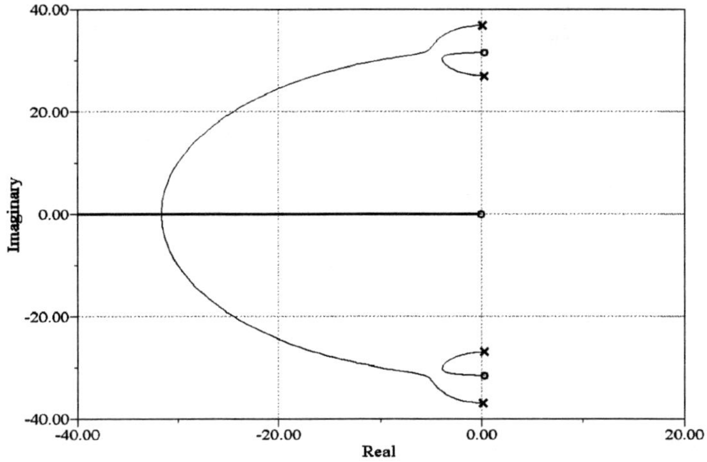

Fig. 3.19 Root loci plot

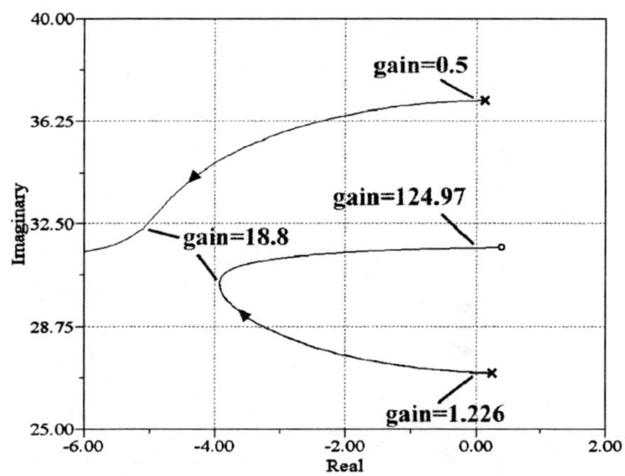

Fig. 3.20 Zoomed root loci plot

The optimum damping is obtained when we have the eigenvalues with the largest possible negative real part. In this case, the corresponding gain is found to be 18.8. Establishing a negative feedback with gain of 18.8, we have a transfer function

$$G_f(s) = \frac{G(s)}{1+18.8G(s)} = \frac{s^3 - 0.8s^2 + 1000s}{s^4 + 18.02s^3 + 2084.94s^2 + 17900s + 1000000}. \quad (3.20)$$

The inverse Laplace transform of $G_f(s)$ is given by

$$y_f(t) = -1.875\cos(30.26t + 1.591)\,e^{-3.925t} + 2.38\cos(32.37 + 1.155)\,e^{-5t}. \quad (3.21)$$

In fact, Equation 3.21 describes the response of the system to unit impulse, which is plotted in Figure 3.21.

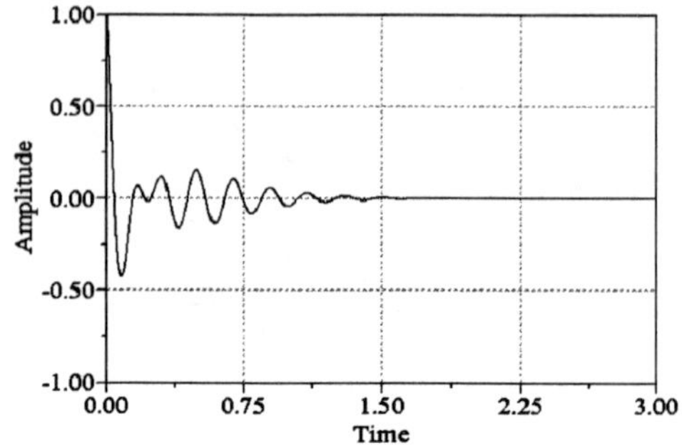

Fig. 3.21 Impulse response of the vigilance system antenna

We leave it to the reader to verify that if the system in Figure 3.16 is represented in block diagram form and is simulated with $R_f = 18.8$ N.s/m and an initial velocity of 1 m/s given to the foundation mass (I:M_f), then the same response given in Figure 3.21 is obtained.

One may now have a closer look at the system dynamics from the point of view of physics of the problem: why does the system become unstable for low as well as high values of foundation damping? The negative damping draws energy from the environment (as opposed to positive damping); the source of energy in this case is the air flowing over the bluff body. This imported energy should be dissipated by other components of the system such that the system remains stable. This is why we need some minimum damping in the foundation. On the other hand, as the foundation damping increases, the entire foundation becomes very hard, just like a jammed shock isolator of a vehicle. This means that the foundation will not be able to provide sufficient energy dissipation; it is almost like mass M_f and everything below it becoming part of the ground structure. Hence the remaining week damper R_e is not enough to stabilize the system. The energetic interactions between various components in this example would become clearer when we introduce the concept of activity index in the next chapter.

Note that if we consider the designed value of the damper in the model and replace the foundation stiffness as a displacement dependent force feedback, then we can also design a proper foundation stiffness (another part of the driving point impedance). Iteratively, we can arrive at the best spring-damper parameters for the foundation. It is worthwhile to mention that the combined spring damper combination acts as a PI (proportional-integral) feedback where velocity of the foundation mass is the controller input and a force on the foundation mass is the controller output.

3.3 Causal Paths

One of the basic concepts used in structural analysis of models is the understanding of the information exchange taking place across a broad spectrum of elements in the

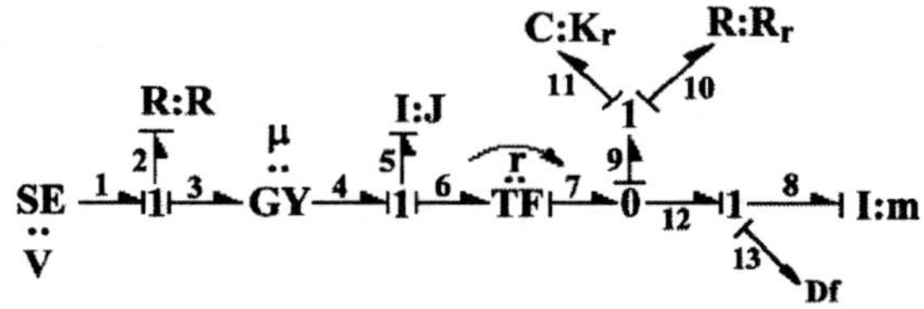

Fig. 3.22 Bond graph model of the electro-mechanical system

model. In the previous chapter we have discussed causality of different elements, two ports and junctions, which may be considered as a localized causal assignment concept in which SCAP was followed to arrive at computable models. However, if one is interested in studying which elements are going to influence which dynamics, then one has to interpret how the information between different segments of the graph are coupled. This would enable us to find if the influence of dynamics of a particular segment would be significant (strong interaction) or minor (weak interaction) on another segment.

Definition 3.1. A causal path [59] tracks the progress of a signal through the model. In bond graph terms, a signal is one of the power variables, *i.e.* effort or flow. When the signal reaches a junction, it is either distributed (*i.e.* follows any one of the weak bonds if the signal is the strong variable of the junction) or goes to the strong bond (if it is the weak variable of the junction). When the signal reaches a passive element (I, C or R), it returns back through it but undergoes a qualitative change (*i.e.* flow variable becomes effort variable and *vice versa*). Similarly, if the signal passes through a gyrator, then there is a qualitative change. When a signal reaches a source or a sensor, it is terminated there.

Definition 3.2. When a causal path starts from a particular variable in a particular bond and returns to the same variable in the same bond again, it is called a closed path and the path is terminated there.

Definition 3.3. A closed causal path in which no passive elements are encountered (*i.e.* no I,C or R) is a causal loop [173, 263, 264].

Definition 3.4. A closed causal path in which all passive elements encountered are either resistive elements or differentially causalled storage elements is an algebraic loop [43, 173].

3.3.1 Transfer Functions from Bond Graph Models

Consider the bond graph of the system described earlier in Figure 3.3a, whose bond graph model is reproduced here in Figure 3.22 for ready reference. We have used this bond graph model earlier to illustrate the procedure of constructing signal flow graphs.

In Figure 3.22, let us try to find the causal paths from the source to the output, *i.e.* the velocity of the mass being lifted (f_{13}). The starting signal in this case is e_1. Because e_1 is the weak variable in the next 1-junction, it goes to the strong bond, *i.e.* bond number 2. We will write this as $e_1 \rightarrow e_2$. In fact, this traversal can be visually done: bond number 1 has a causality towards the one junction and so it can take a path along any bond having causality away from the junction. In this case, because we have a 1-junction, there is only one bond having causality away from the junction. If it were a 0-junction, then all other bonds would have causality away from the junction and the causal path could have traversed along any one of them. All those possibilities are counted as different causal paths.

In the next step, we encounter a passive element ($R:R$ at bond number 2) through which the causal path returns with a qualitative change. This is because the R-element in this case takes effort information and returns flow information. Thus the causal path will now be written as $e_1 \rightarrow e_2 \rightarrow R:R \rightarrow f_2$.

The flow information f_2 is now distributed by the 1-junction: one goes back to the source and the causal path terminates there, *i.e.*, $e_1 \rightarrow e_2 \rightarrow R:R \rightarrow f_2 \rightarrow f_1$ (which is called a source loading) and the other path is $e_1 \rightarrow e_2 \rightarrow R:R \rightarrow f_2 \rightarrow f_3$.

The causal path now encounters a gyrator element and hence it will further undergo a qualitative change as: $e_1 \rightarrow e_2 \rightarrow R:R \rightarrow f_2 \rightarrow f_3 \rightarrow GY:\mu \rightarrow e_4$. Then we reach another 1-junction through which the path reaches and returns after qualitative change from an I-element as follows: $e_1 \rightarrow e_2 \rightarrow R:R \rightarrow f_2 \rightarrow f_3 \rightarrow GY:\mu \rightarrow e_4 \rightarrow e_5 \rightarrow I:J \rightarrow f_5$. The flow variable f_5 can follow two paths from the 1-junction; we will proceed along the forward path through the TF element (where the signal quality remains same) and reach $e_1 \rightarrow e_2 \rightarrow R:R \rightarrow f_2 \rightarrow f_3 \rightarrow GY:\mu \rightarrow e_4 \rightarrow e_5 \rightarrow I:J \rightarrow f_5 \rightarrow f_6 \rightarrow TF:r \rightarrow f_7 \rightarrow f_9$.

Bond number 9 is the strong bond of the next 1-junction and it is divided into two paths there, each one of them reaching f_{13} as follows:

$$\begin{aligned} e_1 &\rightarrow e_2 \rightarrow R:R \rightarrow f_2 \rightarrow f_3 \rightarrow GY:\mu \rightarrow e_4 \rightarrow e_5 \rightarrow I:J \rightarrow f_5 \rightarrow f_6 \\ &\rightarrow TF:r \rightarrow f_7 \rightarrow f_9 \rightarrow f_{10} \rightarrow R:R_r \rightarrow e_{10} \rightarrow e_9 \rightarrow e_{12} \\ &\rightarrow e_8 \rightarrow I:m \rightarrow f_8 \rightarrow f_{13}, \end{aligned} \tag{3.22}$$

and

$$\begin{aligned} e_1 &\rightarrow e_2 \rightarrow R:R \rightarrow f_2 \rightarrow f_3 \rightarrow GY:\mu \rightarrow e_4 \rightarrow e_5 \rightarrow I:J \rightarrow f_5 \rightarrow f_6 \\ &\rightarrow TF:r \rightarrow f_7 \rightarrow f_9 \rightarrow f_{11} \rightarrow K:K_r \rightarrow e_{11} \rightarrow e_9 \rightarrow e_{12} \\ &\rightarrow e_8 \rightarrow I:m \rightarrow f_8 \rightarrow f_{13}. \end{aligned} \tag{3.23}$$

From Equations 3.22 and 3.23, we find that there are two causal paths from the source to the sensor. In fact, if we consider appropriate element gains (R and $1/R$ for R-elements in resistive and conductive causality, respectively; $1/Is$ and K/s for integrally causalled I and C elements, respectively; and the transformer and gyrator moduli as are appropriate) and signs (due to power directions), we can arrive at the same forward path gains as in Equation 3.3.

Similarly, one may proceed to identify the loops in the model. For example, if one starts from variable f_2, then one can have one of the causal paths as

$$\begin{aligned} &f_2 \rightarrow f_3 \rightarrow GY:\mu \rightarrow e_4 \rightarrow e_5 \rightarrow I:J \rightarrow f_5 \rightarrow f_6 \rightarrow f_4 \\ &\rightarrow GY:\mu \rightarrow e_3 \rightarrow e_2 \rightarrow R:R \rightarrow f_2, \end{aligned} \tag{3.24}$$

which is in fact a loop because it starts and ends with the same variable and its loop gain is same as that of $L_1 = -\mu^2/(RJs)$ in Equation 3.2; the minus sign appearing because of the power direction change during traversal of effort variable from bond number 3 to bond number 2. This way, all loops in the model (five in total) can be found and Mason's rule can be applied directly on the bond graph model to arrive at the transfer function [37, 142].

If a causal path from a source to an output does not contain any integrally causalled storage elements, then the input directly influences the output. In other words, this is called a direct input feed through and, in this case, the numerator and denominator polynomials of the transfer function will have the same order.

Remark 3.1. Signal flow graphs can be used to find causal paths in linear models whereas bond graph causal paths are more general in the sense that there is no restriction on the linearity of the model. However, in models of hybrid systems, some causal paths may get truncated in the middle due to specific changes in the parameter values of some modulated elements. Therefore, causal paths in hybrid system models are themselves time-varying.

3.3.2 Delay and Attenuation Dynamics

One of the fundamental uses of causal paths is to study the influence of dynamics of a segment on the dynamics of another segment. If the causal path from one segment reaches another segment by passing through many storage elements, then the interaction between the segments is considered to be weak. This is because of the fact that storage elements act like low-pass filters and thus a disturbance at one segment would be mildly felt at the other segment. The number of storage elements encountered between the segments decides the filter order, *i.e.* first order, second order and so on. If these filters are additionally coupled with resistive elements which form local loops containing parts of the variables in the forward path, they would damp-out the disturbance. If there are no R-elements forming loops which contain parts of the variables in the forward path, then the disturbance would not be attenuated, but reach the destination after some delay whose order is defined by the number of storage elements in integral causality in the forward causal path.

3.4 Augmented Controller and Observer Design

In feedback control, such as those discussed in previous examples, a suitable feedback gain is chosen to stabilize the system. As the feedback gain changes, the poles of the system move along the branches of the root-loci. However, a desired performance criterion may not be achieved by selecting any particular gain, *e.g.* a system which is unstable for all gains would remain unstable. Moreover, one cannot specify a desired pole location and design a specific feedback gain to achieve it; the poles are restricted to move along the root-loci. Furthermore, all the poles shift with changing feedback gain and one has no control over where the poles should shift along each branch. In this section, we introduce the concepts of pole placement for single-input-single-output (SISO) systems which also needs the concepts of controllability and observability from a classical viewpoint. This will help in understanding the importance of the results given in the next section.

3.4.1 Pole Placement

For pole placement of SISO systems, the system must be completely controllable and observable. Therefore, study of these two control properties assumes an important role from the inception of the design stage. Similarly, for full order observer design accounting for sensor noise and process disturbances by using a Kalman filter, and for optimal control applications, full controllability and observability of the system is a prerequisite. How these control properties can be determined from a system model is discussed in the next section.

The poles of the system are placed at the desired location by a state feedback, *i.e.* $u' = u + \mathbf{K}x$, where u' is the new input vector, $\mathbf{K}$ is a matrix of feedback gains and x is the state vector. Consequently, the state space form of the system becomes

$$\begin{aligned} \dot{x} &= \mathbf{A}x + \mathbf{B}(u + \mathbf{K}x) = (\mathbf{A} + \mathbf{BK})x + \mathbf{B}u, \\ y &= \mathbf{C}x + \mathbf{D}(u + \mathbf{K}x) = (\mathbf{C} + \mathbf{DK})x + \mathbf{D}u. \end{aligned} \tag{3.25}$$

The new poles of the system are the eigenvalues of matrix $\mathbf{A} + \mathbf{BK}$. Finding a proper gain matrix constitutes the pole placement problem, which can be solved in various ways. One of the commonly used methods is to bring the system to a controllable canonical form through similarity transformation, and the other commonly used method is the Ackermann's formula. We will not go into details of those methods. What is of interest is to know that all the poles of the system can be moved to desired locations if the system is completely state controllable. If the system is partially state controllable, only the controllable poles may be shifted. Alternatively, if there exists a matrix $\mathbf{K}$ such that poles of matrix $\mathbf{A} + \mathbf{BK}$ are sufficiently different from poles of matrix $\mathbf{A}$, then the system is controllable.

To implement the controller which places the poles of the system at desired locations, we need to know the states of the system. However, we only have one output but no information about the actual states of the system. Then how do we implement the feedback $\mathbf{K}x$? This is the central reason to develop an observer system [186].

An observer is a model of the plant which receives the same input as the plant. The output of the actual plant and the observer are compared and the error is fed back through a gain such that the error dynamics is asymptotically stable.

Consider the plant dynamics described by

$$\begin{aligned} \dot{x} &= \mathbf{A}x + \mathbf{B}u, \\ y &= \mathbf{C}x + \mathbf{D}u; \end{aligned} \tag{3.26}$$

and its corresponding observer dynamics given by

$$\begin{aligned} \dot{x}_o &= \mathbf{A}x_o + \mathbf{B}u + \mathbf{L}(y - y_o), \\ y_o &= \mathbf{C}x_o + \mathbf{D}u; \end{aligned} \tag{3.27}$$

where subscript 'o' refers to observer and all other variables have their usual meaning.

Then the state error dynamics may be written as

$$\begin{aligned} \dot{x}_e &= \dot{x} - \dot{x}_o \\ &= \mathbf{A}x + \mathbf{B}u - \mathbf{A}x_o - \mathbf{B}u - \mathbf{L}(y - y_o) \\ &= \mathbf{A}(x - x_o) - \mathbf{L}(\mathbf{C}x + \mathbf{D}u - \mathbf{C}x_o - \mathbf{D}u) \\ &= (\mathbf{A} - \mathbf{LC})(x - x_o) = (\mathbf{A} - \mathbf{LC})x_e. \end{aligned} \tag{3.28}$$

Obviously, if all eigenvalues of matrix $\mathbf{A} - \mathbf{LC}$ lie on the left half of the complex plane, *i.e.* have negative real part, then the error between the plant and the observer states would decrease asymptotically. The objective then is to chose a gain matrix $\mathbf{L}$ such that the observer starts tracking the plant as soon as possible. This is again a pole placement problem.

Then the general scheme of pole placement may be as shown in Figure 3.23, where $y(t)$ and $y_0(t)$ are, respectively, the plant and the observer outputs. The individual gains in this augmented controller and observer system (**K** and **L**) can be independently designed, thanks to the separation principle [14].

In Simulink®, this architecture can be represented as shown in Figure 3.24.

In Figure 3.24, we have used the readily available Simulink® state-space block by performing the following transformations:

$$\begin{aligned} \dot{x}_o &= \mathbf{A}x_o + \mathbf{B}u + \mathbf{L}(y - y_0), \\ &= \mathbf{A}x_o + \left[\mathbf{B} \mid \mathbf{I}_{n\times n}\right] \left\{ \begin{matrix} u \\ \mathbf{L}(y - y_0) \end{matrix} \right\} \\ &= \mathbf{A}_o x_o + \mathbf{B}_o u_0, \end{aligned} \tag{3.29}$$

Fig. 3.23 General schematics for pole placement

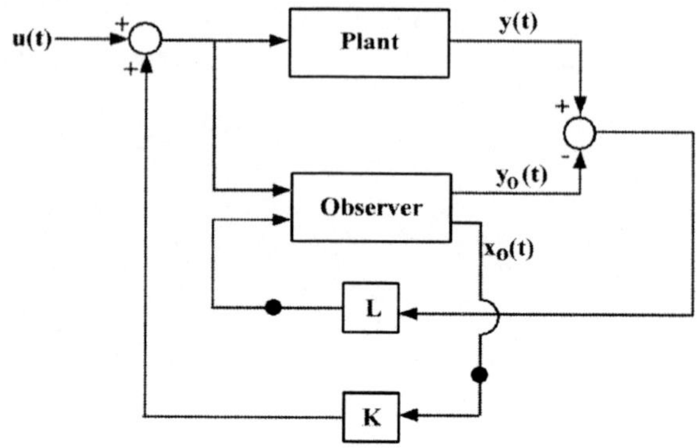

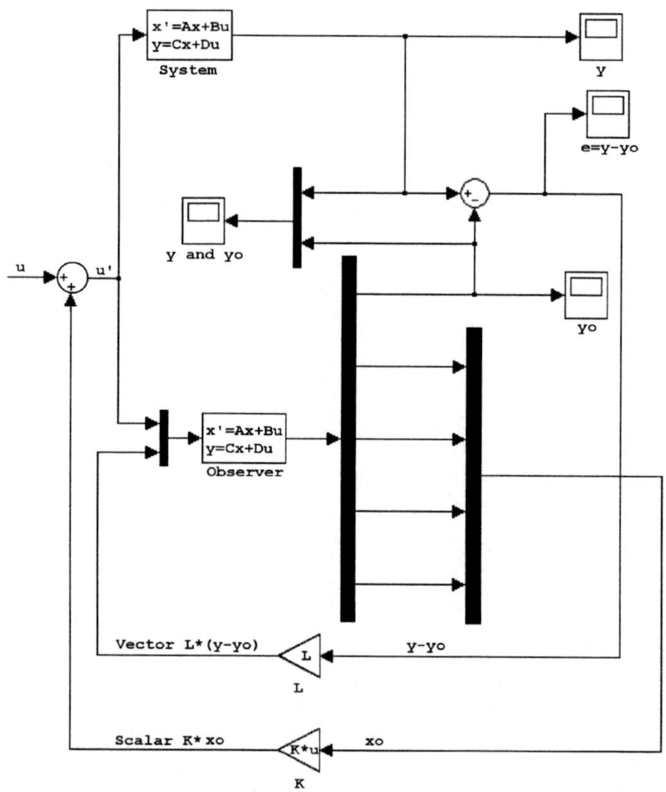

Fig. 3.24 Pole placement architecture in Simulink®

where $\mathbf{A}_o = \mathbf{A}$ and $\mathbf{B}_o = \left[\, \mathbf{B} \mid \mathbf{I}_{n\times n} \,\right]$ are used to define the state-space form of the observer, and the observer input vector is considered as $u_o = [u \;\; \mathbf{L}(y - y_0)]^T$, where $\mathbf{I}_{n\times n}$ is an identity matrix and n is the number of states.

Similarly, from the observer, we not only need output y, but also the states for incorporating feedback in the plant, *i.e.* $y_0 = \begin{bmatrix} \mathbf{C} \\ \mathbf{I}_{n\times n} \end{bmatrix} x_o$. Therefore, in Figure 3.24, the observer block has the following variables in the state-space representation:

$$\mathbf{A}_o = \mathbf{A}, \mathbf{B}_o = \begin{bmatrix} \mathbf{B} \vdots \mathbf{I}_{n\times n} \end{bmatrix}, \mathbf{C}_o = \begin{bmatrix} \mathbf{C} \\ \mathbf{I}_{n\times n} \end{bmatrix}, \mathbf{D}_o = \begin{bmatrix} \mathbf{D} & 0 & \cdots & 0 \\ 0 & 0 & & \\ \vdots & & \ddots & \\ 0 & & & 0 \end{bmatrix}. \tag{3.30}$$

The mux and demux blocks (thick vertical lines) are used in Figure 3.24 to inflate the input vector dimension (u_o) and to dissociate observer output into actual output y_0 (for comparison with plant output) and observer states (x_o).

3.4.2 Example: Active Flow-induced Vibration Isolation

We will continue with the example taken in the preceding section (Figure 3.17). In that example we designed a foundation damper with the aid of root-loci diagram. If we want better control over location of the poles, then we may replace the damper by a controller as shown in Figure 3.25a. Note that the design in the earlier example was for a passive damper whereas the new system incorporates an active foundation. Active systems have gained recent importance in many areas, *e.g.* active suspensions, active vibration dampers, active magnetic bearings, ... An active system requires four essential components as compared to a passive system: (1) sensor(s), (2) actuator(s), (3) controller and (4) power supply with power modulator. Almost all mechatronic devices are active systems.

We will consider the open loop system with the data given in Figure 3.25a. The corresponding bond graph model is given in Figure 3.25b. The numerical matrices in state-space representation are found from Equation 3.17 with the state vectors $x = \begin{bmatrix} P_8 \; P_1 \; Q_9 \; Q_6 \end{bmatrix}^T$ and output vector $y = V_f$ as follows:

$$\mathbf{A} = \begin{bmatrix} -0.02 & 0.2 & -1000 & -100 \\ 0.02 & 0.8 & 0 & 100 \\ 1 & 0 & 0 & 0 \\ 1 & -10 & 0 & 0 \end{bmatrix}, \mathbf{B} = \begin{bmatrix} 1 \\ 0 \\ 0 \\ 0 \end{bmatrix},$$
$$\mathbf{C} = \begin{bmatrix} 1 \; 0 \; 0 \; 0 \end{bmatrix}, \text{ and } \mathbf{D} = 0. \tag{3.31}$$

Note that the eigenvalues of matrix $\mathbf{A}$ are $0.131651 \pm 37j$ and $0.25834 \pm 27j$, which mean that all the eigenvalues of the open loop system have a positive real part and hence the open loop system is unstable. If we choose to place the poles of the system at some desired location, say at $-0.2 \pm 20j$ and $-0.3 \pm 10j$, then we need to determine the gain matrix $\mathbf{K}$. One may use MATLAB® to perform this pole placement by using commands *Place* or *Acker*. Note that MATLAB® places poles

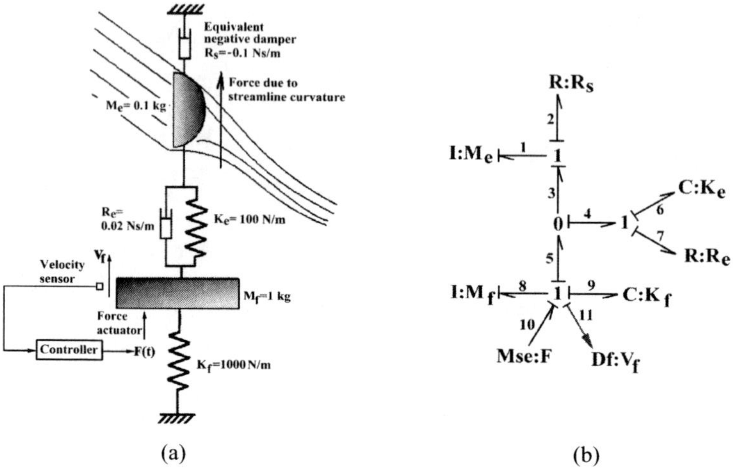

Fig. 3.25 Controller for vigilance system antenna and its open-loop model

of matrix $\mathbf{A}-\mathbf{BK}$ instead of matrix $\mathbf{A}+\mathbf{BK}$; thus negative of the gains obtained from MATLAB® will be considered here: $\mathbf{K}=\left[-1.78\ 20.05\ 956\ 638\right]$.

Observer poles are usually placed about two to ten times to the left of the system poles. There is no direct MATLAB® command to place poles of the observer. However, one may follow this simple trick: eigenvalues of matrix $(\mathbf{A}-\mathbf{LC})$ are the same as that of $(\mathbf{A}-\mathbf{LC})^T=\left(\mathbf{A}^T-\mathbf{C}^T\mathbf{L}^T\right)$. Then one may define matrices $\mathbf{A}'=\mathbf{A}^T$, $\mathbf{B}'=\mathbf{C}^T$ and $\mathbf{K}'=\mathbf{L}^T$. This allows us to place the poles of $\mathbf{A}'-\mathbf{B}'\mathbf{K}'$ by using available commands and then we take a reverse transform to obtain $\mathbf{L}=\mathbf{K}'^T$. For the given system and pole locations, we have $\mathbf{L}=\left[2.78\ -1.3475\ 0.6397\ 1.5629\right]^T$.

Naturally, the matrices used in observer's state space representation are given by

$$\mathbf{A}_o=\mathbf{A},\mathbf{B}_o=\left[\mathbf{B}\vdots\mathbf{I}_{n\times n}\right],\mathbf{C}_o=\begin{bmatrix}\mathbf{C}\\ \mathbf{I}_{n\times n}\end{bmatrix},\mathbf{D}_o=\begin{bmatrix}\mathbf{D} & 0 & \cdots & 0\\ 0 & 0 & & \\ \vdots & & \ddots & \\ 0 & & & 0\end{bmatrix},$$

where $\mathbf{I}_{n\times n}$ is a square identity matrix and $n=4$ is the number of states. These matrices can be easily created by using the following MATLAB® commands: Ao=A, Bo=[B,eye(4)], Co=[C;eye(4)] and Do=zeros(5). Consequently, the full-order augmented controller observer of the active suspension system is represented in block diagram form in Figure 3.24 with $u(t)=0$. Note that in Figure 3.24 the matrix gain block $\mathbf{L}$ is an element-wise gain (has a scalar input and vector output) whereas the matrix gain block $\mathbf{K}$ is a matrix gain (has a vector input and scalar output). Simulation results with unit momentum given to the foundation in the plant model, *i.e.* with plant initial conditions as $\left[1\ 0\ 0\ 0\right]^T$ and observer initial conditions as $\left[0\ 0\ 0\ 0\right]^T$ (because it is assumed that the observer has no prior knowledge of the plant states), are shown in Figure 3.26 . From these results, we find that (1) the system is stable (Figure 3.26a) because its poles are placed properly by the controller and (2) the observer tracks the plant with sufficient accuracy (Figure 3.26b).

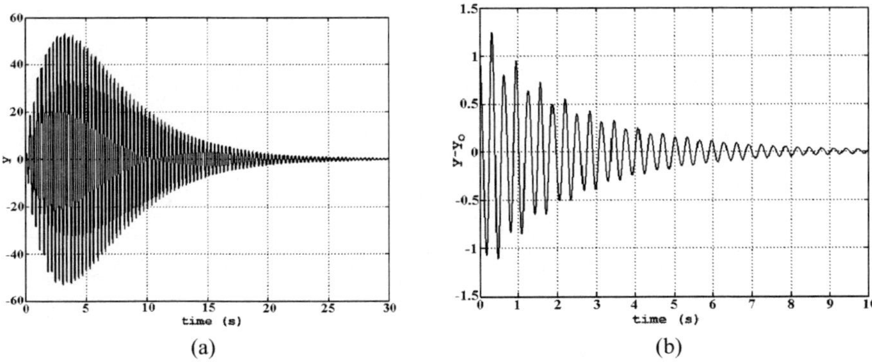

Fig. 3.26 Results of pole placement. **a** Plant response. **b** Error dynamics

3.4.3 Pole Placement Architecture in Bond Graph Models

So far, the discussion in this section is centered around block-diagram models. If we go for a bond graph model based observer [122, 202, 203], we have to bear a few things in mind: (1) in the observer, the feedback $\mathbf{L}(y-y_o)$ adds to every term in $\dot{x}_o$ and (2) unlike block diagrams, we have to take care of the type of signals, *i.e.* effort or flow.

Let us consider an integrally causalled I-element connected to a 1-junction by a bond, having bond number m. Then the associated state variable is P_m and $\dot{P}_m = e_m$. The relevant feedback term of $\mathbf{L}(y-y_o)$ should thus be an effort variable. Then we should use a modulated effort source (MSe) at the 1-junction connected to bond number m.

On the other hand, consider an integrally causalled C-element connected to a 1-junction by a bond, having bond number l. Then the associated state variable is Q_l and $\dot{Q}_l = f_l$. Then the relevant feedback term of $\mathbf{L}(y-y_o)$ should be a flow variable. But if we use a modulated flow source (MSf) at the 1-junction connected to bond number l, it will directly set the value of f_l and flows of all other bonds at that junction. We need to break bond number l by introducing a 0-junction in between and then connect the MSf element to it.

Similarly, if an integrally causalled C-element is connected to a 0-junction, a modulated flow source can be directly added to the 0-junction. On the other hand, if an integrally causalled I-element is connected to a 0-junction, we need to break the bond connecting the 0-junction to the I-element by introducing a 1-junction in-between and then add the modulated effort source to that 1-junction.

Furthermore, besides the actual output, we also need the state variables as outputs of the observer so that we can create the input for the plant and the observer as $u'(t) = u(t) + \mathbf{K}x_0$. However, there is no means for creating such direct state measurements. Instead, we can specify the actual process input as a state modulated source.

On the basis of the above developments, the augmented controller observer architecture of the system in bond graph form is given in Figure 3.27.

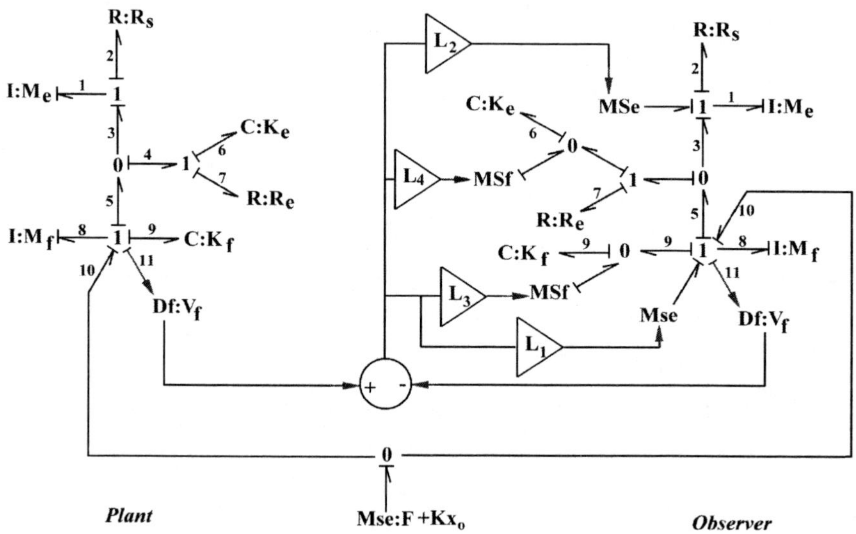

Fig. 3.27 Pole placement using bond graph models

The model and observer input is a modulated effort source where states $x_o = \left[P_8\ P_1\ Q_9\ Q_6\right]_0^T$ are taken from the observer side. Instead of using states themselves, one may use virtual sensors to construct the feedback $\mathbf{K}x_o$. For example, one may measure the force in the 0-junction connected to element $C:K_e$ on the observer side by adding an effort detector and then multiply the signal by a gain $1/K_e$ to get the state Q_6. Similarly, flow sensors may be added to one junctions connected to I-elements and then the acquired signals may be multiplied with corresponding mass to obtain the corresponding momentum states. These states may be used to construct the common input of the plant and the observer.

Note that pole-placement can be performed by directly analyzing the bond graph structure as developed in [203, 210].

Many model based fault diagnosis tools are based on observers. In Chapter 5, we will deal with some more observer types and their use in fault diagnosis.

3.4.4 Discrete-time Augmented Controller and Observer

Modern control systems are generally implemented in digital computers. This allows for implementation of efficient control laws, which otherwise cannot be implemented in analog controllers, *e.g.* neuro-fuzzy control, overwhelming control, *etc.* Signals acquired by the sensors in the system are sampled by an A/D data-acquisition card and then they are used in the control algorithm with a suitable hold. The controller outputs are sent to the power modulators of the actuators in the process through the D/A port of the same data-acquisition card. Design of digital control systems is fundamentally different from that of continuous time control systems. We start from the basics here.

The solution of a linear system may be written in continuous-time as

$$x(t) = \Phi(t,t_0)x(t_0) + \int_{t_0}^{t} \Phi(t,\tau)\mathbf{B}(\tau)u(\tau)\,d\tau, \tag{3.32}$$

where the state-transition matrix $\Phi(t,\tau) = e^{\int_\tau^t \mathbf{A}(\xi)d\xi}$ and $x(t_0)$ is the vector of initial conditions.

Let us assume that we discretize this system by sampling it at intervals of T seconds. From Shannon's sampling theorem, we understand that to avoid aliasing and frequency folding, the sampling frequency must be at least twice that of the Nyquist frequency, *i.e.* the largest frequency component in the signal. We assume that the sampling frequency is far larger than Shannon frequency. Then, in each sampling interval, we can assume that $u(t)$ does not vary much, *i.e.* $u(t) = u(kT) \forall kT \leq t < (k+1)T$ under a zero-order-hold (zoh) assumption. We further assume that the system is time invariant, *i.e.* an LTI system. Instead of writing sample instances as kT, $(k+1)T$ ..., for constant sampling rate we will write them simply as k, $k+1$, ... Then by treating the just previous sample as the initial state, we can write Equation 3.32 as

$$x(k+1) = \Phi(k+1,k)x(k) + \int_{k}^{k+1} \Phi(k+1,\tau)\mathbf{B}(\tau)\,d\tau\, u(k). \tag{3.33}$$

For the discretized LTI system, $\Phi(k+1,k) = e^{\mathbf{A}((k+1)T-kT)} = e^{\mathbf{A}T}$ and the second term $\int_k^{k+1} \Phi(k+1,\tau)\mathbf{B}(\tau)\,d\tau = \mathbf{A}^{-1}e^{\mathbf{A}T}\mathbf{B}$. Then we can write the final discretized version of Equation 3.33 in a discrete-time state-space form as

$$\begin{aligned} x(k+1) &= \mathbf{A}_d x(k) + \mathbf{B}_d\, u(k), \\ y(k) &= \mathbf{C}_d x(k) + \mathbf{D}_d\, u(k), \end{aligned} \tag{3.34}$$

where $\mathbf{A}_d = e^{\mathbf{A}T}$, $\mathbf{B}_d = \mathbf{A}^{-1}e^{\mathbf{A}T}\mathbf{B}$, $\mathbf{C}_d = \mathbf{C}$ and $\mathbf{D}_d = \mathbf{D}$ may be formulated from direct inspection. Matrix exponents may be calculated by using MATLAB® command *expm* or MATLAB® command *c2d* may be used to directly discretize a continuous time state-space model with various options, such as the zoh option used here. If $G(s)$ is a transfer function matrix corresponding to the continuous time model and $G(z)$ is its z-transform, then poles of $G(z)$ are eigenvalues of $e^{\mathbf{A}T}$. Furthermore, if an eigenvalue of matrix **A** is in the left half of the complex plane, then the corresponding eigenvalue of matrix $e^{\mathbf{A}T}$ appears inside a unit circle about the origin in the complex plane. In digital systems, the unit circle in the complex plane acts as the stability boundary.

The basic discrete time full order controller and observer is similar to the continuous time version.

Controller:

$$\begin{aligned} x(k+1) &= (\mathbf{A}_d + \mathbf{B}_d\mathbf{K})x(k) + \mathbf{B}_d u(k), \\ y(k) &= (\mathbf{C}_d + \mathbf{D}_d\mathbf{K})x(k) + \mathbf{D}_d u(k). \end{aligned} \tag{3.35}$$

Observer:

$$\begin{aligned} x_o(k+1) &= \mathbf{A}_d x_o(k) + \mathbf{B}_d u(k) + \mathbf{L}\left(y(k) - y_o(k)\right), \\ y_o(k) &= \mathbf{C}_d x_o(k) + \mathbf{D}_d u(k); \end{aligned} \tag{3.36}$$

giving error dynamics

$$\begin{aligned} x_e(k+1) &= x(k+1) - x_o(k+1), \\ &= (\mathbf{A}_d - \mathbf{L}\mathbf{C}_d)x_e(k). \end{aligned} \tag{3.37}$$

Thus the pole placement of discrete-time systems is exactly similar to that for the continuous time system. However, the poles must be placed inside the unit circle in the complex plane to ensure stability of the system.

3.4.5 Current Estimator

The discrete-time observer defined in Equation 3.36 uses one step previous measurements to establish the feedback in the term $\mathbf{L}\left(y(k) - y_o(k)\right)$. It would be better if we could use a feedback term $\mathbf{L}\left(y(k+1) - y_o(k+1)\right)$ in Equation 3.36, because it improves both the observer performance and also that of the controller, where we have a feedback term $\mathbf{K}x(k)$ [182]. However, while $y(k+1)$ and $u(k+1)$ are available at time $t = (k+1)T$ from the real process, we do not have the value of $y_o(k+1)$. One of the solutions is to use an estimate of $y_o(k+1)$ by following a two step procedure, called a current estimator.

State update: here we create an intermediate state-space form as

$$\begin{aligned} \hat{x}(k+1) &= \mathbf{A}_d x_o(k) + \mathbf{B}_d u(k), \\ \hat{y}(k) &= \mathbf{C}_d \hat{x}(k) + \mathbf{D}_d u(k). \end{aligned} \tag{3.38}$$

Measurement update: the intermediate form estimates the state variables of the next step and we can obtain an estimate $\hat{y}(k+1) = \mathbf{C}_d \hat{x}(k+1) + \mathbf{D}_d u(k+1)$. Then we may rewrite the observer's state space equation as

$$\begin{aligned} x_o(k+1) &= \hat{x}(k+1) + \mathbf{L}\left(y(k+1) - \hat{y}(k+1)\right), \\ y_o(k) &= \mathbf{C}_d x_o(k) + \mathbf{D}_d u(k). \end{aligned} \tag{3.39}$$

The error dynamics is then found to be

$$\begin{aligned} x_e(k+1) &= x(k+1) - x_o(k+1) \\ &= x(k+1) - \hat{x}(k+1) - \mathbf{L}\left(y(k+1) - \hat{y}(k+1)\right) \\ &= (\mathbf{A}_d - \mathbf{L}\mathbf{C}_d\mathbf{A}_d)x_e(k). \end{aligned} \tag{3.40}$$

This means that poles of matrix $(\mathbf{A}_d - \mathbf{L}\mathbf{C}_d\mathbf{A}_d)$ should be placed at desired locations inside a unit circle in the complex plane. By defining a matrix $\mathbf{C}_d^* = \mathbf{C}_d\mathbf{A}_d$, the pole placement problem is similar to that of the continuous time observer. This two step procedure is used to design discrete-time Kalman filters, which have widespread use in process control as well as diagnosis.

3.5 Structural Analysis of Control Properties

The choice of sensors, actuators and their position plays the vital role in design of FDI systems. However, from a bond graph model representing the true system behavior along with its actuators and observers, feasibility of control strategies can be verified through the analysis of structural properties of the model. Structural properties relate to properties depending only on the model structure (junction structure) and the elements (components) it is composed of, and they do not depend on the actual values of the parameters. Structural analysis was first developed in [147] for SISO systems and it was later extended to MIMO systems [231]. Thereafter, different graphical analysis methods using digraphs [247] and bond graphs were developed. In this chapter, we are simply presenting the algorithms for structural analysis without going through their proofs. Interested readers may refer to [61, 211, 245–247] for the detailed background of the theory.

3.5.1 Structural Rank

Consider a linear time-invariant (LTI) system represented by the following state space model:

$$\begin{aligned} \dot{x} &= \mathbf{A}x + \mathbf{B}u, \\ y &= \mathbf{C}x + \mathbf{D}u. \end{aligned} \tag{3.41}$$

If n is the number of states, then the characteristic polynomial of the system or the denominator of transfer functions from the control input to the observed outputs may be written as

$$D(s) = |s\mathbf{I} - \mathbf{A}| = s^p(s^q + a_{q-1}s^{q-1} + ... + a_1 s + a_0), \qquad (3.42)$$

where s is the Laplace operator, $\mathbf{I}$ is an $n \times n$ identity matrix, $p + q = n$ and the coefficients $a_{q-1}...a_0$ are functions of the system parameters. The number of structurally null modes in the system is p, *i.e.* there are at least p number of eigenvalues which are zero. However, the actual null modes in the system may be higher owing to parameter dependence, *e.g.* when $a_0 = 0$, the number of null modes is $p+1$.

Since the term s^pcan be separated out as a factor from $|s\mathbf{I} - \mathbf{A}|$, the $s\mathbf{I} - \mathbf{A}$ matrix contains p rows with diagonal elements containing s and other terms in those rows cancel out during determinant calculation. Thus for $s = 0$, or $s\mathbf{I} - \mathbf{A} = -\mathbf{A}$, these rows are linearly dependent on other rows. This implies that matrix $\mathbf{A}$ can be brought to a form where p rows will contain all elements equal to zero. Thus, the structural rank of matrix $\mathbf{A}$ is $n - p = q$.

To find q, the system equations need not be derived and tested. They can be obtained by using the following simple rules applied to the bond graph model:

1. The order of a system (dimension of matrix $\mathbf{A}$ or number of states) is the number of integrally causalled storage elements (I and C), when a preferred integral causality is assigned to the model.
2. The structural rank of the model (rank of matrix $\mathbf{A}$ without parameter dependence) is the number (q) of storage elements (I and C) that can be brought to differential causality, when a preferred differential (or derivative) causality is assigned to the bond graph model.
3. The number of structurally null modes is the number ($p = n - q$) of storage elements (I and C) that cannot be brought to differential causality, when a preferred differential/ derivative causality is assigned to the model.

3.5.2 Structural Controllability

Definition 3.5. If there exists an input $u(t)$ such that a linear system at any arbitrary initial state $x(t_0)$ can be brought to a final state $x(t_f) = 0$ then the linear system is said to be controllable in the interval $[t_0, t_f]$.

The numerical method to determine controllability of a system is to check whether the controllability matrix has full rank n, where n is the number of states. The controllability matrix($\mathbf{S}$) of a continuous time system is defined as

$$\mathbf{S} = \begin{bmatrix} \mathbf{B} \ \mathbf{AB} \ ... \ \mathbf{A}^{n-1}\mathbf{B} \end{bmatrix}. \qquad (3.43)$$

The controllability matrix for a discrete-time system has exactly the same form except that the matrices $\mathbf{A}$ and $\mathbf{B}$ correspond to those appearing in the discretized state-space equation (see Equation 3.34). Consider a recursive execution of Equation 3.34 with $t_0 = 0$ and $t_f = kT$ (T is the sampling time) as follows:

$$\begin{aligned}
x(1) &= \mathbf{A}x(0) + \mathbf{B}u(0),\\
x(2) &= \mathbf{A}x(1) + \mathbf{B}u(1) = \mathbf{A}^2x(0) + \mathbf{AB}u(0) + \mathbf{B}u(1),\\
x(3) &= \mathbf{A}x(2) + \mathbf{B}u(2) = \mathbf{A}^3x(0) + \mathbf{A}^2\mathbf{B}u(0) + \mathbf{AB}u(1) + \mathbf{B}u(2),\\
&\vdots\\
x(k) &= \mathbf{A}^kx(0) + \mathbf{A}^{k-1}\mathbf{B}u(0) + \ldots + \mathbf{AB}u(k-2) + \mathbf{B}u(k-1).
\end{aligned} \tag{3.44}$$

According to the definition, the final state $x(k) = 0$ implies

$$-\mathbf{A}^kx(0) = \begin{bmatrix}\mathbf{B}\ \mathbf{AB}\ \ldots\ \mathbf{A}^{k-2}\mathbf{B}\ \mathbf{A}^{k-1}\mathbf{B}\end{bmatrix}\begin{bmatrix}u(k-1)\\ u(k-2)\\ \vdots\\ u(1)\\ u(0)\end{bmatrix}, \tag{3.45}$$

and thus the sequence of inputs required to bring the system to final desired state are found as

$$\begin{aligned}
[u(k-1)\ldots u(0)]^T &= -\begin{bmatrix}\mathbf{B}\ \mathbf{AB}\ \ldots\ \mathbf{A}^{k-2}\mathbf{B}\ \mathbf{A}^{k-1}\mathbf{B}\end{bmatrix}^{-1}\mathbf{A}^kx(0)\\
&= -\mathbf{P}^{-1}\mathbf{A}^kx(0).
\end{aligned} \tag{3.46}$$

Equation 3.46 describes a set of simultaneous linear equations, which are solvable if the $\mathbf{P}$ matrix having $n \times k$ dimension has rank n. The minimum number of samples required to reach the desired state becomes $k = n$, for which $\mathbf{P} = \mathbf{S}$. Because the definition of controllability neither restricts the initial state nor the number of samples, we should consider the most stringent case ($k = n$), which means that the controllability matrix must be invertible (consequently, $\mathbf{S}$ must have full rank) to find the input sequence and thus results in the same controllability condition as that for continuous-time systems.

The rank of the controllability matrix depends on the numerical values of the parameters and hence it cannot be considered as a measure of the robustness of the control strategy. This method also does not identify the uncontrollable modes of the system. The alternative numerical method is through feedback input injection, where the shift in the eigenvalues of matrix $\mathbf{A} + \mathbf{BK}$ ($\mathbf{K}$ is a gain matrix of small random values) from the eigenvalues of $\mathbf{A}$ is considered as a measure of controllability of particular modes (eigenvalues). This method, used earlier for pole placement, too, is parameter dependent and is not a robust measure of controllability.

The structural controllability based on bond graph structure is however robust and parameter independent. The conditions needed for a system to be structurally controllable are:

1. *Attainability/reachability condition (necessary condition)*: every storage element (integrally causalled I and C) in the bond graph model with preferred integral causality must have at least one causal path from a control source (MSe or MSf).
2. *Sufficient condition*: every integrally causalled storage element (I and C) in the bond graph model (with preferred integral causality) can be differentially causalled (assigned derivative causality) when preferred differential causality is assigned on the bond graph model. If some integrally causalled elements in preferred integral causality mode cannot be assigned differential causality in the preferred differential causality mode, then dualisation of some or all control sources (control MSe to control MSf and *vice versa*) or inversion of their causalities may be performed to allow those storage elements to accept differential causality. Note that causalities of uncontrolled sources, *e.g.* force due to gravity, are not to be altered.

If the necessary and sufficient conditions are satisfied, then the rank of matrix $\mathbf{S}$ is n. If matrix $\mathbf{A}$ is of full rank, the system is controllable with a single actuator whose position is determined by the attainability condition (such that matrix $\left[\mathbf{B}\ \mathbf{A}\right]$ has full rank) and other design considerations.

If the sufficient condition is not satisfied and only q ($q < n$) number of storage elements, originally in integral causality, could be differentially causalled, then the rank of matrix $\mathbf{S}$ is q. Then for the model to be controllable, $p = n - q$ actuators are needed and their location has to be determined by attainability condition and their participation has be determined by control source dualisation.

3.5.3 Structural Observability

Definition 3.6. If the initial state $x(t_0)$ of a linear system can be calculated from the knowledge of inputs $u(t)$ and outputs $y(t)$ in the interval $[t_0, t_f]$ then the linear system is said to be observable in that interval.

The numerical method to determine observability of a system is to calculate the rank of the observability matrix. The observability matrix ($\mathbf{O}$) of a continuous time system is defined as

$$\mathbf{O} = \begin{bmatrix} \mathbf{C} \\ \mathbf{CA} \\ \mathbf{CA}^2 \\ \vdots \\ \mathbf{CA}^{n-1} \end{bmatrix}. \tag{3.47}$$

It can be proved through recursive expansion of output equations in Equation 3.34 ($y(0)$ to $y(k)$ in the time interval $[0, kT]$), in an analogous manner to Equation 3.44, that the same observability matrix and its full rank condition are reached for discrete-time systems.

The rank of the observability matrix depends on the numerical values of the parameters and hence it cannot be considered as a robust measure of full-state observability. This method also does not identify the modes of the system that are not observable. The alternative numerical method is through feedback output injection, where the shift in the eigenvalues of matrix $\mathbf{A}-\mathbf{LC}$ ($\mathbf{L}$ is a gain matrix of small random values) from the eigenvalues of $\mathbf{A}$ is considered as a measure of observability of particular modes (eigenvalues). This method too is parameter dependent and is not a robust measure of observability.

The robust structural observability is based on the satisfaction of the following conditions:

1. *Attainability/reachability condition (necessary condition)*: every storage element (integrally causalled I and C) in the bond graph model with preferred integral causality must have at least one causal path to a sensor (De or Df element).
2. *Sufficient condition*: every integrally causalled storage element (I and C) in the bond graph model with preferred integral causality can be differentially causalled (assigned derivative causality) when preferred differential causality is assigned on the bond graph model. When some integrally causalled elements in preferred integral causality mode cannot be assigned differential causality in the preferred differential causality mode, some or all sensors may be dualized (De to Df and *vice versa*) or their causalities may be inverted such that those storage elements can be differentially causalled.

If the necessary and sufficient conditions are satisfied, then the rank of matrix $\mathbf{O}$ is n. If rank of matrix $\mathbf{A}$ is n, then only one suitably placed observer, determined by attainability condition (so that matrix $\begin{bmatrix}\mathbf{C}^T & \mathbf{A}^T\end{bmatrix}^T$ has full rank), is sufficient to guarantee full-state observability. If the rank of $\mathbf{O}$ is q, then $p = n - q$ number of additional observers are needed to assure observability and the position of these sensors has to be determined by the attainability condition and design considerations.

Remark 3.2. Structural controllability and observability tests, at times, require dualisation of sources and sensors, respectively. During structural controllability tests, one may dualize control sources (MSe and MSf elements) to assign differential causalities to storage elements. This means the ports attached to these elements can take up any causality. One of the alternative methods to check structural controllability is to replace all the control sources by R-elements, which can take any causality, and then complete the preferred differential causality assignment.

Remark 3.3. During structural observability tests, one may dualize sensors (De and Df elements) so that maximum number of storage elements are brought to differential causality. This means that, in this case, ports connected to sensors can be assigned any causality. Alternatively, one may simply replace all sensors by R-elements and then proceed with preferred differential causality assignment.

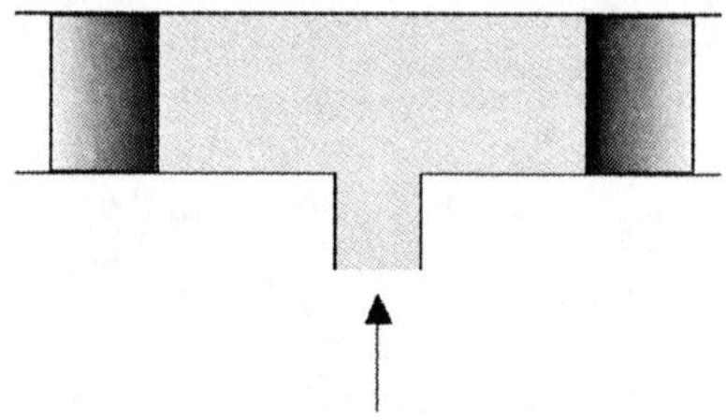

Fig. 3.28 Two spools in a friction-less guide

Remark 3.4. Most commercial bond graph software are designed to build simulation models, *i.e.* they output behavioral models in preferred integral causality. These software try to minimize differentially causalled storage elements. To test structural controllability and observability properties, one may dualize all storage elements (I to C and *vice versa*). Thereafter, depending on which structural property is to be tested (controllability or observability), either control sources or sensors are to be substituted by R-elements. Now the software would automatically try to assign preferred integral causality to storage elements. When it does that, say to a C-element in the revised bond graph, the corresponding element (I-element) in the original bond graph is assigned differential causality. Thus, the number of elements which cannot be assigned integral causality in the revised bond graph model corresponds to the number of elements which cannot be assigned differential causality in the original model. This trick is suitable for all most all available bond graph software.

3.5.4 Example I: Two Spools in a Cylinder

Let us consider a system shown in Figure 3.28. The two massive spools can be moved by injection of compressible fluid at the inlet port. The resistance between the cylinder and spools is neglected.

The bond graph model of the above system is shown in Figure 3.29 for both preferred integral causality and preferred derivative causality. The causal path linking the source to one of the spool masses is shown in the integrally causalled model.

From the preferred integral causality model, the number of states of the system is 3. However, the preferred differential causality model shows that only two storage elements could be assigned differential causality and one storage element is left in integral causality. Thus the structural rank of the system is 2. Furthermore, all three storage elements cannot be brought to differential causality even through control source dualisation (converting MSf to MSe would cause violation of causal structure at the 0-junction) and hence the system is not controllable.

Let us consider the state space representation of the integrally causalled model (the states are P_1, P_2 and Q_3).

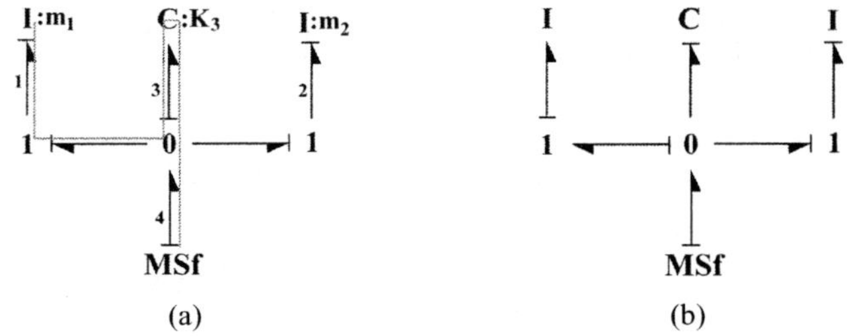

Fig. 3.29 Bond graph model of two spools in a friction-less guide. **a** In integral causality. **b** In differential causality

$$\frac{d}{dt}\begin{Bmatrix} P_1 \\ P_2 \\ Q_3 \end{Bmatrix} = \begin{bmatrix} 0 & 0 & K_3 \\ 0 & 0 & K_3 \\ -1/m_1 & -1/m_2 & 0 \end{bmatrix} \begin{Bmatrix} P_1 \\ P_2 \\ Q_3 \end{Bmatrix} + \begin{Bmatrix} 0 \\ 0 \\ 1 \end{Bmatrix} SF_4. \tag{3.48}$$

Thus the system matrices are given by

$$\mathbf{A} = \begin{bmatrix} 0 & 0 & K_3 \\ 0 & 0 & K_3 \\ -1/m_1 & -1/m_2 & 0 \end{bmatrix} \text{ and } \mathbf{B} = \begin{Bmatrix} 0 \\ 0 \\ 1 \end{Bmatrix}. \tag{3.49}$$

The rank of **A** matrix is 2, which tallies with the results for the structural rank obtained earlier.

Similarly, the controllability matrix $\mathbf{S} = \begin{bmatrix} \mathbf{B} \; \mathbf{AB} \; \mathbf{A}^2\mathbf{B} \end{bmatrix}$ is given by

$$\mathbf{S} = \begin{bmatrix} 0 \; K_3 & 0 \\ 0 \; K_3 & 0 \\ 1 \;\; 0 & -K_3/m_1 - K_3/m_2 \end{bmatrix}, \tag{3.50}$$

whose rank is 2. This proves that the system is uncontrollable.

Let us now consider damping in one of the spools. The bond graph model for the damped system in preferred integral causality and preferred derivative causality are shown in Figure 3.30.

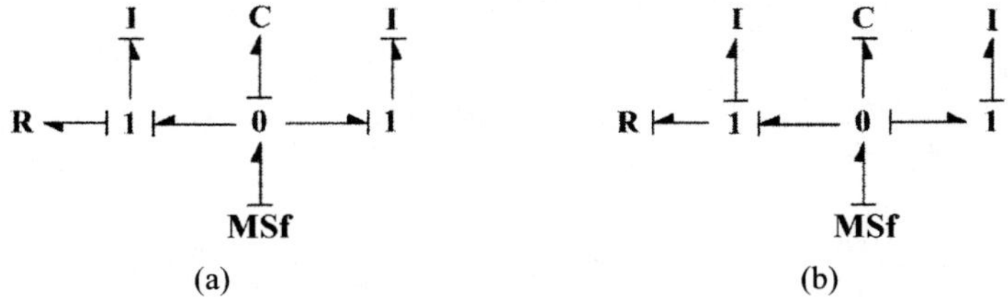

Fig. 3.30 Bond graph model of two spools where one spool has friction. **a** In integral causality. **b** In differential causality

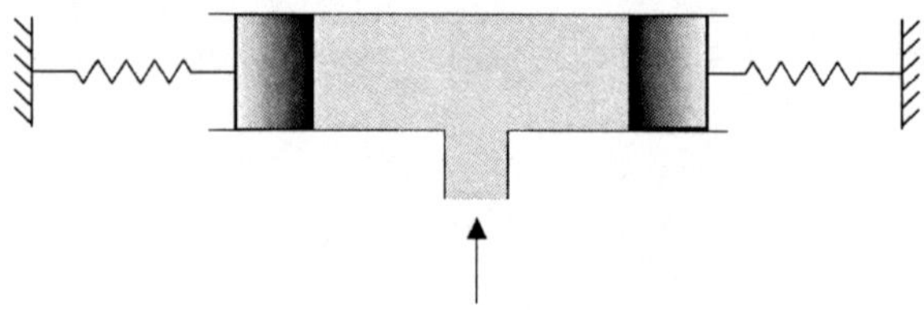

Fig. 3.31 Two anchored spools in a friction-less guide

In this case, all the storage elements with integral causality in the preferred integral causality model could be brought to preferred differential causality. Thus the damped system is controllable.

Let us now consider a modified system as shown in Figure 3.31, where the massive spools have been anchored to the inertial frame by using two springs.

The bond graph model for this system in preferred integral causality and preferred differential causality are shown in Figure 3.32.

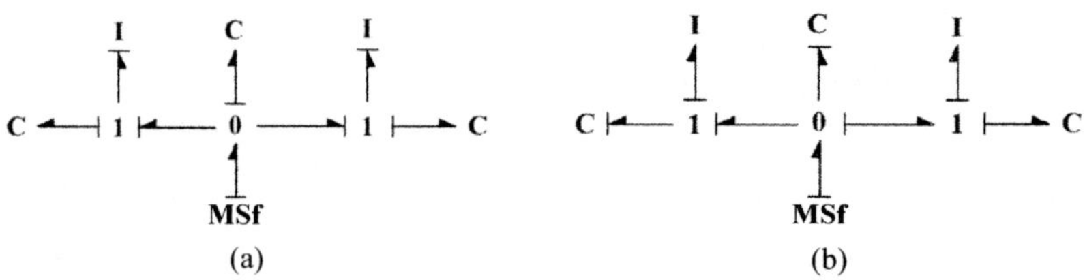

Fig. 3.32 Bond graph model of two anchored spools in a friction-less guide. **a** In integral causality. **b** In differential causality

The preferred differential causality model contains one state that was integrally causalled in the preferred integral causality model and could not be brought to differential causality in preferred differential causality model. However, dualisation of the control input can bring it to differential causality as shown in Figure 3.33. Thus, the anchored system is structurally controllable.

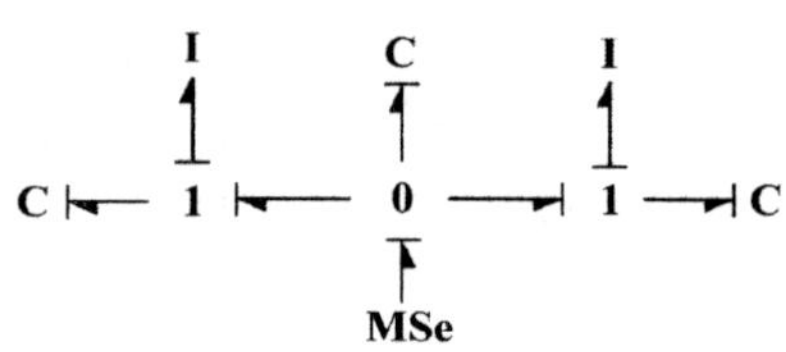

Fig. 3.33 Bond graph model of two anchored spools in a friction-less guide with dualized source and in preferred differential causality

3.5.5 Example II: A Hybrid Two-tank System

Consider a hybrid two-tank system shown in Figure 3.34.

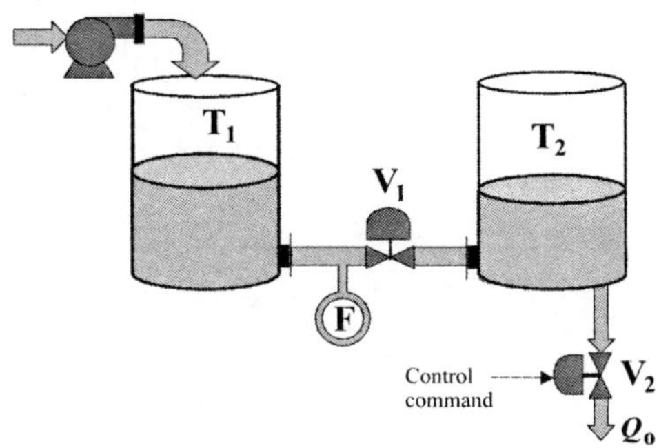

Fig. 3.34 Two-tank system as an example of a hybrid system

The system consists of two tanks, T_1 and T_2, with a valve, V_1, between them. Valve V_2 is used to provide output from T_2. A flow sensor, F, measures the flow between tanks T_1 and T_2. The valve stem position of V_2 is controlled by the user, whereas V_1 is always open at a constant position. The pump feeding T_1 is controlled based on the level in that tank. The corresponding bond graph model in integral causality is shown in Figure 3.35. Note that in the open-loop model, actuators are represented as modulated sources (without any modulating signal).

Here only observability part is discussed; controllability study can be done in a similar way. From Figure 3.35, we find that the attainability/ reachability condition is always satisfied. The preferred derivative causality can be assigned as shown in Figure 3.36 and hence the sufficient condition is also satisfied. Thus, the system is observable with the given instrumentation.

This analysis is not valid when V_2 is completely closed by the user. This system is indeed hybrid with two distinct configurations and hence it requires another model for the latter configuration [226]. The corresponding bond graph model in preferred derivative causality is shown in Figure 3.37.

In Figure 3.37, only one of the storage elements can be assigned derivative causality. Thus, the system is unobservable. In fact, the flow sensor can be used

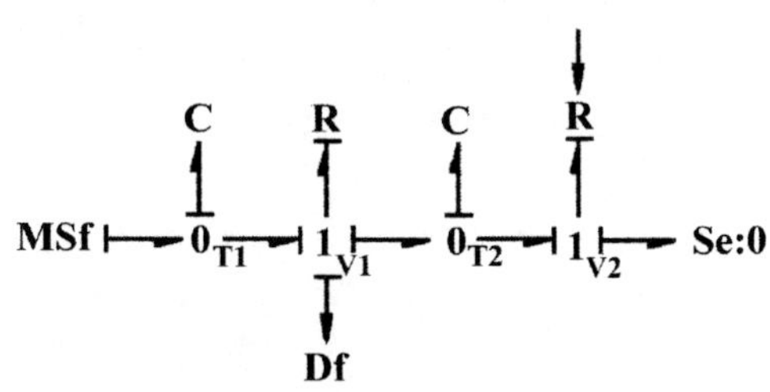

Fig. 3.35 Integrally causalled bond graph model of the two-tank system

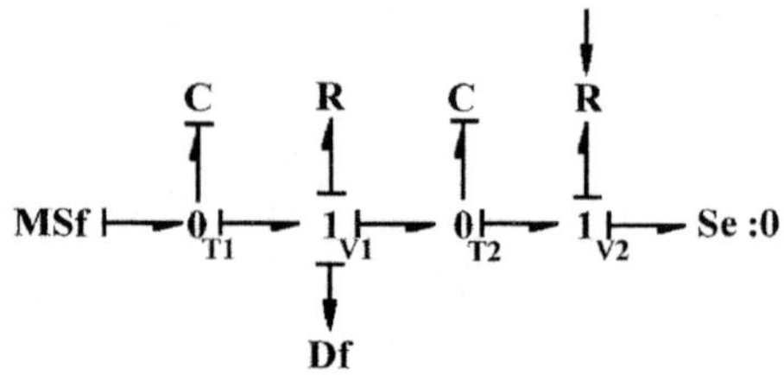

Fig. 3.36 Differentially causalled bond graph model of the two-tank system

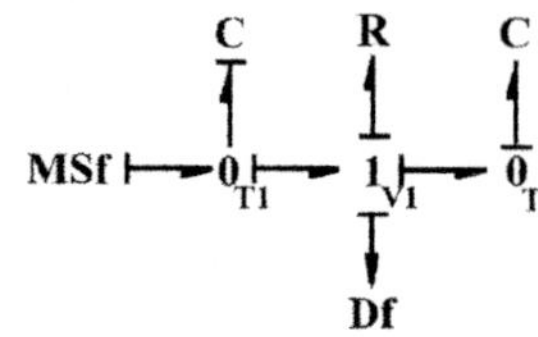

Fig. 3.37 Preferred differential causality in the bond graph model of the two-tank system with closed outlet

to calculate the difference between the pressures in T_1 and T_2, but not their absolute values. This explains why only one eigenvalue (internal mode) of the system is observable, while the other eigenvalue (rigid body mode) is not observable. The same happens when a differential pressure sensor is used. From the analysis it can be seen that the storage element which is not in derivative causality is connected to a 0-junction belonging to component T_2. From this, one can conclude that an additional sensor is required in component T_2. Therefore, an effort sensor (De), which may be a pressure or level sensor, is needed at the corresponding 0-junction. With that sensor added to the process, the corresponding bond graph model in preferred derivative causality is shown in Figure 3.38a. It is seen that the causality of the new effort sensor needs inversion to assign derivative causality to all the storage elements. Also, it is seen that removal of the flow sensor still maintains the derivative causality and satisfies the attainability condition, as shown in Figure 3.38b.

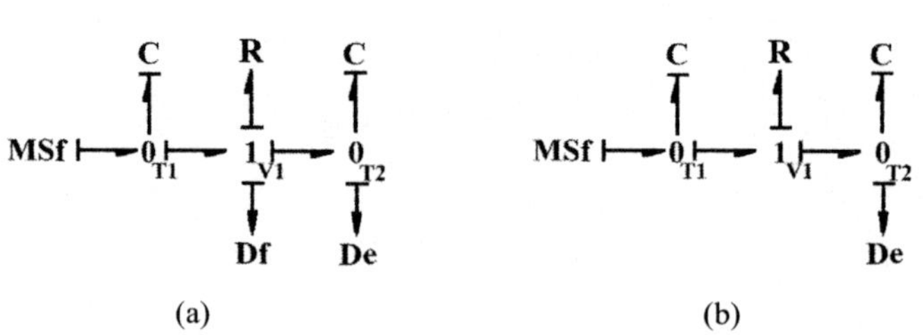

Fig. 3.38 Preferred differential causality in the bond graph model of the two-tank system with closed outlet. **a** With two sensors. **b** Without redundant sensor

In fact, the causality of the flow sensor (Df) in Figure 3.38a cannot be inverted (it leads to integral causality in one of the storage elements) in the given configuration when causality of the effort sensor (De) is already inverted. Such constraint between the causalities of two sensors is termed a *direct hardware redundancy*. Note

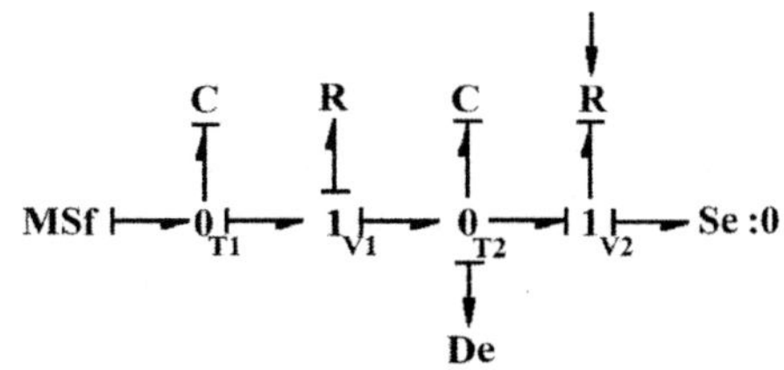

Fig. 3.39 Bond graph model of the two-tank system in differential causality with minimal sensor architecture

that in the given configuration (*i.e.* V_2 closed) and assuming the tanks have constant cross-section, the flow through V_1 is equal to area times the rate of change of level in T_2 (measured by De) and thus, the flow sensor is redundant. Before finalizing the sensor instrumentation, the new sensor location must be tested for compatibility with the other hybrid configuration of the system, *i.e.* when V_2 is open. The preferred derivative causality model of the system with only the effort sensor is shown in Figure 3.39.

It is seen that the effort detector is suitable to assign preferred derivative causality to storage elements in models for both hybrid configurations of the system. Thus a single sensor renders the hybrid system observable. The method discussed here is useful in selection of the sensor, its location and optimization of the number of sensors.

In fact, if we assume that the effort detector measures the level L_2 in tank T_2, then we know from Definition 3.6 that the system is observable when we can find level L_1 in tank T_1 from known inputs and outputs. If C_{d1} and C_{d2} are taken as the discharge coefficients of the two valves V_1 and V_2, respectively, and A_1 and A_2 are the cross-sectional areas of the two tanks T_1 and T_2, respectively, then we can write the relation for conservation of mass in tank T_2 as

$$C_{d1}\sqrt{(\rho g)(L_1 - L_2)} = \rho A_2 \dot{L}_2 + C_{d2}\sqrt{\rho g L_2}. \tag{3.51}$$

Note that Equation 3.51 can be solved to find L_1 for any value of C_{d2}, *i.e.* both when the valve is open and when the valve is closed. However, when $C_{d1} = 0$, we cannot determine L_1. This is due to the fact that if flow through V_1 is restricted then we have two disjoint systems (two separate tanks) and there will be no causal path from one of the storage elements to the sensor, thereby violating the reachability condition.

Remark 3.5. The complexity of structural analysis of systems involving many hybrid components and controllers increases exponentially due to the requirement of testing structural properties of all configuration combinations. However, since each step involves only a causality assignment procedure and no numerical evaluation, this sensor placement algorithm is computationally very efficient.

3.5.6 Example III: A Biomechanics Problem

An inchworm, also called a geometrid caterpillar, is shown in Figure 3.40a. It has a highly complex motion and, for our modeling purpose, we will simplify the problem to a one-dimensional motion form as given in Figure 3.40b. For movement, the worm bends and stiffens itself in a Ω-like shape and then lifts one of its ends to move in that direction. It then anchors the just moved end, relaxes its muscles, releases the other end and finally pulls it to a location to again form the Ω-like shape. Thus the worm follows two distinct steps during its motion, which constitute the two hybrid configurations of the system.

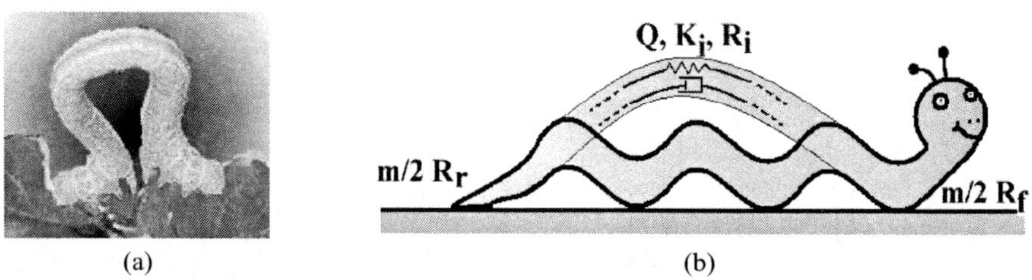

Fig. 3.40 **a** An inchworm. **b** Its schematic representation

We will assume that the mass of the insect is equally lumped at its two ends. We further assume that there are two resistances (R_f and R_r) anchoring these masses to the ground and there is no ground resistance at any other place. These resistances should be modeled as Coulomb friction, but we will show that a damper gives reasonable results. The stiffening and relaxing of the muscles is assumed to be a flow source, Q, which charges a spring (bending stiffness) and damper combination (K_i and R_i). The equivalent mechanical system is then represented in Figure 3.41a with the corresponding bond graph model in Figure 3.41b.

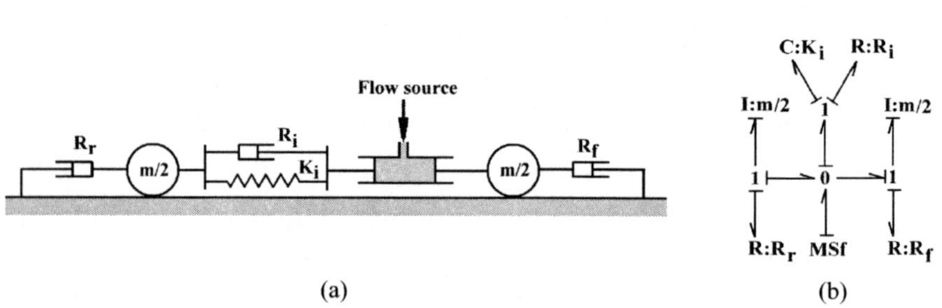

Fig. 3.41 **a** Equivalent mechanical model of an inchworm. **b** The corresponding integrally causalled bond graph model

We find that the control source (MSf) has causal paths to all three storage elements. We consider three separate cases here:

1. If there is no ground damping, *i.e.* if the surface is friction less or the worm cannot grip it properly, then we can drop the R-elements at two ends of the bond graph model. When the corresponding bond graph model is assigned preferred differential causality, as in Figure 3.42a, we find that the system is uncontrollable. In fact, in this configuration the internal mode of vibration is controllable whereas the rigid body mode (translation of the worm as a whole) is uncontrollable.
2. If there is sufficient friction at the contact surface between the ground and the worm, then we can show that all the storage elements in the full bond graph model in Figure 3.41b can be completely assigned differential causality and thus the system is controllable. However, the motion of an inchworm is hybrid in nature: in any given posture, it grips the ground at one of its ends and pulls/pushes the other end. Therefore, we should check controllability of the two hybrid modes, each one of them being represented as LTI system models. From Figures 3.42b,c, we find that both hybrid modes of the system are controllable.
3. If we consider that the micro-grippers on one end of the worm are damaged and the worm is able to grip only on its remaining one end, then too the overall system remains controllable (either Figure 3.42b or Figure 3.42c, depending upon which end is damaged).

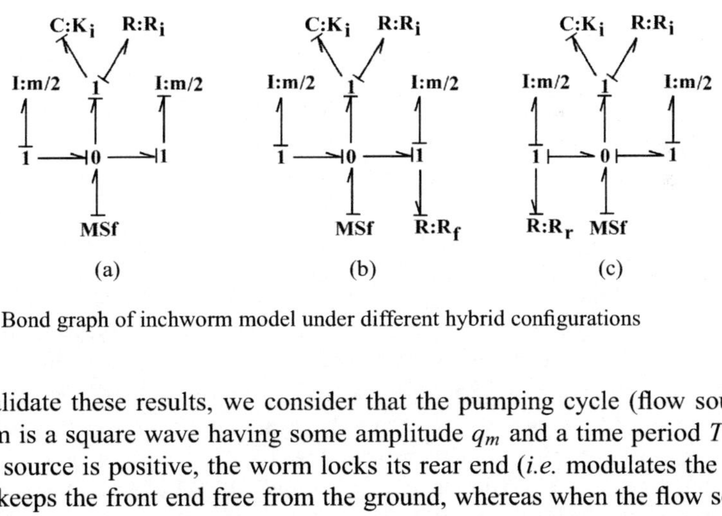

Fig. 3.42 Bond graph of inchworm model under different hybrid configurations

To validate these results, we consider that the pumping cycle (flow source) of the worm is a square wave having some amplitude q_m and a time period T. When the flow source is positive, the worm locks its rear end (*i.e.* modulates the damper R_r) and keeps the front end free from the ground, whereas when the flow source is negative, it locks its front end (*i.e.* a high damping value R_f) and releases its rear end. These periodic modulations are shown in Figure 3.43 where R_g is the maximum value of damping.

We consider the following representative data: $m = 0.05$ kg, $K_i = 10$ N/m, $R_i = 0.01$ N.s/m, $T = 10$ s, $Q = 0.01$ m/s, $R_g = 0.1$ Ns/m. In the first case, we assume that both $R : R_f$ and $R : R_r$ are absent in the bond graph model. The corresponding simulation results are shown in Figure 3.44, where the displacement of the front end is given by the curve with $\star$ marks and the displacement of the rear end is given by

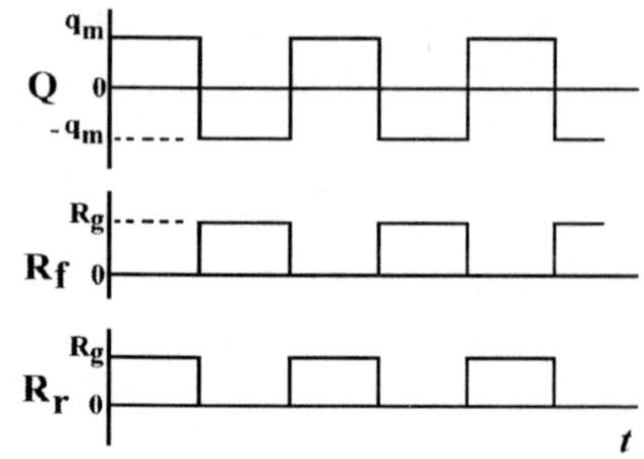

Fig. 3.43 Modulation of inchworm's locomotion parameters

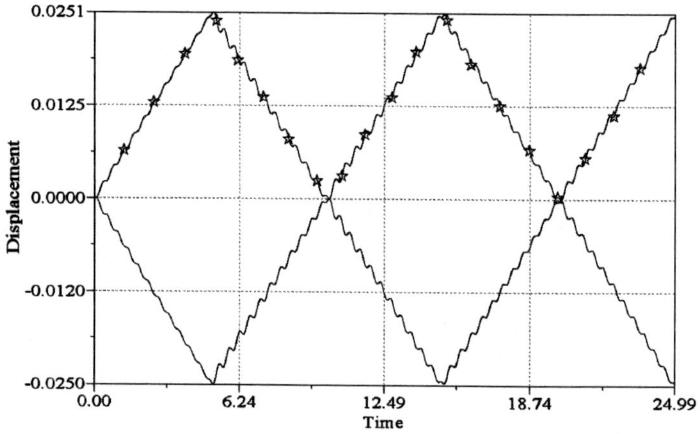

Fig. 3.44 Motion of inchworm without grip

the plain curve. We find that when the front end of the worm goes ahead, the rear end goes back and *vice versa*; thereby meaning that the mass-center of the worm does not move but the internal mode or relative displacement of the ends is controllable. This confirms our earlier results predicting overall uncontrollability of the system, which in this case refers to the uncontrollability of the rigid body mode.

When we include damping at both ends of the worm and consider the modulations shown in Figure 3.43, we find from the simulation results in Figure 3.45 that the worm is able to move without appreciable slip.

As a special case, we consider that the maximum damping between the worm and the ground is reduced to $R_g = 0.01$Ns/m, *i.e.* the surface becomes a bit slippery. In this case, the simulation results shown in Figure 3.46 reveal that the worm is able to move with heavy initial slip, which gradually reduces. Obviously, the worm travels less distance in the same time.

In the last simulation, we consider high damping in the rear end of the worm and assume that the front end does not give damping. The results in Figure 3.47 reveal that the worm is able to move and cover slightly less distance than when it has damping at both its ends. This proves that the system is controllable with

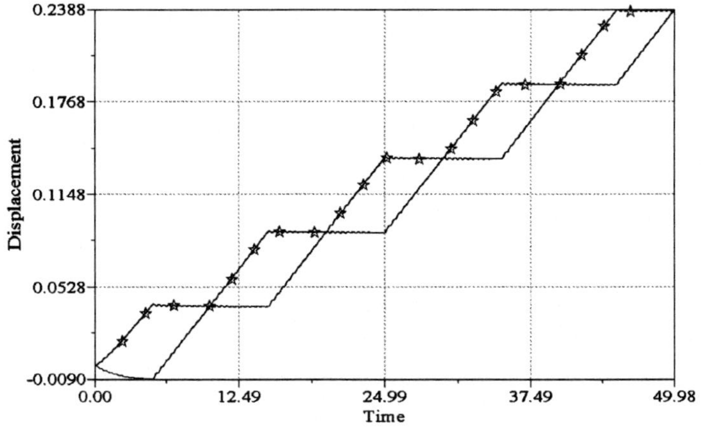

Fig. 3.45 Motion of inchworm with proper grip

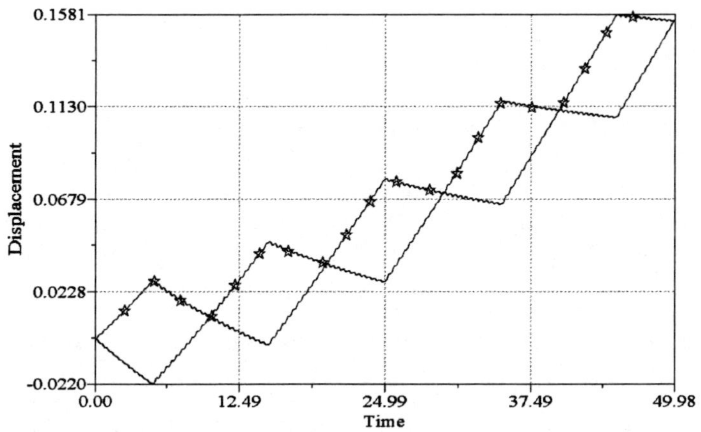

Fig. 3.46 Motion of inchworm with little grip

damping at one of its ends. When one interprets the simulation results, one finds that, interestingly, when the flow source is negative and the rear end is being dragged forward (*i.e.* no damping at both ends during this phase), the front end does not retract significantly. This is attributed to the momentum gained by the front end during the previous half-cycle of the motion.

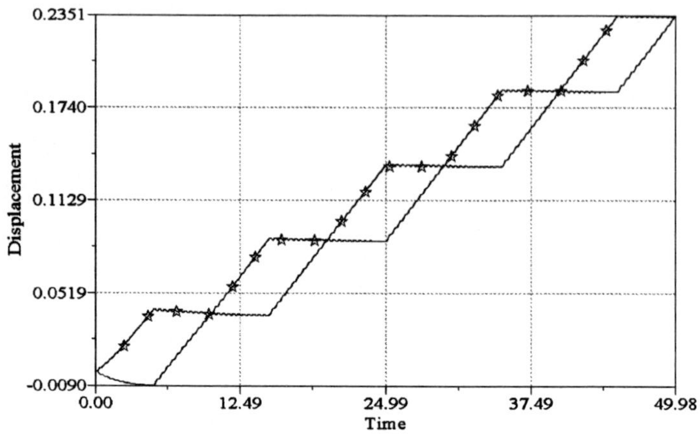

Fig. 3.47 Motion of inchworm with grip at one end

3.5.7 Infinite Zeroes and Relative Degree

While designing a supervision system, one needs to study how easily the system is controllable. This aspect, which is a broader concept when compared to simple study of structural controllability properties, is important from the FTC point of view. In [83], the author ascribes two reasons for why a particular input/output pair of a linear system described by a rational transfer function may be hard to control:

1. The relative degree of the transfer function is large.
2. The zeroes of the transfer function are lightly damped or have positive real parts.

We will limit our discussion to SISO systems. The relative degree of a transfer function is the number of poles (states) – number of zeroes. The zeroes of a transfer function are sometimes referred to as finite zeroes. The number of infinite zeroes of a system is the relative degree of the transfer function, *i.e.* the number of asymptotes in the root locus. Naturally, the ease with which LTI systems can be controlled is related to the number and location of the system's zeroes. Like the number of poles of the transfer function is the number of storage elements in integral causality in the behavior model (also called forward model), the number of zeroes the of the transfer function is the number of storage elements in integral causality in the inverse model [179, 180].

If the transfer function between a source and a sensor is $G_f(s)$, then in the inverse model the original source becomes the sensor and the original sensor becomes the source and the new source to new sensor transfer function is $G_i(s)$. Naturally, $G_i(s) = 1/G_f(s)$, meaning that the zeroes of $G_f(s)$ become poles of $G_i(s)$. Then, because it is easy to find the number of poles of the inverse model, *i.e.* the order of denominator polynomial of a transfer function, which is equal to the number of states or the number of storage elements in integrally causality in the inverse model, the inverse model acts as a suitable conduit to estimate the number of finite zeroes of the forward model.

When we look back at the flow induced vibration problem in Section 3.2.5 and the root loci in Figure 3.19, we find that the zeroes of that system are on the right half of the complex plane. This means that if one uses a high feedback gain, the system will be unstable. In some other systems, zeroes may be on the left half plane, but they may be too lightly damped, *i.e.* very close to the imaginary axis. In contrast, when zeroes of the transfer function of the system are highly damped, then designing a control system for the system is much simpler. As an example, one may use a simple high-gain observer [202] where the plant and observer error is fed back through a high proportional gain, instead of the elaborate observer with feedback to all the states as developed earlier in Section 3.4. Moreover, systems with highly damped zeroes show low noise sensitivity.

Furthermore, if we look at the transfer function in Equation 3.19 and the root loci in Figure 3.19, we find that the system has one infinite zero and the asymptote is along 0° or 180° line. Higher relative degree results in a higher number of asymptotes and break points. In Figure 3.19, if we consider the longer path, we find that if we continue to increase the gain after the break-in point gain, then one root goes to the higher damping side while the other goes to the lower damping side. This reduces the designer's control on the tuning parameters. Similarly, when there are too many asymptotes, adding any physically realizable controller or filter (*e.g.* a lead lag compensator) having more poles than zeroes introduces further infinite zeroes into the system. Overall, the control problem then becomes very complicated.

As an example, consider a transfer function $G(s) = (s+2)/(s^3 + 5s^2 + 100s + 1000)$. The relative degree of this transfer function is 2. Its root loci for negative feedback gains is plotted in Figure 3.48, where we find that the system becomes stable with some feedback gain. The root loci for positive feedback is not considered because in positive feedback, the system is unstable for all gains. However, when the gain is increased, two roots change along the asymptote. This means their damping (real part of eigenvalues) becomes fixed whereas their frequency (imaginary part of the eigenvalues) increases; thereby, effectively reducing the damping coefficient, ξ.

Let us now consider two cases. In case I, a lead compensator having transfer function $G_1 = 2(s+15)/(s+30)$ is added to the system (transfer functions get multiplied) and in case II, a lag compensator having transfer function $G_2 = 0.5(s+30)/(s+15)$ is added to the system. The corresponding root loci are superimposed on the original (leaving the poles and zeroes of the compensator outside the window) root loci in Figure 3.48. We find that these compensators, as is expected, change the centroid of the root loci. The lead compensator gives more high gain damping, but then the damping ratio does not change appreciably and continues to reduce as the asymptotes tend to infinite zeroes. This example shows the difficulty in designing control strategies for systems having many infinite zeroes.

One of the reasons to study the relative degree of a system is to choose the proper type of sensor and its location such that the relative degree of the obtained transfer function is minimized.

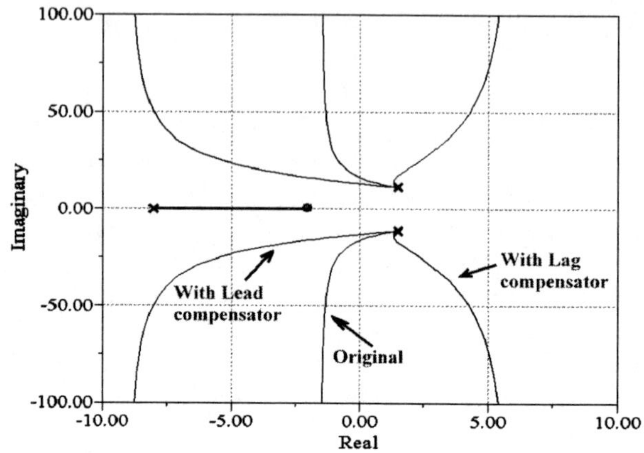

Fig. 3.48 Root loci showing asymptotes and how they change by adding compensators

The inverse bond graph model is created by imposing both effort and flow information from the sensor and receiving both at the source. This cannot be done through normal causality; we need bicausality as introduced in the last chapter. To repeat, in bicausality notation, the effort and flow causalities are split into two half-causal strokes. The side of the bond with the half-arrow for power direction is used to impose flow causality and the effort causality is imposed on the other side. When flow and effort causalities are counter-oriented, the two half causal strokes appear at one end of the bond and merge together to form a single causal stroke; which is called unicausality or normal causality. When flow and effort information are co-oriented, their causal strokes appear at two different ends. The source element (Se or Sf) is represented as a source sensor (SS) element [83, 177] and the sensor (De or Df) is represented as a sensor source (again SS) element, the only difference between them being that the source sensor element receives information of both power variables whereas the signal source element supplies both. The bicausality is propagated from the original sensor to the original source. In the process, some internal bonds in the model are assigned bicausality such that it is propagated to the receiving source sensor. At every junction, two bonds must be bicausalled, which will proved later in Chapter 7. In this chapter, this simple idea of bicausality is sufficient to proceed.

Consider the simple spring-mass-damper system given in Figure 3.49a, and its forward and inverse bond graph models given in Figures 3.49b,c, respectively. From Figure 3.49b, the number of integrally causalled storage elements, consequently meaning the number of poles of the system, is found to be 2. In the inverse model Figure 3.49c, the sensor (Df) and source (Se) have been replaced with source sensor (SS) elements. The SS element in bond number 5 imposes both effort and flow information, thereby causing differential causality in the I-element. The SS element at

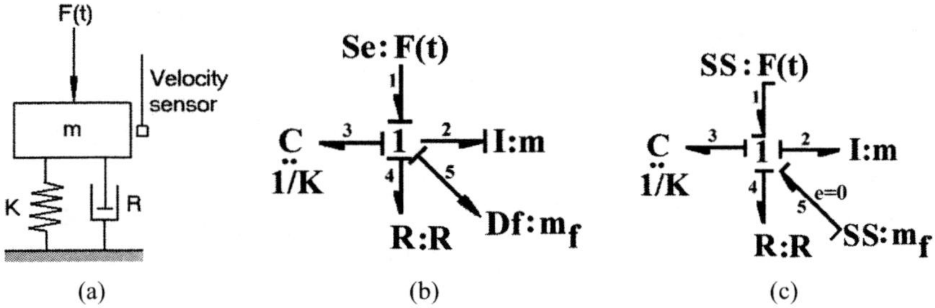

Fig. 3.49 Relative degree of a spring-mass-damper system with a velocity sensor

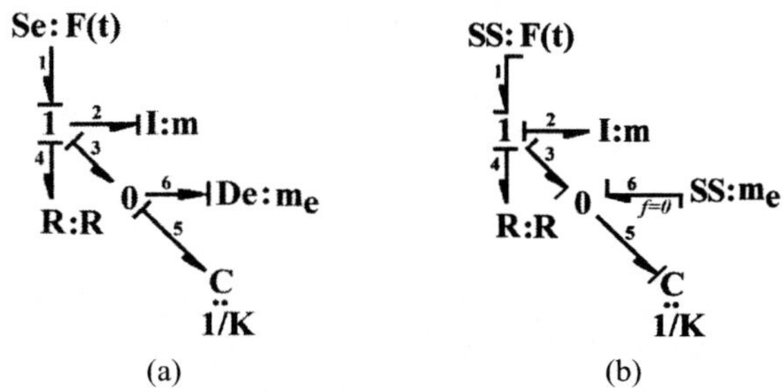

Fig. 3.50 Relative degree of a spring-mass-damper system with a displacement sensor

bond number 1 receives both flow and effort information. Consequently, the inverse model has only one integrally causalled element and then the number of poles of the inverse model is 1. This implies that the number of finite zeroes of the forward model is 1 and the relative degree of the system is $2-1=1$.

Let us now use a force sensor in the spring instead of measuring the velocity of the mass point. Then the forward and inverse bond graph models can be represented as shown in Figures 3.50a,b, respectively.

From Figure 3.50a, the number of poles of the system is found to be 2, and from Figure 3.50b, the number of finite zeroes is found to be 0 (no storage element is in integral causality). Thus the relative degree of the system is $2-0=2$. In fact, one may examine the two transfer functions, $G_1(s)$ and $G_2(s)$ from the bond graph models in Figure 3.49b and Figure 3.50a, respectively:

$$G_1(s) = \frac{s}{ms^2+Rs+K} \qquad \text{and} \qquad G_2(s) = \frac{K}{ms^2+Rs+K}.$$

Let us consider the example of an electromechanical system shown in Figure 3.51 and suppose that the velocity of the mass being lifted is measured. Note that this system was earlier considered in Section 3.2 as an example. The forward bond graph model of the system is given in Figure 3.52. In the forward model, there are three storage elements in integral causality and hence the transfer function has three poles.

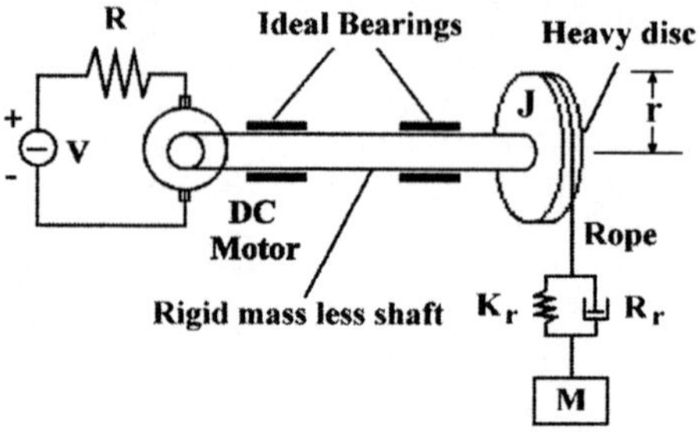

Fig. 3.51 An electro-mechanical system with a velocity sensor

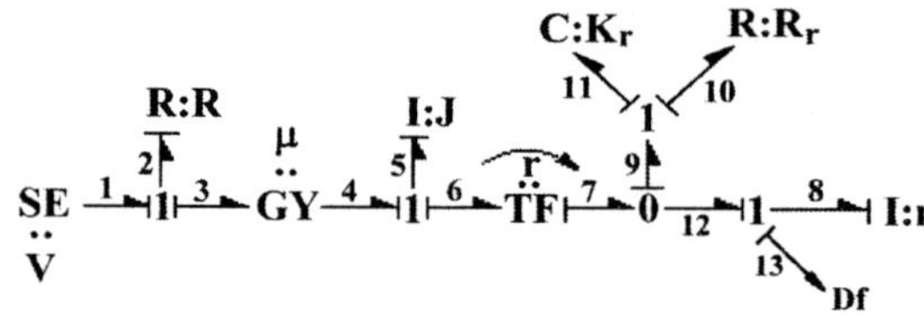

Fig. 3.52 Model of the electro-mechanical system with a velocity sensor

The inverse bond graph model of the system is given in Figure 3.53. When bi-causality is propagated from the detector to the source, it is found that two storage elements are forced into differential causality and only one storage element remains in integral causality. Therefore, the order of the numerator of the transfer function (number of finite zeroes) is 1 and the relative order of the transfer function is 2.

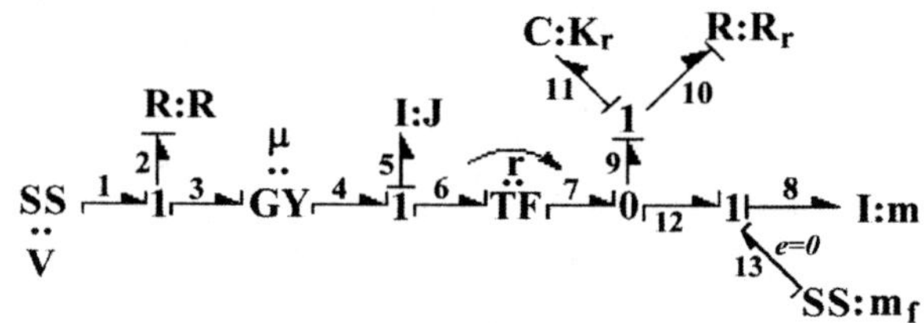

Fig. 3.53 Inverse model of the electro-mechanical system

Indeed, when the transfer function between the input and the output is derived from the signal flow graph (see Figure 3.9), it is given by

$$\frac{f_8(s)}{e_1(s)} = \frac{\mu\, r\, R_r s + \mu\, r\, K_r}{\begin{array}{c} R\, m\, J\, s^3 + \left(\mu^2 m + r^2 R_r R\, m + R_r R\, J\right) s^2 \\ + \left(\mu^2 R_r + r^2 K_r R\, m + K_r R\, J\right) s + \mu^2 K_r \end{array}}, \tag{3.52}$$

which is the same as Equation 3.4 multiplied by s (because Equation 3.4 is a transfer function up to displacement). The transfer function in Equation 3.52 confirms the results derived from structural analysis.

Note that the actual relative degree may be different from the one obtained through structural analysis because we do not consider parameter values during structural analysis, *e.g.* if one considers $R_r = 0$ in Equation 3.52 then the relative degree of the transfer function will be 3. On the other hand, if we knew earlier that $R_r = 0$, then we could have dropped it from the bond graph model. It is evident from the inverse model in Figure 3.53 that without the R-element at bond number 10, the C-element at bond number 11 could not have been assigned integral causality; thereby giving the correct relative degree of the transfer function as 3.

Remark 3.6. We have seen earlier that additional damping can make systems structurally controllable and observable. We also see in this section that damping can at times reduce the relative degree of the system. None of these guarantee that any damping would improve structural properties. As an example, if we consider aerial damping on the disk and the lifted mass in the example just considered, we can show that the relative degree would still remain the same. We leave it to the reader to verify this by answering this simple question: can the causalities of the two I-elements be changed in the inverse model?

3.5.8 Zero Dynamics

Zero dynamics is an important feature in system analysis and controller design. Zero dynamics refers to the internal dynamics of a system when initial conditions and inputs are so applied that the output remains zero. This dynamics arises from the numerator polynomial of a system's transfer function. Therefore, zero dynamics cannot be influenced by feedback compensation or pole-placement. The zero dynamics determines the performance limit of many feedback control systems. An LTI system is said to be minimum-phase if the system and its inverse are causal and stable. Thus even if a system is asymptotically stable (BIBO or Bounded-Input Bounded-Output stable, *i.e.* all its poles are stable), it may become non-minimum-phase due to the location of its zeroes. It is well-known that if the zero dynamics of a system is unstable (*i.e.* a non-minimum-phase system) then it may not be possible to develop a controller to obtain perfect trajectory tracking. The command generation for trajectory tracking (and also for fault tolerant control) is an inverse problem which is discussed in Chapter 11. Unstable zero-dynamics means the inverse system becomes unstable.

Therefore, physical systems must be designed in such a way that they possess desired zero dynamics. Moreover, the calculation of the zero dynamics from the structural model of a physical system (*i.e.* not using numerical values) offers insights into which subsystems of the system are contributing to which parts of the zero-dynamics and thereby allow structural level design modifications. Bond graph model based studies on structural analysis of zero-dynamics for SISO and MIMO systems have been developed in [110, 111, 277]. Note that the developed principles, based on specific causality assignment procedures and use of causal paths to determine subsystem interactions are applicable to both linear and non-linear systems.

We have used bicausality to determine the relative degree of a system in the previous section. The same can be obtained by detecting the shortest causal path between the input and output of an SISO system. If the system is MIMO then the vector relative degree can be determined [111].

Consider the system given in Figure 3.51 and its bond graph model in Figure 3.52. We have derived the transfer function of the system in Equation 3.52 and shown through bicausality assignment in Figure 3.53 that the relative degree of the system is two. In the behavioral bond graph model of the system, the shortest causal path linking the input to the output (Se_1 to Df_{13}) is shown in Figure 3.54 by dotted lines. A shortest causal path contains minimum number of integrally causalled storage elements. There are two integrally causalled storage elements in the shortest causal path. This means that the relative degree of the input-output transfer function is two and the order of the numerator polynomial is one (the number of integrally causalled storage elements not appearing in the shortest causal path). The proof is straightforward: the order of the denominator polynomial is fixed (in this case, three, due to three integrally causalled storage elements). The shortest forward path gain contains a factor $1/s^2$(due to two storage elements in the shortest forward path) which will be multiplied with s^3 (after application of Mason's rule). Obviously, the highest power of s in the numerator polynomial corresponds to the difference between the number of states of the system and the number of states encountered in the shortest causal path.

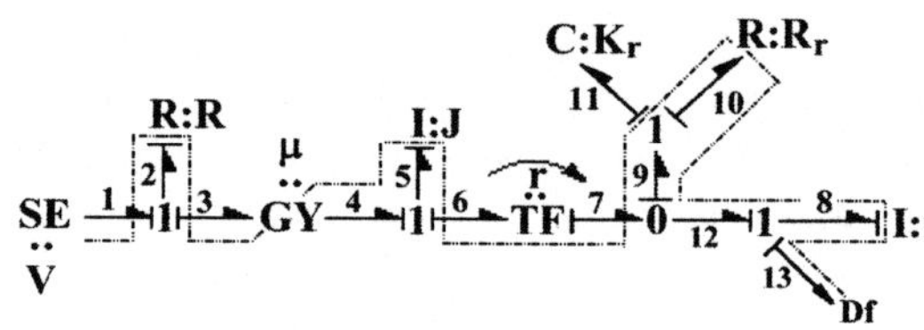

Fig. 3.54 Shortest causal path between the input and the output

Note that the subsystem responsible for zero-dynamics can be identified as the R-C element pair at 1-junction connected to bond number 9 (see bicausalled model in Figure 3.53). The first-order transfer function resulting from this subsystem is $R_r s + K_r$, which happens to be the deciding factor in the numerator polynomial in Equation 3.52 and gives the system's zero at $s = -K_r/R_r$. Then for all positive values of K_r and R_r, the zero-dynamics of the system is stable.

Let us consider a collocated spring-mass-damper system shown in Figure 3.55. In this system, the friction between the masses and the ground is neglected. A force is applied as shown in the figure and velocity of one of the masses is the output variable. The bond graph model of the system in integral causality is given in Figure 3.56.

The shortest causal path between the input and the output is shown by dotted lines in Figure 3.56. The number of integrally causalled storage elements in the shortest causal path is three, which means the relative degree of the input-output transfer function is three. The number of integrally causalled storage elements not appearing in the shortest causal path is four, which is the order of the numerator polynomial of the input-output transfer function.

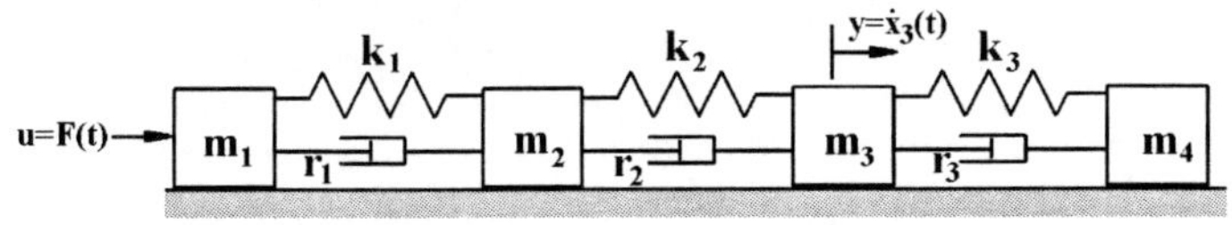

Fig. 3.55 A collocated mechanical system

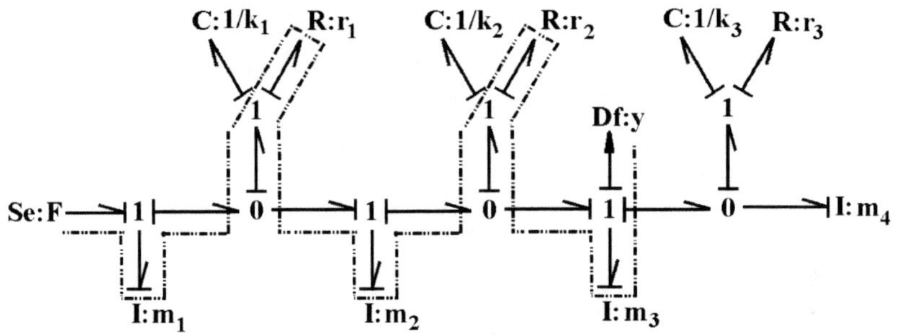

Fig. 3.56 Bond graph model of the collocated mechanical system

The bond graph model of the system may be bicausalled as shown in Figure 3.57, which shows three elements in differential causality and thus the relative degree of the system is three. Needless to say, the shortest causal path analysis and bicausality assignment-based analysis are equivalent.

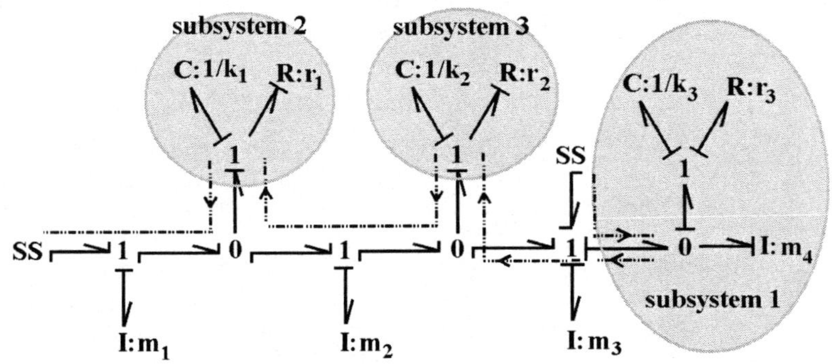

Fig. 3.57 Bicausalled bond graph model of the collocated mechanical system

The subsystems contributing to zero-dynamics are determined in [110, 111] by a complicated algorithm: (1) fixing the measured variable to zero at the junction(s) to which the sensor(s) is/are connected, which in this case means assigning zero-flow to all bonds connected to the 1-junction to which Df element is connected, (2) inverting causality of the source(s), (3) inverting causality of all energy elements on the shortest causal path, while (4) maintaining integral causality of the storage elements not in the shortest causal path. After completing causality assignment, separate subsystems which are not in the shortest causal path, as identified before, and are interacting with other subsystems through zero-power bonds (because of step (1)) are determined. This algorithm violates the causality of the junction(s) to which sensors are connected. This is because of the fact that an inverse system cannot be properly modeled within uni-causality framework.

However, a bicausalled bond graph model is itself sufficient to identify the subsystems contributing to zero-dynamics. Note that the causalities of all energy elements appearing in Figure 3.56 are reversed in the bicausalled model in Figure 3.57. The source causality is inverted and those storage elements which are not in the shortest causal path in Figure 3.56 are kept in integral causality in Figure 3.57. The bicausalled model satisfies all the steps in the aforementioned algorithm and yet there is no violation of causality at any junction.

The shortest causal path between the inverted output to the inverted input may now be identified from Figure 3.57 (which is simply the sequence of bicausalled bonds passing through various junctions to the source sensor SS:V) and the subsystems having integrally causalled storage elements and not appearing in this shortest path can be identified. Three such subsystems, which contribute to the zero-dynamics are marked in Figure 3.57.

These three subsystems are causally coupled as shown by dotted lines in Figure 3.57. The first subsystem is a second order system with a flow input and an effort output (in the same bond). The subsystem transfer function may be obtained as $\left(R_3 m_4 s^2 + m_4 K_3 s\right) / \left(m_4 s^2 + R_3 s + K_3\right)$. The denominator of this transfer function, *i.e.* $m_4 s^2 + R_3 s + K_3$, appears in numerator of the forward system's transfer function. Likewise, the other two subsystems give two first order polynomials, $R_1 s + K_1$ and $R_2 s + K_2$, respectively, for subsystem 2 and subsystem 3. We leave it to the reader to find the transfer function of the actual system from a signal flow graph and verify that its numerator polynomial is actually $\left(m_4 s^2 + R_3 s + K_3\right)\left(R_1 s + K_1\right)\left(R_2 s + K_2\right)$.

Problems

Simple problems

3.1. In the system shown in Figure 3.58a, a sensitive piece of equipment is being hoisted by a pulley. Assume that the shaft is rigid and J is the total equivalent rotary inertia of the motor, the shaft and the pulley. If weights of different bodies are considered in the bond graph model, they should not be treated as control sources; only the voltage source is the control input. (a) Through structural analysis, determine whether this system is completely controllable and observable with the given instrumentation architecture (actuators and sensors) and also find any null-modes in the system. (b) Find the relative degree of the input output transfer function through bicausality assignment and verify the result by deriving the transfer function from a signal flow graph model. Further show that the denominator of the transfer function contains as many free ss as the null modes of the system.

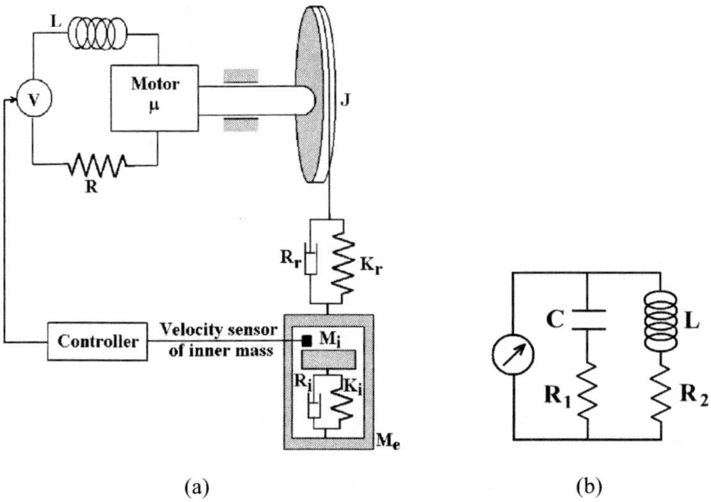

Fig. 3.58 **a** A sensitive piece of equipment being hoisted by a pulley. **b** An electrical network

3.2. The circuit shown in Figure 3.58b contains a current source. The current through the resistance R_2 is measured. (a) Show that the system is structurally controllable and structurally observable. (b) Consider that resistances R_1 and R_2 have same value R. Derive the state equations of the system and formulate the controllability and observability matrices. If all parameters are assumed to be non-zero and finite, show that the system is observable for all parameters whereas for a specific relation between the parameters, $R = \sqrt{L/C}$, the system becomes uncontrollable. (c) Through bicausality assignment, show that the relative degree of this system is one. Verify this by deriving the input output transfer function from a signal flow graph. (d) Also show that input-output causal path analysis on the integrally causalled bond graph model exposes the same relative degree.

3.3. Prove the equivalences shown in Figure 3.59a. Construct a bond graph model of the system shown in Figure 3.59b by neglecting the lever's inertia and assign causalities. Detect the zero-order-causal-path (causal loop) in the model. By using the equivalences shown in Figure 3.59a, reduce the bond graph model to a compact form which does not contain causal loops. Hint: start from two junctions, $1_{\dot{\theta}}$ and $1_{\dot{y}}$, for rotational and linear velocities of the rod, respectively, at the fulcrum point.

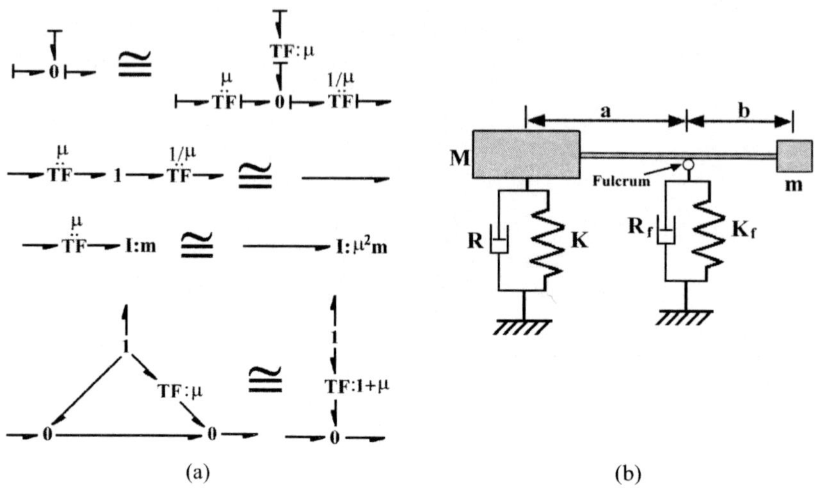

Fig. 3.59 **a** Some bond graph equivalences. **b** A mechanical system

Project Type Problems

3.4. The Shannon's sampling theorem states that a signal is complete characterized if it is sampled at frequencies higher than twice the Nyquist frequency, *i.e.* the sampled output in this case contains all the frequencies of the input signal and the input signal can be reconstructed through the use of an appropriate low-pass filter. Consider the Simulink® block diagram in Figure 3.60 where two sine waves each of amplitude 1 and frequencies 1 rad/s and 5 rad/s are mixed and then sampled. Finite pulse width sampling is done as given in picture (sampling period=0.06 s) and then a fourth order Chebyshev filter with break frequency of 50 rad/s was used to reconstruct the analog signal.

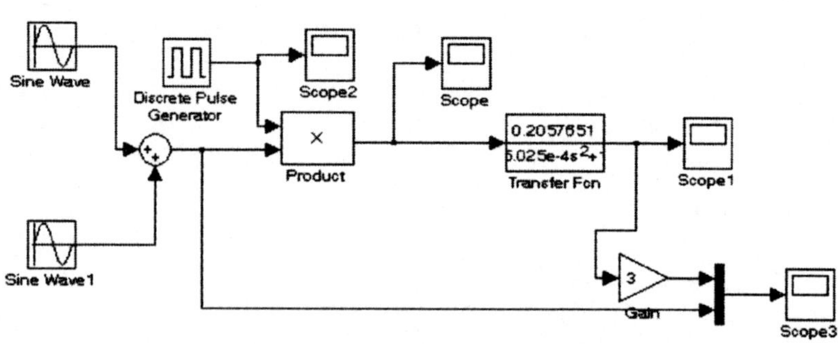

Fig. 3.60 Simulink® model of finite pulse-width sampler

Specify the data for discrete pulse generator as amplitude=1, period=3, pulse width=1 and pulse frequency=0.02 (*i.e.* 0.06/3, this factor of 3 is later considered as a gain). The transfer function of the fourth order Chebyshev filter is

$$G_f(s) = \frac{0.02057651}{\begin{array}{c}1.6\times 10^{-7}s^4 + 5.72972\times 10^{-6}s^3 + 5.025928\times 10^{-4}s^2\\ +1.033596\times 10^{-2}s + 0.2057651\end{array}}. \tag{3.53}$$

Simulate this model for 10 s with good precision and show that the output in block 'Scope3' is indeed a reconstruction of the input signal.

3.5. For a discrete-time system with sampling time of $T = 1$s and defined by

$$x(k+1) = \begin{bmatrix} 0.7 & 0.1 \\ 0.1 & 0.8 \end{bmatrix} x(k) + \begin{Bmatrix} 0.5 \\ -2 \end{Bmatrix} u(k), \tag{3.54}$$

$$y(k) = \begin{bmatrix} 1 & 0 \end{bmatrix} x(k), \text{ with } x(0) = \begin{Bmatrix} 2 \\ -2 \end{Bmatrix} \quad \text{and} \quad u(k) = \sin^2(0.2k),$$

develop a simple observer and a current estimator (either as MATLAB® codes or in Simulink®), in both of which the poles of the error dynamics are to be placed within the unit circle at $0.1 \pm 0.4j$. Compare the output of the plant with the output from the two observers over a 40 s time window by initializing the observer states $x_o(0) = \begin{bmatrix} 0 & 0 \end{bmatrix}^T$ and discuss how the current estimator gives better performance.

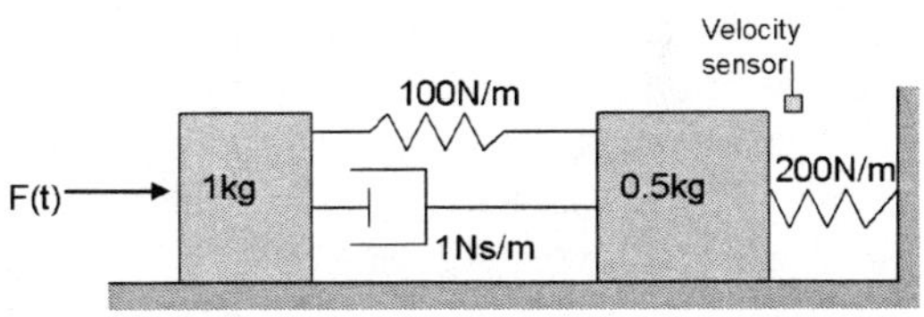

Fig. 3.61 A mechanical system for which an observer is to be designed

3.6. Design a Luenberger observer for the system shown in Figure 3.61 where the surface is considered frictionless. All four observer poles are to be placed at $-6.5 \pm 0i$ on the real axis. Simulate the plant and the observer simultaneously for 3 s duration with all observer states initialized to zero and the plant states as follows: momentum of right mass$= -1$, momentum of left mass$= 0.5$, displacement of right spring$= -2$ and displacement of middle spring$= 5$; all in some consistent units. Plot any two states, one from plant and the other from observer, *e.g.* deformation of middle spring and show how the observer tracks the plant. Change the observer pole positions to $-100 \pm 0j$ and two double poles at $-50 + 0j$, and then show that the tracking error is compensated very quickly, but there is heavy initial overshoot in the response.

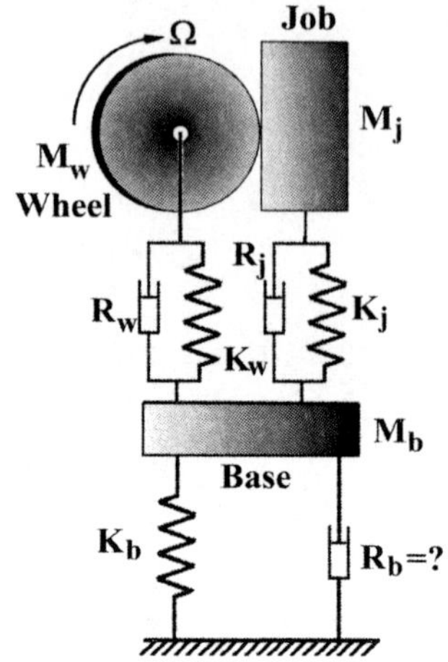

Fig. 3.62 Schematic of a soft handled grinding machine with all fixtures approximated by spring-damper combinations

3.7. A grinding wheel and its job are mounted on a flexible foundation as shown in Figure 3.62. The grinding wheel is driven by a speed controlled motor and it spins at a constant speed Ω. The material removal during grinding is due to friction induced shear. At the wheel-job interface, equal and opposite friction forces develop. The friction force on the wheel produces an equivalent force and a couple moment at the wheel center, the latter balancing the torque generated by the driving motor. The friction force is more or less constant and may be considered as a static load. Here we will consider the dynamics in the vertical direction. At low spinning speeds, the cutting force *vs* the cutting speed characteristics is such that with the increase in cutting speed, the required cutting force is reduced. If a relative velocity is set up between the wheel and the job, due to some disturbance, then the system exhibits flutter instability. This is what is referred to as the tool chatter. Tool chatter is modeled as a negative resistance, R_{ch}, acting on the relative velocity between the grinding wheel and the job. Note that wheel speed Ω times wheel's radius or the circumferential speed is responsible for material removal and does not influence in the vertical dynamics except for static deflections. It is only the slip velocity at the interface which sets up tool chatter. For the following data which have been scaled down and are in consistent units, design a proper foundation damper through driving point impedance design principles and root-loci analysis. The static forces due to interface friction may not be considered in the model. Compare the time responses of the open loop (no foundation damping) and the closed loop (with optimal foundation damping) systems. Also compare the corresponding frequency responses (Bode plots).

Scaled data in consistent units: $M_b = 10$, $K_b = 100$, $R_b =$ to be designed, $K_w = 200$, $R_w = 1$, $M_w = 0.5$, $M_j = 5$, $K_j = 500$, $R_j = 0.5$ and $R_{ch} = -1$.

Chapter 4
Bond Graph Model-based Qualitative FDI

Most of the early approaches for fault diagnosis and isolation were rule based. Such approaches use simple prediction rules to provide possible faults in a system and their causes. These methods suffer from incompleteness and inflexibility. Recent fault diagnosis methods are based on analysis of the underlying model structures and behavior of a system. Models serve as knowledge representation of a large amount of structural, functional and behavioral information and their relationship [8, 16, 50, 93, 139, 197, 198, 237, 239]. This knowledge representation is used to create complex cause-effect reasoning leading to construction of powerful and robust automatic diagnosis and isolation systems [2, 76, 98, 99, 114, 115, 261, 267].

Qualitative reasoning by using bond graphs can be conducted to construct intelligent supervisory control systems [150, 271]. It uses general reasoning strategy through artificial intelligence and bond graph models are employed for the knowledge representation. In this chapter we deal with qualitative model based methods, where model implies an analytical/structural model constructed from a system engineer's perspective.

4.1 Model Order Reduction

Fault diagnosis is aimed at precisely capturing major discrepancies in the process or system behavior. Usually an autonomous system's behavior is a superposition of its steady state dynamics and small transients. For reliable fault diagnosis of steady state behavior, transients are usually neglected. Faults affecting transient behavior are more efficiently handled by analyzing the output frequency response, *e.g.* through Fourier transform, power spectral density, wavelet and Hilbert transforms, *etc.*

Thus, there is a need to separate the steady state behavior from the transient behavior. Most often, one can identify a simpler system model representing only the steady state system dynamics. Moreover, FDI algorithms are always based on simpler models, which have a minimum number of state variables, are easily computable and capture only essential dynamics of the system. Therefore model order reduction or identification of a reduced order model is the starting point in the design of an efficient online model based supervision system. From a control theoretic prospective, higher order systems are difficult to control and it is very difficult to develop optimal control algorithms for them.

Various methods for model order reduction and minimal order system identification are reported in the literature. For LTI systems, mostly projection-based model order reduction is followed. Examples of projection based model order reduction methods are modal truncation or eigenvalue decomposition, balanced reduction through balancing transformations, singular value decomposition and optimal Hankel norm approximation.

Consider the state-space equations of an LTI system

$$\begin{aligned} \dot{x} &= \mathbf{A}x + \mathbf{B}u, \\ y &= \mathbf{C}x, \end{aligned} \tag{4.1}$$

where $x \in \mathbb{R}^n$, $u \in \mathbb{R}^m$, $y \in \mathbb{R}^p$ and there is no direct input feed-through, *i.e.* $\mathbf{D} = 0$. Let $\mathbf{S}_n = \begin{pmatrix} \mathbf{A} & \mathbf{B} \\ \mathbf{C} & 0 \end{pmatrix} \in \mathbb{R}^{(n+p)\times(n+m)}$ be the state-space quadruple. The objective of model order reduction is to find a reduced order quadruple $\mathbf{S}_r = \begin{pmatrix} \mathbf{A}_r & \mathbf{B}_r \\ \mathbf{C}_r & 0 \end{pmatrix} \in \mathbb{R}^{(r+p)\times(r+m)}$ with $r < n$, which for a specific operating range has approximately the same dynamics as that of the original system.

Let $z = \mathbf{T}x$ be a transformation applied to the states, where $\mathbf{T} \in \mathbb{R}^{n\times n}$ is nonsingular. Then the transformed state-space form is

$$\begin{aligned} \dot{z} &= \mathbf{TAT}^{-1}z + \mathbf{TB}u = \hat{\mathbf{A}}z + \hat{\mathbf{B}}u, \\ y &= \mathbf{CT}^{-1}z = \hat{\mathbf{C}}z. \end{aligned} \tag{4.2}$$

The matrices in this state-space form may be partitioned as

$$\hat{\mathbf{A}} = \mathbf{TAT}^{-1} = \begin{bmatrix} \mathbf{A}_{11} & \mathbf{A}_{12} \\ \mathbf{A}_{21} & \mathbf{A}_{22} \end{bmatrix}, \quad \hat{\mathbf{B}} = \mathbf{TB} = \begin{bmatrix} \mathbf{B}_1 \\ \mathbf{B}_2 \end{bmatrix}, \tag{4.3}$$

$$\hat{\mathbf{C}} = \mathbf{CT}^{-1} = \begin{bmatrix} \mathbf{C}_1 & \mathbf{C}_2 \end{bmatrix} \text{ and } z = \begin{bmatrix} z_1 \\ z_2 \end{bmatrix}. \tag{4.4}$$

The reduced order quadruple is then $\mathbf{S}_r = \begin{pmatrix} \mathbf{A}_{11} & \mathbf{B}_1 \\ \mathbf{C}_1 & 0 \end{pmatrix}$ with $z_1 \in \mathbb{R}^r$ as the states and the remaining parts of the full quadruple are called residual. Following singular perturbation method [138], Equation 4.3 must satisfy

$$\begin{aligned} \dot{z}_1 &= \mathbf{A}_{11}z_1 + \mathbf{A}_{12}z_2 + \mathbf{B}_1 u, \\ 0 &= \mathbf{A}_{21}z_1 + \mathbf{A}_{22}z_2 + \mathbf{B}_2 u, \\ \Rightarrow z_2 &= -\mathbf{A}_{22}^{-1}\mathbf{A}_{21}z_1 - \mathbf{A}_{22}^{-1}\mathbf{B}_2 u. \end{aligned} \tag{4.5}$$

Then the reduced order model (state-space quadruple) is given by

$$\mathbf{S}_r = \begin{pmatrix} \left(\mathbf{A}_{11} - \mathbf{A}_{12}\mathbf{A}_{22}^{-1}\mathbf{A}_{21}\right) & \left(\mathbf{B}_1 - \mathbf{A}_{12}\mathbf{A}_{22}^{-1}\mathbf{B}_2\right) \\ \mathbf{C}_1 - \mathbf{C}_2\mathbf{A}_{22}^{-1}\mathbf{A}_{21} & -\mathbf{C}_2\mathbf{A}_{22}^{-1}\mathbf{B}_2 \end{pmatrix}. \tag{4.6}$$

In modal truncation method, one considers the transformation matrix $\mathbf{T}$ to be the eigenvectors of matrix $\mathbf{A}$. Then

$$\hat{\mathbf{A}} = \mathbf{TAT}^{-1} = \left[\begin{array}{c|c} \Lambda_{11} & 0 \\ \hline 0 & \Lambda_{22} \end{array}\right] = \left[\begin{array}{ccc|ccc} \lambda_1 & & & & & \\ & \ddots & & & & \\ & & \lambda_r & & & \\ \hline & & & & \ddots & \\ & & & & & \lambda_n \end{array}\right], \tag{4.7}$$

where the eigenvalues are ordered according to the specifications, *i.e.* which modes need to be retained, *e.g.* in ascending frequency (lower to higher mode), *i.e.* $\lambda_1 < \lambda_2 < ... < \lambda_n$. The modal partitioning can now be related to Equation 4.3 to obtain the reduced order modal state-space equations. The state variables associated with the modal equations are referred to as modal momentums and modal displacements.

In [60], the fast and slow dynamics of a multi-rate system were identified from various loop gains, which can be obtained from a bond graph model by analyzing the causal paths in the model (see Chapter 3), and then they were used in a singular perturbation based model order reduction performed on a bond graph model. Another bond graph model based model order reduction in frequency domain may be consulted in [183].

Balanced reduction techniques, such as singular value decomposition, are based on the evaluation of controllability and observability Grammians and then solution of the matrix Lyapunov equation which yields the Hankel singular values. Those methods are out of the scope of this book.

In the following, we present a model order reduction method based on physical system models, especially by considering energy as a metric for model order reduction [60, 79, 153, 154, 243]. We assume that if the original model is both controllable and observable, such that the system can be brought to its steady state operating regime in a controlled manner, then the reduced order model should retain those properties.

The nature of steady state dynamics is usually dependent on the input frequency range. Therefore, essential peaks with larger bandwidth in the frequency response are used to identify the reduced order model. Different frequencies in the response arise from causal coupling between storage elements in the bond graph model [79]. However, identifying the causal paths linking various storage elements and their combinations in a moderately large bond graph model is extremely difficult. Moreover, there is no generalized method for non-linear systems. Significant peaks in the frequency response are characterized by their amplitudes and bandwidths which, in other words, represent the energy content in the corresponding modes. In bond graph models, it is much easier to handle the energy being transferred to different elements of the system.

We start by associating an activity index [153, 154] with every passive element of the system. Controlled elements (*e.g.* R-elements for controlled valves) are not taken into account. We define the energy index of a passive element (only in the steady state operation) as

$$E_i = \int_{ts}^{t_f} \left(\sum_{k=1}^{n} |e_k.f_k| \right) dt, \tag{4.8}$$

where e_k and f_k are the efforts and flows in the k-th bond connected to i-th element, n is the total number of bonds connected to i-th element, and t_s is the time after which one may assume steady state system behavior and $t_f \gg t_s$ is chosen to include a sufficient contribution of all frequencies (*i.e.* up to several times of the Nyquist frequency) in the observation window. Then the activity index of the i-th element is defined as the weighted contribution of each component as

$$A_i = \frac{E_i}{\sum_{j=1}^{m} E_j}, \tag{4.9}$$

where m is the total number of passive elements in the model.

Thereafter, one may decide a threshold activity index below which the contribution of the corresponding passive element to the dynamics of the system may be considered insignificant. Those elements may now be removed from the model, provided that the model still remains both controllable and observable; there are no problems in causality assignment, no artificial singularity, and the reduced order model produces comparable results both in the time and frequency domains to that of the original model. Note that even if some storage elements cannot be integrally causalled in the reduced order model, it is acceptable for FDI system development; the reasons for which are partially discussed later in this chapter and discussed in details in the next chapter.

As an example, we consider a two-degrees-of-freedom system shown in Figure 4.1a and its bond graph model given in Figure 4.1b.

Note that gravity is not considered in the model because we are concerned about motion over and above the static deflections. For the parameter values $M_e = 1$, $K_e = 100$, $R_e = 1$, $M_f = 2$, $K_f = 200$, $R_f = 2$ and $F(t) = A\cos(\omega t)$, where A =0.01 and $\omega = 100$, and considering that all variables are in some consistent unit, the simulation results were used to study the activity indexes of all I, C and R elements over 10 s observation window. The results are given in Table 4.1, where the subscripts in the activity index refer to the bond numbers and the ordering is done according to descending activity indices.

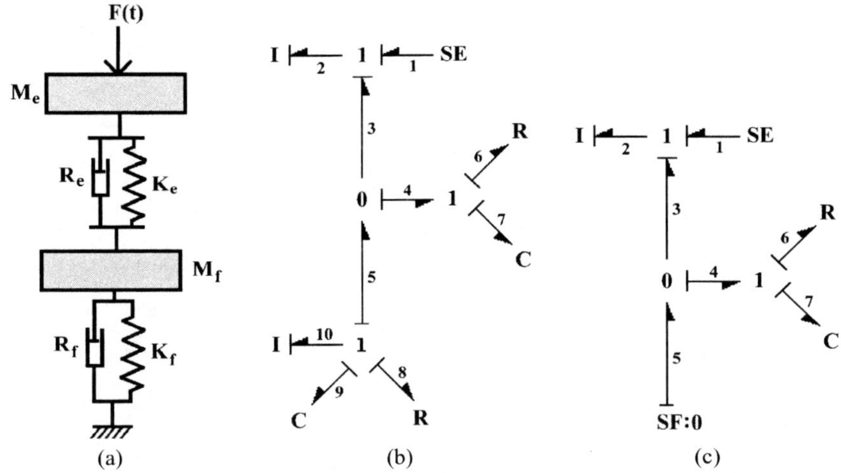

Fig. 4.1 A two degrees-of-freedom system and its bond graph models

Table 4.1 Activity indices of bond graph elements

Bond number (i)	E_i (Joules)	A_i
2	0.0977	0.9745
6	0.00155	0.0154
7	0.00099	0.00987
10	1.41×10^{-5}	1.39×10^{-4}
9	1.88×10^{-6}	1.86×10^{-5}
8	3.84×10^{-7}	3.83×10^{-6}

The technological specification may be given depending upon the process under consideration. Let us assume that the reduced order model must account for 99% of the energy in the model. Then one starts summing the activity indices in descending order untill it exceeds the assumed threshold, *i.e.* 0.99 in this case. From Table 4.1, we find that the sum of activity indices of the top three elements just exceeds the threshold while the sum of the top two activity indices falls short of the threshold. Therefore, the three corresponding elements must be retained in the model. Due to small activity indexes in the last three bonds, they may be removed from the model. Physically this means that not much energetic interaction is taking place in the foundation mass-spring-damper sub-system. A rigid support will give rise to the same steady state response. Therefore, the reduced order model may be represented as given in Figure 4.1c. Note that the reduced order model corresponds to $\omega = 100$, which may be considered large for this system. When the full-order and reduced-order models were simulated and their responses were compared (momentum of the top mass), they we found to be indistinguishable to the naked eye. Even when the excitation frequencies are reduced to about $\omega = 30$ (the original reduction was based on $\omega = 100$), the responses of two systems were found to be in good agreement with each other (see Figure 4.2a for $\omega = 50$). As the input frequencies are reduced further and approach the natural frequencies of the system shown in the bode plot in Figure 4.2b, the error between the two models starts to build up.

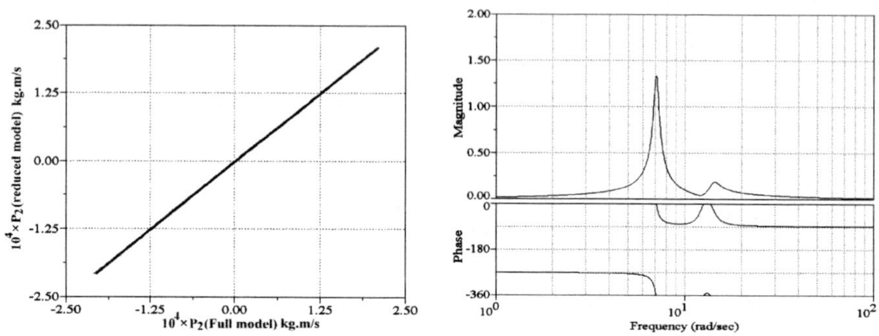

(a) (b)

Fig. 4.2 **a** Comparison of full and reduced order model outputs. **b** Frequency response plot of full order model

For low frequency excitation, *e.g.* $\omega = 1$, one can proceed in a similar way and find that only the two R-elements for damping can be removed to construct the reduced order model. Thus we have three different models: one of the full system for near resonance excitation, one for the low frequency excitation and one for high frequency excitation. This breakdown method is often referred to as one model for one frequency range. In this example system, the low frequency model really does not reduce model order, but may reduce computation time in some specific cases; such reductions are not much help in FDI system design.

Remember that in Example II in Section 2 of the previous chapter, we designed a foundation damper to control flow induced vibrations. One of the apparently contradicting results obtained there was that high foundation damping does not stabilize the system. In reality, high foundation damping reduces the capacity of the foundation to dissipate energy, which in other words mean that activity indices of all elements in the foundation become negligible compared to rest of the elements of the system.

Therefore, reduced order models approximate behavior of the system in a given operating range, *i.e.* input characteristics, relative parameter values, *etc.* The proper threshold for minimum activity index is chosen according to the technological specifications.

Models of distributed parameter systems, such as beam vibration problems, are often treated through finite element reticulations. Finite element (or a finite difference) formulation approximates the system behavior of a distributed parameter system described in terms of partial differential equations by spatially lumping mass, stiffness and damping at/between a few nodes. To obtain good results for higher modes, one needs to consider large number of finite elements. It may not be possible to apply this model reduction technique to finite element bond graphs by checking activity indices of bond graph elements because dropping any inertia or stiffness element would change the element properties. Usually, in such problems, certain modes of vibration are more excited (draw more energy) than other modes. For example, if one considers unbalance response of a rotor shaft with a centrally mounted disk then under steady-state synchronous whirling, only the first bending mode is excited. On the other hand, if a perfectly balanced vertical rotor with a heavy central disk undergoes out of the plane vibrations, then the gyroscopic couplings predominantly excite the even modes (second, fourth, ...) of vibration. The actual system response in linear distributed parameter systems is the superposition of several modes of vibration. In such cases, it is often advisable to use a modal bond graph instead of a bond graph developed from finite elements. To develop a modal bond graph, one needs a modal decomposition of the problem to obtain the modal stiffness, mass and damping parameters; and then first few vibration modes are represented in the bond graph form [121, 219]. When model order reduction technique is applied to a modal bond graph, it can find out which modal oscillators are energetically least active and part of the bond graph representing those oscillators may be dropped from the model [152]. This is an energy metric based modal truncation method. Frequency domain model order reduction may be done directly in the physical domain without converting the bond graph model to its modal form [184].

Consider a quarter car model as shown in Figure 4.3a and its bond graph model as shown in Figure 4.3b. The parameters in the model are $M_s = 300$ kg, $K_s = 2\times10^4$ N/m, $R_s = 200$ Ns/m, $M_u = 50$ kg, $K_u = 2\times10^5$ N/m and $R_u = 1500$ Ns/m, which respectively stand for the sprung mass, the suspension stiffness, the suspension damping, the unsprung mass (wheel), tyre stiffness and tyre damping. The ground excitation may be considered as a harmonic function, $V(t)$.

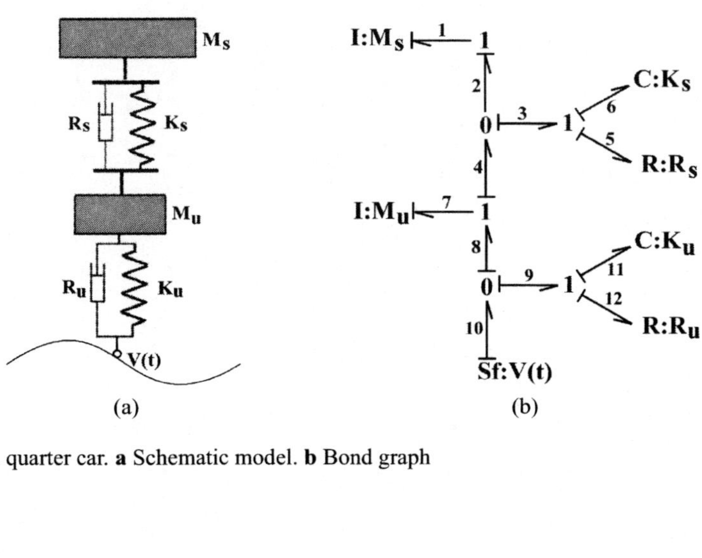

Fig. 4.3 A quarter car. **a** Schematic model. **b** Bond graph

The modal decomposition of the system needs evaluation of the eigenvalues of the system [160]. Instead, let us consider a transfer function between the ground excitation and the force transmitted to the ground, *i.e.*

$$g(s) = \frac{e_{10}(s)}{f_{10}(s)}. \tag{4.10}$$

The reason for considering this transfer function will be apparent later. For the given values of system parameters, the transfer function may be obtained by direct matrix method or from a signal flow graph as follows:

$$g(s) = \frac{1500s^4 + 207000s^3 + 1633333s^2 + 9.333333 \times 10^7 s}{s^4 + 34.666667s^3 + 4486.667s^2 + 4666.667s + 266666.7}. \tag{4.11}$$

This transfer function may be written by factorizing the denominator as

$$g(s) = \frac{1500s^4 + 207000s^3 + 1633333s^2 + 9.333333 \times 10^7 s}{(s^2 + 0.591s + 60.5235)(s^2 + 34.0756s + 4406)}. \tag{4.12}$$

Each of the terms in the denominator may now be thought of as a modal oscillator of the kind defined by the equation $M\ddot{x} + R\dot{x} + Kx$ or in the frequency domain as $(Ms^2 + Rs + K)x(s)$. By comparison, the modal oscillator parameters are: $M_1 = 1$, $R_1 = 0.591$, $K_1 = 60.5235$, $M_2 = 1$, $R_2 = 34.0756$ and $K_2 = 4406$, where M, K and R are respectively modal mass, modal stiffness and modal damping parameters and subscripts 1 and 2 stand for the mode number.

Then the modal bond graph equivalent of the bond graph model in Figure 4.3b can be constructed as shown in Figure 4.4, where each modal oscillator is decoupled from the other. The two transformers with moduli μ_1 and μ_2 are used to represent modal participation.

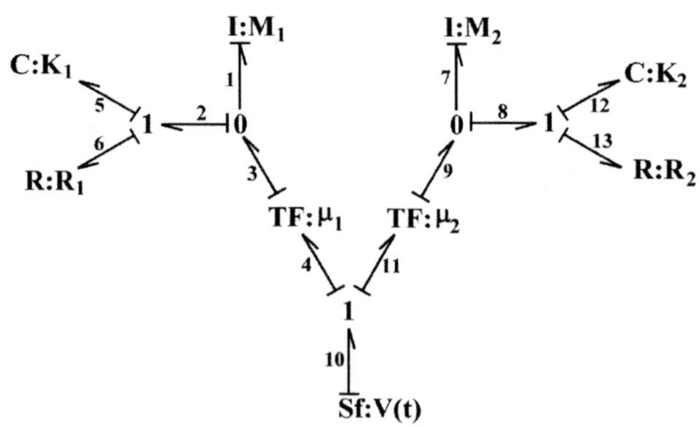

Fig. 4.4 Modal bond graph model of the quarter car system

Note that the state variables of the modal bond graph are modal momentums and modal displacements; they have no relation to the state variables in the original model.

However, the only variable which remains unaffected is the source. This is why we had derived the transfer function of the source to the source loading in the earlier model. If both the models have to be equivalent, then the transfer functions from the source to source loading must be the same for both.

Let us first consider two transfer functions from the bond graph model in Figure 4.4 as

$$g_1(s) = \frac{e_3(s)}{f_3(s)}$$

and

$$g_2(s) = \frac{e_9(s)}{f_9(s)}.$$

It follows that

$$g(s) = \frac{e_{10}(s)}{f_{10}(s)} = \mu_1^2 g_1(s) + \mu_2^2 g_2(s). \tag{4.13}$$

Substituting values of modal parameters, we obtain

$$g_1(s) = \frac{s\,(0.591s + 60.5235)}{s^2 + 0.591s + 60.5235} \tag{4.14}$$

and

$$g_2(s) = \frac{s\,(34.0756s + 4406)}{s^2 + 34.0756s + 4406}. \tag{4.15}$$

By equating $\mu_1^2 g_1(s) + \mu_2^2 g_2(s)$ to $g(s)$ in Equation 4.11, we get identical denominator polynomial and the numerator polynomial gives

$$\begin{aligned}
&\left(0.591s^4 + 80.66s^3 + 4666.6s^2 + 266666.6s\right)\mu_1^2 + \\
&\left(34.08s^4 + 4426.14s^3 + 4666.6s^2 + 266666.6s\right)\mu_2^2 \\
&= 1500s^4 + 207000s^3 + 1633333s^2 + 9.333333 \times 10^7 s.
\end{aligned} \tag{4.16}$$

Now one can separately arrange powers of s and obtain four linear equations. However, from observation it is apparent that only two linearly independent equations can be obtained. Let us consider the largest and the lowest powers of s, which give

$$\begin{aligned}
&0.591\mu_1^2 + 34.08\mu_2^2 = 1500, \\
&266666.6\left(\mu_1^2 + \mu_2^2\right) = 9.333333 \times 10^7, \text{ or } \mu_1^2 + \mu_2^2 = 350;
\end{aligned} \tag{4.17}$$

whose solution yields $\mu_1 = 17.646$ and $\mu_2 = 6.214$.

Once these values are obtained, the modal bond graph can be simulated for any input. The true states of the system model in Figure 4.3b may be represented in the modal bond graph as shown in Figure 4.5. The gains, γ_1 and γ_2, in Figure 4.5 are vectors. Their values are calculated by comparing transfer functions obtained from Figure 4.5 with those obtained from 4.3b. This process is similar to the transformation defined in Equation 4.4.

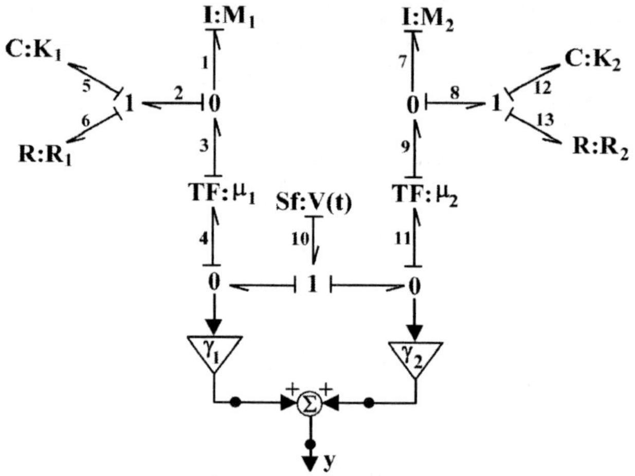

Fig. 4.5 Modal bond graph with true states as outputs

The model reduction on the modal bond graph may be based on modal activity index or element activity index; the element here means the modal element and it has no relation to the actual system. The modal activity index is based on the energy exchanged to different modes, *i.e.* the energy transfer through the transformer elements modeling modal participation. Modal energy index can be defined as

$$E_i = \int_{t_s}^{t_f} |e_i(t) f_i(t)|\, dt \tag{4.18}$$

where $e_i(t)$ and $f_i(t)$ are respectively the effort and flow in any one of the bonds connected to the i-th modal participation transformer. Then the modal activity index is defined as

$$A_i = \frac{E_i}{\sum\limits_{j=1}^{m} E_j}, \tag{4.19}$$

where m is the total number of modes in the model. This formulation allows elimination of entire modes from the model if certain modal activities are found to be insignificant with respect to others. On the other hand, if element-wise activity indices are considered, then it is possible both to remove entirely some modes or to remove some elements from some modes. Which is the better option depends on the application in hand.

For example, at extremely low excitation frequencies, activities of all modal elements except modal masses are so low that the corresponding elements may be removed from the model. Then the bond graph model can be reduced to a single equivalent mass excited by the velocity, where the equivalent mass is $M = \mu_1^2 M_1 + \mu_2^2 M_2 = \mu_1^2 + \mu_2^2 = 350$. Note that this is the same as the sum of two masses in the original model, *i.e.* $M = M_s + M_u = 350$. This means that the reduction of the original model and reduction of the modal model are equivalent.

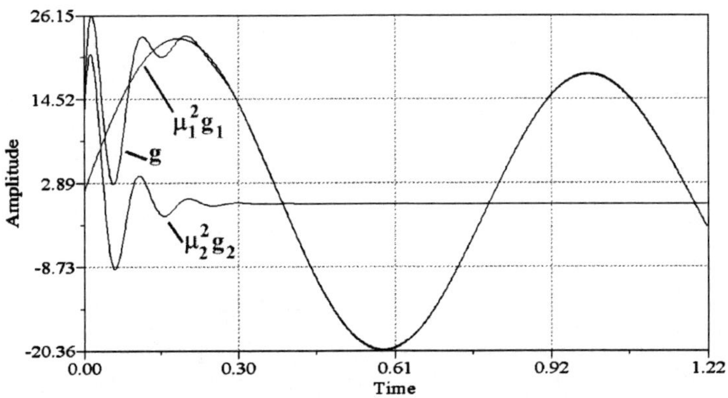

Fig. 4.6 Time responses of full system and the two modal oscillators to low frequency excitation

Activity analysis shows that, for moderately low frequency excitation in the considered example, its higher mode sub-model may be removed from the modal bond graph model whereas, for high frequency excitation, its lower mode sub-model may be removed from the modal bond graph model. For low frequency excitation ($V(t) = 0.01\cos(0.1t)$) or $V(s) = 0.01s/(s^2 + 0.01)$, the time response of the original model is compared with the corresponding responses obtained from the modal-decomposed models in Figure 4.6, by considering each mode separately. The overall system response is superposition of two modal responses. These results show that only the lower mode (*i.e.* $\mu_2 = 0$) is sufficient to approximate the system behavior at low frequency excitations.

Similarly, results shown in Figure 4.7 are for high frequency excitation $V(t) = 0.01\cos(1000t)$ and it is observed that consideration of only high frequency mode dynamics (*i.e.* $\mu_1 = 0$) in the model gives good enough approximation of the system behavior.

The frequency response plots of two modes and the full system are compared in Figure 4.8, where the magnitude axis is in decibels scale. It is evident that, except a small frequency range (the shaded are), consideration of one suitable mode is sufficient to describe the system behavior.

Modal decomposition based model reduction is applicable to only LTI systems. Their utility is found mainly while models of distributed parameter systems (such as beams, columns) are considered. In process engineering systems, structural mod-

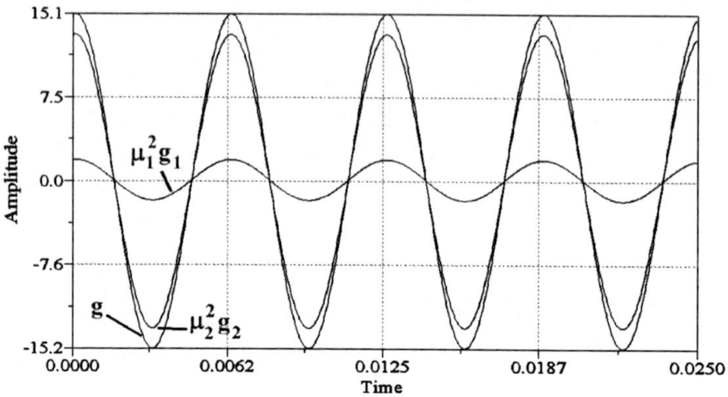

Fig. 4.7 Time responses of full system and the two modal oscillators to high frequency excitation

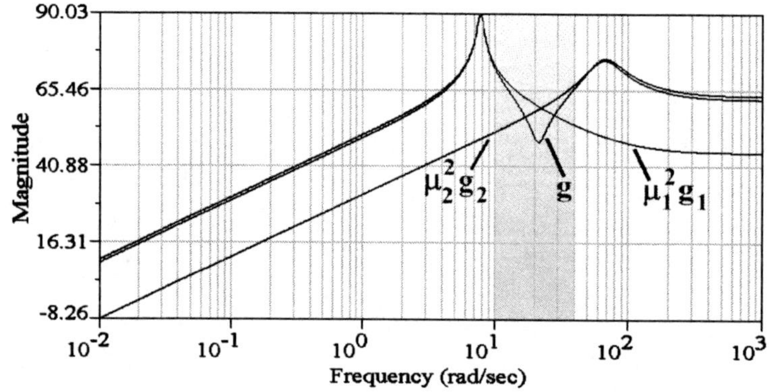

Fig. 4.8 Comparison of frequency response of the full model with those of individual modes

els of long pipelines, tall columns, such as risers in petrochemical plants, fall into this category. Distributed parameter system models are described in terms of partial differential equations which involve both time and space as independent variables. The usual solution to such systems is through separation of variables to formulate an initial value problem with spatial boundary conditions and their solution is usually done through finite difference or finite element scheme. If the problem is broken into finite segments, *i.e.* mass, stiffness and damping are lumped at various nodes, then too many finite elements are needed to obtain an accurate model. This is where a modal formulation produces a significantly low dimensional and yet accurate model. Further reduction of the model retains only essential dynamics and fault detection of such systems can be done in frequency domain based on Fourier or wavelet analysis of measured signals, such as accelerometer readings. More discussion on these aspects is outside the scope of this book.

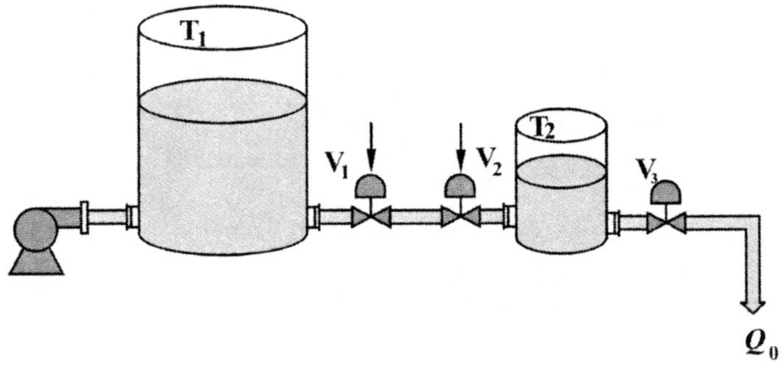

Fig. 4.9 A two-tank system with adjacent non-linear control valves

Fig. 4.10 Bond graph model of two-tank system showing algebraic loop

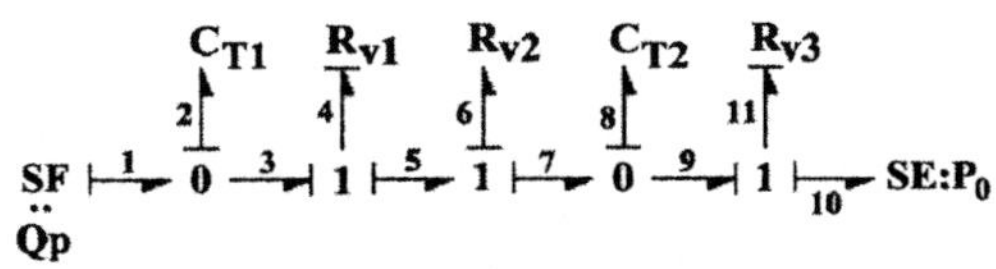

Fig. 4.11 Breaking algebraic loop with a coupling capacitor

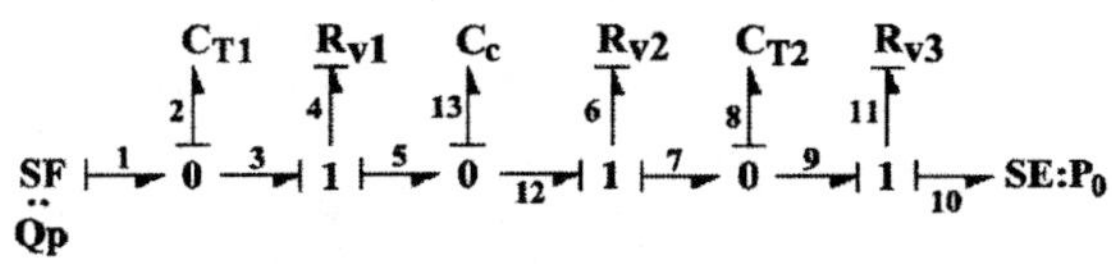

While building models for simulation, we often add pads or coupling capacitors to avoid differential causalities. Coupling capacitors are also used to avoid algebraic/differential algebraic loops involving non-linear R-elements.

For example, consider the system given in Figure 4.9 and its bond graph model given in Figure 4.10. Causal path analysis shows that there is an algebraic loop between elements R_4 and R_6. Considering that the flow through the valves is given by a non-linear relation, such that R_4 and R_6 cannot be merged together and represented as a single R-element, the model leads to implicit differential algebraic state equations. Implicit equations require special solution techniques. Therefore, usually a pad like structure called a coupling capacitor is added between the two valves, as shown in Figure 4.11. This coupling capacitor breaks the algebraic loop. It may be physically thought of as a tank with very small cross-sectional area such that a small flow into it produces enough pressure to drive almost an equal amount of flow out.

However, pads and coupling capacitors add extra states to the model. Therefore, for FDI analysis, the first model reduction step is to remove all such spurious elements used in the behavioral model to remove loops and differential causalities.

4.2 FDI Using Bond Graphs and Qualitative Reasoning

Models for dynamic systems may be classified as qualitative and quantitative. When model parameters are unknown or uncertain, but the model structure is well defined, qualitative equations from bond graph models instead of the differential equations may be used for FDI. Such qualitative analysis is done using available measurements, *i.e.* the history of past data[155, 240, 266]. Therefore, the qualitative behavior equations are always written in differential causality.

For passive elements, qualitative behavior equations are defined as

$$\begin{aligned} \text{R-element} &: e(k) = Rf(k), \\ \text{C-element} &: f(k) = C(e(k) - e(k-1)), \\ \text{I-element} &: e(k) = I(f(k) - f(k-1)), \end{aligned} \tag{4.20}$$

where k denotes the current time and $k-1$ denotes the time at the previous sample. The actual sampling time T, which is assumed to be a constant, does not appear in these equations (while taking derivatives) because we are not considering actual values in a qualitative analysis. The qualitative state of a parameter is defined as

$$\begin{aligned} [+] \text{ or } [1] &\quad : \text{higher than normal, increase;} \\ [0] &\quad : \text{normal, no change;} \\ [-] \text{ or } [-1] &\quad : \text{lower than normal,decrease.} \end{aligned} \tag{4.21}$$

The qualitative equations thus derived assume that the nominal values of parameters are positive and when they deviate from their nominal values, *i.e.* after a fault, they still remain positive. If there are some parameter or measurements in the model whose qualitative values are unknown then they are removed from the qualitative equations. When an abnormal deviation in the signature of a particular measured quantity is observed, its qualitative value is entered into the appropriate equation. Fault is recursively assumed in each parameter of the system (both $[+]$ and $[-]$ faults) and the qualitative value of the power variable associated with the parameter is explicitly calculated using qualitative operators. This procedure, which results in a set of fault candidates and their associated qualitative states, is called fault hypothesis generation. Then the fault propagation or qualitative time evolution corresponding to each fault candidate is constructed. The parameter for which all the equations match without any conflict is considered as the actual cause of the fault. The latter step is referred to as fault hypothesis validation.

One of the simplest fault hypothesis generation scheme [150, 271], which is based on causal structure analysis on a bond graph model, is presented here. Initially, a table of antecedents and consequences is constructed from the bond graph model of the system.

All power variables, parameters and two-port moduli involved in a relation are divided into two classes, namely antecedents and consequences [150, 271]. For any equation, say $x = f(y,z)$, where $f(\,)$ is a function, the variable(s) on the left side, *e.g.* x, is(are) consequence(s) whereas the variables on the right side, *e.g.* y and z, are antecedents. Construction of antecedents and consequences follows directly from the computational causalities imposed on a bond graph model. When causal structure of each element and junction in the graph is considered, parameters always appear in the antecedent (cause) class.

For example, let us consider the case of 1-junction. The strong bond (one which is causalled away from the 1-junction) decides the flow variable of the junction and is the antecedent. Flows in all other bonds connected to that junction are consequences. The 1-junction is also an effort sum junction. The effort in the strong bond is the signed sum (taking power directions into account) of efforts in other bonds. Thus, the effort in the strong bond is consequence and the efforts in all other bonds are antecedents. Similar logic applies to the 0-junction and two-port elements. The antecedent and consequence classes for all junction types and elements are given in Table 4.2.

For each model, the table of antecedents and consequences are prepared first. In this table, each power variable appears only once in the antecedent and once in the consequence classes; whereas all parameters appear as antecedents.

4.2.1 Determination of Initial Fault Set

Let us start with a single fault hypothesis. The initial fault set is determined through traversal of causal path. The faulty variable is first located in the consequences list and assigned a qualitative value ($[+]/1$, 0 or $[-]/-1$). The back propagation is done by assigning qualitative values to corresponding antecedents. For instance, if a variable e_1 is below normal (-1) and its antecedents are e_2, $-e_3$; then e_2 must be assigned qualitative value -1 (below normal) and e_3 must be assigned qualitative value $-(-1) = 1$ (or above normal). Then these newly assigned antecedents are treated as consequences and antecedents for them are searched. A tree like structure [120] thus evolves, which terminates at outer vertices, where either the parameters are detected as antecedents or the qualitative value of a detected antecedent is in conflict with an earlier value (which gets higher precedence or rank). The following example illustrates this approach.

Let us consider a two-tank hydraulic system (Figure 4.12a) and its pseudo-bond graph model (Figure 4.12b). In the bond graph model, pressure is the effort variable and mass flow rate is taken as the flow variable. A pump supplies flow Q_p to a tank T_1. The outflow from tank T_1 to tank T_2 is regulated by a valve V_b. Finally, the fluid is discharged to atmosphere through a valve V_o. There are two pressure sensors, P_1

Table 4.2 Derivation of antecedents and consequences from bond graph causality

Component	Antecedents	Consequences
⇀| I:m	$1/m$ or I, e	f
|⇀ I:m	m or $1/I$, f	e
|⇀ C:K	K or $1/C$, f	e
⇀| C:K	$1/K$ or C, e	f
|⇀ R:R	R, f	e
⇀| R:R	$1/R$, e	f
|⇀(1) TF(μ) |⇀(2)	e_2, f_1, μ	e_1, f_2
⇀|(1) TF(μ) ⇀|(2)	$e_1, f_2, 1/\mu$	e_2, f_1
⇀|(1) GY(μ) |⇀(2)	$e_1, e_2, 1/\mu$	f_1, f_2
|⇀(1) GY(μ) ⇀|(2)	f_1, f_2, μ	e_1, e_2
1 junction (bonds 1, 2, 3)	f_2 $e_1, -e3$	f_1, f_3 e_2
0 junction (bonds 1, 2, 3)	e_2 $f_1, -f_3$	$e_1, e3$ f_2
SE ⇀|	Se	e
SF |⇀	Sf	f

and P_2 in the two tanks. We assume that the pipes are connected at the bottom of these tanks. The pressure produced at the bottom of the tanks is due to the storage of the fluid. Therefore, the capacity of the tanks is given by $C_{Ti}=A_i/g$, where $i = 1,2$ and A refers to the cross-sectional area.

The antecedents and consequences for this model are given in Table 4.3.

Let us consider a case where the observed level in T_1 is higher than nominal while the level in T_2 is nominal, *i.e.* $P_1{}^+$ and $P_2{}^0$. We start by propagating the fault, *i.e.* $P_1{}^+$, and construct the fault tree shown in Figure 4.13. To construct the fault tree, we start by finding variable P_1 in the consequences and find its antecedent to be e_4. Then using e_4 as consequence, we find e_2 as its antecedent, and so on.

In the fault tree, qualitative values are denoted using '+' for higher and '$-$' for lower. The conflicting nodes are represented by a cross ($\times$) near them. Thus the initial fault set detected is C_{T1}^-, $Q_p{}^+$, R_{Vb}^+, C_{T2}^-, $R_{V0}^+$and P_0^+. Since we are working

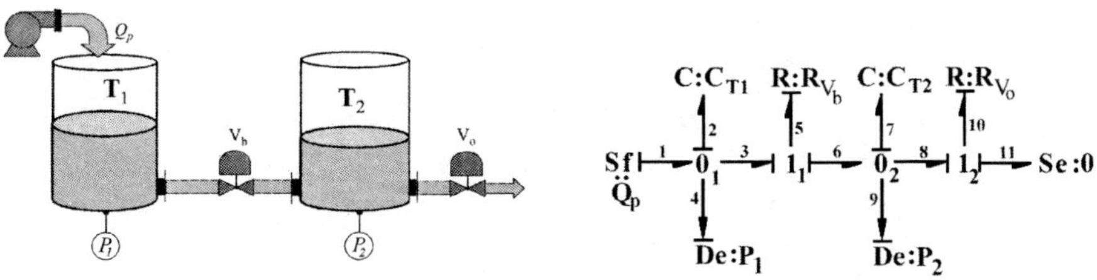

(a) (b)

Fig. 4.12 A two-tank systems and its bond graph model

Table 4.3 Causal interpretations

Antecedents		Consequences
Parameter	Variables	
Q_p		f_1
$1/C_{T1}$	f_2	e_2
	e_2	e_1, e_3, e_4
	e_4	P_1
	$f_1, -f_3$	f_2
$1/R_{Vb}$	e_5	f_5
	$e_3, -e_6$	e_5
	f_5	f_3, f_6
$1/C_{T2}$	f_7	e_7
	e_7	e_6, e_8, e_9
	e_9	P_2
	$f_6, -f_8$	f_7
$1/R_{V0}$	e_{10}	f_{10}
	f_{10}	f_8, f_{11}
	$e_8, -e_{11}$	e_{10}
	P_0	e_{11}

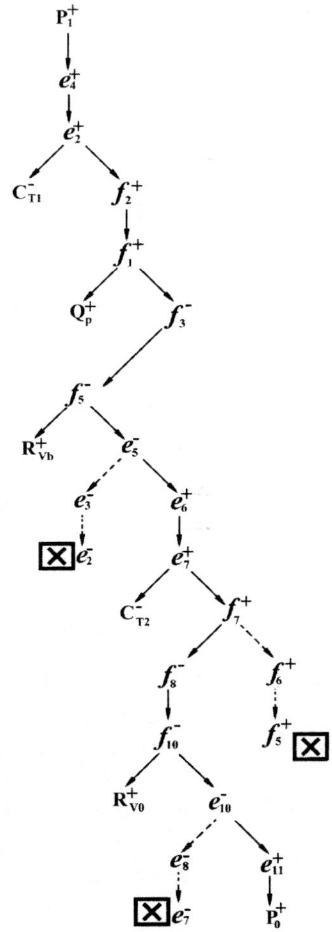

Figure 4.13 Fault tree

with a single fault hypothesis, among these faults any one parameter could be the cause of the fault.

4.2.2 Fault Disambiguation

Now constraints may be added to the tree or suggestions for qualitative measurement of the most appropriate variable/signal can be made. We have the qualitative state for P_2, which is normal. By looking at antecedents and consequences table, one finds that e_7 should be normal. Thus a conflict appears at node for e_7^+. Then the tree can be reconstructed after conflict resolution [120] and finally represented as shown in Figure 4.14.

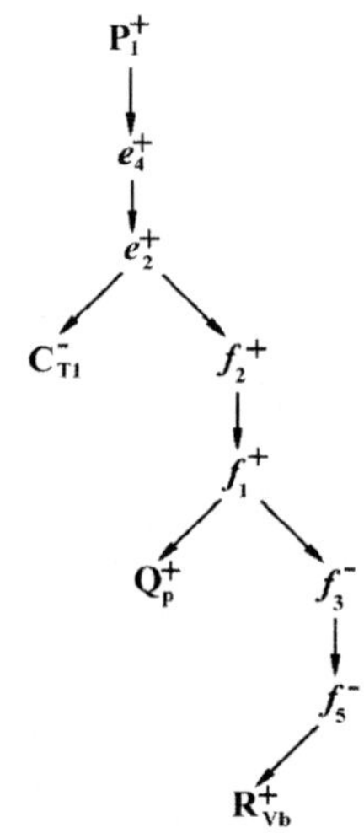

Fig. 4.14 Reduced fault tree obtained after conflict resolution

From the final fault tree, three fault candidates are obtained: C_{T1}^-, $Q_p{}^+$ and R_{Vb}^+. In a single fault hypothesis, only one of them can be faulty at a time. Further isolation of fault is possible only when more qualitative information about some other signals or temporal evolution of the measured signals (*e.g.* does the level continue to rise?) is available. Assuming that temporal logic is not to be applied, qualitative reasoning (QR) using artificial intelligence (AI) is the best solution. A supervisory system using QR uses the information about the likelihood (probability) of failure of each component, the past history of failure of components, robustness of control system, *etc.* For example, the fault C_{T1}^- corresponds to a decrease in the capacity of T_1, which is possible (by diagnosis) due to something falling into the tank. Considering that there are covers (with a small hole for air passage) on each tank, this possibility is ruled out. Similarly, if the pump is controlled to deliver a fixed rate of flow and the control system is robust, or the pumps flow output is measured, then the pump's failure too can be ruled out. This leaves us with one isolated fault candidate, R_{Vb}^+, which corresponds to blockage of the valve between two tanks. Note that this process of elimination of fault candidates using QR and rules in an AI based supervisory system does not always lead to an isolated fault candidate. Most often, there are more than one fault candidates, which may belong to a particular module or component of the system or a few components of the system. At least this helps in carrying out quick repair/maintenance work by concentrating only on the susceptible subsystems.

Fault trees grow exponentially with increase in the system's complexity. For larger systems, it is difficult to construct a fault-tree manually. Normally such trees are represented as linked lists in a computer program. Automated fault diagnosis programs prune some branches of the tree by accepting qualitative inputs for other suggested (most critical) signals and resolving the conflicts. The tree ultimately converges to a critical parameter list.

So far we have considered all variables to be positive. This works out well for most process engineering systems. The power variables too are generally positive if proper power direction is chosen in the bond graph model. If the power directions are assigned in such a way that all effort and flow variables in the model remain positive at all times, then diagnosis discussed in this section is applicable. However, in some systems, there is no fixed effort (pressure) gradient and flow direction may change. Therefore, in addition to the qualitative value of the signal $([+],[0],[-])$, another qualifier for its sign $([+],[-])$ is needed. These two different and independent attributes allow representing qualitative state of a variable in six different ways. More details on how to use such qualitative classification for FDI and also its extension to handle multiple faults are discussed in [151, 220].

Note that the qualitative value of a signal $([+],[0],[-])$ is with reference to its nominal steady state value. In systems having oscillating steady state response, there is no way to define whether the output is higher or lower than the nominal. If one specifies the amplitude or frequency as being higher/lower than nominal, the preceding analysis is not applicable. Therefore, this qualitative FDI method may be well-suited to many process engineering (hydraulic, thermo-fluid and chemical) systems, but is generally unsuitable for the majority of mechanical, mechatronic and electrical systems.

4.3 Qualitative Analysis Using Tree Graphs

Tree graphs [135, 136] are a qualitative representation of system equations derived from a model. It follows both qualitative and quantitative failure effect analysis for systems operating in steady state. Qualitative analysis of the tree graph predicts the final effects (new steady state) under assumed failure conditions. On the other hand, quantitative analysis evaluates the transient effect leading to the final effect by utilizing detailed characteristics and information of system components.

At the steady state condition, it is assumed that the rate of states is zero. Thus, their fluctuations in qualitative terms remain 0 at steady state. This assumption means that the following analysis would be applicable to process engineering systems in static steady state. This includes systems whose reduced order bond graph model contain no coupling between I and C-elements (also two I-elements or two

C-elements coupled through a gyrator); if such couplings are there then they should affect only transient dynamics.

A tree graph is a tree-structured network connection between the input nodes and the output nodes through the constraints represented by a structure of the internal nodes. A tree graph is a forward model which could be traversed backwards for FDI. In the forward tree graph model, the states of the system and the external excitations are the input nodes and the outputs are the time derivatives of the states. There are two types of internal nodes, namely algebra and function nodes, as shown in Figure 4.15. These two types of internal nodes are used to represent the qualitative model structure, *i.e.* the qualitative state equations, in a tree graph.

Fig. 4.15 Algebra and function nodes of a tree graph

Initially, all assumed faults are included in the system model. The assumed faults are obtained from the fault hypothesis generation step, as discussed in the previous section. The state equations for the global bond graph model is derived and represented as the tree graph. Reverse traversal of a tree graph generates symptoms corresponding to each fault. These symptoms are recorded and later used for diagnosis.

Let us consider the model of the two-tank system shown in Figure 4.16a and its pseudo-bond graph model shown in Figure 4.16b. The possible faults in this model are malfunctioning of valves, storage devices (leakages) and the input. Valve blockage/leakage is attributed to variation in its resistance. Faults to simulate leakage from tanks are added explicitly in the model by introducing two valves to discharge fluid to environment. In the fault-free operation, these valves are closed.

The state equations obtained from the bond graph model in Figure 4.16 are

$$\begin{aligned}\frac{dQ_2}{dt} &= Sf_1 - \frac{K_2Q_2 - K_7Q_7}{R_5} - \frac{K_2Q_2 - SE_{13}}{R_{12}},\\ \frac{dQ_7}{dt} &= \frac{K_2Q_2 - K_7Q_7}{R_5} - \frac{K_7Q_7 - SE_{11}}{R_{10}} - \frac{K_7Q_7 - SE_{15}}{R_{14}}.\end{aligned} \tag{4.22}$$

Although the R-elements have been assumed to be linear in Equation 4.22, we will show later that the following analysis is valid for non-linear constitutive relations. Qualitative equations are written in terms of functions. Function expression like $C_1(Q_1)$ returns K_1Q_1 for linear springs. As an example, relation $f_5 = 1/R_5\,(e_5)$,which is applicable to a linear resistor, is written in functional terms

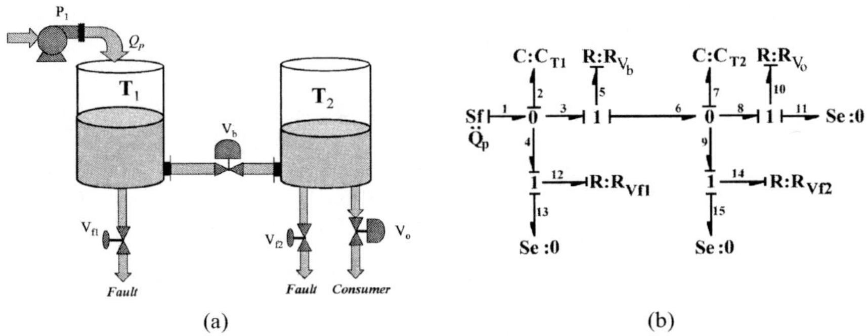

Fig. 4.16 System considered for tree graph analysis and its bond graph model

as $f_5 = \mathrm{R}_5^{-1}(e_5)$ where R_5^{-1} corresponds to the inverse function for a resistance, *i.e.* conductance. These functions are often termed as characteristic functions, which define behavior of elements. The qualitative equation for the two-tank system in function form is then given as

$$\begin{aligned}\frac{d\mathrm{X}_1}{dt} &= \mathrm{U}_1 - \mathrm{R}_{\mathrm{Vb}}^{-1}(\mathrm{C}_2(\mathrm{X}_1) - \mathrm{C}_7(\mathrm{X}_2)) - \mathrm{R}_{\mathrm{Vf1}}^{-1}(\mathrm{C}_2(\mathrm{X}_1) - \mathrm{U}_3),\\ \frac{d\mathrm{X}_2}{dt} &= \mathrm{R}_{\mathrm{Vb}}^{-1}(\mathrm{C}_2(\mathrm{X}_1) - \mathrm{C}_7(\mathrm{X}_2)) - \mathrm{R}_{\mathrm{Vo}}^{-1}(\mathrm{C}_7(\mathrm{X}_2) - \mathrm{U}_2)\\ &\quad - \mathrm{R}_{\mathrm{Vf2}}^{-1}(\mathrm{C}_7(\mathrm{X}_2) - \mathrm{U}_4),\end{aligned} \tag{4.23}$$

where X_1 and X_2 are mapped to states Q_2 and Q_7, U_1 to U_4 are mapped to inputs SF_1,SE_{11},SE_{13} and SE_{15} and the functions correspond to various constitutive relations. The tree graph representing Equation 4.23 is given in Figure 4.17. The starting nodes are the states of the system and the inputs. The terminating nodes are the derivatives of the states, *i.e.* $\frac{d\mathrm{X}_1}{dt}$ and $\frac{d\mathrm{X}_2}{dt}$.

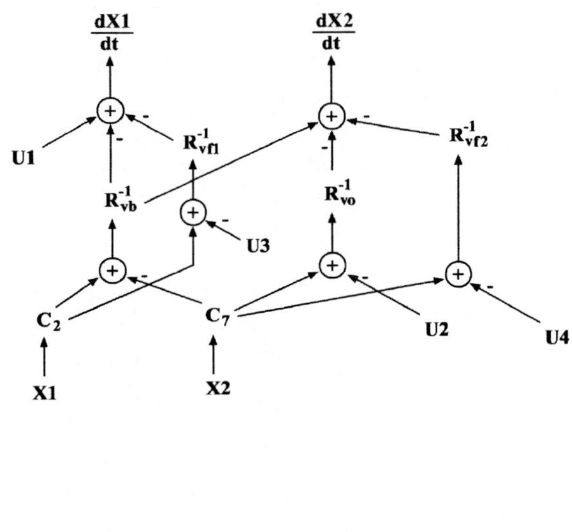

Fig. 4.17 Tree graph of two-tank system

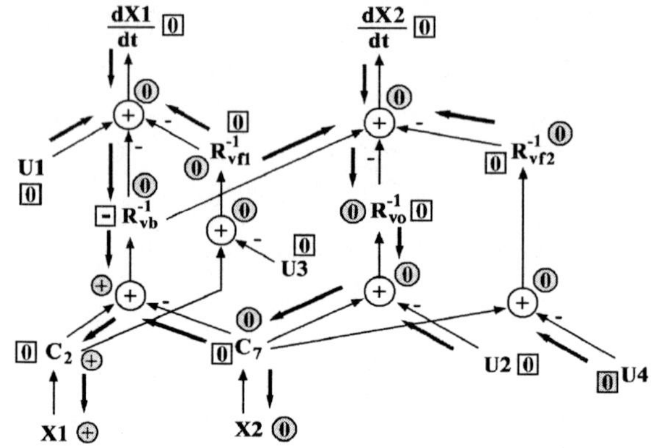

Fig. 4.18 Fault propagation on the tree graph of two-tank system

Let us now assume a fault in resistance R_{vb}, which may be due to a blockage, *i.e.* R_{vb} takes a very high value. This is represented as R_{vb}^{+} qualitatively or denoted by a '+' sign within a box near the function node R_{vb}.

However, we are dealing with inverse of the function and thus the qualitative state is denoted by a '−' sign within a box near the function node R_{vb}^{-1}. The steady state conditions imply both dX_1/dt and dX_2/dt are zero (constant). Also at steady state, the inputs are not changing. So all these nodes and those representing static parameters are qualitatively assigned 0. The fault is then propagated according to the qualitative equations. The qualitative deviation of a signal (output of each internal node) is shown within a circle near the node. The fault tree graph is shown in Figure 4.17.

The fault propagation shown in Figure 4.18 starts from fixed nodes. Since qualitative change in $dX1/dt$ is zero, the nodes connected to it also experience no qualitative change. That implies the output from function node R_{vb}^{-1} is unchanged or qualitatively 0. However, the magnitude of R_{vb}^{-1} has qualitatively decreased. So the magnitude of input signal has increased (+). If we start from qualitative change in $dX2/dt = 0$, then we observe that qualitative change of the output from function node C_7 is 0. At the node depicting the difference between output from function node C_2 and output from C_7, it is evident that qualitative value of output from function node C_2 has increased (+), because the output from function node C_7 has not changed. This implies an increase in value of state variable X_1 (state variable Q_2 in the bond graph model).

Furthermore, the tree graph shows that in a steady-state process operation, the output from node R_{vo}^{-1} (flow output to consumer) and level in tank T_2 would remain unchanged whereas the level in the first tank, T_1, would increase. To confirm these results, a numerical simulation was performed. The flow through the valve (f) was modeled as a non-linear function of the pressure difference (e) as $f = R_v^{-1}(e) = C_d\sqrt{e}.\text{sign}(e)$,where C_d is the overall discharge coefficient. The representative parameter values were taken as $Q_p = 1$ kg/s, $K_2 = g/A_1 = 19.62$ N.m^{-1}.s^{-2}, $K_7 = g/A_2 = 32.7$ N.m^{-1}.s^{-2}, $C_{dvb} = 0.2$ kg$^{-1/2}$m$^{-1/2}$, $C_{dvo} = 0.1$ kg$^{-1/2}$m$^{-1/2}$ and

$C_{dvf1} = C_{dvf2} = 0\ \mathrm{kg}^{-1/2}\mathrm{m}^{-1/2}$ (no leakage), where C_{dvb} and C_{dvo} are discharge coefficients of valves V_b and V_o, respectively. During the simulation, a fault was introduced by changing the value of C_{dvb} to $0.1\ \mathrm{kg}^{-1/2}\mathrm{m}^{-1/2}$ after the system has reached the steady state. The results in Figure 4.19 show an increase in the value of state X_1 and a temporary decrease in the value of state X_2, which slowly returns to its earlier magnitude in the new steady state. The quantitative simulation results thus validate the results obtained for steady state conditions through qualitative fault tree analysis even though non-linear constitutive relations have been used.

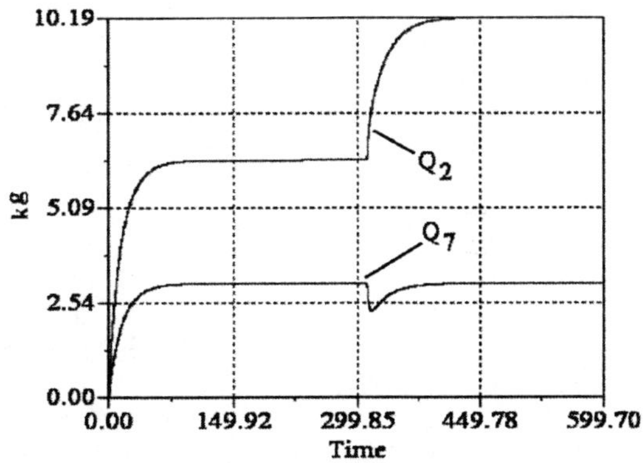

Fig. 4.19 Two-tank system response to step change in a valve's discharge coefficient

The transient trends in the response obtained through quantitative analysis are manifestation of different time-constants. They can be compared to DNA fingerprinting of individuals. These trends may be used to take corrective measures *via* intelligent supervisory control schemes. Trends of other faults, such as leakages from the tanks, can likewise be obtained and their features recorded in a knowledgebase. The online FDI problem then reduces to how intelligently the implemented software can interpret the symptoms and match them with fault signatures stored in a knowledgebase to diagnose the fault.

4.4 Qualitative FDI Using Temporal Causal Graphs

A causal mathematical model is one which is already written in proper sequence so that all the unknown variables in it are computable. An acausal mathematical model is a set of equations which need to be ordered and organized (*i.e.* a causal assignment) before computation; different ways of ordering correspond to different ways of causality assignment. Graphical representations of causal mathematical models are often called causal graphs (*e.g.* linear graph, signal flow graph, block diagram). On the other hand, graphical representation of acausal models is a non-oriented graph, *e.g.* a digraph. Bond graph models can represent both acausal and causal properties. Linear causal graphs represent the model structure by linking various nodes (variables) in the model. The directed edges in the graph are used to represent constraints corresponding to various passive elements in a bond graph model (I,C,R and also TF and GY), whereas additive constraints are represented at the nodes.

Temporal causal graph (TCG) [72, 73, 171] is a special class of causal graph, in which temporal evolution of variables are represented qualitatively by accounting for the storage elements encountered in the causal path. Temporal causal graphs are almost identical to signal flow graphs and thus they are easily constructed from bond graph models using the same procedure for creating signal flow graphs. The frequency domain Laplace variable s, which stands for derivative with respect to independent variable t (time) in time-domain, is represented as $1/dt$ in temporal causal graphs. The integration with respect to time corresponding to integrally causalled storage elements introduces delay in the system's response. This delay is depicted as dt in a temporal causal graph. Similarly, $1/dt$ term is used to represent differentiation in edges corresponding to storage elements in derivative causality.

A temporal causal graph can be traversed in both forward and backward directions from an observed or hypothesized fault (in a signal or a parameter). The backward propagation is used to construct the list of fault candidates, whereas the forward propagation derives predictions for posteriori behavior (time evolution) corresponding to each of the hypothesized faults [118].

Let us consider a system shown in Figure 4.20 and its bond graph model shown in Figure 4.21.The temporal causal graph for the system is shown in Figure 4.22 (note the similarity with SFG of the same system in Chapter 3).

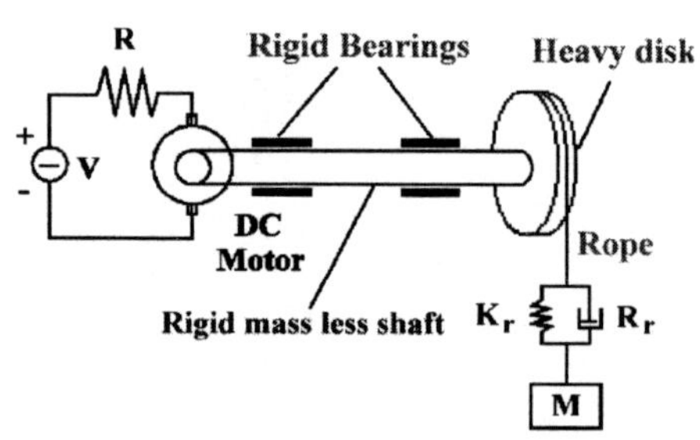

Fig. 4.20 An electro-mechanical hoisting mechanism

4.4.1 Fault Hypothesis Generation

Backward propagation, *i.e.* tracing the path opposite to the signal flow direction, is used to generate the fault hypothesis, *i.e.* a list of fault candidates. Backward

Fig. 4.21 Bond graph model of electro-mechanical hoisting mechanism

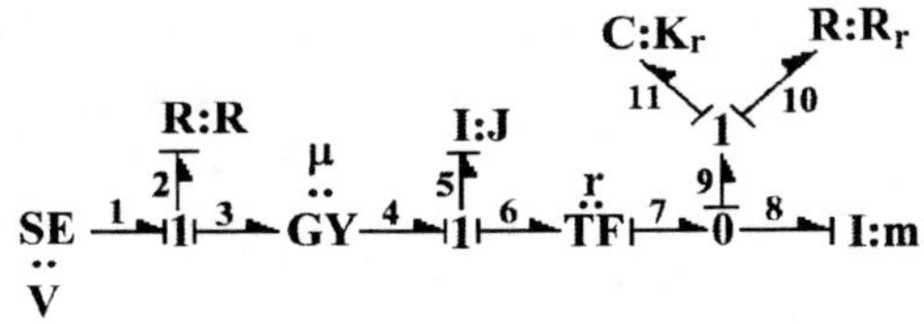

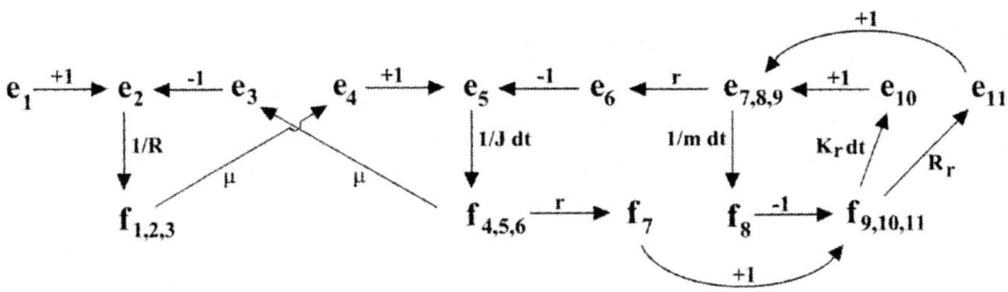

Fig. 4.22 Temporal causal graph of electro-mechanical hoisting mechanism

propagation starts from the measurement node whose qualitative state is available and terminates at parameters or conflicting nodes.

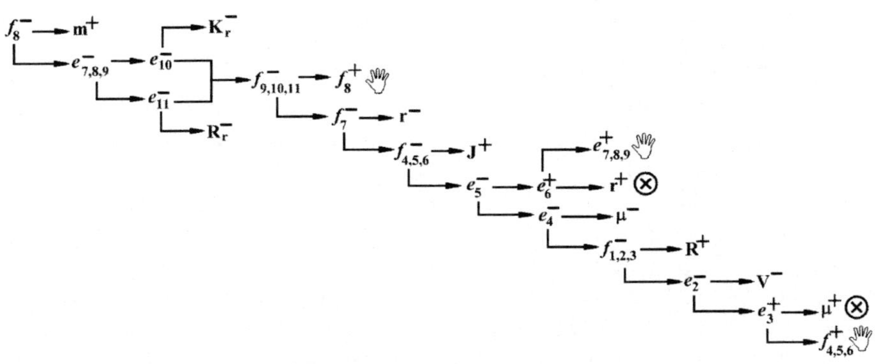

Fig. 4.23 Back propagation in a temporal causal graph and conflict resolution

Let us consider a fault scenario whereby the measured lifting velocity (at the mass point) is found to be below its nominal value, *i.e.*, in qualitative terms, f_8^-. When we traverse back from f_8^-, there is a single edge with gain $1/m.dt$ to the node $e_{7,8,9}$. Since f_8 is qualitatively lower, it results that load mass is higher (m^+) and/or $e_{7,8,9}$ is lower ($e_{7,8,9}^-$). When a parameter is encountered, the propagation is terminated for that branch. Then the back propagation continues for remaining nodes, *e.g.* from $e_{7,8,9}^-$. The tree representing the back propagation is given in Figure 4.23. Note that during back propagation, instantaneous edges are given precedence. This means that if there are two backward edges from a node with two different orders of delays and then those two branches meet somewhere else, *i.e.* they are parallel, then the branch with least order of delays is propagated first. This is why, in Figure 4.23, we continue propagation from node e_{11}^{-} but terminate at e_{10}^{-}.

The generated fault hypothesis contains the list of fault candidates and their likely qualitative state. Under single fault hypothesis, it is assumed that only one parameter from the set $(m^{+}, K_r^{-}, R_r^{-}, r^{-}, J^{+}, \mu^{-}, R^{+}, V^{-})$ has deviated from its nominal value. Then the objective is to find which one?

The very first step in isolating the fault candidate is reduction of dimension of the set of initial fault candidates. This requires qualitative input from other sensors and follows the principle similar to the fault tree analysis, discussed in Section 4.2. In the next step, there are some parameters that can be considered robust or at least robust to drift in a specific direction. For example, one may consider that while $r^{+}$is possible, *i.e.* increase in pulley radius due to continuous winding, $r^{-}$is not expected. Moreover, both the states $r^{+}$and r^{-}cannot be abrupt faults, but progressive faults taking place over a long time of operation. Considering that the observed deviation in measured signal is abrupt, the initial fault set can be heuristically reduced to $(K_r^{-}, R_r^{-}, \mu^{-}, R^{+}, V^{-})$, which correspond to the set of parameters belonging to the most likely components to fail, *i.e.* faults in the power supply, motor, and the rope.

4.4.2 Fault Hypothesis Validation

Among the fault candidates, only one is faulty. We generate the qualitative trend (QT), *i.e.* magnitude, slope, and so on, of the output after inception of the fault, considering each fault candidate one by one. The forward propagation of the temporal causal graph yields predictions for future behavior for the set of measurements for a postulated fault. This prediction takes account of temporal delays encountered during forward propagation. Let us consider the fault hypothesis that the supply voltage has fallen (V^{-}) for some reason. Starting forward propagation (along the direction of signal only) from the edge containing parameter V leads to $f_{1,2,3}^{-} \rightarrow e_4^{-} \rightarrow e_5^{-}$. At this node, temporal edge (*dt* in forward path) is encountered and therefore the effect on subsequent nodes is time delayed (due to integration). This implies derivative of f_4 is affected instead of its magnitude. This is represented as $f_4 \downarrow$, where the number of arrows represents the number of time delays encountered during forward propagation to that node and the arrow direction represents the qualitative magnitude, *i.e.* ‘ $\downarrow$’ for ‘$-$’ and ‘ $\uparrow$’ for ‘$+$’. Note that arrows are always co-oriented.

Normally, forward propagation is terminated when predictions are available for a sufficiently higher order. The predicted temporal qualitative trend, also called temporal signatures, corresponding to hypothesized fault V^{-} is given in Figure 4.24.

Temporal signatures for each hypothesized fault are listed as a table and are compared with the observed measurements to reduce the set of hypothesized faults to the actual ones. From Figure 4.24, we find that the second derivative of measured signal f_8 would decrease, and thus it will settle somewhere. We have the information of

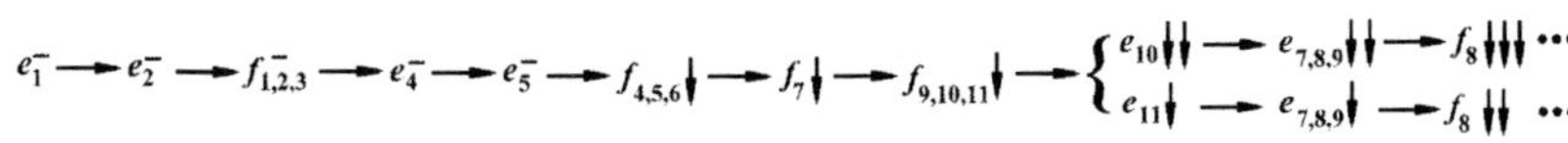

Fig. 4.24 Forward propagation of fault hypothesis

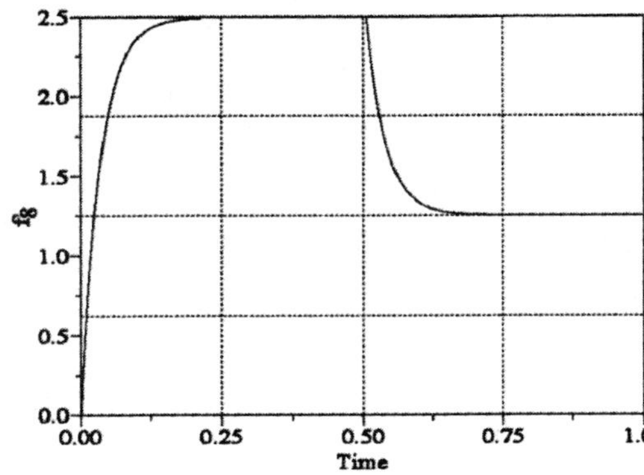

Fig. 4.25 Trend generation for TCG based diagnosis

zeroth order and second order evolutions, but the information of first order evolution is missing. This means there is not sufficient qualitative information to proceed for FDI and there should be some more sensors placed in the system.

If the pulley is considered sufficiently light, such that delay in accelerating it is insignificant (no fly-wheeling effect), then we can neglect delay in that edge and obtain an initial negative first order and second order deviation. This result is shown in Figure 4.25 for the following representative data: $V = 50$ V initially (at $t = 0$) and changed to 25 V (at $t = 0.5$ s) to simulate abrupt fault, $R = 10\ \Omega$, $\mu = 10$ N.m/A, $J = 0.1$ kg.m^2, $r = 0.5$ m, $m = 1$ kg, $K_r = 10^6$ N/m^2 and $R_r = 200$ N.s/m.

External disturbances and apprehended faults can be added to the model to study their qualitative effect on the system. For the system under consideration, failures like field collapse, ambient resistance on the rotating pulley and load, increase in pulley radius due to continuous winding, *etc.* can be added to the model and a new temporal causal graph (TCG) can be created and studied to generate global fault signatures. In fact, there is a specialized software, called *Transcend*, to perform this qualitative analysis with temporal causal graphs.

As another example, let us consider the two-tank system (Figure 4.16a) discussed earlier in Section 4.3. Note that, in this system, we have inputs for qualitative states of two signals. From the bond graph model of the system given in Figure 4.16b, one may construct the TCG given in Figure 4.26.

Let us now assume the same fault scenario considered in Section 4.3, *i.e.* the measured level in tank T_1 is found to be higher than nominal whereas the level in tank T_2 is found to be unchanged. This corresponds to e_2^+ and $e_{6,7,8}^0$ in the TCG. First of all, we start generating the fault hypothesis through back-propagation from the node e_2^+, as shown in Figure 4.27. During this process, we encounter the node

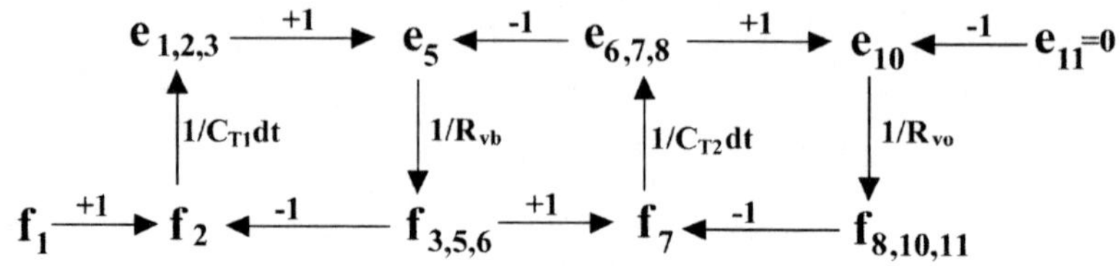

Fig. 4.26 Temporal causal graph of the two-tank system

$e^{-}_{6,7,8}$, which contradicts one of the known inputs, *i.e.* $e^{0}_{6,7,8}$. Therefore, the rest of the expansion in Figure 4.27 (marked within a box) will be considered as conflicts.

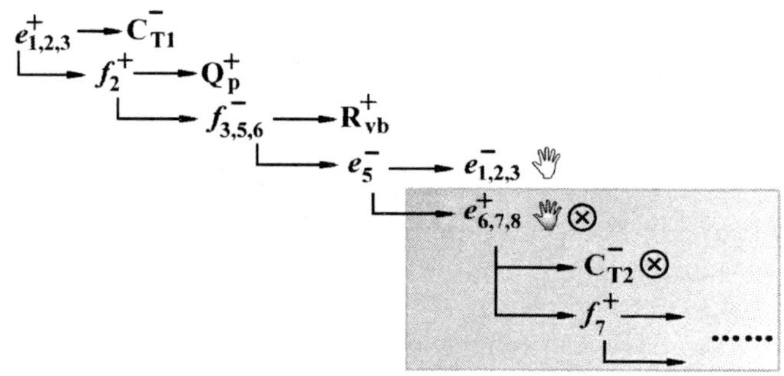

Fig. 4.27 Fault hypothesis generation for two-tank system

Thus we arrive at the list of fault candidates as C^{-}_{T1}, Q^{+}_{p} and R^{+}_{vb}. To isolate the fault further, one has to take transient trends into account. Note that forward propagation (fault hypothesis validation) corresponding to the three fault hypotheses gives identical results and hence there is no means of qualitatively differentiating between these faults. However, one may refer to the transient trend corresponding to fault hypothesis R^{+}_{vb} as shown in Figure 4.19. An exactly similar trend is obtained for fault hypothesis C^{-}_{T1}. However, if the fault hypothesis is Q^{+}_{p}, we get different transient trends as shown in Figure 4.28 where levels in both the tanks increase. Note that forward propagation from fault hypothesis Q^{+}_{p} has given wrong transient predictions. Obviously, Q^{+}_{p} should not be a fault candidate and we are left with two remaining fault candidates, namely C^{-}_{T1} and R^{+}_{vb}. Further fault isolation is impossible in this case with the given instrumentation architecture. In the next chapter, we will show that, with two level sensors, the two faults C^{-}_{T1} and R^{+}_{vb} would be isolatable from each other by following quantitative FDI methods.

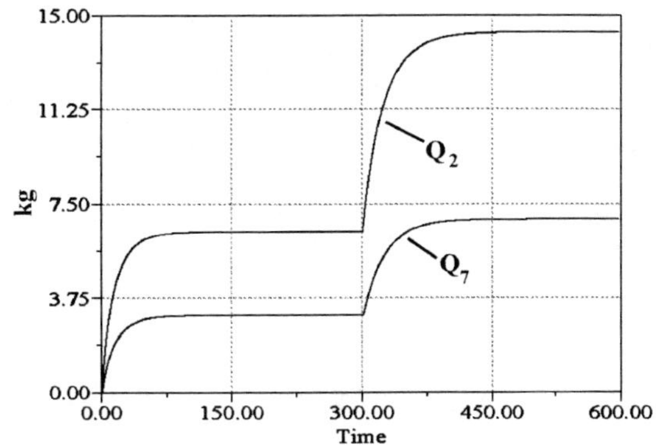

Fig. 4.28 Response of two-tank system to step change in input flow rate

4.5 Hybrid Diagnosis with Temporal Causal Graphs

The disadvantages of the aforementioned qualitative FDI methods may be summarized as follows:

1. They handle only abrupt faults.
2. Fault-tree and qualitative causal analyses are applicable to non-linear continuous-time systems, but the effect of response time is not considered in the qualitative model. It is assumed that cause and effect relations are instantaneous. TCG based diagnosis considers the temporal delays, but it is only applicable to linear time invariant (LTI) systems.
3. The application of these methods is constrained to only open-loop systems. This is because these approaches rely on temporal trends of systems evolution obtained from the measurements, which may not show any deviation if they are controlled. It is well known that controllers try to hide the fault effects.
4. All these methods are based on static steady-state assumptions, which may be ill-suited to processes with dynamic steady-state and processes having large response time.
5. For some systems, a TCG does not provide information about all consecutive orders of trend.
6. TCG may at times predict wrong qualitative trends.
7. Noise and robustness issues are not considered in such diagnosis — many orders of derivatives would exacerbate the noise effects.
8. Most often, one is given a large list of fault candidates, which could be very confusing to the field engineers.

Inclusion of some quantitative information may be considered as a solution to problems faced in purely qualitative analysis. Such a diagnosis is often referred to as hybrid diagnosis, where full/partial quantitative analysis improves the prediction quality.

One of the starting points in quantitative analysis is the estimation of parameters. The approach followed in [159] uses the standard recursive least-square minimization method to identify an LTI system model with a specific fault. Models are constructed for each hypothesized fault and then they are simulated and their responses are compared with those obtained from the actual process. The parameter corresponding to the fault model giving the best match with the process response is isolated as the actual fault. However, there are some unanswered questions: how to initialize the model states and how to account for the delay between the actual observation and model response? These questions are answered in Chapter 9.

4.6 Remarks on Model Linearization

Most of the qualitative FDI schemes are applicable to LTI systems. For this purpose, a non-linear system model has to be linearized before qualitative FDI is applied to it with the assumption that the LTI model would remain valid even after the fault. Usually, industrial processes are controlled about some chosen set-points and their models can be linearized about the corresponding operating points.

A non-linear dynamical system's mathematical model is given by

$$\begin{aligned} \dot{x}(t) &= f(x(t), u(t)), \\ y(t) &= g(x(t), u(t)), \end{aligned} \tag{4.24}$$

where $x \in \mathbb{R}^n$, $u \in \mathbb{R}^m$, $y \in \mathbb{R}^p$ and the partial derivatives $\frac{\partial f(x,u)}{\partial x}$, $\frac{\partial f(x,u)}{\partial u}$, $\frac{\partial g(x,u)}{\partial x}$, and $\frac{\partial g(x,u)}{\partial u}$ exist and are continuous.

This system may have multiple equilibrium points. Let us consider an equilibrium point defined by (x_e, u_e) such that $f(x_e, u_e) = 0$. The nominal outputs are then given by $y_e = g(x_e, u_e)$. Taylor series expansion of the non-linear functions around this equilibrium point up to the first order gives

$$\begin{aligned} x = x_e + x_\Delta, \;\; u &= u_e + u_\Delta, \; y = y_e + y_\Delta, \\ f(x,u) &= f(x_e, u_e) + \left.\frac{\partial f(x,u)}{\partial x}\right|_{x_e,u_e} x_\Delta + \left.\frac{\partial f(x,u)}{\partial u}\right|_{x_e,u_e} u_\Delta, \\ g(x,u) &= g(x_e, u_e) + \left.\frac{\partial g(x,u)}{\partial x}\right|_{x_e,u_e} x_\Delta + \left.\frac{\partial g(x,u)}{\partial u}\right|_{x_e,u_e} u_\Delta, \end{aligned} \tag{4.25}$$

where x_Δ, u_Δ and y_Δ are small deviations of corresponding variables from the equilibrium point. Substitution of the linearized functions in Equation 4.24 results in the linearized state-space form

$$\begin{aligned} \dot{x}_\Delta(t) &= \mathbf{A}x_\Delta + \mathbf{B}u_\Delta, \\ y_\Delta &= \mathbf{C}x_\Delta + \mathbf{D}u_\Delta, \end{aligned} \tag{4.26}$$

where $\mathbf{A} = \left.\frac{\partial f(x,u)}{\partial x}\right|_{x_e,u_e}$, $\mathbf{B} = \left.\frac{\partial f(x,u)}{\partial u}\right|_{x_e,u_e}$, $\mathbf{C} = \left.\frac{\partial g(x,u)}{\partial x}\right|_{x_e,u_e}$ and $\mathbf{D} = \left.\frac{\partial g(x,u)}{\partial u}\right|_{x_e,u_e}$.

As an example, we consider linearization of the model of a spring-mass-damper system with a hardening spring, as shown in Figure 4.29a.

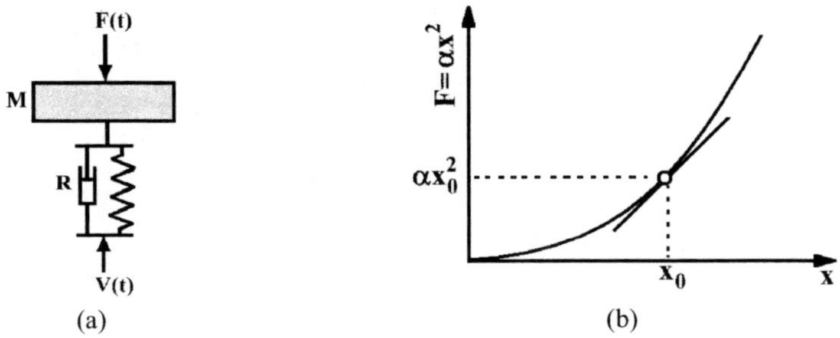

Fig. 4.29 A spring-mass-damper system and hardening spring characteristic

The self-weight of the mass causes some deflection of the spring. The spring force is given by $F = \alpha|x|x$ which may be written as $F = \alpha x^2$, where x is the spring's compression and it is assumed that $x > 0 \forall t \geq 0$. It is assumed that the excitations are weak, *i.e.* they do not cause large deviations from the static equilibrium point. The nominal inputs are taken as $F(t) = V(t) = 0$ and the equilibrium states are momentum of the mass $P = 0$ and displacement of the spring $x = Mg = \alpha x_0^2 \Rightarrow x_0 = \sqrt{Mg/\alpha}$. The linearization about the equilibrium point is shown in Figure 4.29b. The linearized stiffness becomes the local slope of the curve, *i.e.* $K = \partial F/\partial x|_{x=x_0} = 2\alpha x_0$.

The bond graph model of the non-linear system is given in Figure 4.30a, where element C_n models the non-linear spring. When one linearizes the spring about the equilibrium point then the spring force $F = \alpha(x_0 + x_\Delta)^2 \simeq \alpha x_0^2 + 2\alpha x_\Delta$ (by neglecting higher powers of x_Δ). This may be represented as shown in Figure 4.30b, where the C-element is linear and the term αx_0^2 is added through a pseudo-source (Se-element). Instead of using this pseudo-source, one may assign an appropriate initial condition to the state of the linear C-element, which would produce the same effort, *i.e.* $Q|_{t=0} = x_0/2$. However, the state variable Q would then neither represent the actual spring deformation not the deformation over and above the equilibrium deformation. The proper reduction of the non-linear bond graph to its linear form is as shown in Figure 4.30c, where small perturbations x_Δ are the states and they are given zero-initial conditions.

When one performs FDI by using a linearized model and the fault is isolated as the change in a certain parameter's value, then it is the linearized parameter whose value has changed. For example, if one identifies a fault in the stiffness of the system in Figure 4.29a by using the bond graph model in Figure 4.30c, one cannot conclude if parameter α has changed or the equilibrium point x_0 has shifted. Moreover, if the equilibrium point has shifted, the linearized model itself is invalid. This along with the problems associated with qualitative FDI are the primary reasons to go for quantitative FDI methods as developed in the next chapter.

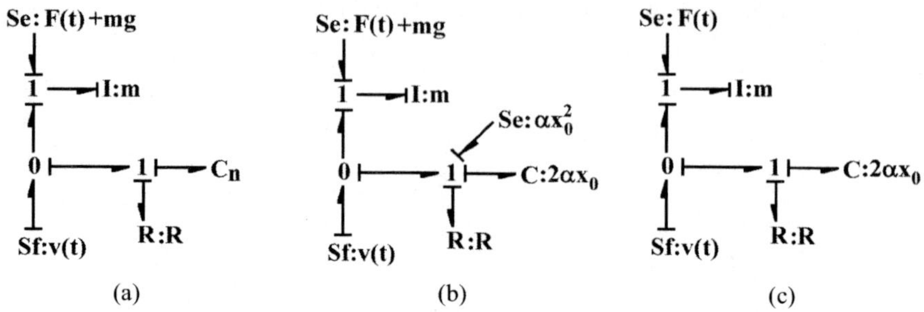

Fig. 4.30 Non-linear and linearized bond graph models

Problems

4.1. Draw the bond graph model of the inverted pendulum system shown in Figure 4.31. The forces due to ground friction and aerial resistance are approximated with the use of a damper, R, hooking the vehicle to the inertial reference frame. Derive the state equations and linearize them with respect to the unstable equilibrium point, *i.e.* the upright pendulum position. Express the linearized equations in state-space representation and draw the corresponding linear bond graph showing appropriate expressions for different parameters.

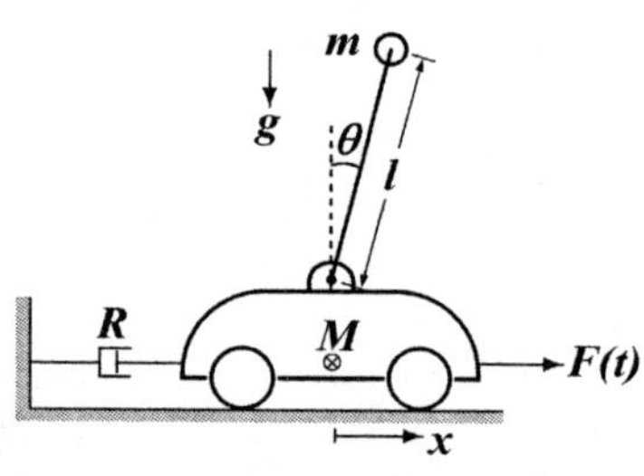

Fig. 4.31 Inverted pendulum on a vehicle

4.2. A two-tank system is shown in Figure 4.32. It has two valves with discharge coefficients C_{d1} and C_{d2}, respectively. The two conical funnel-like tanks are identical and they have the cone angle α. In steady state operation, the flow through each valve is the same as the supply from the pump. Therefore, one may obtain the steady state level in two tanks, say L_1 and L_2, from the equation $C_{d1}\sqrt{\rho g(L_1 - L_2)} = C_{d2}\sqrt{\rho g L_2} = Q_p$. Consider the masses of the liquid in the two tanks as state variables and linearize the system about its steady-state operating point. Draw the corresponding linearized bond graph model and temporal causal graph. If the outflow to the user is found to be slightly higher than the nominal, then what are the possible fault candidates? Select any one of the fault candidates and predict the corresponding transient trend up to the second order. Simulate this system with suitable data

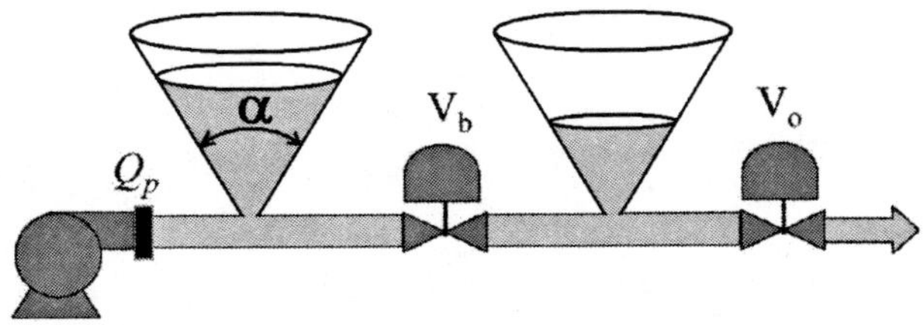

Fig. 4.32 A two-tank system with conical tanks

(*e.g.* $\alpha = 1$ rad and other data as considered in Section 4.3) and introduce the fault to check if the predicted trend matches the actual trend.

4.3. Consider the electrical circuit shown in Figure 4.33. If the current through inductor L_1 is measured and found to be higher than the nominal value, determine the fault candidates by performing qualitative analysis and then by constructing a fault tree. Further assume that the inductances cannot change suddenly and then minimize the fault candidates through a tree-graph analysis if it is specified that the current through both the inductors are higher than nominal in the post-fault steady-state operation.

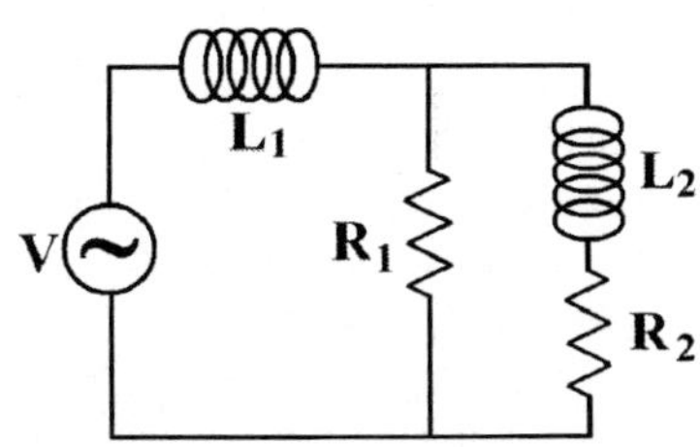

Fig. 4.33 An electrical circuit considered for qualitative analysis

4.4. A simple heat exchanger system is shown in Figure 4.34. The cold water at temperature T_w is supplied with a constant rate (a flow source) of Q_w and the hot water is taken out at the same rate in steady-state operation. The water outlet is through a long pipe, which discharges water to the environment. The discharge coefficient of the outlet pipe is C_d. The hot steam enters the heat exchanger through a valve having discharge coefficient C_s at the nominal valve position. The inlet side steam pressure is constant at P_s and its temperature is T_s. The valve position is controlled to maintain a constant water temperature in the bath.

For modeling purpose, we assume that the water temperature in the bath is uniform and the water level in the tank remains constant (consequently the mass) in

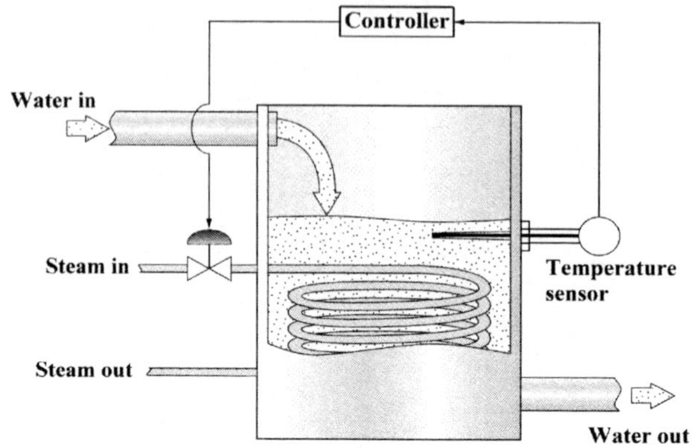

Fig. 4.34 A simple heat-exchanger system with valve control

steady state process operation. Moreover, the mass of the steam inside the pipes may be considered as a single storage element. The steam returns from the storage element to the outlet side through some resistance R_e. The heat stored in the metallic pipes may be neglected and then the overall heat exchange coefficient between the steam and water is given by R_{HEXA}. Nomenclature for other variables (specific heats, density, ...) may be suitably assumed.

Consider that the controller was inactive and the valve controlling the steam flow was at the nominal position. The process was operating in a steady-state condition. Suddenly, it was observed that due to some fault, the water temperature gradually dropped from its nominal value to another steady-state temperature. Perform a fault tree analysis to determine the fault candidates. If more information is available about the process state and it was found that the water level has increased, the steam control valve is in proper position, the steam and water flow rates are also normal, then what are the remaining fault candidates?

4.5. A towing boat is commissioned to bring ashore a motorized yacht (see Figure 4.35) whose motor has failed. They are fastened together by a rope, which is assumed to remain taut at all times. The yacht has also developed a leak and, to stop it from sinking, a pump is installed in it to pump out the water. Although this becomes a variable mass system, we will assume that the pumping out process produces a forward thrust force F_y on the yacht. The towing boat's propellers produce a forward propelling force F_b. In fact, this a controlled force and it is the output of a complex optimization process with the aim to control the tension in the connecting rope such that the two bodies do not collide with each other. The wind and water currents are in fixed directions and they resist the motion of the boat and the yacht. The wind resistance on the boat, R_b, is higher than that on the yacht, R_y. The wind and water resistances are two sky-hook dampers connected to the two rigid bodies. The rope may be modeled as a spring-damper combination in mechanical parallel.

Fig. 4.35 A towing boat and a yacht travelling against the wind

In steady state sailing, the towing boat and the yacht have the same velocity. (a) Suddenly, it was found that the towing boat starts to sail much faster. Diagnose the possible causes by constructing a temporal causal graph of the system. If the fall in wind resistance is considered as the fault hypothesis, generate the trends. (b) If it was found that the speed of the two boats have simultaneously decreased, then generate the trends with a fault hypothesis that the pump installed in the yacht is not functioning properly (*i.e.* the mass of the yacht is increasing).

4.6. A controlled fluid supply unit is shown in Figure 4.36.

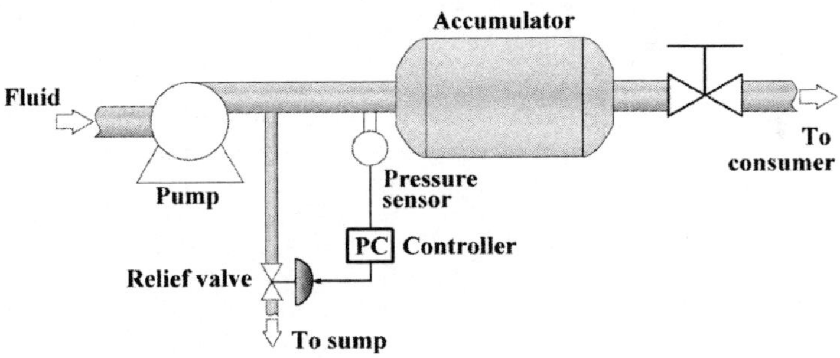

Fig. 4.36 A controlled fluid supply unit

The working fluid is assumed to be compressible. The pump delivers a constant flow rate and, to maintain a constant pressure in the accumulator, a relief valve is used to bypass the pumped fluid to the sump. During steady state operation, it was found that suddenly the pressure in the accumulator started to fall. The controller automatically tried to maintain the accumulator pressure and thus completely closed the relief valve. Then the operator found that the accumulator pressure sensor readings were falling and gradually reached a steady state value. The operator first checked the relief valve to ascertain that there was no leakage through it and thereafter he went to the output side and found that the flow rate to the consumer had reduced significantly. The pump was very new and known to be working properly. Use any of the qualitative analysis methods discussed in this chapter to diagnose the fault.

4.7. A passive level control system is shown in Figure 4.37. The level indicator bar actuates two spools to move a puppet valve and thus achieves the on-off control action. The thickness of the two spools decides the dead-zone between the *on* and the *off* action of the controller.

The actuating piston area is A_p and the cross-sectional area of the cylindrical storage tank is A_t. The hydraulic valve controller's actuator is powered from a constant pressure (P_{psl}) source. All fluids in the circuit are assumed to be incompressible. Other geometrical parameters are shown in the figure. Resistances may be included at appropriate locations in the actuating mechanism. The flow through the puppet valve is assumed to be proportional to the valve position.

When the tank was full up to its maximum (*i.e.* the puppet valve was closed), the consumer completely opened the outlet valve. It was found that the water level dropped below its minimum level but the controller failed to open the puppet valve. Use a fault tree analysis to determine the possible causes for this behavior.

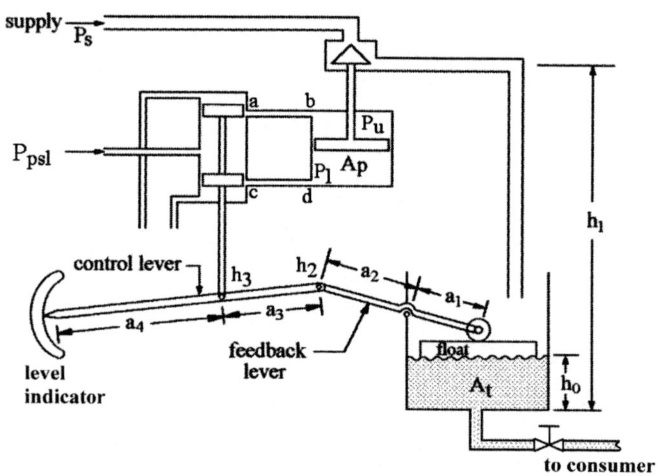

Fig. 4.37 A level control system

Chapter 5
Bond Graph Model-based Quantitative FDI

5.1 Introduction

Recent methods of fault diagnosis are based on the analysis of the underlying structure and behavior of a system [240]. Bond graph models serve as knowledge representation of a large amount of structural, functional and behavioral information and their relationship. This knowledge representation is used to create complex cause-effect reasoning leading to construction of powerful and robust automatic diagnosis and isolation systems. We have discussed a few of such qualitative reasoning methods in the previous chapter and reasoned that, in most cases, proper fault isolation is impossible with simple qualitative analysis. Full or partial quantitative analysis helps in better fault isolation. We introduce a few of the model based quantitative FDI methods in this chapter.

A model-based FDI method makes use of a mathematical model of the system. Formally, model-based fault diagnosis can be defined as the detection, isolation and characterization of faults in components of a system by comparing the available system measurements with *a priori* information represented by the mathematical model of the system. Quantitative FDI methods check the consistency between the actual process and its model. They check the differences generated from the comparison of different model variables with those of the process. These differences are called residuals or symptom signals [8, 16, 50, 93, 139, 197, 198, 237, 239]. Consistency checking is normally achieved through a comparison between the measured signals with their estimations, which are generated by a mathematical model of the considered process.

Every residual should be normally zero or close to zero when no fault is present, but should be distinguishably different from zero when a fault occurs. This is sometimes referred to as the *fundamental problem of residual generation*. An ideal residual is characteristically independent of system inputs, outputs and the system itself [78]. Fault detection concerns the process of differentiating between a normal operation and a faulty operation. Faults are detected by monitoring the trend of the residuals, which usually consists of setting a fixed threshold on a residual quantity generated by difference between real measurements and estimates of these measurements using the mathematical model. A number of residuals can be designed with each having special sensitivity to individual fault occurring in different locations in the system. The subsequent analysis of each residuals, once threshold is exceeded, then leads to fault isolation. Such a simple monitoring technique is called geometrical analysis. It is simple to implement but has some serious drawbacks: in the case of noises, input variations and change of operating points of the monitored process, the transients set-up in the process lead to transients in the residuals and during these transient periods, residuals may intermittently reach their fixed thresholds and thus give rise to false alarms [45, 69].

If a perfect mathematical model of the process under consideration is used (*i.e.* accounting for the transients, noises, *etc.*) then the resulting residuals will be such that there will be no false alarms. Such a perfect model then has to take care of minor process dynamics, *i.e.* it will have higher model order and will take more computation time. Such large and complex models are unsuitable for real-time process control and real-time FDI. Moreover, a perfectly accurate mathematical model of a physical system is never available. Usually, the parameters of the system may vary with time and the characteristics of the disturbances and noises are unknown; thus they cannot be modeled accurately. Therefore, there is always a mismatch between the actual process and its mathematical model even under no fault conditions. Such discrepancies directly influence residuals and hence they are the major causes of false alarms and misdetections.

To overcome these problems, a model-based FDI scheme needs to be insensitive to modeling uncertainties. Note that reducing the residuals sensitivity to modeling uncertainties is not necessarily a solution to this problem because this would also reduce the sensitivities to faults. The correct approach is to increase insensitivity to modeling uncertainties while simultaneously increasing sensitivity to faults. This means that fault sources should be decoupled from modeling uncertainties and disturbances right during the model formulation [234]. A number of such methods have been developed, for example the Unknown Input Observer (UIO), eigenstructure assignment and parity relation methods [12, 17, 92, 94, 109, 148, 195, 206, 269].

Abrupt fault or hard failures are easily detected because their effects on the residuals are usually larger than modeling uncertainties and thus a simple fixed residual threshold is usually sufficient to detect them. However, a good FDI scheme should as well be able to diagnose incipient faults in a process. With respect to abrupt faults, incipient faults have a small effect on residuals and they can be hidden by disturbances. An incipient fault may not necessarily degrade the performance of the plant. However, it should be detected soon so that the faulty component may be replaced before the probability of more serious malfunctions increases. This is a form of predictive maintenance.

Along with sensitivity to incipient faults, promptness of detection, fewer misdetections and low false alarm rate are other constraints on an FDI system. Out of these constraints, low false alarm rate and few missed detections are contradictory and a trade-off between them is required. On the other hand, promptness of detection requires that the time between the onset of a fault and its detection, usually called detection delay, must be small. This means that when a residual slightly exceeds its threshold, albeit for a sort duration, there will be an alarm. Therefore, promptness of detection and low false alarm rate are again contradictory constraints. All these constraints are optimized with appropriate weight given to each item, which depend on the FDI requirements of the specific system under consideration.

The primary requirements of the FDI system are a residual generator and a residual evaluator or decision support system. The general schematics of a model-based FDI system architecture is shown in Figure 5.1. One of the major advantages in following a model-based FDI approach is that, usually, no additional hardware are required. The measurements necessary for process control are usually sufficient to implement the FDI algorithm. Furthermore, a model-based FDI algorithm can be implemented in software on the same computer used to control the process. Note that process control and FDI are interrelated activities, as will be shown in later chapters of this book.

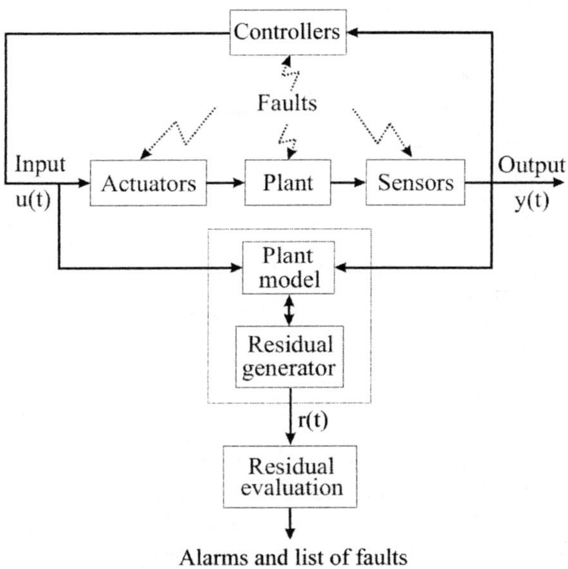

Fig. 5.1 A general scheme of model-based FDI

In Figure 5.1, it is assumed that faults can affect process components, sensors, controllers and actuators. Modern processes are usually computer controlled and thus controller faults are hypothetical. The two distinct blocks in Figure 5.1 are:

1. *Residual generation*: this block generates residuals (or fault symptoms) [77, 78, 91, 132, 144] which should indicate any fault occurrence. Residual generators can be developed using different methods, *e.g.* observer [49, 55, 56, 71, 74, 78, 131, 148, 196, 269, 272], parity relations [92, 94, 206] and parameter estimation [17, 44, 113, 114, 244, 283] based methods.

2. *Residual evaluation*: this block determines whether any faults have occurred or are going to occur in the process by examining the residuals and their trends. The occurrences or the likelihood of faults are found by applying certain decision rules which may be specific for the process. The decision rule may be based on a geometric method such as a simple threshold test on the instantaneous residual values or moving averages of the residuals, adaptive thresholds [69, 213, 230], interval models [6, 7, 62, 103, 170, 212, 252, 254], or on cumulative sums [13] of residuals. Some decision rules are based on statistical methods, *e.g.* generalized likelihood ratio test or sequential probability ratio test [28, 273].

Note that if residuals are well-designed (*i.e.* robust and structured) then the decision-making becomes relatively easy. Therefore, proper residual generation is the main objective in quantitative model-based FDI.

5.2 Classical Quantitative FDI and Residual Generation

Observer based FDI is a well known analytical model based FDI scheme, which compares the actual output from a system with reference output from an analytical model and the difference between the predicted and observed values, called a residual, which should be ideally zero during normal operation of the actual system, is monitored every time.

Another analytical FDI procedure works by evaluating physical constraint laws (described by the model) using input data, sensor data and parameter values from the monitored system. Constraint laws, when written symbolically in such a way that they contain only known variables (measurements and inputs), are called Analytical Redundancy Relations (ARRs) [53, 76, 96, 130, 209, 242]. In other words, ARRs are static or dynamic constraints which link the time evolution of the known variables when a system operates according to its normal operation model. Each constraint relation should always be valid within a certain bound of error, when evaluated using measured data from the real system. This error, which is theoretically zero during the normal operation of a process, is called a residual.

These two approaches of residual generation for FDI in real-time are shown in Figure 5.2. When there is a fault in the system, some residuals must deviate sufficiently such that the deviation is distinguishable from the residual response during normal operation. Usually, due to process and sensor noise, and their derivatives appearing in the ARR, it is very difficult to see this deviation with the naked eye. This requires some post processing, namely signal treatment techniques such as averaging, to extract relevant features from the residuals.

Observer based residuals use a behavioral model and compare the observer output to the measurements from the sensors installed in the process. At every sampling instance, the inputs are fed into the model and the outputs are calculated. The convergence of the solution for the model can take a significant amount of time depending on the model size, non-linearities and complexities, and the solution stiffness (due to large differences in time constants in the different parts of the system), *etc.* When the solution time is significant with respect to the sampling interval, the performance of the supervision platform can get seriously degraded. In contrast, in the case of ARR based FDI, residual evaluation is in derivative form and hence it is easily evaluated in real-time by making use of the values of the current and a few previous samples. In this book, we will follow an ARR based FDI approach.

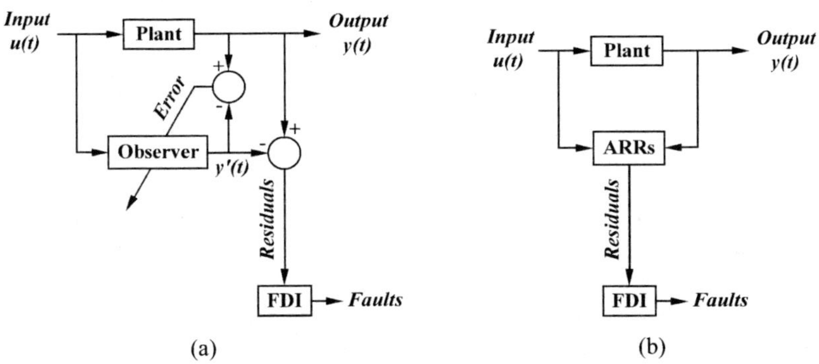

Fig. 5.2 Residual generators

The most important issue in model-based fault detection concerns the accuracy of the model describing the behavior of the monitored system. Modeling uncertainty arises from the impossibility of obtaining complete knowledge and understanding of the monitored process, which include unmodeled dynamics, loss of accuracy due to simplifying assumptions, un-quantifiable noises in process measurements, errors in estimating process parameter values and unknown disturbances.

5.2.1 Observer-based Methods

The residual generation may be done by using one or several estimators [55, 74, 78, 131, 196, 270, 272]. An estimator/observer is a quantitative process model which outputs an estimate of the actual process output. Most often, only one estimator is sufficient for fault detection. However, a bank of estimators is usually needed for fault isolation. The bank of estimators can be arranged in different schemes.

5.2.1.1 Dedicated Observer Scheme

This scheme may be used for sensor or actuator FDI. For sensor fault isolation, one may simply design a number of dedicated Luenberger observers where each observer is connected to a single output. For actuator fault detection, each observer is connected to a single input. The basic idea is to have a bank of observers where each observer only depends on one fault and is approximately decoupled from all other faults and from all disturbances; design of observers with disturbance decoupling is discussed later in this section. The observers should be so designed that the k-th residual generated by the k-th observer depends only on the k-th fault. Ideally, this scheme should be able to detect the same number of faults simultaneously as the number of observers. The schematic representation of a dedicated observer scheme is given in Figure 5.3.

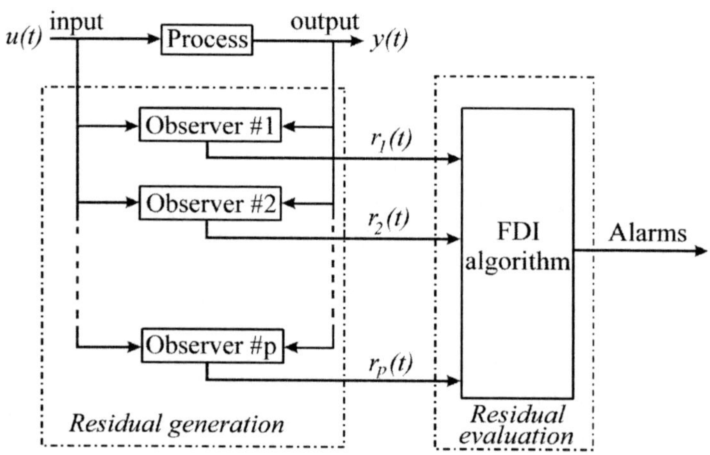

Fig. 5.3 Dedicated observers scheme for FDI

5.2.1.2 Generalized Observer Scheme

This scheme is similar to the dedicated observer scheme. In this scheme, for sensor fault isolation, each observer is connected to all sensors except the one whose fault will be detected by that observer. Likewise, for actuator fault isolation, each observer is connected to all inputs except the one whose fault will be detected by that observer. Naturally, if the residual obtained from a particular observer is a null vector and all other observers give non-null residual vector, the sensor/actuator which is not connected to that particular observer is the faulty one. The observers should be so designed that the k-th residual generated by the k-th observer depends on all faults except the k-th fault.

5.2.2 Observer-based Residuals

We have already discussed observers in the previous chapter. Let us consider an LTI system model given by

$$\begin{aligned}\dot{x}(t) &= \mathbf{A}x(t) + \mathbf{B}u(t),\\ y(t) &= \mathbf{C}x(t) + \mathbf{D}u(t),\end{aligned} \tag{5.1}$$

and a Luenberger observer of it as

$$\begin{aligned}\dot{x}_o(t) &= \mathbf{A}x_o(t) + \mathbf{B}u(t) + \mathbf{L}\left(y(t) - y_o(t)\right),\\ y_o(t) &= \mathbf{C}x_o(t) + \mathbf{D}u(t).\end{aligned} \tag{5.2}$$

Then the error in estimating the state from the observer is $e(t) = x(t) - x_o(t)$. The error dynamics is given by

$$\dot{e}(t) = (\mathbf{A} - \mathbf{LC})\,e(t), \tag{5.3}$$

with $e(0) = x(0) - x_o(0)$, which means if the real part of eigenvalues of matrix $\mathbf{A} - \mathbf{LC}$ are sufficiently negative by choosing matrix $\mathbf{L}$ suitably, then the error will quickly converge to zero and we will get a good estimate of the plant state from the observer. If an open-loop observer is used, it transpires that real part of eigenvalues of matrix $\mathbf{A}$ must be sufficiently negative, *i.e.* the open-loop plant must be highly stable; otherwise an open-loop observer with incorrect initial conditions will not track the plant states. Indeed, in process engineering installations, assumption of asymptotic stability holds good for a large number of processes.

However, using open-loop observers is not a good idea for lightly damped systems, where choice of initial conditions for observer's states plays an important role, and also in those systems containing feedback of noisy signals. Therefore, let us consider a Linear Time Invariant (LTI) system defined by the following state-space model:

$$\begin{aligned}\dot{x}(t) &= \mathbf{A}x(t) + \mathbf{B}u(t) + \mathbf{E}_d d(t) + \mathbf{E}_f f(t),\\ y(t) &= \mathbf{C}x(t) + \mathbf{D}u(t) + \mathbf{G}_d d(t) + \mathbf{G}_f f(t),\end{aligned} \tag{5.4}$$

where $x(t)$ is the state vector, $y(t)$ is the output vector, $u(t)$ is the vector of known inputs, $d(t)$ is the vector of unknown inputs such as disturbances, $f(t)$ is the vector of faults, and matrices $\mathbf{A},\mathbf{B},\mathbf{C},\mathbf{D}$ *etc.* are constants. In normal operation, $f(t) = 0, \forall t$.

Considering faults in the model (*i.e.* $\mathbf{E}_f \neq 0$, $\mathbf{G}_f \neq 0$, and $\mathbf{E}_d = \mathbf{G}_d = 0$), the error dynamics (state estimation error) can be written as

$$\dot{e}(t) = (\mathbf{A} - \mathbf{LC})\,e(t) + \left(\mathbf{E}_f - \mathbf{LG}_f\right) f(t), \tag{5.5}$$

and the output error can be written as

$$\dot{e}_o(t) = y(t) - y_o(t) = \mathbf{C}e(t) + \mathbf{G}_f f(t). \tag{5.6}$$

Taking Laplace transform of the two error equations,

$$\begin{aligned}
s\,e(s) &= (\mathbf{A} - \mathbf{LC})\,e(s) + \left(\mathbf{E}_f - \mathbf{LG}_f\right) f(s),\\
\text{or}\;\; e(s) &= (s\mathbf{I} - (\mathbf{A} - \mathbf{LC}))^{-1}\left(\mathbf{E}_f - \mathbf{LG}_f\right) f(s);\\
e_o(s) &= \mathbf{C}e(s) + \mathbf{G}_f f(s)\\
&= \left(\mathbf{C}\,(s\mathbf{I} - (\mathbf{A} - \mathbf{LC}))^{-1}\left(\mathbf{E}_f - \mathbf{LG}_f\right) + \mathbf{G}_f\right) f(s);\\
\mathbf{X}(s) &= \frac{e_o(s)}{f(s)} = \mathbf{C}\,(s\mathbf{I} - (\mathbf{A} - \mathbf{LC}))^{-1}\left(\mathbf{E}_f - \mathbf{LG}_f\right) + \mathbf{G}_f,
\end{aligned} \tag{5.7}$$

where $\mathbf{X}(s)$ is the transfer function matrix between faults and the output error. As is evident, if none of the components of $\mathbf{X}(s)$ is null, any fault (non-zero) will affect the observer output error, which after the transients are over can be used for FDI. Thus, in the observer based approach to FDI, $e_o(t)$ is considered as a residual. These principles are also extendable to non-linear system models [131]. Different faults in the system affect the outputs differently. The deviations in sensor output (both magnitudes and sense) are used in the decision procedure for fault isolation.

5.2.2.1 Factorization Approach

In the frequency domain observer design [77, 132], the factorization technique is commonly followed for fault isolation, which requires a bank of residual generators. Each residual generator is designed in such a way that the corresponding residual becomes sensitive to a particular fault and insensitive to all other faults and disturbances. If disturbances are considered in the model in Equation 5.4; *i.e.* $\mathbf{E}_f \neq 0$, $\mathbf{G}_f \neq 0, \mathbf{E}_d \neq 0$ and $\mathbf{G}_d \neq 0$, then the error dynamics given in Equation 5.7 can be written as

$$e_o(s) = \mathbf{H}_f(s)f(s) + \mathbf{H}_d(s)d(s), \tag{5.8}$$

where $\mathbf{H}_f(s)$ and $\mathbf{H}_d(s)$ are transfer function matrices given as

$$\begin{aligned}
\mathbf{H}_f(s) = \mathbf{X}(s) &= \mathbf{C}\,(s\mathbf{I} - (\mathbf{A} - \mathbf{LC}))^{-1}\left(\mathbf{E}_f - \mathbf{LG}_f\right) + \mathbf{G}_f,\\
\mathbf{H}_d(s) &= \mathbf{C}\,(s\mathbf{I} - (\mathbf{A} - \mathbf{LC}))^{-1}\left(\mathbf{E}_d - \mathbf{LG}_d\right) + \mathbf{G}_d.
\end{aligned} \tag{5.9}$$

Instead of considering $e_o(s)$ as the residual, let us reformulate the residual as

$$r(s) = \mathbf{T}(s)e_o(s) = \mathbf{T}(s)\left(\mathbf{H}_f(s)f(s) + \mathbf{H}_d(s)d(s)\right), \tag{5.10}$$

where $\mathbf{T}(s)$ is a transformation matrix. If this transformation is properly designed, then one may obtain perfect decoupling of fault effects from the residual [71, 74, 78, 196]. One requirement for achieving perfect decoupling is that the residual should not depend on the disturbance, *i.e.* $\mathbf{T}(s)\mathbf{H}_d(s) = 0$. If each residual should be influenced by only one fault, then an additionally required condition is that $\mathbf{T}(s)\mathbf{H}_f(s)$ must be a diagonal transfer function matrix. Thus the fault detection filter, $\mathbf{T}$, must be designed in such a way that the reachable subspace of each fault is placed into invariant subspaces which do not overlap each other. So, if the process contains a fault, then the residual $r(s) = \mathbf{T}(s)e_o(s)$ is non-zero and the fault is detected. Since each fault is projected onto invariant subspaces, the fault can also be isolated.

This means that each transformed measurement must be sensitive to one fault. Such a perfect decoupling would then require a lot of sensors in the process. Therefore, instead of aiming for perfect decoupling, the usual practice is to achieve some sort of approximate decoupling. Approximate decoupling means that the dependence between the residual and the faults should be maximized and the influence of the disturbances on the residual should be minimized. An approximate decoupling is usually obtained by maximizing H_2 or H_∞ norm of a performance index defined as $J = \left\|\mathbf{T}(s)\mathbf{H}_f(s)\right\| / \|\mathbf{T}(s)\mathbf{H}_d(s)\|$.

5.2.3 Unknown Input Observers

An observer is defined as an unknown input observer if its state estimation error vector approaches zero asymptotically regardless of the presence of the unknown inputs, *i.e.* disturbances, in the system [49, 148, 269].

Consider a linear time-invariant system described by the following equations:

$$\begin{aligned} \dot{x}(t) &= \mathbf{A}x(t) + \mathbf{B}u(t) + \mathbf{D}v(t), \\ y(t) &= \mathbf{C}x(t) \end{aligned} \tag{5.11}$$

where $x(t) \in \mathbb{R}^n$ is the state vector, $u(t) \in \mathbb{R}^m$ is measurable input vector, $v(t) \in \mathbb{R}^q$ is the unknown input vector, $y(t) \in \mathbb{R}^p$ is the output vector and $\mathbf{A}, \mathbf{B}, \mathbf{C}$ and $\mathbf{D}$ are matrices of suitable dimensions. The unknown input observer of this system [56] may be given as

$$\begin{aligned} \dot{z}(t) &= \mathbf{N}z(t) + \mathbf{L}y(t) + \mathbf{G}u(t), \\ x_o(t) &= z(t) - \mathbf{E}y(t), \end{aligned} \tag{5.12}$$

where $z(t) \in \mathbb{R}^n$ is the state vector, and the matrices $\mathbf{N}$, $\mathbf{L}$, $\mathbf{G}$ and $\mathbf{E}$ are of appropriate dimensions which are to be determined in such way that $x_o(t)$ converges asymptotically to $x(t)$.

The error dynamics is then given by

$$
\begin{aligned}
e(t) &= x(t) - x_o(t), \\
\dot{e}(t) &= \dot{x}(t) - \dot{x}_o(t) \\
&= \mathbf{A}x(t) + \mathbf{B}u(t) + \mathbf{D}v(t) - \dot{z}(t) + \mathbf{E}\dot{y}(t) \\
&= \mathbf{A}x(t) + \mathbf{B}u(t) + \mathbf{D}v(t) - \mathbf{N}z(t) - \mathbf{L}y(t) - Gu(t) \\
&\quad + \mathbf{EC}\left(\mathbf{A}x(t) + \mathbf{B}u(t) + \mathbf{D}v(t)\right) \\
&= \mathbf{A}x(t) + \mathbf{B}u(t) + \mathbf{D}v(t) - \mathbf{N}x_o(t) + \mathbf{NE}y(t) - \mathbf{L}y(t) - \mathbf{G}u(t) \\
&\quad + \mathbf{EC}\left(\mathbf{A}x(t) + \mathbf{B}u(t) + \mathbf{D}v(t)\right) \\
&= \mathbf{A}x(t) + \mathbf{B}u(t) + \mathbf{D}v(t) - \mathbf{N}x_o(t) + \mathbf{NEC}x(t) - \mathbf{L}y(t) - \mathbf{G}u(t) \\
&\quad + \mathbf{EC}\left(\mathbf{A}x(t) + \mathbf{B}u(t) + \mathbf{D}v(t)\right) \\
&= \mathbf{N}x(t) - \mathbf{N}x_o(t) - (\mathbf{NP} + \mathbf{LC} - \mathbf{PA})x(t) - (\mathbf{G} - \mathbf{PB})u(t) \\
&\quad + \mathbf{PD}v(t) \\
&= \mathbf{N}e(t) - (\mathbf{NP} + \mathbf{LC} - \mathbf{PA})x(t) - (\mathbf{G} - \mathbf{PB})u(t) + \mathbf{PD}v(t),
\end{aligned}
\tag{5.13}
$$

where $\mathbf{P} = \mathbf{I} + \mathbf{EC}$ and $\mathbf{I} \in \mathbb{R}^{n \times n}$ is an identity matrix. The error converges to zero iff

$$\mathbf{PD} = \mathbf{D} + \mathbf{ECD} = 0, \tag{5.14}$$

$$\mathbf{NP} + \mathbf{LC} - \mathbf{PA} = 0, \tag{5.15}$$

$$\mathbf{G} - \mathbf{PB} = 0. \tag{5.16}$$

The solution of Equation 5.14 for $\mathbf{E}$ may be written as

$$\mathbf{E} = -\mathbf{D}(\mathbf{CD})^+ + \mathbf{Y}(\mathbf{I}_p - (\mathbf{CD})(\mathbf{CD})^+), \tag{5.17}$$

where $(\mathbf{CD})^+$ is the generalized inverse of $\mathbf{CD}$, $\mathbf{I}_p \in \mathbb{R}^{p \times p}$ is an identity matrix and $\mathbf{Y}$ is an arbitrary matrix of appropriate dimension.

The matrix $\mathbf{G}$ can be obtained from Equation 5.16 as follows:

$$\mathbf{G} = \mathbf{PB} = \mathbf{B} + \mathbf{ECB}. \tag{5.18}$$

The state error converges to zero if the poles of $\mathbf{N}$ are stable. Let us introduce a matrix $\mathbf{K}$ such that

$$\mathbf{N} = \mathbf{PA} - \mathbf{KC}. \tag{5.19}$$

Then Equation 5.15 is satisfied iff

$$\mathbf{K} = \mathbf{L} + \mathbf{NE}. \tag{5.20}$$

This derivation is straightforward: $\mathbf{N} = \mathbf{PA} - \mathbf{KC} = \mathbf{PA} - (\mathbf{L} + \mathbf{NE})\,\mathbf{C} = \mathbf{PA} - \mathbf{LC} - \mathbf{NEC}$, $\Rightarrow \mathbf{N} + \mathbf{NEC} = \mathbf{PA} - \mathbf{LC}$ or $\mathbf{N}\,(\mathbf{I} + \mathbf{EC}) = \mathbf{NP} = \mathbf{PA} - \mathbf{LC}$. Because $\mathbf{P}$ is already known, the poles of $\mathbf{N}$ can be placed by properly choosing a matrix $\mathbf{K}$ (see Equation 5.19). Thus the designed observer feedback gains are obtained as

$$\begin{aligned}\mathbf{L} &= \mathbf{K} - \mathbf{NE} \\ &= \mathbf{K} - (\mathbf{PA} - \mathbf{KC})\mathbf{E} \\ &= \mathbf{K}(\mathbf{I}_p + \mathbf{CE}) - \mathbf{PAE}.\end{aligned} \tag{5.21}$$

Calculation of matrices **N**, **G** and **L** completes the construction of the full order unknown input observer.

If we consider a UIO for fault detection, then we need to distinguish disturbances from faults. The UIO will be designed such a way that disturbance effects are decoupled whereas fault will affect the estimation error. Let us redefine the state space form in Equation 5.11 differently as

$$\begin{aligned}\dot{x}(t) &= \mathbf{A}x(t) + \mathbf{B}u(t) + \mathbf{E}_d d(t) + \mathbf{E}_f f(t), \\ y(t) &= \mathbf{C}x(t) + \mathbf{G}_d d(t) + \mathbf{G}_f f(t),\end{aligned} \tag{5.22}$$

where $x(t)$ is the state vector, $y(t)$ is the output vector, $u(t)$ is the vector of known inputs, $d(t)$ is the vector of unknown inputs such as disturbances, $f(t)$ is the vector of faults, and matrices $\mathbf{A}, \mathbf{B}, \mathbf{C}, \mathbf{D}, \mathbf{E}_d, \mathbf{E}_f, \mathbf{G}_d$ and $\mathbf{G}_f$ are constants. A commonly used unknown input observer for fault detection and the corresponding residual generator are given by

$$\begin{aligned}\dot{x}_0(t) &= \mathbf{H}_x x_0(t) + \mathbf{H}_u u(t) + \mathbf{H}_y y(t), \\ r(t) &= \mathbf{M}_x x_0(t) + \mathbf{M}_y y(t),\end{aligned} \tag{5.23}$$

where the matrices $\mathbf{H}_x$, $\mathbf{H}_u$, $\mathbf{H}_y$, $\mathbf{M}_x$ and $\mathbf{M}_y$ are of appropriate dimensions. Let the error be defined as $e(t) = x_0(t) - \mathbf{T}x(t)$, where **T** is a transformation matrix. Then the estimation error dynamics can be written as

$$\begin{aligned}\dot{e}(t) &= \dot{x}_0(t) - \mathbf{T}\dot{x}(t) \\ &= \mathbf{H}_x x_0(t) + \mathbf{H}_u u(t) + \mathbf{H}_y y(t) - \mathbf{TA}x(t) - \mathbf{TB}u(t) \\ &\quad - \mathbf{TE}_d d(t) - \mathbf{TE}_f f(t) \\ &= \mathbf{H}_x x_0(t) - \mathbf{H}_x\mathbf{T}x(t) + \mathbf{H}_x\mathbf{T}x(t) + \mathbf{H}_u u(t) \\ &\quad + \mathbf{H}_y\left(\mathbf{C}x(t) + \mathbf{G}_d d(t) + \mathbf{G}_f f(t)\right) - \mathbf{TA}x(t) - \mathbf{TB}u(t) \\ &\quad - \mathbf{TE}_d d(t) - \mathbf{TE}_f f(t) \\ &= \mathbf{H}_x\left(x_0(t) - \mathbf{T}x(t)\right) + \left(\mathbf{H}_x\mathbf{T} + \mathbf{H}_y\mathbf{C} - \mathbf{TA}\right)x(t) + \left(\mathbf{H}_u - \mathbf{TB}\right)u(t) \\ &\quad + \left(\mathbf{H}_y\mathbf{G}_d - \mathbf{TE}_d\right)d(t) + \left(\mathbf{H}_y\mathbf{G}_f - \mathbf{TE}_f\right)f(t). \\ &= \mathbf{H}_x e(t) + \left(\mathbf{H}_x\mathbf{T} + \mathbf{H}_y\mathbf{C} - \mathbf{TA}\right)x(t) + \left(\mathbf{H}_u - \mathbf{TB}\right)u(t) \\ &\quad + \left(\mathbf{H}_y\mathbf{G}_d - \mathbf{TE}_d\right)d(t) + \left(\mathbf{H}_y\mathbf{G}_f - \mathbf{TE}_f\right)f(t).\end{aligned} \tag{5.24}$$

The residual vector may be likewise derived as

$$r(t) = \mathbf{M}_x e(t) + (\mathbf{M}_y\mathbf{C} - \mathbf{M}_x\mathbf{T})x(t) + \mathbf{M}_y\mathbf{G}_d d(t) + \mathbf{M}_y\mathbf{G}_f f(t). \tag{5.25}$$

Asymptotic error convergence and disturbance decoupling may then be achieved if the following system of equations is satisfied:

$$\begin{aligned}
&\mathbf{H}_x \text{ is stable,}\\
&\mathbf{TA} = -\left(\mathbf{H}_x\mathbf{T} + \mathbf{H}_y\mathbf{C}\right),\\
&\mathbf{TB} = \mathbf{H}_u,\\
&\mathbf{H}_y\mathbf{G}_d - \mathbf{TE}_d = 0,\\
&\mathbf{M}_y\mathbf{C} - \mathbf{M}_x\mathbf{T} = 0,\\
&\mathbf{M}_y\mathbf{G}_d = 0.
\end{aligned} \tag{5.26}$$

If the matrices are properly designed then the estimation error and residual become

$$\begin{aligned}
\dot{e}(t) &= \mathbf{H}_x e(t) + \left(\mathbf{H}_y\mathbf{G}_f - \mathbf{TE}_f\right) f(t),\\
r(t) &= \mathbf{M}_x e(t) + \mathbf{M}_y\mathbf{G}_f f(t).
\end{aligned} \tag{5.27}$$

Because of the fact that $\mathbf{H}_x$ is stable, in the fault free case $e(t) \to 0$ and $r(t) \to 0$ as $t \to \infty$. The conditions for existence of a UIO and the rank conditions to ensure that the faults do not cancel each other are out of the scope of this book. However, note that if the dimension of the unknown disturbance vector $d(t)$ is larger (consequently dimensions of matrices $\mathbf{E}_d$ and $\mathbf{G}_d$), such that Equation 5.26 cannot be solved with proper inversion, then the residual vector $r(t)$ will be non-zero even in the absence of faults. For proper fault detection, the number of unknown disturbances cannot exceed the number of sensors. Fault isolation is usually based on a bank of unknown input observers, each having a single hypothesized fault. Different faults become unknown inputs (disturbances) to different observers and the one giving zero residual is the fault candidate.

5.2.3.1 Example

Consider a two-tank system shown in Figure 5.4a. We assume that the system is linear, *i.e.* the flow through each valve is proportional to the pressure difference across it. The linear bond graph model is given in Figure 5.4b.

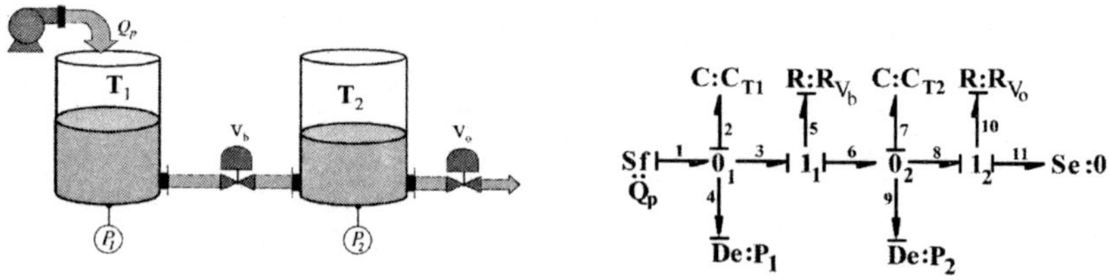

(a) (b)

Fig. 5.4 A two-tank system and its bond graph model

The state-space model of the process with the state variables as $x = \begin{bmatrix} Q_2 & Q_7 \end{bmatrix}^T$ is given by

$$\begin{Bmatrix} \dot{Q}_2 \\ \dot{Q}_7 \end{Bmatrix} = \begin{bmatrix} -\frac{1}{C_{T1}R_{vb}} & \frac{1}{C_{T2}R_{vb}} \\ \frac{1}{C_{T1}R_{vb}} & -\frac{1}{C_{T2}R_{vb}} - \frac{1}{C_{T2}R_{vo}} \end{bmatrix} \begin{Bmatrix} Q_2 \\ Q_7 \end{Bmatrix} + \begin{bmatrix} 1 \\ 0 \end{bmatrix} Q_p. \tag{5.28}$$

Let us consider four fault candidates as $C : C_{T1}$, $R : R_{vb}$, $C : C_{T2}$ and $R : R_{vo}$; which correspond to leakages from the storage tanks and leakages/blockages in the valves. Sensor and actuator faults are not considered in this example. If we consider the fault hypothesis that the element $R : R_{vb}$ is faulty then we see from Figure 5.4b that this element gives flow to the system. Then this element can be replaced by a flow source, as shown in Figure 5.5a. If we create a UIO and treat this newly added source as a disturbance and decouple it, then the residual will be zero if and only if $R : R_{vb}$ is actually the fault. Similarly, we can construct UIOs by considering other elements in our fault hypothesis; the modified models in Figure 5.5b,c correspond to fault hypotheses $C : C_{T2}$ and $R : R_{vo}$, respectively. Residuals from the bank of UIOs are compared in a decision procedure to isolate the fault.

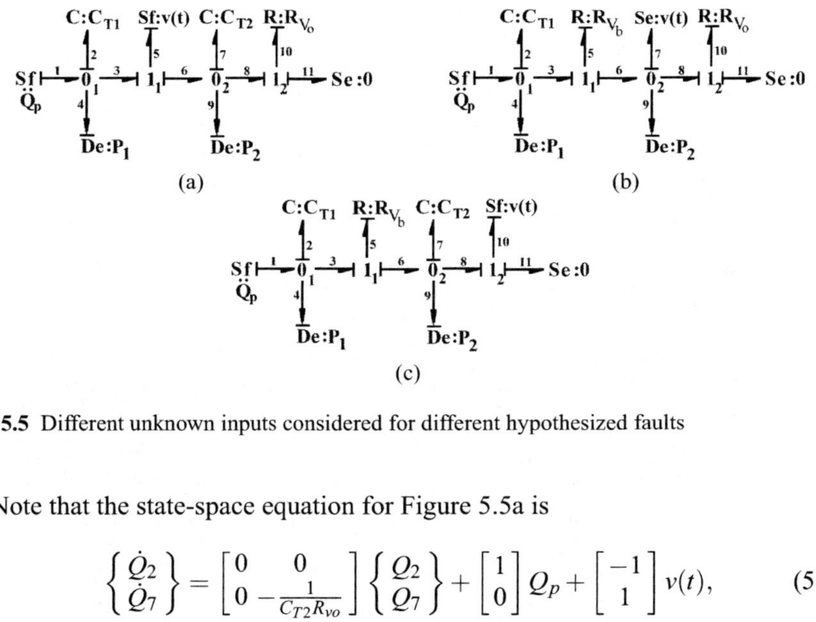

Fig. 5.5 Different unknown inputs considered for different hypothesized faults

Note that the state-space equation for Figure 5.5a is

$$\begin{Bmatrix} \dot{Q}_2 \\ \dot{Q}_7 \end{Bmatrix} = \begin{bmatrix} 0 & 0 \\ 0 & -\frac{1}{C_{T2}R_{vo}} \end{bmatrix} \begin{Bmatrix} Q_2 \\ Q_7 \end{Bmatrix} + \begin{bmatrix} 1 \\ 0 \end{bmatrix} Q_p + \begin{bmatrix} -1 \\ 1 \end{bmatrix} v(t), \tag{5.29}$$

where $v(t)$ is the unknown input and $\mathbf{D} = \begin{bmatrix} -1 & 1 \end{bmatrix}^T$ (see Equation 5.11). Similar equations can be derived for other faults. From Figure 5.5b we have

$$\begin{Bmatrix} \dot{Q}_2 \\ \dot{Q}_7 \end{Bmatrix} = \begin{bmatrix} -\frac{1}{C_{T1}R_{vb}} & 0 \\ \frac{1}{C_{T1}R_{vb}} & 0 \end{bmatrix} \begin{Bmatrix} Q_2 \\ Q_7 \end{Bmatrix} + \begin{bmatrix} 1 \\ 0 \end{bmatrix} Q_p + \begin{bmatrix} -\frac{1}{R_{vb}} \\ -\frac{1}{R_{vo}} \end{bmatrix} v(t), \tag{5.30}$$

and from Figure 5.5b we have

$$\left\{\begin{array}{c}\dot{Q}_2\\ \dot{Q}_7\end{array}\right\}=\left[\begin{array}{cc}-\frac{1}{C_{T1}R_{vb}} & \frac{1}{C_{T2}R_{vb}}\\ \frac{1}{C_{T1}R_{vb}} & -\frac{1}{C_{T2}R_{vb}}\end{array}\right]\left\{\begin{array}{c}Q_2\\ Q_7\end{array}\right\}+\left[\begin{array}{c}1\\ 0\end{array}\right]Q_p+\left[\begin{array}{c}0\\ -1\end{array}\right]v(t). \tag{5.31}$$

Although it seems that we can detect and isolate all faults, it is not so in reality. This issue will be handled in the next section.

Note that the **A** matrix is different in Equations 5.29–5.31 from Equation 5.28. This means that each observer will have to be designed differently. However, if we consider change in a particular parameter value, say $R : R_{vo}$, then we can rewrite Equation 5.28 as

$$\begin{aligned}\left\{\begin{array}{c}\dot{Q}_2\\ \dot{Q}_7\end{array}\right\}&=\left[\begin{array}{cc}-\frac{1}{C_{T1}R_{vb}} & \frac{1}{C_{T2}R_{vb}}\\ \frac{1}{C_{T1}R_{vb}} & -\frac{1}{C_{T2}R_{vb}}-\frac{1}{C_{T2}(R_{vo}+\Delta R_{vo})}\end{array}\right]\left\{\begin{array}{c}Q_2\\ Q_7\end{array}\right\}\\ &\quad+\left[\begin{array}{c}1\\ 0\end{array}\right]Q_p\\ &=\mathbf{A}x+\mathbf{B}u+\Delta\mathbf{A}x\\ &=\mathbf{A}x+\mathbf{B}u+\mathbf{D}v,\end{aligned} \tag{5.32}$$

where $\Delta\mathbf{A}$ is the change in matrix **A** due to parametric fault and its effect is recast as an unknown input. In the general case, there may be change to matrix **B**, which can also be shifted to unknown input vector. One advantage of this formulation is that the **A** matrix of all the observers remains the same as that of the system, only the unknown input part changes.

The formulation given in Equation 5.32 may be implemented in bond graph models by adding appropriate flow and effort sources. This would generally require insertion of an appropriate junction. Note that there can be no generalization of UIOs for fault representation in a bond-graph field element because this would simultaneously require as many unknown inputs as the number of bonds connected to the field element. The modified system models with unknown inputs for fault hypotheses in $R : R_{vb}$ and $C : C_{T2}$ are shown in Figure 5.6a,b, respectively.

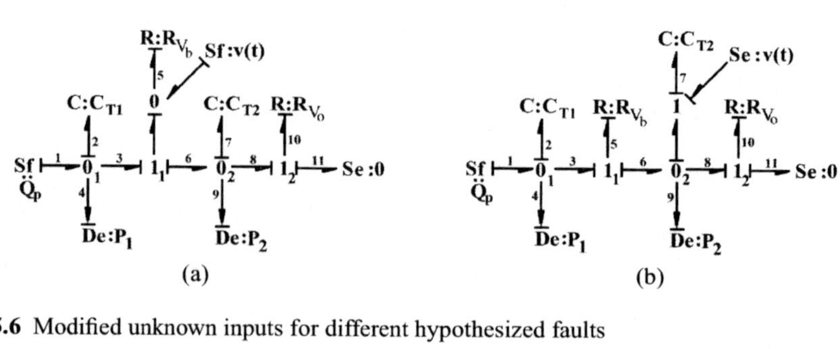

Fig. 5.6 Modified unknown inputs for different hypothesized faults

5.2.4 Parity Space Residuals

Let us consider a Linear Time Invariant (LTI) system defined by the following state-space model:

$$\begin{aligned} \dot{x}(t) &= \mathbf{A}x(t) + \mathbf{B}u(t) + \mathbf{E}_d d(t) + \mathbf{E}_f f(t), \\ y(t) &= \mathbf{C}x(t) + \mathbf{D}u(t) + \mathbf{G}_d d(t) + \mathbf{G}_f f(t), \end{aligned} \tag{5.33}$$

where $x(t)$ is the state vector, $y(t)$ is the output vector, $u(t)$ is the vector of known inputs, $d(t)$ is the vector of unknown inputs such as disturbances, $f(t)$ is the vector of faults, and $\mathbf{A}, \mathbf{B}, \mathbf{C}, \mathbf{D}$ *etc.* are constant matrices of appropriate dimensions. In normal operation, $f(t) = 0, \forall t$.

Let us consider the following form of differential equations, which can be used to model a large class of non-linear dynamic systems:

$$\begin{aligned} \dot{x}(t) &= \Phi_x\left(x(t), u(t), d(t), f(t)\right), \\ y(t) &= \Phi_y\left(x(t), u(t), d(t), f(t)\right), \end{aligned} \tag{5.34}$$

where Φ_x and Φ_y are two smooth non-linear functions, *i.e.* they have continuous partial derivatives up to any order with respect to their arguments.

When the non-linear system described in Equation 5.34 works around a nominal operating regime, which is characterized by a stable equilibrium state, it was shown in Chapter 4 that these non-linear differential equations can be linearized around the operating point and the dynamics for deviations, *i.e.* when $x(t)$ represents deviations from equilibrium state, $y(t)$ represents deviations from equilibrium state output, and so on, can be expressed in the form of the LTI system model described in Equation 5.33. We will consider those linearized equations in this section.

Furthermore, we will assume that the faults considered in the LTI state-space model are additive, *e.g.* a leakage, sensor bias, blockage, as opposed to multiplicative faults, which change the parameters of system, *i.e.* the entries in the matrices.

Let us assume a case, where there are no unknown inputs, *i.e.* $d(t) = 0$. In the faultless mode, *i.e.* $f(t) = 0$, the first derivative of the output can be expressed as

$$\begin{aligned} y(t) &= \mathbf{C}x(t) + \mathbf{D}u(t), \\ \dot{y}(t) &= \mathbf{C}\dot{x}(t) + \mathbf{D}\dot{u}(t). \end{aligned} \tag{5.35}$$

By substituting $\dot{x}(t) = \mathbf{A}x(t) + \mathbf{B}u(t)$ in the above equation and taking further successive derivatives with respect to time up to m-th order of the output in the LTI model, one obtains

$$
\begin{aligned}
\dot{y}(t) &= \mathbf{CA}x(t) + \mathbf{CB}u(t) + \mathbf{D}\dot{u}(t),\\
\ddot{y}(t) &= \mathbf{CA}^2x(t) + \mathbf{CAB}u(t) + \mathbf{CB}\dot{u}(t) + \mathbf{D}\ddot{u}(t),\\
&\ \vdots\\
y^{(m)}(t) &= \mathbf{CA}^m x(t) + \mathbf{CA}^{m-1}Bu(t) + \cdots + \mathbf{CB}u^{(m-1)}(t) + \mathbf{D}u^{(m)}(t),
\end{aligned}
\tag{5.36}
$$

where exponents inside braces indicate the order of the derivatives with respect to time.

Considering that matrix $\mathbf{A}$ is of full rank (rank$=n$), taking the characteristic equation $\phi(s) = |s\mathbf{I} - \mathbf{A}| = s^n + a_{n-1}s^{n-1} + \cdots + a_1 s + a_0$ and by applying Caley-Hamilton theorem (every matrix satisfies its own characteristic equation, *i.e.* $\phi(\mathbf{A}) = 0$), at $m = n$, we have

$$
\begin{aligned}
y^n(t) &= \mathbf{CA}^n x(t) + \mathbf{CA}^{n-1}\mathbf{B}u(t) + \cdots + \mathbf{CB}u^{(n-1)}(t) + \mathbf{D}u^{(n)}(t),\\
&= -a_{n-1}\mathbf{CA}^{n-1}x(t) - a_{n-2}\mathbf{CA}^{n-2}x(t) - \cdots - a_1\mathbf{CA}x(t)\\
&\quad -a_0\mathbf{C}x(t) + \mathbf{CA}^{n-1}\mathbf{B}u(t) + \cdots + \mathbf{CB}u^{(n-1)}(t) + \mathbf{D}u^{(n)}(t),\\
&= -a_{n-1}\begin{pmatrix} y^{(n-1)}(t) - \mathbf{CA}^{n-2}\mathbf{B}u(t) - \cdots \\ -\mathbf{CB}u^{(n-2)}(t) - \mathbf{D}u^{(n-1)}(t)\end{pmatrix}\\
&\quad -a_{n-2}\begin{pmatrix} y^{(n-2)}(t) - \mathbf{CA}^{n-3}\mathbf{B}u(t) - \cdots \\ -\mathbf{CB}u^{(n-3)}(t) - \mathbf{D}u^{(n-2)}(t)\end{pmatrix}\\
&\quad -\cdots\\
&\quad -a_0\left(y(t) - \mathbf{D}u(t)\right)\\
&\quad +\mathbf{CA}^{n-1}Bu(t) + \cdots + \mathbf{CB}u^{(n-1)}(t) + \mathbf{D}u^{(n)}(t);
\end{aligned}
\tag{5.37}
$$

or

$$
\Phi_{\mathrm{ARR}}\left(\left(y(t), u(t)\right), \left(\dot{y}(t), \dot{u}(t)\right), \cdots \left(y^{(n)}(t), u^{(n)}(t)\right)\right) = 0. \tag{5.38}
$$

Note that the above equation defines a set of constraints between inputs and outputs, which are, indeed, the ARRs.

For general multi-input-multi-output (MIMO) case, one may write

$$
Y_m(t) = \mathbf{O}_m x(t) + \mathbf{T}_m U_m(t), \tag{5.39}
$$

where

$$
\mathbf{O}_m = \left[\mathbf{C}^T \ \ (\mathbf{CA})^T \cdots (\mathbf{CA}^m)^T\right]^T,
$$

$$
\mathbf{T}_m = \begin{bmatrix}
\mathbf{D} & 0 & \cdots & \cdots & 0\\
\mathbf{CB} & \mathbf{D} & & \cdots & 0\\
\vdots & & \ddots & \ddots & \\
\vdots & & & \ddots & \ddots \\
\mathbf{CA}^{m-1}\mathbf{B} & \cdots & \cdots & \mathbf{CB} & \mathbf{D}
\end{bmatrix}^T
$$

is a lower triangular block Toepliz matrix, and $Y_m(t), U_m(t)$ are vectors obtained by stacking $y(t)$, $u(t)$ and their derivatives up to m-th order. Note that for $m = n-1$, $\mathbf{O}_n$ is the observability matrix. When m is sufficiently large, let there be a matrix $\mathbf{R}_m$, which is constructed by the row vectors of $\mathbf{O}_m$, such that $\mathbf{R}_m\mathbf{O}_m = 0$, *i.e.* $\mathbf{R}_m\left(Y_m(t) - \mathbf{T}_m U_m(t)\right) = 0$.When fault is introduced in the model,

$$Y_m(t) = \mathbf{O}_m x(t) + \mathbf{T}_m U_m(t) + \mathbf{T}_f F_m(t), \tag{5.40}$$

where $\mathbf{T}_f$ is another lower triangular block Toepliz matrix (obtained by substituting $\mathbf{E}_f$ and $\mathbf{G}_f$, respectively for $\mathbf{B}$ and $\mathbf{D}$ in $\mathbf{T}_m$) and $F_m(t)$ is a vector spanning $f(t)$ and its derivatives up to m-th order. Therefore, one obtains

$$\mathbf{R}_m\left(Y_m(t) - \mathbf{T}_m U_m(t)\right) = \mathbf{R}_m\mathbf{T}_f F_m(t). \tag{5.41}$$

The left side of the above equation is zero in the absence of faults and is indeed an ARR. The residuals generated using this analysis lead to what are called parity space residuals [92, 94, 195, 206], whereby the residuals are projected to different invariant subspaces [146] such that they become sensitive to only specific faults. The parity space approach can be extended to non-linear systems with polynomial type non-linearity.

5.2.4.1 Example

Consider the single-degree-of-freedom spring-mass-damper system shown in Figure 5.7a and its bond graph model shown in Figure 5.7b. The mass in the system is excited by a known force and the velocity of the mass-point is measured. The system is, naturally, observable.

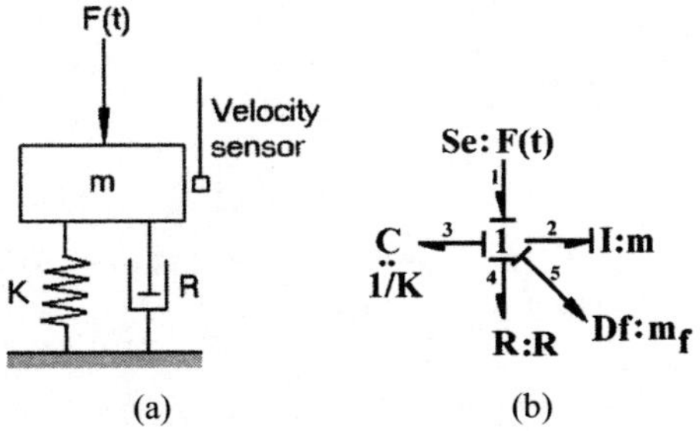

Fig. 5.7 A spring-mass-damper system and its bond graph model

From the bond graph model, the state space model for this LTI system is given by

$$\begin{Bmatrix} \dot{x}_1 \\ \dot{x}_2 \end{Bmatrix} = \begin{bmatrix} -R/m & -K \\ 1/m & 0 \end{bmatrix} \begin{Bmatrix} x_1 \\ x_2 \end{Bmatrix} + \begin{bmatrix} 1 \\ 0 \end{bmatrix} u(t),$$

$$\{y_1\} = [1/m \quad 0] \begin{Bmatrix} x_1 \\ x_2 \end{Bmatrix}, \tag{5.42}$$

where $x_1 = P_2$, $x_2 = Q_3$, $u(t) = F(t)$ and y_1 is the output from the velocity sensor. Taking successive derivatives of y_1,

$$\begin{aligned} \dot{y}_1 &= \dot{x}_1/m = -\left(R/m^2\right)x1 - (K/m)x_2 + u(t)/m \\ &= -(R/m)y_1 - (K/m)x_2 + u(t)/m, \\ \ddot{y}_1 &= -(R/m)\dot{y}_1 - (K/m)\dot{x}_2 + \dot{u}(t)/m \\ &= -(R/m)\dot{y}_1 - \left(K/m^2\right)x_1 + \dot{u}(t)/m \qquad (5.43) \\ &= -(R/m)\dot{y}_1 - (K/m)y_1 + \dot{u}(t)/m, \qquad (5.44) \end{aligned}$$

which leads to an ARR

$$m\ddot{y}_1 + R\dot{y}_1 + Ky_1 - \dot{u}(t) = 0. \tag{5.45}$$

If we consider one more sensor to measure the displacement, then the output equation is

$$\begin{Bmatrix} y_1 \\ y_2 \end{Bmatrix} = \begin{bmatrix} 1/m & 0 \\ 0 & 1 \end{bmatrix} \begin{Bmatrix} x_1 \\ x_2 \end{Bmatrix}, \tag{5.46}$$

from which we obtain

$$\begin{aligned} \dot{y}_1 &= \dot{x}_1/m = -\left(R/m^2\right)x1 - (K/m)x_2 + u(t)/m \\ &= -(R/m)y_1 - (K/m)y_2 + u(t)/m, \qquad (5.47) \\ \dot{y}_2 &= \dot{x}_2 = (1/m)x_1 = y_1. \qquad (5.48) \end{aligned}$$

This leads to two ARRs:

$$\begin{aligned} m\dot{y}_1 + Ry_1 + Ky_2 - u(t) &= 0, \\ y_1 - \dot{y}_2 &= 0. \end{aligned} \tag{5.49}$$

Note that two sensors lead to two ARRs. Furthermore, the maximum order of derivatives in the ARRs is one. This means that the maximum order of derivatives (m) chosen to construct matrices Y_m, R_m, *etc.* must be suitably chosen such that ARRs can be written using minimum orders of derivatives. Consider the two ARRs, among which the first one is sensitive to faults arising out of static bias in sensor readings y_1 and y_2, and then the second ARR, which is sensitive to a static bias in sensor reading y_1. If these ARR are written using higher order derivatives (derivatives of an ARR is an ARR, but not its integration), they will not be sensitive to these sensor errors. Moreover, higher order derivatives of noisy signals appearing in ARRs will cause more problems in implementing a suitable decision procedure.

5.3 Analytical Redundancy Relations and Fault Signature

Analytical Redundancy Relations (ARRs) represent various compatibility conditions or constraints between a set of known process variables [242]. The general form of an ARR is given by

$$f_1(K_1) = f_2(K_2), \tag{5.50}$$

where f_1 and f_2 are two functions relating K_1 and K_2, which are sets of known process variables. In a bond graph based approach, the known variables are the sources (Se and Sf), the modulated sources (MSe and MSf), the measurements from sensors (De and Df), the model parameters (θ) and the controller outputs (u). An ARR is then written as

$$ARR : f(De, Df, Se, Sf, MSe, MSf, u, \theta) = 0. \tag{5.51}$$

An ARR, which has been obtained from a physical law, represents some conservation phenomena, *e.g.* Bernoulli equation in hydraulic domain; Newton's law in mechanical domain; Kirchoff's law in electrical domain .... In inter-domain couplings (*e.g.* an electric motor), dynamics of one domain is usually mapped to the other and a combined conservation law is formulated. Note that ARRs may not always have some physical meaning; nevertheless they represent some constraints to be satisfied during normal operation of a system.

5.3.1 Residual and Decision Procedure

A residual, r, is evaluation of an ARR, *i.e.*

$$r = \text{Eval}[f_1(K_1) - f_2(K_2)]. \tag{5.52}$$

For the FDI procedure to work properly, as mentioned in [16], ARRs should be:

1. Robust, *i.e.* insensitive to unknown inputs and unknown parameters, so that they are satisfied when no fault is present
2. Sensitive to faults, so they are not satisfied when faults are present
3. Structured, so that in the presence of a given fault, only a subset of the ARRs is not satisfied, thus allowing recognition (from the subset of satisfied and the subset of not satisfied ARRs) of the fault, which occurred

Structurally independent ARRs are considered as basis ARRs. An ARR which can be algebraically constructed from the basis ARRs is not structurally independent. There are various ways of selecting basis ARRs whose number equals the number of sensors. Alternatively, it is said that each structurally independent residual has a unique (fault) signature.

The residuals (evaluation of the ARRs by using the actual sensor data and the process parameters) are used to detect the faults in the process. The next phase of isolating the faulty component(s), along with the signal treatment to decide which residuals have deviated from normal operation, is called the decision procedure.

The type of decision procedure used to supervise a system varies from case to case, *i.e.* it is usually system specific. The elements, $c_i\,(i=1\ldots n)$ of the binary coherence vector $C=[c_1,c_2,...,c_n]$ are determined from one or more decision procedures, Θ_i. These decision procedures generate the alarm conditions. Hence, $C=[\Theta_1(r_1),\Theta_2(r_2),\cdots,\Theta_n(r_n)]$. Robust decision procedures minimize misdetection and false alarms by treating the residual noises. A simple decision procedure (Θ) for all residuals, whereby each residual, r_i, is tested against a corresponding threshold, δ_i, fixed *a priori*, can be written as

$$c_i=\Theta(r_i)=\begin{cases}1,\text{if } |r_i|\rangle\,\delta_i;\\ 0,\text{ otherwise.}\end{cases} \tag{5.53}$$

Note that to account for modeling uncertainties, process and measurement noises, *etc.* these thresholds may be adaptive, *i.e.* functions of inputs, measurements and time [6, 7, 62, 64, 65, 69, 103, 213, 230, 252, 254]. The coherence vector is calculated at every step. A fault is detected, when $C\neq[0,0,\cdots,0]$, *i.e.* at least one element of the coherence vector is non-zero (alternatively, at least one residual exceeded its threshold). The isolation of the faulty component is done by matching the coherence vector to a binary fault signature matrix.

5.3.2 The Fault Signature Matrix

The Fault Signature Matrix (FSM), *S,* describes the structural sensitivity of each residual to various faults in physical devices, sensors, actuators and controllers [240, 275]. Thus, matrix **S** forms a structure that links the discrepancies in components to changes in the residuals. The elements of matrix **S** are determined from the following analysis:

$$S_{ji}=\begin{cases}1,\text{if the } i\text{-th residual is sensitive to faults in the } j\text{-th component;}\\ 0,\text{ otherwise.}\end{cases} \tag{5.54}$$

Note that while constructing the elements of S_{ji}, where $r_i=f_i(K_i)$, component j may involve one or more sub-components.

A component's (variable) fault is monitorable (in other words its monitorability index $\mathrm{M}_b=1$) if at least one residual is sensitive to it. The alarm interpretation or fault isolation is done using the binary FSM. A fault in a component/variable can be isolated (or its isolability index $I_b = 1$) only when it is monitorable and its fault signature is different from fault signatures of all other variables/components. After fault detection, the coherence vector is matched with the fault signatures in **S** and the component, which has a unique match, is isolated as the fault candidate.

Each residual is sensitive to some faults and insensitive to some others. It is fundamental that different residuals become sensitive to different faults. When residuals are so designed that each residual is sensitive to only one fault and no other residual is sensitive to that same fault, then the resulting set of residuals are called structured residuals. In practice, it is not always possible to obtain structured residuals, especially when number of faults modeled is more than the number of sensors, or sensors are not placed appropriately. If the set of faults affecting a particular residual is different from the set of faults affecting any other residual, then the former residual is termed structurally independent of the others. Theoretically, the number of the structurally independent residuals that can be obtained from the model of a system is equal to the number of sensors in the model.

The FDI designer can decide that some equipment are more robust as compared to others and leave those robust components out of the FDI analysis. For example, if the atmospheric conditions (pressure and temperature, say) are not going to affect the dynamics of the system significantly, then its parameters need not be included in the fault signature matrix. Similarly, if a controller is implemented in a digital computer and has been tested sufficiently to be free from programming errors, then it may be considered robust. These technical specifications enhance FDI ability.

A component's fault need not be explicitly included in models used for ARR generation. Faults in any physical component can be mapped to the undesirable changes in the values of the parameters of the component. As an example, a leakage from a liquid storing tank can be mapped to a hypothetical increase in the cross-sectional area that brings the liquid level down; a blockage in a pipe can be mapped to a change in its coefficient of discharge, *etc.* Thus, a residual is sensitive to faults in a component, when the parameters or the measurements belonging to that component appear in the corresponding ARR. The elements of the fault signature matrix can as well be constructed through experimentation by introducing various faults, one at a time, in the real plant.

Furthermore, it is important to distinguish between the two forms of the fault signature matrix: a theoretical one, which is derived from structural analysis as given in Equation 5.54, and a practical one, which is used in the real-time supervision platforms. The practical fault signature matrix takes into account the actual range of values of the parameters and the measurements. Obviously, the relative magnitudes of different influences (parts of the ARR) play an important part. This is expected because a residual may be sensitive to several faults, but quantitative sensitivities to those faults are usually different. In general, the practical fault signature matrix is a refined form of the theoretical fault signature matrix (unless some dynamics is neglected in the modeling stage), taking the actual numerical values of the dimensions, the environment, the operating setpoints, and the resulting relative sensitivities into consideration. This matrix is easily constructed from the off-line data obtained through the experiments by introducing various faults in the process, and also through simulation studies by introducing additional faults, which cannot be introduced manually in the real process.

Obtaining ARRs in closed symbolic form is difficult because the elimination of unknown variables from the model is not a trivial task [63, 141, 228]. Most often, compatibility conditions or ARRs relate to well-known physical laws. However, ARRs need not always represent physical laws and even if they do so, it is not easy to write them down directly from a mathematical model of the system. This is why many structured approaches [141] (by using digraph, bipartite graph, bond graph *etc.*) to ARR derivation have been developed. Some of the relevant ARR derivation approaches are discussed in the following section.

5.4 Structured Approach to ARR Derivation

A bipartite graph [41] provides a structured approach for derivation of ARRs. The structure of a system can be represented by a bipartite graph [16, 242] with two sets of vertices: the constraints and the variables, and edges between the constraints and the variables. A bipartite graph is a non-oriented graph, in which each constraint vertex is connected to some variables and parameters used in the constraint equation, *e.g.* differential equations and measurement equations. In the following, we consider a pedagogical example to show bipartite graph representation and ARR derivation procedures.

5.4.1 Behavior Model

A controlled two-tank system, shown in Figure 5.8, is considered. It consists of two tanks, T_1 and T_2, connected by a pipe with a valve, V_1. Two level sensors, L_1 and L_2, are installed in the tanks, T_1 and T_2, respectively. The fluid level in tank T_1 is controlled by a PI-level controller, which acts on a pump to maintain the water level in tank T_1 (*i.e.* L_1) at some constant predefined setpoint, S_{Pt}.

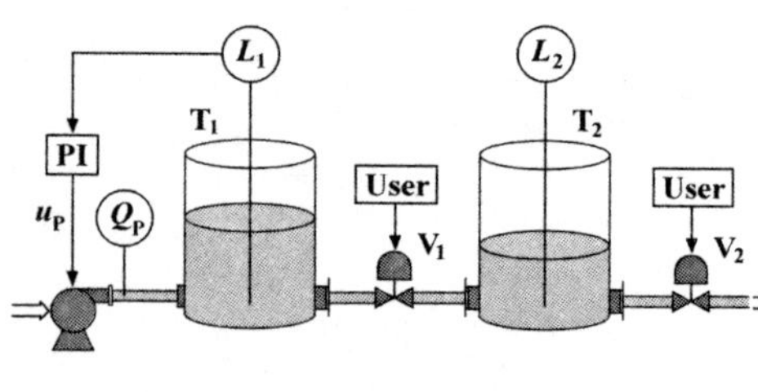

Fig. 5.8 Process and instrument diagram (P&ID) of a two-tank system

The outputs from the PI controller u_P and the pump flow (Q_P), given in Equation 5.55 and Equation 5.56, respectively, are measured:

$$u_P = K_p \left(S_{Pt} - \rho.g.L_1(t)\right) - K_I \int \left(S_{Pt} - \rho.g.L_1(t)\right) dt = \Phi_{PI}\left(L_1(t)\right), \tag{5.55}$$

$$Q_P = \begin{cases} u_P & \text{if } 0 \le u_P \le f_{\max} \\ 0 & \text{if } u_P \le 0 \\ f_{\max} & \text{if } u_P > f_{\max} \end{cases} = \Phi_P(u_P), \tag{5.56}$$

where K_p and K_I are, respectively, the proportional and integral gains, and S_{Pt} is a predefined setpoint for the PI controller. The pump function is represented by Φ_P and the PI controller function is defined by Φ_{PI}.

The quantity of water outflow to the consumer, Q_O, is manually controlled by a valve, V_2. The purpose of this system is to provide a continuous fluid flow to the consumer. The atmospheric pressure is taken as reference, *i.e.* zero.

The mass flow rate, $\dot{m}$, through a valve is usually given by the non-linear relation

$$\dot{m} = C_d(x)\sqrt{\Delta P}.sign(\Delta P) \tag{5.57}$$

where ΔP is the pressure difference across the valve, x is the valve stem position (between 0 to 1, where 0 means fully closed state and 1 means fully open state), C_d is a function representing the coefficient of discharge depending on the features of the valve, (linear, equal percentage, quick opening, square root, *etc.*) and 'sign' is a function used to adjust the direction of flow. Variations of flow with respect to the stem position for different features of valves are shown in Figure 5.9. If the stem positions of the valves are considered to be fixed, then their corresponding discharge coefficients can be taken as constants.

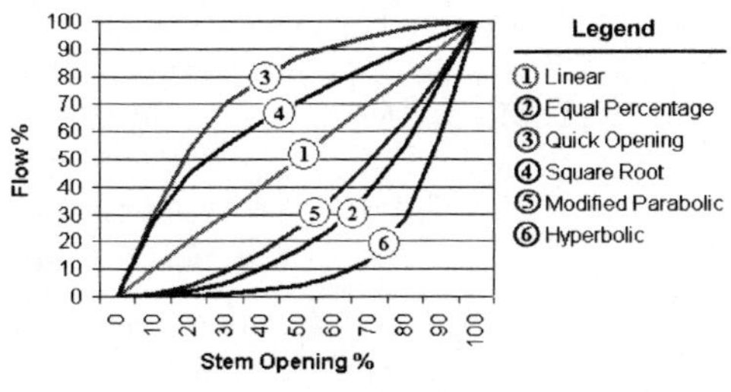

Fig. 5.9 Control valve flow characteristics

The volume flow rate Q through an orifice during incompressible steady state flow is also considered to be proportional to the square root of the pressure drop (ΔP) across the orifice [22, 23, 258]:

$$Q = C_d.A.\sqrt{\frac{2}{\rho}|\Delta P|}.\text{sign}(\Delta P), \tag{5.58}$$

where A is the area of the orifice. The expression for laminar flow (Q) of an incompressible viscous fluid flowing under steady state conditions through a horizontal pipe or tube can be obtained from the Hagen-Poiseuille equation [278]:

$$Q = \frac{\pi d^4}{128.\mu.L}.\Delta P = C_d.\Delta P, \tag{5.59}$$

where μ is the viscosity of the fluid, and L is the length of the pipe. So, in this case, the flow Q is directly proportional to the pressure drop ΔP across the length of the pipe. These formulae are usually modified by using Reynolds number dependent friction factors to account for flow separation and turbulence. The coefficient of discharge C_d itself depends on the Reynolds number. It has been shown experimentally that the value of C_d varies sharply at low Reynolds number (*i.e.* for laminar flow) and thereafter, with the increase in the Reynolds number, it converges to a steady value asymptotically, which is known as Von Mises asymptote [173].

In the case of a controlled valve, the flow through the valve is modulated by the controller output (u). The effective coefficient of discharge, $\hat{C}_d$, through a control valve is given by [258]

$$x = \begin{cases} 0 \text{ for } u < 0, \\ u \text{ for } 0 \leq u \leq 1, \\ 1 \text{ for } u > 1; \end{cases} \tag{5.60}$$

$$\phi_e(x) = R_n.e^{x-1},$$
$$\hat{C}_d = C_d.\phi_e(x),$$

where x is the valve stem position, ϕ_e is the equal percentage function (the valve feature considered for this case) and R_n is the rangeability factor.

Considering compressible fluid flow (*e.g.* for saturated steam), the flow characteristics for the turbulent and laminar flow through a valve is governed by the following equation [258]:

$$\dot{m} = \begin{cases} \frac{1}{2}\hat{C}_d.P_u, \text{for } P_d \leq \frac{1}{2}P_u \,(\text{turbulent flow}); \\ \hat{C}_d.\sqrt{P_d}.\sqrt{P_u - P_d}, \text{for} P_d \geq \frac{1}{2}P_u \,(\text{laminar flow}) \end{cases} \tag{5.61}$$

where P_u and P_d are the upstream and downstream pressures, respectively, and $\dot{m}$ is the mass flow rate of the fluid flowing through the valve. This relation (Equation 5.61) is continuously differentiable, because at $P_d = \frac{1}{2}.P_u$, both turbulent as well as the laminar considerations give $\dot{m} = \frac{1}{2}.\hat{C}_d.P_u$.

Likewise, various symbolic and empirical relations for superheated steam, gas *etc.* are available in the literature. In bond graphs, valves, pipes and orifices can be modeled by an R-element [188], whose constitutive relation is determined by the flow condition as well as the device characteristics. Note that the discharge coefficient depends upon various factors such as the inside roughness, Reynolds number, type of valve and pipe fittings *etc.* and so it is usually determined from experimental study.

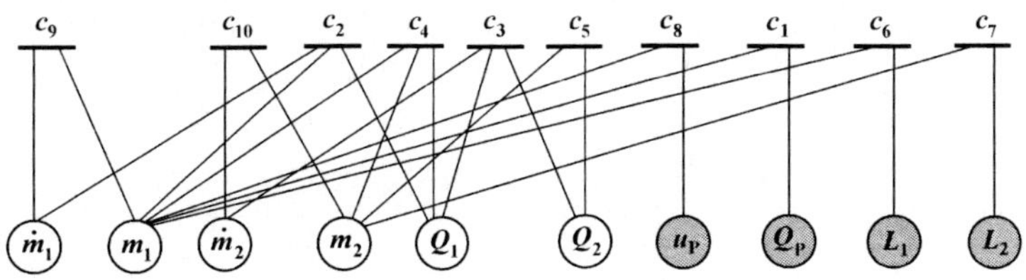

Fig. 5.10 Constraints and process variables linked in a bipartite graph

5.4.2 Constraints and Variables

The behavioral equations, *i.e.* various constraints, for the two-tank system are given in Table 5.1. The effective discharge coefficients for the valves V_i are denoted as C_{di}, where $i = 1,2$.

Table 5.1 Constraints for different components of the two-tank process

Component	Constraint	Model (equations)
Pump	c_1	$Q_P(t) = \Phi_P(u_P(t)) = \Psi_P(m_1(t))$
Tank T_1	c_2	$\dot{m}_1(t) = Q_P(t) - Q_1(t)$ $= \Psi_P(m_1(t)) - Q_1(t)$
Tank T_2	c_3	$\dot{m}_2(t) = Q_1(t) - Q_2(t)$
Valve V_1	c_4	$Q_1(t) = C_{d1} \cdot \sqrt{\left(\frac{m_1(t)}{A_1} - \frac{m_2(t)}{A_2}\right) \cdot g}$
Valve V_2	c_5	$Q_2(t) = C_{d2} \cdot \sqrt{\frac{m_2(t).g}{A_2}}$
Sensor L_1	c_6	$L_1(t) = \frac{m_1(t)}{\rho.A_1}$
Sensor L_2	c_7	$L_2(t) = \frac{m_2(t)}{\rho.A_2}$
Controller	c_8	$u_P(t) = \Phi_{PI}(L_1(t)) = \Psi_{PI}(m_1(t))$

Derivatives with respect to the independent variable t are added explicitly as two constraints:

$$c_9 : \dot{m}_1(t) = \frac{d}{dt}(m_1(t)),$$

$$c_{10} : \dot{m}_2(t) = \frac{d}{dt}(m_2(t)).$$

The set of constraints $c = [c_1, c_2, c_3, c_4, c_5, c_7, c_8, c_9, c_{10}]$ links variables $Z = [m_1(t), \dot{m}_1(t), m_2(t), \dot{m}_2(t), Q_1(t), Q_2(t), u_P(t), Q_P(t), L_1(t), L_2(t)]$ in a bipartite graph as shown in Figure 5.10.

The structural analysis, *i.e.* the analysis of the constraints, is performed through matching on a bipartite graph [16, 240]. A matching is a causal assignment, which associates some system variables with the constraints from which they can be calculated; *i.e.* it converts a bipartite graph to a directed graph. Variables which cannot be matched cannot be calculated. Variables, which can be matched in several ways, can be calculated by different (redundant) means, thus providing a means for system reconfiguration. Any finite-dimensional graph can be decomposed into three sub-graphs with specific properties [41, 66]. This decomposition is canonical, *i.e.* for a given system, it is unique. These sub-graphs are: an over-constrained subsystem which means that the variables (let n be their number) have to satisfy more than n constraints; a just-constrained subsystem, because it introduces as many new variables as constraints; and an under-constrained subsystem which introduces more new variables than constraints [66].

5.4.3 Derivation of ARRs

ARRs are obtained from the over-constrained subsystem in two different ways: direct redundancy and deduced redundancy. Direct redundancy is any constraint, which involves only known variables; whereas, deduced redundancy is any constraint, which contains both known and unknown variables. If the unknown variables are observable then they can be calculated from the known ones and putting the results in the constraint leads to an ARR.

To eliminate unknown variables, we first construct an incidence matrix [16, 240] shown in Table 5.2, which depicts involvement of different variables in different constraints. In Table 5.2, entries marked '1' in the matrix indicate that the variable in corresponding column is involved in the constraint in the corresponding row; whereas, items marked '×' mean the same thing with the additional information that the corresponding variable cannot be calculated from the corresponding constraint. For example, by using constraint c_9, one may calculate $\dot{m}_1(t)$ from $m_1(t)$, but not *vice versa*, because it is assumed that initial condition $m_1(0)$ is unknown. Similarly, from constraint c_8, one cannot calculate the mass stored in the tank T_1 from known variable u_P, because the PI controller function is not invertible. Note that controllers also involve integration of variables, *e.g.* error integral in the PI controller, but if we assume that all controllers are implemented in computer software then the corresponding initial conditions are known variables.

In the next step, causality is assigned to eliminate the unknown variables. The outer vertices of the structure are the known variables: $u_P(t)$, $Q_P(t)$, $L_1(t)$ and $L_2(t)$. A causal matching starting from $L_1(t)$ and $L_2(t)$ is shown in Table 5.3 and its corresponding graphical representation is shown in Figure 5.11.

The steps in causal assignment propagation are numbered sequentially in Table 5.3. In the first (and third) column of Table 5.3 the mark '✓' means by using c_6 (and c_7), $m_1(t)$ (and $m_2(t)$) can be calculated from $L_1(t)$ (and $L_2(t)$) in step 1. Then the expressions for $m_1(t)$ (and $m_2(t)$) are used to calculate $\dot{m}_1(t)$ (and $\dot{m}_2(t)$) in step 2. Finally, these are used with constraints c_4 and c_5 to calculate $Q_1(t)$ and $Q_2(t)$, respectively, in step 3. Note that each unknown variable is determined only once

Table 5.2 Incidence matrix of the two-tank process

	m_1	$\dot{m}_1$	m_2	$\dot{m}_2$	Q_1	Q_2	u_P	Q_P	L_1	L_2
c_1	×							1		
c_2	1	1			1					
c_3				1	1	1				
c_4	1		1		1					
c_5			1			1				
c_6	1								1	
c_7			1							1
c_8	×						1			
c_9	×	1								
c_{10}			×	1						

Table 5.3 Causal matching for elimination of unknown variables

	$m_1(t)$	$\dot{m}_1(t)$	$m_2(t)$	$\dot{m}_2(t)$	$Q_1(t)$	$Q_2(t)$	$u_P(t)$	$Q_P(t)$	$L_1(t)$	$L_2(t)$
c_1	×							1		
c_2	1	1			1					
c_3				1	1	1				
c_4	1		1		✓ (3)					
c_5			1			✓ (3)				
c_6	✓ (1) ←	‐ ‐ ‐	‐ ‐ ‐	‐ ‐ ‐	‐ ‐ ‐	‐ ‐ ‐	‐ ‐ ‐	‐ ‐ ‐	1	
c_7			✓ (1) ←	‐ ‐ ‐	‐ ‐ ‐	‐ ‐ ‐	‐ ‐ ‐	‐ ‐ ‐	‐ ‐ ‐	1
c_8	×						1			
c_9	×	✓ (2)								
c_{10}			×	✓ (2)						

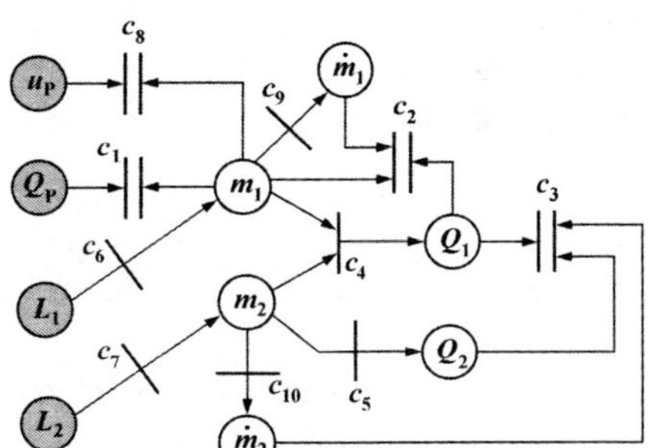

Fig. 5.11 Graphical representation of causal matching

(only one '✓' in each column corresponding to unknown variables). Finally, there are some constraints, namely c_1, c_2, c_3 and c_8, which have all variables determined (characterized by no '✓' in those rows). This means that there are more constraints than required to calculate the unknown variables. The four unused constraints correspond to the over-constrained subspace and they lead to formation of four ARRs. The ARRs are the constraint relations themselves in which the involved variables are already determined from other constraints. The four ARRs of the considered example system are

$$
\begin{aligned}
&\mathrm{ARR}_1\,(c_1) : Q_P\,(t) - \Phi_P\,(u_P\,(t)) = 0,\\
&\mathrm{ARR}_2\,(c_8) : u_P\,(t) - \Phi_{PI}\,(L_1\,(t)) = 0,\\
&\mathrm{ARR}_3\,(c_2) : Q_P\,(t) - \rho.A_1.\frac{d}{dt}\,(L_1\,(t)) - C_{d1}.\sqrt{\rho.g.\,(L_1\,(t) - L_2\,(t))} = 0,\\
&\mathrm{ARR}_4\,(c_3) : C_{d1}.\sqrt{\rho.g.\,(L_1\,(t) - L_2\,(t))} - \rho.A_2.\frac{d}{dt}\,(L_2\,(t)) -\\
&\qquad C_{d2}.\sqrt{\rho.g.L_2\,(t)} = 0.
\end{aligned}
\tag{5.62}
$$

Note that corresponding to four measured variables, there are four ARRs. Further, note that ARR_1 and ARR_2 which do not involve parameters of the system, *i.e.* A_1, A_2, C_{d1} and C_{d2} (ρ and g are considered exogenous or medium parameters, which are never faulty), correspond to functional redundancies, *i.e.* outputs of two sensors can be obtained directly from other sensors.

One may assign different ways of matching on the incidence matrix to eliminate the unknown variables. These lead to use of different constraints to calculate all unknown variables and thus result in a different set of unused constraints (four in this example). Obviously, the form of the ARRs change. However, ARRs obtained from different ways of matching can be always brought to another form. Note that linear combinations of ARRs are also ARRs.

5.5 ARR Generation from Bond Graph Models

A system model, M, may be described by a set of constraints, F (which represents the algebraic constraints and constitutive relations of elements in the model), a set of variables, Z and a set of parameters θ. Each variable may be known, or unknown:

$$
M = M(F, Z, \theta). \tag{5.63}
$$

5.5.1 Constraints and Variables

5.5.1.1 Constraints

The constraints, F, can be seen as the set of relations which link the system variables and the parameters. It has to include information about the structure (F_J), the behavior (F_B), the measurement (F_Y), the control system (F_C) and the controlled sources (F_A):

$$
F = \{\, F_J,\, F_B,\, F_Y,\, F_C,\, F_A \,\}.
$$

- *Structural equations* F_J: they represent a set of conservation laws (of mass, energy, *etc.*) and/or equilibrium equations, *i.e.* the weak and the strong laws of the junctions. They are deduced from the junction equations:

$$F_J = \{F_{J0}\} \cup \{F_{J1}\} \cup \{F_{TF}\} \cup \{F_{GY}\},$$

where $F_J \in \mathbb{R}^{n_j}$ and the number of structural equations, n_j, is equal to the sum of the number of bonds connected to each junction (0-junctions and 1-junctions) and twice the number of 2-port elements (Transformers and Gyrators, *i.e.* TF and GY elements).

- *Behavioral equations* F_B: The physical laws expressing how the energy is transformed are mathematically described by the behavior model. In a bond graph model, they describe the physical phenomena which are represented in lumped-parameter bond graph elements (I, C and R). These equations are called "constitutive laws":

$$F_B = \{F_C\} \cup \{F_I\} \cup \{F_R\}, \tag{5.64}$$

where $F_B \in \mathbb{R}^{n_e}$ and n_e is the total number of power bonds in the bond graph connected to I, C and R elements. These equations are schematically represented as shown in Figure 5.12, which is often referred to as the Paynter's tetrahedron [127] or the Thoma's carousal [256].

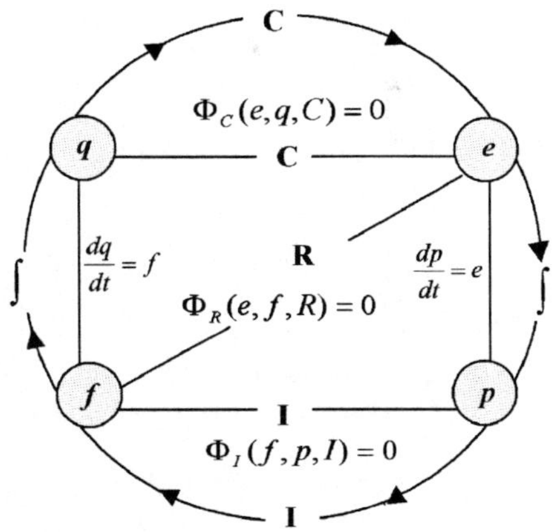

Fig. 5.12 Behavioral constraints in bond graph

- *Measurement equations* F_Y: they describe the measurements which are available, *i.e.* the way in which the sensors transform some state variables of the process into output signals which can be used for FDI and control purposes. In bond graph models, the sensors are represented as detectors of flow (Df) and detectors of effort (De).

$$F_Y = \{F_{De}\} \cup \{F_{Df}\},$$

where $F_Y \in \mathbb{R}^{n_s}$ and n_s is the number of sensors.

- *Control algorithms* F_C: they describe the controller algorithms. The controller inputs are the setpoints and the sensor outputs. A controller output acts on the actuator which is represented as a modulated effort (MSe) or flow (MSf) source:

$$F_C(u, u_{ref}, Y_m, \theta_{reg}) = 0, \tag{5.65}$$

where θ_{reg} is a set of controller parameters (gain, sampling time, *etc.*); u, u_{ref} and Y_m represent, respectively, the controller output, the setpoint value, and the sensor outputs. $F_C \in \mathbb{R}^{n_r}$ and n_r is the number of regulators.
Unlike structural and behavioral equations, which use the power variables (efforts and flows) as input-output variables; measurement and control algorithms use information signals as variables.

- *Source modulation equations* F_A: they describe modulated power sources (controlled pump, controlled valve, *etc.*). The input signals, u, are provided by the controllers, and the outputs are the regulated variables, MSe and MSf:

$$F_{A1}(MSf, u) = 0, F_{A2}(MSe, u) = 0,$$

where $F_A \in \mathbb{R}^{n_a}$ and n_a is the number of the modulated sources.
Thus,

$$F \in \mathbb{R}^{2n_j + n_e + n_s + n_a + n_r}.$$

5.5.1.2 Variables

The set of constraints, F, map to a set of variables, Z: known (K) and unknown (X):

$$Z = X \cup K. \tag{5.66}$$

The unknown variables are the power variables (flow and effort) that label the bonds. The vector X containing all the power variables is

$$X(t) = \{e_1(t), f_1(t)\} \cup \{e_2(t), f_2(t)\} \ldots \cup \{e_{nc}(t), f_{nc}(t)\}, \tag{5.67}$$

where n_c is twice the number of power bonds plus the number of signal bonds in the model.

The set, K, of known variables contains the control variables, u; the variables whose values are measured by the sensors, Y_m, and the supervision parameters (such as u_{ref}):

$$K = MSe \cup MSf \cup Se \cup Sf \cup Y_m. \tag{5.68}$$

MSe and MSf represent respectively a modulated effort source variable and a modulated flow source variable (controlled pump, controlled heater, *etc.*). They depend on the output, u, of the control system:

$$MSe = F_{C1}(u), MSf = F_{C2}(u). \tag{5.69}$$

Output signal, u, depends on the control algorithm function, F_C; the setpoint value; and the sensor outputs. We suppose that u, which is calculated by the control algorithm, Fc (Equation 5.65), is known.

Se and Sf, represent respectively a constant effort source and a constant flow source (weight, cell voltage, *etc.*).

Y_m is the sensor (De,Df) output vector of dimension n_s, $Y_m \in \mathbb{R}^{n_s}$.

Finally, $K \in \mathbb{R}^l$, where $l = n_S + n_a$ and n_a is the number of sources (including modulated and simple sources).

5.5.1.3 Parameters

$\theta \in \mathbb{R}^p$ is the vector of parameters. In a bond graph model, it is associated with the characteristics of I, C and R elements, *i.e.* the discharge coefficient, the hydraulic capacity, *etc.* and the moduli of two port elements.

A regular bond graph used for simulation is a just constrained system. Consider a junction having n number of power bonds, which are connected to different elements. Because there are n bonds, total number of power variables is $2n$. The junction has one strong bond and we have $n-1$ number of equality constraints (*i.e.* efforts are same or flows are same depending on the junction type) and one sum equation; thereby totaling to n number of equations. Furthermore, the n number of bonds connected to n number of external elements give n number of constitutive laws. Therefore, we always have $2n$ number of power variables to solve from $2n$ number of equations. That is why the model is just constrained and in principle solvable. While creating models for simulation, power variables in sensor elements are treated as unknowns. On the other hand, if we start treating them as known variables, such as in FDI, then the system model becomes over constrained and we can derive ARRs.

The general schematics of bond graph model based supervision is shown in Figure 5.13.

5.5.2 Algorithm for Generation of ARRs

For ARRs generation, the bond graph model of the monitored process should be generated in preferred derivative causality. Note that integral causality is recommended for engineering simulation in order to avoid the numerical problems arising out of differentiation. However, the derivative causality is more suitable in ARR

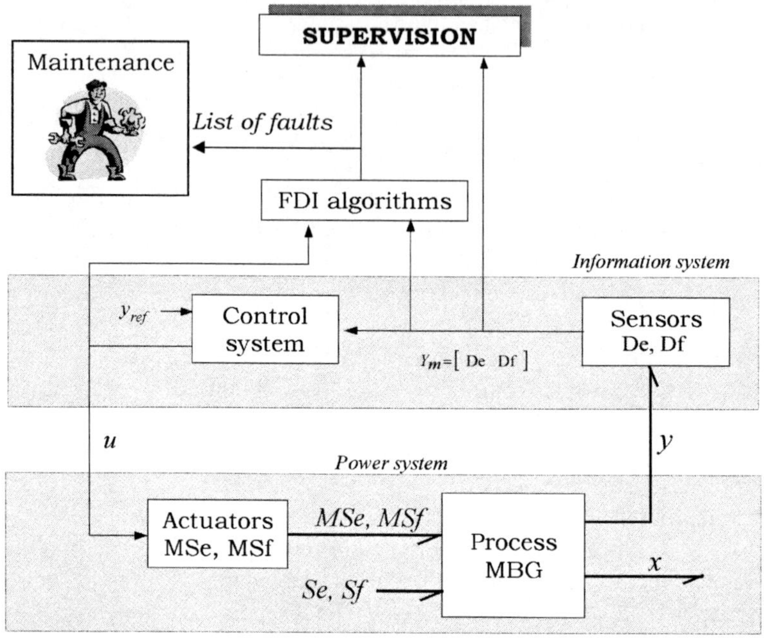

Fig. 5.13 Bond graph model based supervision

expression to avoid influence of the initial conditions. Otherwise, the number of unknown variables increases and it may not be possible to obtain ARRs.

From the bond graph modeling viewpoint, the ARR will be written in the following form:

$$f\left(\begin{array}{c}(De, Df, Se, Sf, MSe, MSf, \theta_m),\\ \left(\frac{dDe}{dt}, \frac{dDf}{dt}, \frac{dSe}{dt}, \frac{dSf}{dt}, \frac{dMSe}{dt}, \frac{dMSf}{dt}, \frac{d\theta_m}{dt}\right), \ldots\end{array}\right) = 0. \tag{5.70}$$

The ARR generation algorithm is a recursive elimination technique [42, 165, 251], which works as follows:

1. *Choose a junction.*
2. *Find the corresponding ARR by writing the structural equation, F_J, of the given junction.* The main idea is to associate two indicators, N_k and N_u, with each constraint relation (constitutive equation). N_k is the number of known variables and N_u is the number of unknown variables:

$$N_k + N_u = n_{var}, \tag{5.71}$$

where n_{var} is the total number of variables. The goal is to decrease N_u and increase N_k so that the ARR can be obtained when N_u becomes zero, or alternately stated, when $n_{var} = N_k$.The unknown variables in symbolic format are directly deduced from the bond graph model by using the covering causal path rules.

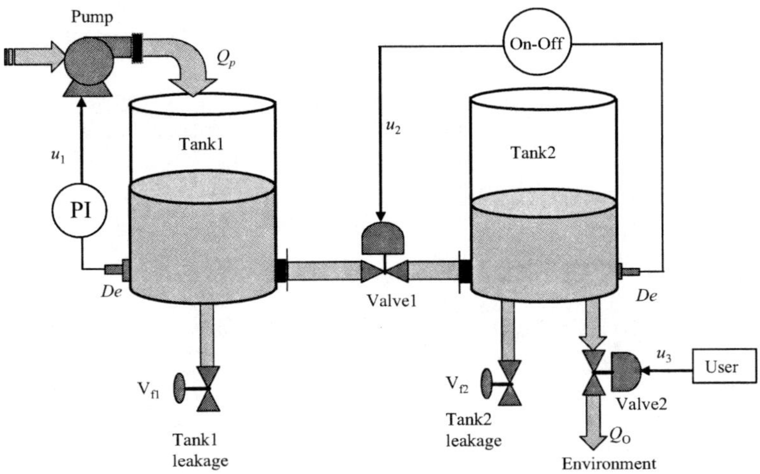

Fig. 5.14 A controlled two-tank system

Recall that a causal path is a sequence of links which have the same causal orientation. Depending on the causality, the followed variable is either the effort or the flow. When one passes through a GY-element or returns through a passive element (I,C, or R) then the followed variable changes.

3. *Consider the next junction and derive the corresponding ARR through step 2.*
4. *If the derived ARR is independent of the others derived previously, then keep it, else consider another junction.*
5. *Repeat the procedure until all the junctions are considered and all possible ARRs with distinctly separate signatures are obtained.*

5.5.3 Example

Consider a controlled two tank system shown in Figure 5.14. This system is similar to the one considered in the previous section (see Figure 5.8) except that valve-1 is controlled by an On-Off controller and two pressure sensors are used instead of two level sensors in the tanks.

The pseudo-bond graph model of the system is given in Figure 5.15, where pressure and mass flow rate are considered as the power variables. The PI controller is supposed to maintain a constant level in the tank. Its model is implemented in the signal domain with a level setpoint h_c. The On-Off controller tries to keep the level in Tank2 within a specific range. The lower and upper bounds of this range are setpoints which are directly implemented in the control law.

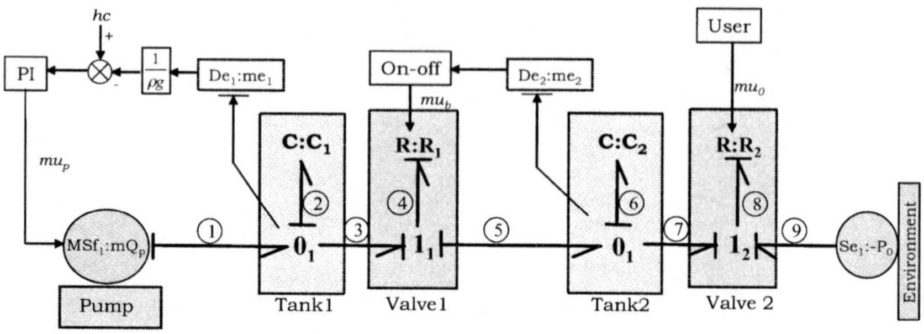

Fig. 5.15 Bond graph model of the controlled two-tank process

5.5.3.1 Used Variables

The vector containing all the variables associated with the bond graph model given in Figure 5.15 is $Z = X \cup K$. We have nine power bonds. Thus, the unknown variables vector X is of dimension 18, *i.e.*

$$X \in \mathbb{R}^{18} : X = \left[e_1 \; f_1 \; e_2 \; f_2 \; \dots \dots \; e_9 \; f_9 \right]^T . \tag{5.72}$$

There are two sources, two sensors, and three controllers. Then the subset K of known variables is of dimension 7, *i.e.*

$$K \in \mathbb{R}^7 : K = \left[mQ_p \; P_0 \; me_1 \; me_2 \; mu_p \; mu_b \; mu_0 \right]^T . \tag{5.73}$$

The parameter vector θ is composed of discharge coefficients (R_1 and R_2 elements) and hydraulic capacities ($C_1$and C_2), *i.e.*

$$\theta = \left[C_{d1} \; C_{d2} \; C_1 \; C_2 \right]^T . \tag{5.74}$$

5.5.3.2 Constraints

The bond graph model of the process contains four junctions. Each of the two 0-junctions (junctions 0_1 and 0_2) give rise to four structural equations, whereas each of the two 1-junctions contribute three structural equations. Therefore, we have 14 structural equations ($n_j = 14$), four bond graph elements ($n_e = 4$), two sensors ($n_s = 2$), two controllers (PI, and On-Off) ($n_c = 2$) and one controlled source/actuator ($n_a = 1$). Thus, 23 constraint relations are available and only 18 variables are unknown, implying that the system is over determined and we should be able to derive 5 ARRs.

- *Structural equations* Φ_J. From junction equations, we have

$$\begin{aligned} &\Phi_1 : mQ_p - f_4 - f_2 = 0,\ \Phi_2 : e_2 - e_4 - e_6 = 0, \\ &\Phi_3 : -f_6 - f_8 + f_4 = 0,\ \ \Phi_4 : -e_8 + e_6 - P_0 = 0. \end{aligned} \tag{5.75}$$

Note that these relations are already written in terms of strong bond variables, *e.g.* instead of writing $\Phi_1 : MQ_p - f_3 - f_2 = 0$, we have substituted f_3 by f_4 in the same step.

- *Behavioral equations* Φ_b. The constitutive equations for C_1, R_1, C_2 and R_2 elements are respectively:

$$\begin{aligned} &\Phi_5 : \begin{cases} s.e_2 = \frac{1}{C_1} f_2, \text{ if } e_2 < \rho g Z_{1\max} \\ s.e_2 = 0, \text{ if } e_2 = \rho g Z_{1\max} \\ e_2 \in [0, \rho g Z_{1\max}] \end{cases} ;\ \Phi_6 : f_4 = C_{d1}.\mathrm{sgn}(e_4)\sqrt{|e_4|}.mu_b; \\ &\Phi_7 : \begin{cases} s.e_6 = \dfrac{1}{C_2} f_6, \text{ if } e_6 < \rho g Z_{2\max} \\ s.e_6 = 0, \text{if } e_6 = \rho g Z_{2\max} \\ e_6 \in [0, \rho g Z_{2\max}] \end{cases} ;\ \Phi_8 : f_8 = C_{d2}.\mathrm{sgn}(e_8)\sqrt{|e_8|}mu_0; \end{aligned} \tag{5.76}$$

where s stands for the derivative operator. Furthermore, the maximum allowable levels in the two tanks (*i.e.* overflow conditions) are considered in the constitutive relations.

- *Measurement equations* Φ_m:

$$\begin{aligned} &\Phi_9 : me_1 = \begin{cases} e_2, \text{ if } e_2 < \rho g Z_{1\max}, \\ \rho g Z_{1\max}, \text{ if } e_2 = \rho g Z_{1\max}. \end{cases} \\ &\Phi_{10} : me_2 = \begin{cases} e_6, \text{ if } e_6 < \rho g Z_{2\max}, \\ \rho g Z_{2\max}, \text{ if } e_6 = \rho g Z_{2\max}. \end{cases} \end{aligned} \tag{5.77}$$

- *Control algorithms* Φ_c. The water level in Tank1 is controlled by a PI level controller. The water level in Tank2 has a switching controller with hysteresis (On_Off controller):

$$\begin{aligned} &\Phi_{11} : mu_p = K_p(h_c - \tfrac{1}{\rho g} me_1) + K_i \int (h_c - \tfrac{1}{\rho g} me_1)dt, \\ &\Phi_{12} : mu_b = On\_Off(me_2, L_{State}, L_{Min}, L_{Max}), \end{aligned} \tag{5.78}$$

where h_c, K_p and K_i are respectively the set-point, the proportional and integral gains of PI controller; and me_2, L_{State}, L_{Min} and L_{Max} are respectively the pressure sensor indication, the hysteretic state, minimum threshold, and maximum threshold.

- *Source modulation equations* Φ_a. For simplicity, we assume that the flow provided by the pump (mQ_p) is proportional to the controller output mu_p. Taking into account that the flow from the pump is limited to Q_{Min} to Q_{Max}, MQp can be written as

$$\Phi_{13}{:}mQ_p = \begin{cases} mu_p, \text{ if } Q_{Min} < mu_p < Q_{Max}; \\ Q_{Min}, \text{ if } mu_p \le Q_{Min}; \\ Q_{Max}, \text{ if } mu_p \ge Q_{Max}; \end{cases}$$
$$= MinMax(mu_p, Q_{Min}, Q_{Max}). \quad (5.79)$$

5.5.3.3 Residual Generation

The ARRs are obtained using the algorithm given earlier. We can start from any arbitrarily selected junction. Here let us start from junction 0_1.

1. **Junction $\mathbf{0_1}$**. Considering the first structural equation (constraint Φ_1 in Equation 5.75), two unknown variables in it are f_4 and f_2. The goal is to decrease the number of unknown variables and to increase the number of known variables. The ARR is then generated when all unknown variables are determined. The unknown variables in symbolic format are directly deduced from the bond graph model by utilizing the covering causal path rules.

- f_2 is calculated by the generic constitutive equation of storage tank modeled by C element (constraint Φ_5). The effort e_2 is measured and determined using measurement equation, Φ_9 ($e_2 = me_1$). We finally obtain $f_2 = C_1.s.me_1$.
- f_4 is given by the constitutive equation for R1 in constraint Φ_6 (Equation 5.76) which introduces a new unknown variable e_4. In Φ_2 ("1" junction equation), e_4 is unknown. This variable is expressed in terms of known variables by covering the causal paths (4-3-De$_1$:me$_1$) and (4-5-De$_2$:me$_2$) which lead to $e_4 = me_1 - me_2$. Consequently, we have $f_4 = Cd_1.\text{sgn}(me_1 - me_2).\sqrt{|me_1 - me_2|}.mu_b$, where 'sgn' is a function used to evaluate sign of its argument. The second variable in constraint Φ_1 is then known.
- Substituting expressions for f_2 and f_4 in "0_1 junction" relation Φ_1, we obtain the first ARR:

$$\begin{aligned} ARR_1 &= mQ_p - C_1 s.me_1 - C_{d1}\text{sgn}(me_1 - me_2)\sqrt{|me_1 - me_2|}.mu_b \\ &= ARR_1(De_1, De_2, MSf_1, C_1, R_1, mu_b). \end{aligned} \quad (5.80)$$

2. **Junction $\mathbf{1_1}$**. This structural equation is given by constraint Φ_2 in Equation 5.75. The unknown variables in this equation are e_2, e_4 and e_6. Covering different causal paths leads to $e_2 = me_1$, $e_6 = me_2$ and $e_4 = \Phi_{R1}^{-1}(f_4, mu_b)$. One may express f_4 in terms of known variables as has been done earlier. Thus, the second ARR is obtained by substituting expressions for e_2, e_6 and e_4 in Φ_2; it can can be written as

$$ARR_2 = me_1 - \Phi_{R1}^{-1}(me_2, me_1, mQ_p, mu_b) - me_2 \quad (5.81)$$
$$= ARR_2(De1, De2, MSf1, C_1, R_1, mu_b). \quad (5.82)$$

ARR_2 and $ARR1$ have identical fault signatures; *i.e.* the residuals evaluated using them would be sensitive to the same set of faults (in De_1, De_2, MSf_1, C_1,

R_1, mu_b). *This result can be deduced without any calculation, because the same causal paths are used in derivation of the second ARR (for calculation of unknown variables) as for the first one.* Thus we can only consider one of them as the basis ARR and remove the other. From a computational point of view, $\Phi_{R1}^{-1}(me_2, me_1, mQ_p, mu_b)$ term in Equation 5.81 is undefined when $mu_b = 0$ and thus the ARR form given in Equation 5.80 is, preferably, retained.

3. In the next step, we consider junction 0_2, *i.e.* the constraint Φ_3 in Equation 5.75. This leads to an ARR (which will be the new second ARR because we have taken only one of the two ARRs derived earlier and removed the other) as

$$\begin{aligned} ARR_2 &= C_{d1}.\text{sgn}(me_1 - me_2).\sqrt{|me_1 - me_2|}.mu_b - C_2 s.me_2 \\ &\quad - C_{d2}.\text{sgn}(me_2 - P_0).\sqrt{|me_2 - P_0|}.mu_0 \\ &= ARR_2(De_1, De_2, C_2, R_1, R_2, mu_b, mu_0). \end{aligned} \tag{5.83}$$

This ARR has different fault signature from the existing ARR (in Equation 5.80) and hence will be considered as a legitimate basis ARR.

4. We leave it to the reader to verify that use of the constitutive relation for junction 1_2 (*i.e.* constraint Φ_4 in Equation 5.75) leads to an ARR which has same fault signature as the ARR derived in Equation 5.83. Thus it does not qualify as a new basis ARR.
Finally, we have obtained two ARRs. These two ARRs are indeed conservation of mass relations when the two tanks are taken as separate control volumes. That is why they are referred to as process ARRs and only they can be used to detect process faults.

5. If we look at the control algorithms and source modulation equations in Equations 5.78 and 5.79, respectively, we find that they are already written in terms of known variables. This is because of the fact that inputs and outputs of controllers and modulated sources are always assumed to be known variables. Therefore, those constraint relations (Φ_{11}, Φ_{12} and Φ_{13} in this example) can always be recast and written as ARRs:

$$\begin{aligned} ARR_3 &= mu_p - K_p(h_c - \frac{me_1}{\rho g}) + K_i \int (h_c - \frac{me_1}{\rho g})dt, \\ ARR_4 &= mu_b - On\_Off(me_2, L_{State}, L_{Min}, L_{Max}), \\ ARR_5 &= mQ_p - MinMax(mu_p, Q_{Min}, Q_{Max}). \end{aligned} \tag{5.84}$$

The fault signatures of these three ARRs are different from each other and also different from the two ARRs existing earlier. Thy do not contain any of the process parameters and hence they can only be used to detect controller, actuator and sensor faults.

Finally, we have arrived at 5 structurally independent ARRs, which were expected from our initial analysis of the model structure (13 available constraint relations having 8 unknown variables).

5.6 Causality Inversion Approach for ARR Derivation

The ARR derivation, from a bond graph model, presented in the last section is not a well structured method. First of all, ARRs are generated using the conservation laws at each junction (1 and 0) and then their structural independence are checked with existing ARRs. Deriving so many ARRs and checking their structural independence with others is computationally intensive. Moreover, the derivation procedure is not based on proper causal analysis. For example, in the example considered in the last section, the equation for one of the measured variables was Φ_9: $me_1 = e_2$ (in Equation 5.77) and the constitutive relation of one of the C-elements was Φ_5:$s.e_2 = \frac{1}{C_1}.f_2$ (in Equation 5.76). Then while evaluating the constraint at junction 0_1, expression for f_2 was required. This was obtained in the following steps: (1) $f_2 = C_1 s.e_2$ and then (2) $me_1 = e_2 \Rightarrow f_2 = C_1 s.me_1$. A finer look at these reveals that step-1 is a causal inversion of Φ_5 and step-2 is a causal inversion of Φ_9. Then one may ask the fundamental question of causality: which relations should be inverted such that unknown variables can be eliminated, *i.e.* which causal matching is required?

We will treat the ARR derivation problem such that the required causal matching are preformulated in the model, *i.e.* we will select a different causality assignment scheme from the very beginning. We understand that all measurement relations need to be inverted (such as Φ_9 in the earlier example). We further understand that storage element relations should be in derivative form. Therefore, we will assign preferred derivative causality to the bond graph model and also invert the causalities of all the sensors. This means, sensors will impose the measured variable on the system, *i.e.* sensors become sources [162, 192, 226, 227]. Then causal propagation (SCAP procedure) would assign appropriate causalities to all the R-elements and the junction structure (1- and 0-junctions, TF and GY elements).

Remark 5.1. Changing causalities of sensors and storage elements does not change the constraints themselves. They remain the same as in the behavioral bond graph model, only that they are written differently. When sensors are replaced by sources, the number of unknown variables is still the same as the number of constraints. When a flow sensor becomes a flow source (or an effort sensor becomes an effort source), the effort (or flow) in the bond connected to this source is an unknown variable and will not be zero. This is called the source loading and the output of the source is not influenced by this load. Because of our knowledge that such a source element is originally a sensor (where the unmeasured variable was zero), one can equate the source loading expression obtained from the modified bond graph to zero. This introduces as many new constraints to the model as the number of sensors. Thus the bond graph model becomes over-constrained and we can derive as many ARRs from it as the number of sensors in the behavioral model.

Note that sometimes it may not be possible to invert causalities of all sensors and make them sources. For example, consider that a pressure and a level sensor (two De elements) are connected to the same storage tank, whose model is represented at 0-junction connected to a C-element, the two sensors and other bonds. When we invert causality of one of the sensors (*i.e.* it imposes effort information at the junction instead of measuring it) and replace it by one Se element, the C-element is automatically forced into differential causality. However, we cannot invert the causality of the other sensor because two bonds cannot determine effort at a 0-junction (*i.e.* two bonds cannot be causalled close to the junction). This means that one of the sensors is redundant. In fact, a level sensor is a functional redundancy to the pressure sensor because of the fact that the readings from one of them can be calculated from the readings of the other.

Similarly, it may not be possible to assign derivative causalities to all storage elements. In that case, some of the resulting ARRs contain initial values of state variables and further derivatives of those ARRs, as a postprocessing step, remove those unknown constants of integration (initial values of states).

5.6.1 Example I: A Mechanical System

We take up a mechanical system shown in Figure 5.16a to illustrate the new ARR derivation approach. The single-degree-of-freedom system has a velocity sensor and a force sensor as shown in the figure. In the first case, we consider only the velocity sensor in the bond graph model in Figure 5.16b and in the second case, both the sensors are considered in the bond graph model in Figure 5.16c.

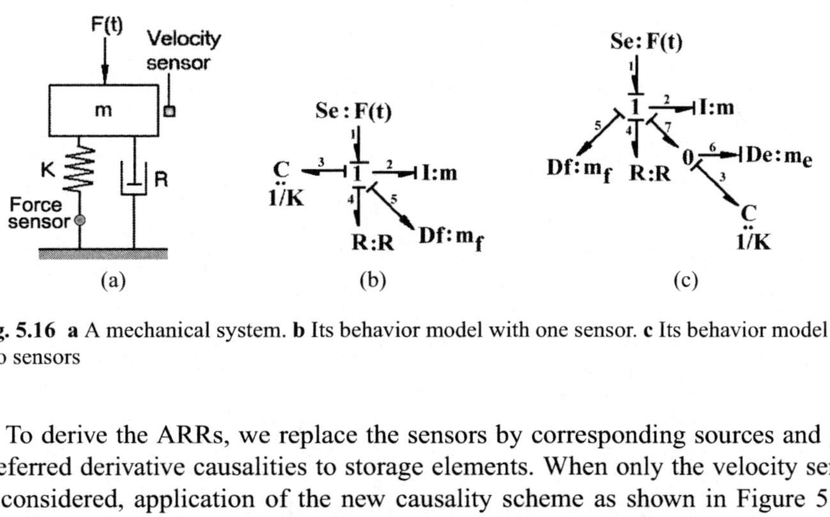

Fig. 5.16 **a** A mechanical system. **b** Its behavior model with one sensor. **c** Its behavior model with two sensors

To derive the ARRs, we replace the sensors by corresponding sources and give preferred derivative causalities to storage elements. When only the velocity sensor is considered, application of the new causality scheme as shown in Figure 5.17a reveals that one of the storage elements (C_3) cannot be assigned derivative causality. However, when both velocity and force sensors are considered, all storage elements can be assigned derivative causality as shown in Figure 5.17b.

From Figure 5.17a, only one ARR can be obtained (because of only one sensor) by equating the effort in bond number 5 (source loading to Sf element) to zero, *i.e.*

$$ARR = e_5 = 0. \tag{5.85}$$

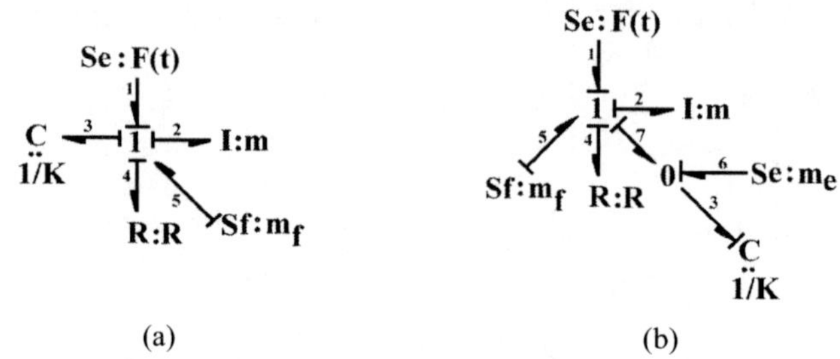

Fig. 5.17 Bond graph models in preferred derivative causality and sensors replaced by sources

Following standard equation derivation principles, we have

$$\begin{aligned} f_1 &= f_2 = f_3 = f_4 = m_f, \\ e_5 &= -e_1 + e_2 + e_3 + e_4, \\ \Rightarrow ARR &= -F(t) + m.\dot{m}_f + K\int m_f dt + Rm_f. \end{aligned} \tag{5.86}$$

This ARR contains an integration with unknown initial values. Therefore, we can take a derivative of it (this corresponds to the integrally causalled C-element in Figure 5.17a) and formulate the ARR as

$$ARR = m.\ddot{m}_f + Km_f + R\dot{m}_f - \dot{F}(t). \tag{5.87}$$

Note that the ARR in Equation 5.87 is the same as the one derived earlier in Equation 5.45 with $y_1 = m_f$ and $u(t) = F(t)$.

From Figure 5.17b, two ARRs can be obtained by equating the two source loadings to zero, *i.e.*

$$\begin{aligned} ARR_1 &= e_5 = 0, \\ ARR_2 &= f_6 = 0. \end{aligned} \tag{5.88}$$

Following normal equation derivation procedure,

$$\begin{aligned} f_1 &= f_2 = f_4 = f_7 = m_f \text{ and } e_3 = e_7 = m_e, \\ e_5 &= -e_1 + e_2 + e_4 + e_7, \\ \Rightarrow ARR_1 &= -F(t) + m.\dot{m}_f + Rm_f + m_e. \end{aligned} \tag{5.89}$$

Likewise, the second ARR is obtained as follows:

$$\begin{aligned} f_6 &= f_3 - f_7, \\ \Rightarrow ARR_2 &= \frac{\dot{m}_e}{K} - m_f. \end{aligned} \tag{5.90}$$

Note that if one considers that $m_e = Ky_2$, where y_2 is the output of a displacement sensor and that $y_1 = m_f$ and $u(t) = F(t)$, then the ARRs given in Equations 5.89 and 5.90 are the same as those derived earlier in Equation 5.49.

5.6.2 Example II: A Two-tank System

Let us consider the controlled two-tank system used in the last section (Figure 5.14). The behavioral bond graph model of the system (Figure 5.15) can be modified by replacing the two sensors by two sources. However, unlike the open-loop system in Example I, this is a closed loop system where sensors are used in the control architecture. In this case, we will keep the sensors as they are in the control circuit and use two modulated sources and then assign derivative causalities to all the storage elements, as shown in Figure 5.18.

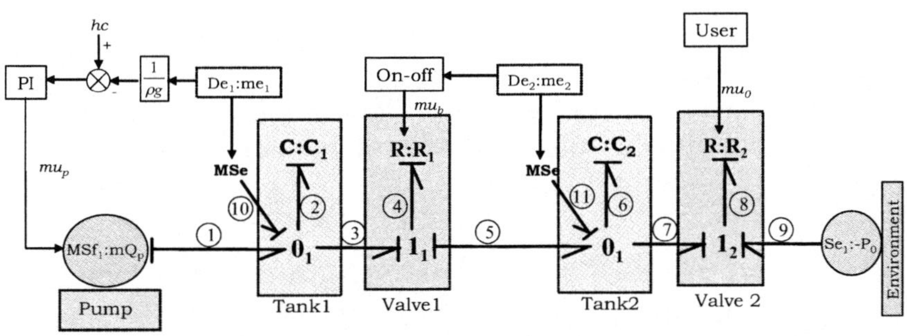

Fig. 5.18 Bond graph model of the controlled two-tank process with inverted sensor causalities

In Figure 5.18, the ARRs are obtained from the source loading to the two effort sources in bond numbers 10 and 11, *i.e.* the constraints $f_{10} = 0$ and $f_{11} = 0$. The ARRs can be derived in terms of the known variables as follows:

$$\begin{aligned} e_1 &= e_2 = e_3 = me_1 \text{ and } e_5 = e_6 = e_7 = me_2, \\ f_{10} &= f_2 + f_3 - f_1 \\ \Rightarrow ARR_1 &= C_1 s.me_1 - mQ_p \\ &\quad + C_{d1}.\text{sgn}(me_1 - me_2).\sqrt{|me_1 - me_2|}.mu_b, \qquad (5.91) \\ f_{11} &= f_6 + f_7 - f_6 \\ \Rightarrow ARR_2 &= C_2 s.me_2 - C_{d1}.\text{sgn}(me_1 - me_2).\sqrt{|me_1 - me_2|}.mu_b \\ &\quad + C_{d2}.\text{sgn}(me_2 - P_0).\sqrt{|me_2 - P_0|}.mu_0. \qquad (5.92) \end{aligned}$$

Note that the ARRs in Equations 5.91 and 5.92 are identical to the two ARRs derived earlier in Equations 5.80 and 5.83. Other ARRs of the system are simply derived from the constitutive relations of the controllers and modulated actuators; they remain the same as those given in Equation 5.84. One of the major contribution of the causality inversion approach has been that we have directly obtained the two structurally independent ARRs. Moreover, the equations used to eliminate unknown variables are the same as those portrayed in the causalities on the bond graph model. Therefore, this approach can be implemented in computer software offering symbolic computation facilities to derive closed form ARR expressions. In fact, the ModelBuilder software [165, 166, 190, 191] uses the algorithm presented here to derive ARRs from bond graph models and then it automatically couples them with a simulation model or the actual process outputs for residual evaluation. In addition, ModelBuilder software generates the fault signature matrix and fault isolation logics from the bond graph model of a process. The C-language code generated by ModelBuilder software can then be implemented directly in most process control software to initiate an FDI application.

5.7 An FDI Application

In this section, we continue with the previous example of the controlled two-tank system and show how the system can be monitored.

5.7.1 Residual Evaluation and Fault Signature Matrix

First of all, we analyze the fault signatures of various residuals. There are five residuals corresponding to the five ARRs given in Equations 5.91, 5.92 and 5.84. The fault candidates are the four process components (Tank1, Tank2, Valve1 and Valve2), five sensors (De1, De2, and those measuring pump flow and controller outputs), two controllers, two sources (pump and environment) and the user.

5.7.1.1 Technological Specifications

Every FDI system is developed upon certain assumptions. These are usually based on experience and observations. In this application, we will assume that the environmental pressure is constant. Moreover, we will assume that the user exactly knows what input it has given to control Valve2. Additionally, controllers are implemented in a process control software and therefore neither controllers nor the measurement of their outputs can be faulty. This reduces the list of faults to be monitored to a manageable set: Tank1, Tank2, Valve1, Valve2, De1, De2, Qp (flow measurement) and the Pump.

5.7.1.2 Construction of the Fault Signature Matrix

Considering the above-mentioned set of faults to monitor, this step constitutes the determination of which faults influence which residuals. The steps for constructing the fault signature matrix are already outlined in Equation 5.54. We will specifically search for variables or functions belonging to a component in different ARRs. For example, C_1 is a parameter associated with Tank1 and this parameter appears only in ARR_1 (Equation 5.91). So, only residual r_1 will be sensitive to faults in Tank1. Similarly, the pump's flow output (mQ_p) appears in ARR_1 and ARR_5 (Equation 5.84), which means that both residuals r_1 and r_5 will be sensitive to a fault in the flow sensor. The pump's model is a functional relationship, which appears only in ARR_5 (Equation 5.84), which means that only residual r_5 will be sensitive to the actuator fault. Note that this is a structural analysis and these sensitivities do not take actual parameter values into consideration. The fault signature matrix of the system is then represented in Table 5.4.

Table 5.4 Fault signature matrix of the two-tank system for the assumed technological specifications

	Residuals						
Components	r_1	r_2	r_3	r_4	r_5	M_b	I_b
Tank1	1	0	0	0	0	1	1
Tank2	0	1	0	0	0	1	0
Valve1	1	1	0	0	0	1	1
Valve2	0	1	0	0	0	1	0
De1	1	1	1	0	0	1	1
De2	1	1	0	1	0	1	1
Q_p	1	0	0	0	1	1	1
Pump	0	0	0	0	1	1	1

In Table 5.4, sensitivity of five residuals (in columns) to eight component faults are shown. When a residual is sensitive to faults in a component, then the corresponding entry in the matrix is 1, otherwise it is 0. The column with heading M_b denotes monitorability of a fault [251] in the component in the corresponding row. If no residual is sensitive to faults in a component, then that fault is not detectable/monitorable ($M_b = 0$). Otherwise the fault can be detected/monitored ($M_b = 1$). This means, if the fault signature of a component is not null (*i.e.* different from the case when all entries under residuals are zero) then $M_b = 1$. Mathematically speaking, it is the result of a binary 'OR' operation on all entries in the fault signature of a component in the fault signature matrix. For the current application, all faults are monitorable.

5.7.2 Single Fault Hypothesis and Fault Isolation

At first, we will assume that the system may either be normal or it may have a single faulty component. This excludes the case when more than one faults can exist at the same time and naturally this also excludes the possibility of one or more fault effects canceling out each other in the residual. This assumption simplifies the fault isolation task. The fault isolation logic is then formulated as follows:

1. At every sampling step, evaluate the ARRs by substituting the parameter values and measurements into them and obtain residuals at that sampling step.
2. Check each residual against a fixed threshold and generate a coherence vector $C = [c_1\ c_2\ ...\ c_n]$. Sometimes a residual preprocessor may be applied before the threshold test. The i-th element of the coherence vector, $c_i = 0$ when the i-th residual is within its threshold, otherwise $c_i = 1$.
3. When the coherence vector is not null, *i.e.* at least one of its elements in none-zero, then raise an alarm. This means some fault is detected. Sometimes, average of coherence vectors in last few samples is used to test if the coherence vector is persistent. This can avoid false alarms, which may be due to some transient effect.
4. If there is an alarm, then match the coherence vector to the fault signatures of different components in the fault signature matrix. If no match is found, then the fault cannot be isolated. If there are more than one matches, then too the fault cannot be isolated, *i.e.* there are more than one fault candidates. If there is a unique match, then the fault is isolated.

This means that the fault signature matrix must be so designed that each residual is sensitive to one and only one fault. Such a fault signature matrix requires as many residuals as the considered faults. The resulting set of residuals are called structured and diagonal. The fault signature matrix of the controlled two-tank system given in Table 5.4 is not structured because the fault signatures of two components, namely Tank1 and Valve2 are identical. All other components have distinctly separate fault signatures and hence faults in them can be isolated under the single fault hypothesis. This is denoted by the isolability index (I_b) in the last column of the fault signature matrix in Table 5.4, where $I_b = 1$ means the fault signature of the component in the corresponding row is distinctly unique from the fault signatures of all other components. Consequently, when a non-null coherence vector matches the fault signature of a component having isolability index $I_b = 1$ then the fault is uniquely isolated.

The reason why faults in certain components cannot be isolated is generally insufficient instrumentation in the concerned parts of the process. In this example, faults in Tank2 and Valve2 cannot be isolated. Consider a leakage from Tank2 (a hole in its bottom plate) and a leakage from Valve2 (its discharge coefficient has increased due to wear and tear). Both of these faults lead to more liquid discharge to the environment and are not distinguishable from each other (by looking at the measurements only) without visual inspection of the process. If we enhance the instrumentation architecture in this part of the system, say by adding a flow sensor to measure the flow through Valve2, then we can show that the faults in Tank2 and

Valve2 will be isolatable from each other. We leave it to the reader to verify this by deriving the additional ARR (due to the additional sensor) and reformulating the fault signature matrix. Therefore, analysis of the fault signature matrix is a structural approach to sensor placement for development of FDI systems.

5.7.2.1 Residual Sensitivity

The residual sensitivities used so far are based on structural analysis, *i.e.* they do not consider the actual values of system parameters and measurements. In practice, some system parameters may be such that the sensitivity of residuals to variations in those parameters may be negligible or may get overshadowed by other influences such as noise. Moreover, each residuals sensitivity to different faults may be different, often the orders of magnitude may vary to a large extent. Therefore, it is necessary to test the residual sensitivities to faults in the real application.

It may not be possible to introduce faults in a real application because often controllers will try to compensate the faults. Consequently the process should be taken offline and one fault should be introduced in it at a time to record all measurements and evaluate the residuals. However, in most cases, the process cannot be taken offline and moreover there may not be provisions in the process to introduce all kinds of faults. Furthermore, some faults may lead to serious consequences and cannot be tried, *e.g.* leakage from a chemical plant. This is why a simulation model of the process is usually used to simulate faults and record residual sensitivities. All kinds of faults can be introduced in a simulation model at no risk. These simulation results help in reformulation of the fault signature matrix (*i.e.* generation of a practical fault signature matrix), determination of residual thresholds and finally classification of faults (as weakly detectable or strongly detectable) [163, 190].

5.7.3 Simulation Results

For validation of fault signatures, we considered the integrally causalled bond graph model of the controlled two-tank system given in Figure 5.15 and modified it by introducing two flow sources ($Sf : Q_{T1}$ and $Sf : Q_{T2}$) at the two 0-junctions. These two flow sources help to introduce leakage in the two tanks. The outputs of this model (which replaces the actual process) were fed to the residual evaluation block (containing expressions of five ARRs). A fault is virtually introduced in the process by altering one of its parameter's value while the same parameter's value in the residual evaluation block is not changed. This generates the discrepancy between the process and the residual model and some residuals change, thereby enabling the supervision platform to monitor the process.

The nominal parameter values chosen for the process are given in Table 5.5 and the corresponding process response is shown in Figure 5.19. Note that the hydraulic capacity of i-th tank is taken as A_i/g; and variables me_1 and me_2 in the following plots represent water levels in the two tanks, respectively.

Table 5.5 Parameter values of the two-tank system used in simulation

Symbol	Description	Value	Unit
$A_i(i=1,2)$	Cross-sectional area of Tank$_i$		
ρ	Density of water	1×10^3	$kg.m^{-3}$
g	Gravity	9.81	$m.s^{-2}$
C_{d1}	Discharge coefficient of Valve1	1×10^{-3}	$kg^{\frac{1}{2}}m^{\frac{1}{2}}$
C_{d2}	Discharge coefficient of Valve2	1×10^{-3}	$kg^{\frac{1}{2}}m^{\frac{1}{2}}$
mu_0	Stem position of Valve2	1	-
h_c	Level setpoint of PI controller	0.5	m
K_p	Proportional gain of PIC	1×10^{-2}	-
K_i	Integral gain of PIC	5×10^{-6}	-
Q_{Min}	Minimum outflow from pump	0	kg/s
Q_{Max}	Maximum outflow from pump	5	kg/s
L_{Min}	Lower setpoint of On-Off controller	0.09	m
L_{Max}	Upper setpoint of On-Off controller	0.11	m
Q_{T1}	Leakage from Tank1	0	kg/s
Q_{T2}	Leakage from Tank2	0	kg/s

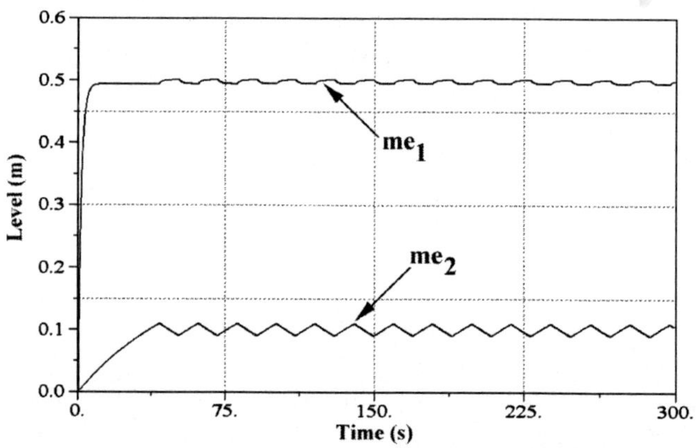

Fig. 5.19 Response of the two-tank process in normal operation

Simulation results in Figure 5.19 show the process start-up period and the steady state operation. The PI-controller maintains the water level in Tank1 at about 0.5m and the On-Off controller maintains the water level in Tank2 between 0.09cm and 0.11 cm.

In the first fault scenario simulation, a leakage fault was introduced in Tank1 by changing the parameter value Q_{T1} from nominal value 0 kg/s to a value -1 kg/s for some duration; after which, it was restored back to its nominal value (*i.e.* the fault was repaired). The corresponding process response is shown in Figure 5.20a. In the fault duration, the water level in Tank1 came down a little whereas the On-Off controller overwhelmed the fault and maintained proper water level in Tank2. Note that if the flow output to the user is considered, it would not be affected by this fault. In closed loop processes, controllers try to hide fault effects and thus it is usually impossible to find faults in such processes by simply observing the products or process outputs. In this process, the process response is not smooth (due to On-Off control action) and thus the raw residual signals are noisy with large amplitudes for very short durations. Use of such raw residuals for diagnosis would lead to a large number of false alarms. Therefore, various residual postprocessors are used. In this application, moving average of the residuals are taken over a time window of 40 s span. The responses of the first two residuals are shown in Figure 5.20b after postprocessing. Other residuals remain zero and are omitted from the figure. We find that during the leakage from Tank1, only residual r_1 diverges from its nominal value (a threshold around mean zero value). Therefore, in the fault duration, we obtain a coherence vector $C = \begin{bmatrix} 1 & 0 & 0 & 0 & 0 \end{bmatrix}$. Because the coherence vector is non-zero, an alarm is raised and a match for this coherence vector is searched in the fault signature matrix. In Table 5.4, we find a unique match with the fault signature of component Tank1. Therefore, the fault can be isolated from the residual responses.

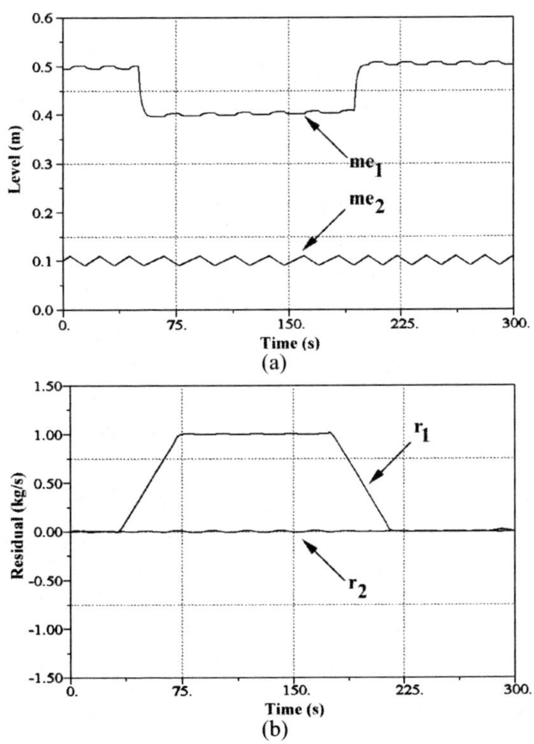

Fig. 5.20 Sensor (**a**) and residual (**b**) responses of the two-tank process showing Tank1 leakage

Remark 5.2. Note that in Figure 5.20b, the residual response is outside the residual threshold only during the fault and once the fault is repaired, the residual response returns back to its nominal value. This is due to the use of ARRs written in the derivative form. If the ARRs were written in integral form then the residual values will not return to their nominal values after repairing the fault, rather they should be manually re-initialized to zero.

In the second fault scenario simulation, blockage fault was introduced in Valve1 by changing the parameter value of C_{d1} from its nominal value $1\times10^{-3}\mathrm{kg}^{\frac{1}{2}}\mathrm{m}^{\frac{1}{2}}$ to a value $1\times10^{-4}\mathrm{kg}^{\frac{1}{2}}\mathrm{m}^{\frac{1}{2}}$ for some duration and then the fault was repaired. The corresponding process response and residual response (after taking moving average over a window of 40 s span) are shown in Figures 5.21a,b, respectively.

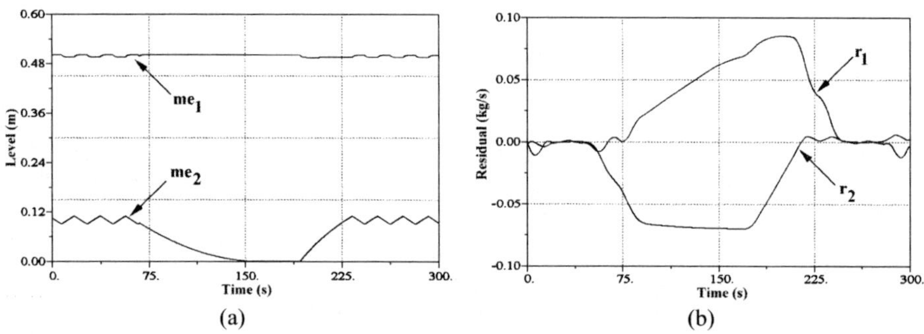

Fig. 5.21 Sensor (**a**) and residual (**b**) responses of the two-tank process showing Valve1 blockage

From the residual response in Figure 5.21b, we obtain a coherence vector $C = \begin{bmatrix} 1 & 1 & 0 & 0 & 0 \end{bmatrix}$ during the fault. This coherence vector has a unique match with the fault signature of component Valve1 in Table 5.4. Therefore, the fault can be isolated from the residual response.

Similarly, other faults are simulated to check the sensitivities of residuals to them and it was found that the practical fault signature matrix is the same as the theoretical one.

Another utility of the simulation is to find how much each residual deviates due to a fault. For example, if the value of C_{d1} is changed from nominal value $1\times10^{-3}\ \mathrm{kg}^{\frac{1}{2}}\mathrm{m}^{\frac{1}{2}}$ to a value $8\times10^{-4}\ \mathrm{kg}^{\frac{1}{2}}\mathrm{m}^{\frac{1}{2}}$ then the residuals change so little (compared to the ones in Figure 5.21b) that it would be impossible to put a threshold on them to catch this fault. If a tight threshold is used, then there will be false alarms during normal process operation. Thus fault scenario simulation aids in determination of proper thresholds and also to quantify the minimum magnitude of fault(s) that will be detected. Faults of very small magnitude will be missed. This is why better a thresholding technique than simple fixed thresholding is required in processes having non-smooth response and uncertainties like sensor noise.

The simulation problem discussed in this section can be modified to a hardware-in-the-loop (HIL) simulation architecture. This requires the process model to be implemented in one computer and the controller algorithms, residual evaluator and decision procedure to be implemented in a separate computer. Each computer should have a data acquisition system for interfacing it to the other computer. After satisfactory level of development, the computer housing the process model may be removed and the actual plant may be interfaced with the supervising computer.

Problems

5.1. A simple level control system is shown in Figure 5.22. Write down the constraints in the mathematical model of this system. Draw its bipartite graph and derive the ARRs with any suitable causal matching on an incidence matrix. Draw the bond graph model of the system and derive the same ARRs. Neglecting controller faults, determine if the sensor architecture is sufficient to isolate all faults.

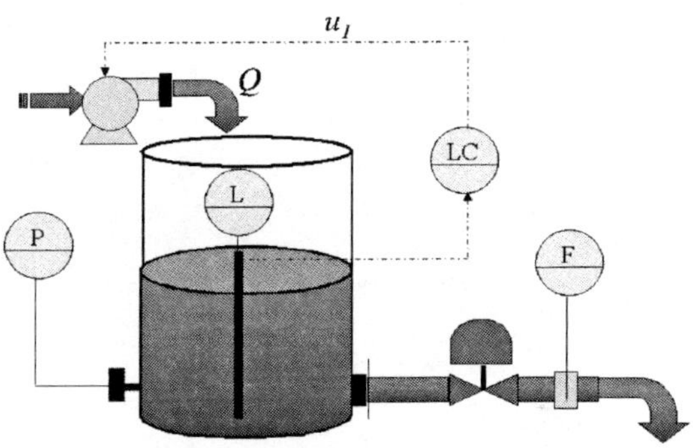

Fig. 5.22 A pump-operated level control system

5.2. The electromechanical system shown in Figure 5.23 is instrumented to measure the motor current, angular velocity of the rotating disk and the tension in the rope. The supply voltage to the motor is controlled through a PI controller to maintain a constant rope tension during the lifting of the load. Find the ARRs from the bond graph model of the system. Assume that controllers and sensors are robust (*i.e.* not faulty) and then determine the faults which cannot be isolated. Devise a new sensor architecture so that all the faults in the mechanical part of the system become isolatable and the order of derivatives of measured signals in the ARRs does not exceed one. The rope as a component includes both its stiffness and damping; there is no need to separately consider faults in these two parameters.

5.3. Consider that Problem 5.1 is extended to a thermo-fluid system,. The temperature of the pumped fluid is T_{in}. A temperature sensor, T, measures the temperature of the liquid stored in the tank. A temperature controller, TC, acts on a submerged

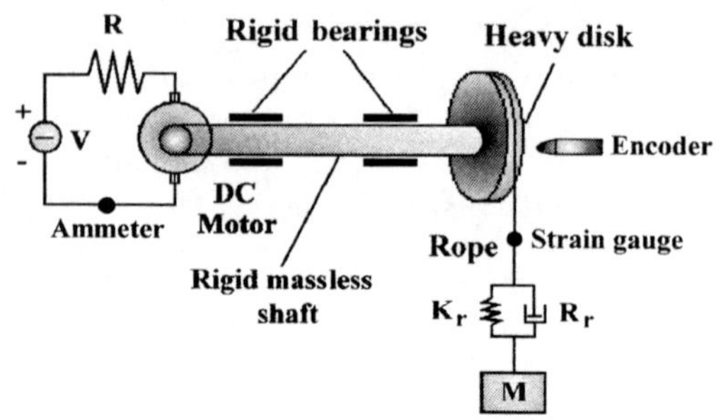

Fig. 5.23 A well-instrumented electromechanical system

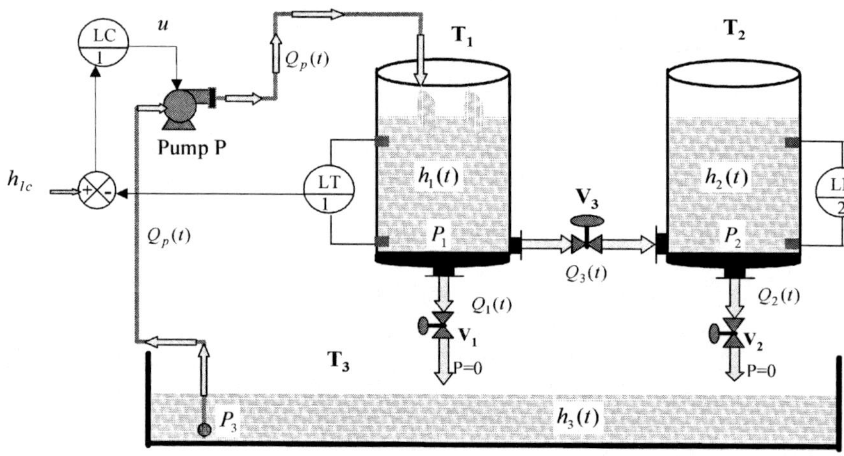

Fig. 5.24 A two-tank system with common sump

electrical heater to control the temperature of the liquid in the tank. The power delivered by the heater is measured by a power meter, W. Derive the ARRs of the thermofluid system and reformulate the fault signature matrix with the heater as the only additional component.

5.4. Consider the system given in Figure 5.24, which has been described earlier in the Problem 2.1. Assume that the level in the sump is constant. Construct a bond graph model of the system. Moreover, the flows through all the valves and the pump are measured. Determine the hardware and functional redundancies. Derive the ARRs of the system and construct the theoretical fault signature matrix.

5.5. The schematic of a simple blending process is shown in Figure 5.25. Two liquids A and B are taken, respectively, from two reservoirs R_1 and R_2, and are mixed in a tank T_2. The reservoirs are flat and big, so they can be considered as constant pressure sources. Two more tanks T_1 and T_3 act as intermediate storages of the fluids. The liquid levels in T_1 and T_3 (level setpoints h_1 and h_3, respectively) are controlled by pumping fluid into them through two controlled pumps, P_1 and P_2.

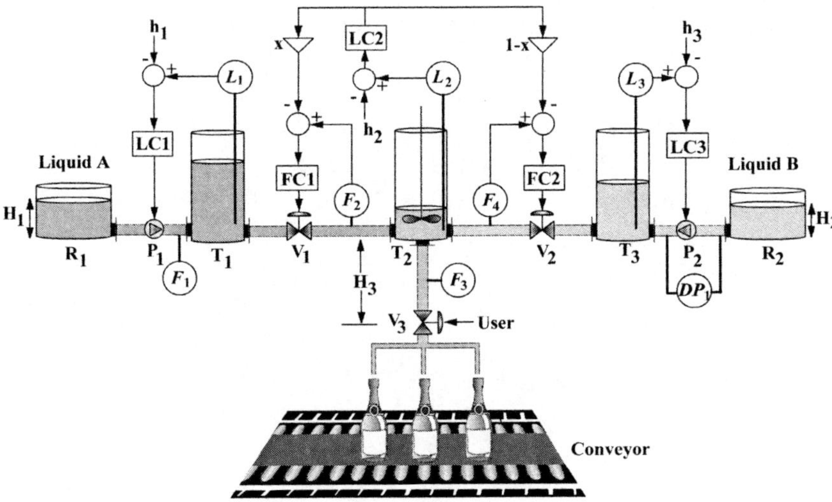

Fig. 5.25 Schematic representation of a simple blending process

The fluid level in T_2 is also controlled by a level controller LC2. However, the mixture must maintain a constant proportion of x parts of liquid A to $1-x$ parts of liquid B. Therefore, cascaded flow controllers are used to transfer the two fluids to T_2 through two valves V_1 and V_2. The output of controller LC2 is appropriately scaled to generate setpoints for the two flow controllers. The product is bottled as shown in the figure through simple gravitational flow. Usually a controller takes the signal from a proximity sensor to open valve V_3 for some specified duration and then closes the valve following which the batch of bottles already filled are transported along the conveyor and a new batch of empty bottles arrive to be filled. However, in this problem, assume that V_3 is manually operated and the fluid is discharged to the atmosphere.

The process instrumentation shows different level sensors (L_i), flow sensors (F_i) and a differential pressure sensor (DP). For modeling purposes, the two pumps may be considered as bond graph R-elements with negative resistance at 1-junctions, *i.e.* they produce positive flows when the effort difference is negative.

Derive the ARRs to monitor this system by assuming that the two fluids have comparable densities. Generate the fault signature matrix. Suggest a solution to diagnose improper mixture quality.

5.6. A preconditioner for a certain fermentation process is shown in Figure 5.26.

The environment is at a pressure P0 and temperature T0. The fluid supplied by the pumps is at environmental temperature. The process is instrumented with a set of level sensors, temperature sensors, flow meters, level controllers and temperature controllers. Assume suitable component naming convention. Further, assume that the output from all actuators (pumps and heater) and controllers are measured.

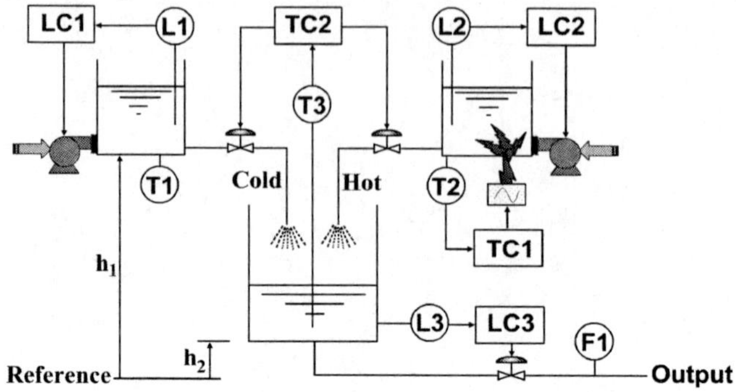

Fig. 5.26 Schematic representation of a preconditioner for a fermentation process

Derive the ARRs of the system. Determine which faults are monitorable and which faults are isolatable. Note that gravity head (with respect to the reference level) should be considered in the model.

Chapter 6
Application to a Steam Generator Process

6.1 Introduction

The methodology developed in the previous chapters is applied in this chapter to supervise an industrial steam generator. The purpose of this application is to design a supervision platform which allows one to control and to monitor the process in normal situations as well as in the presence of faults and failures. On the outset, we generate a bond graph model of the process and validate it so that the model can be used for ARR generation and its online implementation.

6.1.1 Process Description

The steam generator process is a pilot installation which mimics a reduced scale operation of a part of a power plant [140]. It is situated at the University of Lille 1, France.

The plant is mainly composed of a boiler, with a total capacity of 170 L, which performs vaporization of water inside it by means of a 55 kW heater (thermal resistor), a steam expansion system, a condenser coupled with a heat exchanger, a storage tank, and a feed water circuit. In fact, the boiler load is realized by the steam expansion system formed by a set of two modulated valves (V_1 and V_2) and the condenser coupled with a heat exchanger. The level of water in the boiler is kept within ± 3 L of a given setpoint by switching on a pump which draws water from a reservoir tank. An On-Off controller acting on the heater maintains the pressure inside the boiler within ± 0.2 bar of the specified pressure setpoint. The Process and Instrumentation Diagram (P&ID) of the process is given in Figure 6.1.

The following modeling hypotheses are considered: the steam and water are saturated. The thermodynamic properties are calculated in thermodynamic equilibrium, the mixture is at a uniform pressure, the variables are considered as localized parameters, and finally, the liquid phase (water) is incompressible.

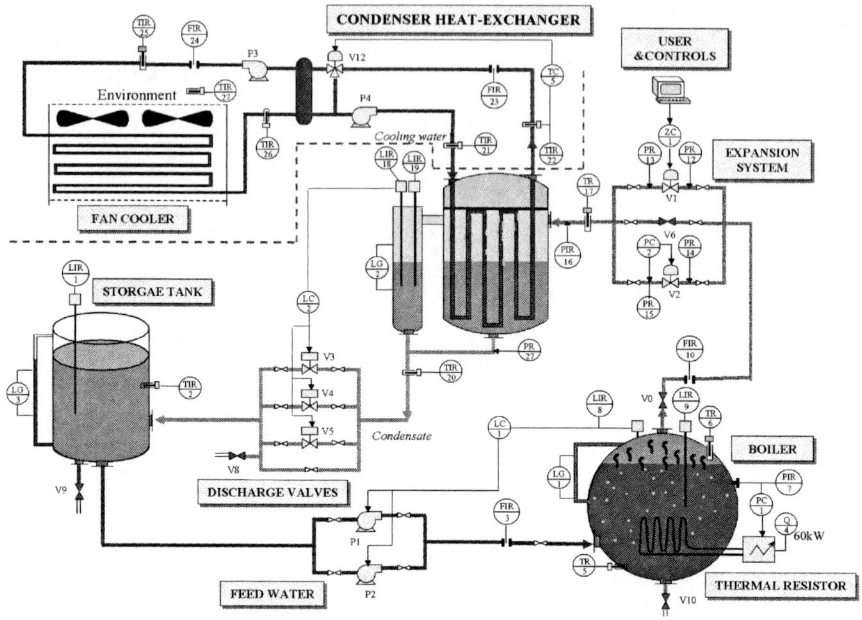

Fig. 6.1 Process and Instrumentation Diagram of the steam generator

The condenser is a complex system. The steam coming from the expansion system to the condenser is condensed over the vertical tubes through which the cooling water flows. The tubes are split into three parts: at the input to the condenser in touch with the steam, within the condensate and then again in contact with the steam at the output from the condenser.

The level of the condensate is controlled by means of three On-Off valves (V_3, V_4 and V_5) placed in parallel between the condenser and the tank. These On-Off valves keep the condensate level (L_{18}) within ± 0.5 L of a given setpoint.

The water supply system is made of two pumps (P_1 and P_2) and a pipe. At any given point of time, only one pump can be in operation; the other pump simply provides equipment redundancy. The pump drives water into the boiler according to the On-Off controller command depending on the level of water in the boiler (sensor L_8).

This installation has been specifically designed to perform as a pilot plant to test FDI applications. It is possible to simulate most of the faults manually or automatically. A leakage from the condenser, the tank or the boiler can be simulated by opening the valves V_8, V_9 or V_{10}, respectively. A boiler output blockage can be contrived by fully or partially closing the valve V_0.

6.1.2 Nomenclature

Symbol	Description	Value	Unit
g	Constant of gravity	9.8	$m.s^{-2}$
ρ	Density of water	1000	$kg.m^{-3}$
c_p	Specific heat of water	4184	$J.kg^{-1}.K^{-1}$
W_p	The power of the heater	5.5×10^4	$J.s^{-1}$
P_0	Atmospheric pressure	1.013×10^5	Pa
T_0	Atmospheric temperature	Sensor	K
A_{Tank}	Tank area	0.43676	m^2
V_{boiler}	Volume of the boiler	0.17	m^3
K_{me}	Overall heat transfer coefficient of the boiler's body	3.6	$J.s^{-1}.K^{-1}$
K_1	Pump characteristic	-8.33×10^{-7}	$m^3 \cdot s^{-1}.Pa^{-1}$
K_2	Pump characteristic	0.972	$m^3 \cdot s^{-1}$
c_{pm}	Specific heat of the metal of the boiler	498	$J.kg^{-1}.K^{-1}$
V_{cond}	Volume of the condenser excluding cooling tubes	7.9×10^{-3}	m^3
A_{cond}	Area of the condenser	1.227×10^{-2}	m^2
c_{pCond}	Specific heat of metal of the condenser	521	$J.kg^{-1}.K^{-1}$
Mt_{Cond}	Mass of the condenser body	17	kg
n_t	Number of turns of coolant tubes	10	
D	Diameter of coolant tubes	0.016	m
ρ_l	Density of condensate	962	$kg.m^{-3}$
μ_l	Dynamic viscosity of condensate	3.0×10^{-4}	$kg.m^{-1}.s^{-1}$
λ_l	Thermal conductivity of condensate	0.675	$J.(m.s.K)^{-1}$
L_v	Latent heat *vs* pressure	Function	$J.kg^{-1}$
l_t	Length of coolant tubes	0.6	m
$V_0\_C_d$	V_0 Discharge valve coefficient	3.6×10^{-8}	$kg.s^{-1}.Pa^{-1}$
$V_1\_C_d$	V_1 Discharge valve coefficient	8.4×10^{-8}	$kg.s^{-1}.Pa^{-1}$
$V_2\_C_d$	V_2 Discharge valve coefficient	5.1×10^{-8}	$kg.s^{-1}.Pa^{-1}$
$V_3\_C_d$	V_3 Discharge valve coefficient	2×10^{-8}	$kg.s^{-1}.Pa^{-1/2}$
$V_4\_C_d$	V_4 Discharge valve coefficient	2×10^{-8}	$kg.s^{-1}.Pa^{-1/2}$
$V_5\_C_d$	V_5 Discharge valve coefficient	2×10^{-8}	$kg.s^{-1}.Pa^{-1/2}$
$V_7\_C_d$	V_7 Discharge valve coefficient	9.5×10^{-4}	$kg.s^{-1}.Pa^{-1/2}$
L_1	Level in the tank	Sensor	m

Symbol	Description	Value	Unit
T_2	Temperature in the tank	Sensor	K
F_3	Flow from pump	Sensor	$m^3.s^{-1}$
L_8, L_9	Volume of water in the boiler	Sensor	m^3
P_7	Pressure in the boiler	Sensor	Pa
P_{16}	Pressure of the steam in the condenser	Sensor	Pa
T_{17}	Temperature of the steam in the condenser	Sensor	K
L_{18}, L_{19}	Condensate level in the condenser	Sensor	m
P_{27}	Pressure at the bottom of the condenser	Sensor	Pa
T_{20}	Temperature of condensate at the bottom of the condenser	Sensor	K
T_{21}	Temperature of the coolant at the inlet to the condenser	Sensor	K
T_{22}	Temperature of the coolant at the outlet from the condenser	Sensor	K
F_{23}	Coolant flow rate to the condenser	Sensor	$m^3.s^{-1}$
Q_4	Heater power output	Sensor	W
T_5	Temperature of the water in the boiler	Sensor	K
T_6	Temperature of the steam in the boiler	Sensor	K
F_{10}	Steam flow from boiler	Sensor	$kg.s^{-1}$
P_{12}	Pressure upstream of V_1	Sensor	Pa
P_{13}	Pressure downstream of V_1	Sensor	Pa
P_{14}	Pressure upstream of V_2	Sensor	Pa
P_{15}	Pressure downstream of V_2	Sensor	Pa
mO_1	Pump command	Sensor	
mO_2	Heater command	Sensor	
mO_5	On-Off valve command	Sensor	
mU_0	V_0 opening (manual)	1(100%)	
mU_1	V_1 command (manual)	0.4(40%)	
mU_2	V_2 command	Sensor	
L_{8_ref}	Boiler level setpoint	Setpoint	m
P_{7_ref}	Boiler pressure setpoint	Setpoint	Pa
P_{15_ref}	P_{15} pressure setpoint	Setpoint	Pa
L_{18_ref}	Condenser level setpoint	Setpoint	m
$\mathbf{OnOff_1}()$	Boiler pressure controller	Function	
$\mathbf{OnOff_2}()$	Pump controller	Function	
$\mathbf{PI}()$	P_{15} pressure controller	Function	

Symbol	Description	Value	Unit
OnOff$_5$()	Condensate discharge controller	Function	
$s()$	Time derivative operator	Operator	
$h_v()$	Specific enthalpy of steam *vs* pressure	Function	$J.kg^{-1}$
$h_l()$	Specific enthalpy of water *vs* pressure	Function	$J.kg^{-1}$
$v_v()$	Specific volume of steam *vs* pressure	Function	$m^3.kg^{-1}$
$v_l()$	Specific volume of water *vs* pressure	Function	$m^3.kg^{-1}$
$\Phi_e()$	Equal percentage valve opening *vs* position	Function	
$\rho_v()$	Density of the steam *vs* pressure	Function	$kg.m^{-3}$
Ps2Ts()	Steam temperature *vs* pressure	Function	K
Ts2Ps()	Steam pressure *vs* temperature	Function	Pa

6.1.3 Word Bond Graph Model of the Process

The word pseudo-bond graph model of the steam generator process is given in Figure 6.2.

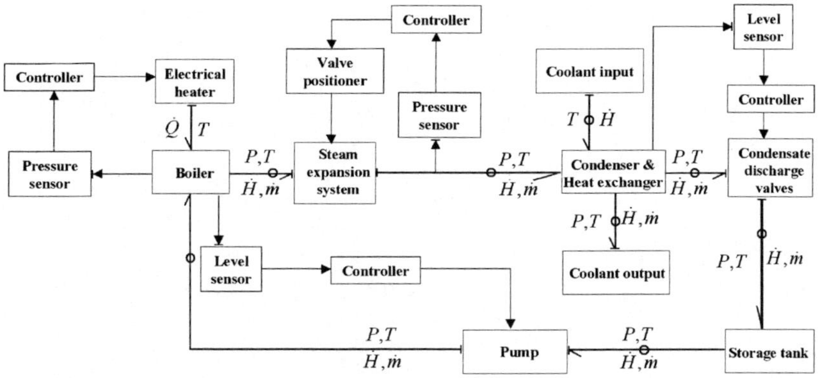

Fig. 6.2 Word pseudo-bond graph of the steam generator process

This representation helps us to derive the bond graph sub-model of each component, to assign the correct causality and, consequently, to complete the bond graph model of the entire process.

6.2 Bond Graph Models of Steam Generator's Components

Each of the components of the steam generator is represented as a sub-model. The sub-models are classified into different process classes, namely storage, transport, transformation, actuator, sensor, *etc.* depending upon the functionalities of each device [187, 226]. This allows storage of different component models in a hierarchical tree structure [188] and also aids the process engineers in easily constructing their process models by connecting various thermo-fluid and block-diagram component models available in the generic process database. The approach presented in this book is implemented in ModelBuilder software developed in the framework of CHEM project. This software includes many common process engineering component sub-models. The component sub-models for steam generator process are added to the ModelBuilder library. Here, we detail only the components present in the steam generator process.

6.2.1 Bond Graph Model of the Storage Tank

The tank in the steam generator is considered as a coupled thermo-fluid storage device. It stores the condensate discharged from the condenser and the pumps draw water from this source to supply water to the boiler. The corresponding pseudo-bond graph model is given in Figure 6.3.

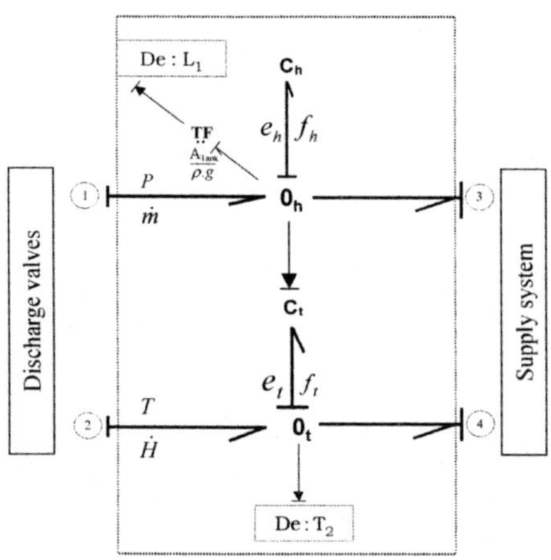

Fig. 6.3 Pseudo-bond graph model of the storage tank

The condensate mass flow ($\dot{m}$) from the condenser is stored in the tank to generate pressure (P). Similarly, the total enthalpy flow ($\dot{H}$) to the tank results in temperature (T_2) variation. The energy coupling between the hydraulic and thermal phenomena is represented by an information bond in the model. For the steam generator application, the storage tank has a constant cross-section A_{Tank}. The capacities of the bond graph storage elements C_h and the state dependent capacity C_t in Figure 6.3 are given by

$$\begin{aligned} C_h &= A_{Tank}/g, \\ C_t &= c_p Q_h = c_p \int f_h dt, \end{aligned} \tag{6.1}$$

where Q_h is the state variable representing the total mass of the liquid in the tank and c_p is the specific heat capacity of the liquid.

6.2.2 Bond Graph Model of the Supply System

Water supply system of the steam generator is composed of two pumps and a pipe. At any point of time, only one pump is used. The other pump serves as a material redundancy for reconfiguration in emergency situations. The mass flow rate from the tank to the boiler is a function of the pressure head across the pump. From the bond graph point of view, the pump is a non-linear resistance with an effective negative resistance. The pipe too is another non-linear resistance and the series coupling between them leads to an algebraic loop. The characteristics of the impeller pumps used in the steam generator is given by

$$f_{pump} = (K_1.\Delta P_{pump} + K_2).mO_2, \tag{6.2}$$

where K_1 and K_2 are the characteristic parameters of the pump, ΔP_{pump} is the pressure gain across the pump, and mO_2 is a binary signal from the output of a controller (in this case, boiler level controller).

The constitutive relations for the flow through the pipe is given by

$$f_{pipe} = V_7\_C_d.\sqrt{|\Delta P_{pipe}|}.sign(\Delta P_{pipe}).mO_2, \tag{6.3}$$

where ΔP_{pipe} stands for the pressure drop across the pipe and V_{7_Cd} represents the pipe discharge coefficient.

In this special case, equating both the flows (see Figure 6.4), the algebraic loop can be symbolically resolved [226] by setting $f_{pump} = f_{pipe}$.

The resulting equivalent resistance, R_{tot}, is shown in Figure 6.5 where the CETF stands for *Coupling Element for Thermofluids* and it is equivalent to the RC element introduced in Chapter 2. The constitutive relation for the equivalent resistance is given as

$$f_9 = \frac{V_7\_C_d\left(V_7\_C_d + \sqrt{(V_7\_C_d)^2 - 4.K_1.K_2 - K_1.e_9}\right)}{2.K_1}.mO_2. \tag{6.4}$$

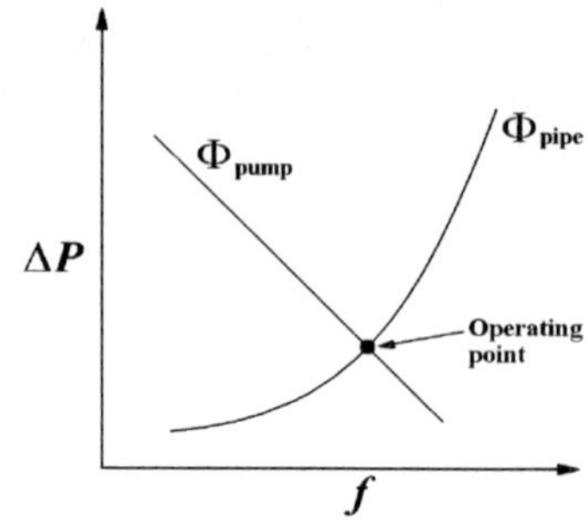

Fig. 6.4 Resolution of operating point for pump-pipe sub-system

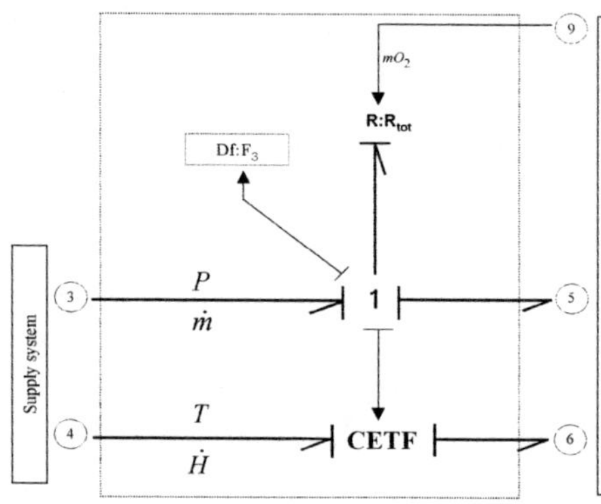

Fig. 6.5 Pseudo-bond graph model of the water supply system

6.2.3 Bond Graph Model of the Boiler

The bond graph model of the boiler is given in Figure 6.6. The storage of hydraulic and thermal energies is modeled by the two-port bond graph C-field element C_b. During boiling, it is assumed that the water and the steam are saturated and are in thermal equilibrium [9]. The resident time and volume of steam bubbles inside the liquid phase is neglected. Since the flow output from the boiler is controlled by controlling the downstream pressure of the steam expansion sub-system, the load does not vary too much and thus the *shrink and swell* phenomenon is not considered in the dynamics. The thermal energy stored by the metal of the boiler's body is modeled by a storage element, C_m. The heat transfers from the water-steam mixture to the boiler's metallic body and from the boiler's body to the environment are modeled by two resistive elements R_{fm} and R_{me}, respectively. These resistances for the heat transfer are expressed in terms of the overall heat transfer coefficients, K_{fm} and K_{me}. The ambient temperature of the environment is given by a source of effort, $Se : T_0$. The pressure sensor, P_7, output is used to control the heater, represented by the modulated thermal source, MSf. The mechanical work in the total internal energy change is given by a source, $Sf : -V_b.\dot{P}_b$. In fact, since the pressure inside the boiler is controlled, this work may be neglected.

For the saturated water-steam mixture inside the boiler, the temperature and pressure are dependent. Their dependency is governed by the relations in the steam ta-

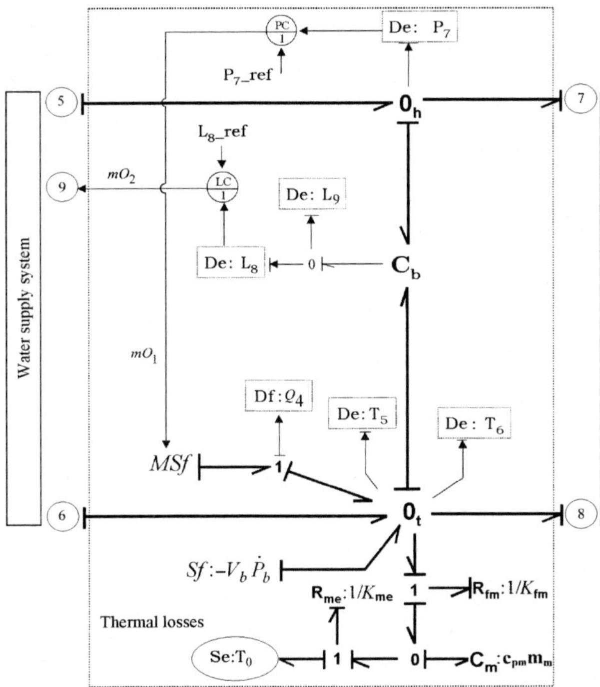

Fig. 6.6 Pseudo-bond graph model of the boiler

bles or Mollier charts. So, if one of these two related variables is known the other can be calculated. The Steam quality, X, is defined as

$$X = \frac{m_v}{m_l + m_v}, \tag{6.5}$$

where m_l and m_v are the masses of saturated liquid and dry saturated steam, respectively. The total enthalpy and total mass inside the boiler are given by the state variables, H and m, of the C_b field. Since the volume of the boiler (V_b) is a known constant parameter, the specific volume v_m and the specific enthalpy h_m of the mixture are calculated as follows:

$$\begin{cases} v_m = \dfrac{V_b}{m} = (1-X).v_l(P_b) + X.v_v(P_b), \\ h_m = \dfrac{H}{m} = (1-X).h_l(P_b) + X.h_v(P_b), \end{cases} \tag{6.6}$$

where v_l and v_v, h_l and h_v are respectively specific volumes and the specific enthalpies of the liquid and vapour given as functions of the steam pressure, P_b. These functions are usually obtained using polynomial interpolation of the steam table data. The set of equations in Equation6.6 can be solved iteratively using different numerical methods (*e.g.* bisection) to determine the pressure and steam quality. Steam tables are again used to determine the saturation temperature, T_b, from the saturation pressure, P_b. A non-iterative formulation of a multi-phase thermal C-field may be obtained by differently choosing the power variables [40].

The level in the boiler (L_b), is measured by sensors, L_8 and L_9, which are calibrated to measure the volume of saturated water. It is calculated from the states of the C_b-field using the solutions for the steam pressure and steam quality, as given below:

$$L_b = (1-X).v_l\,(P_b)\,.m. \tag{6.7}$$

The actual level, in height, of saturated water inside the boiler depends on the geometry and other constructional features.

6.2.4 Bond Graph Model of the Steam Expansion System

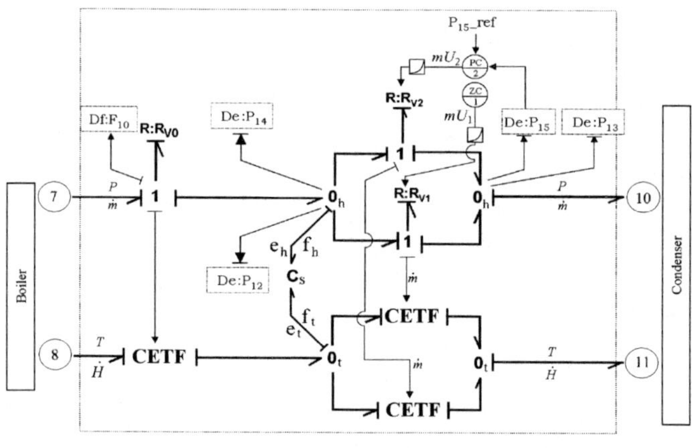

Fig. 6.7 Pseudo-bond graph model of the steam expansion system

The steam expansion system in the steam generator installation is used to simulate the load on the boiler. The effective load can be varied by using different downstream pressure setpoints and also through manual opening of the valves. The bond graph model of the steam expansion system is given in Figure 6.7. The valves in parallel are represented by two resistances R_{v1} and R_{v2}. R_{v1} is controlled manually (signal mU_1 given from a computer) whereas R_{v2} is regulated by the controller output mU_2 according to the downstream pressure P_{15}. The relations between the controller output and the flow through each valve is given by the valve characteristics. In the steam generator installation, the valves have equal percentage characteristics which are represented in the model using block-diagram form. The effective coefficient of discharge, $\hat{C}_d$ through these valves is given as

$$x = \begin{cases} 0 \text{ for } u < 0, \\ u \text{ for } 0 \le u \le 1, \\ 1 \text{ for } u > 1; \end{cases},$$

$$\Phi_e(x) = R.e^{x-1},$$

$$\hat{C}_d = C_d.\Phi_e(x), \tag{6.8}$$

where u is the controller output, x is the valve stem position, Φ_e is the equal percentage function, R is the rangeability factor, and $\hat{C}_d$ is the effective coefficient of discharge of the valve. The first valve at the output of the boiler, R_{v0} has a linear characteristics and is usually completely open. Since saturated steam flows through these valves, the flow characteristic is governed by the following equation:

$$f = \begin{cases} \frac{1}{2}\hat{C}_d.P_u, \text{ for } P_d \le \frac{1}{2}P_u; \\ \hat{C}_d.\sqrt{P_d}.\sqrt{P_u - P_d}, \text{ for } P_d \ge \frac{1}{2}P_u; \end{cases} \tag{6.9}$$

where P_u, P_d and f are the upstream pressure, the downstream pressure and the mass flow rate, respectively.

The two-port element C_s is a non-linear coupling capacitor used to avoid algebraic loops between the non-linear resistive elements (R_{v1} and R_{v0}, and R_{v2} and R_{v0}). By assuming saturated dry steam flow ($X = 1$), the constitutive relations for this element are given as

$$e_h = P = v_v^{-1}\left(\frac{V_\varepsilon}{\int f_h dt}\right),$$

$$e_t = T = \mathbf{Ps2Ts}(P), \tag{6.10}$$

where V_ε is an infinitesimal volume, v_v^{-1} is a thermodynamic function that calculates saturation pressure from the specific volume of dry steam by using steam tables, **Ps2Ts** is a thermodynamic function used to calculate saturated steam temperature from known pressure, P is the steam pressure, and T is the steam temperature.

6.2.5 Bond Graph Model of the Condenser

The vertical condenser is a complex sub-system in the steam generator installation. The steam is condensed over vertical U-tubes carrying the coolant. The condensate is stored at the bottom of the condenser. The condensate level is controlled by using three on-off valves at the discharge. The U-tubes carrying the coolant are partially submerged in the condensate. Thus, there are heat transfers both from the steam and liquid phases. The changes in the condensate level results in variation of steam phase volume and produces compression or expansion of the steam.

The film thickness grows along the path of the condensate flow and there is further heat transfer from the flowing condensate to the cooling tubes. The local film thickness, effective total heat transfer coefficient, and rate of condensation can be calculated by using the well known Nusselt's formulae [181] for film-condensation over vertical tubes.

The condenser is also part of the loading system. Different condensate level setpoints change the length of the U-tubes in contact with the steam and thus change the rate of condensation. Thus the condensate level setpoints have direct effect on the steam flow rate (load) from the boiler and also on the frequency of condensate discharge. The heat transfer from the condensate to the coolant ensures that the condensate is at a safe temperature during the discharge and avoids *pool boiling* of the condensate.

For modeling purposes, the tubes are split into three parts [164]: at the input to the condenser in touch with the steam (section 1), within the condensate (section 2) and then again in contact with the steam at the output from the condenser (section 3). It is assumed that the condenser is insulated, the parameters are localized and the steam in the condenser is saturated. Usual Nusselt's assumptions [48] are made:

1. The steam is pure and saturated. Advection is neglected.
2. The temperature of the condensation surface is constant.
3. The condensate flows down due to gravity force and the flow is laminar.
4. The maximum film thickness is small compared to the diameter of the tubes.

The assumptions that the steam is pure and the condensate flow is laminar does not hold good in practice. Due to deposition of sediments over the tubes carrying the coolant, the condensate flow is wavy and the heat transfer coefficients are different as compared to the theoretical one. Furthermore, the steam is not pure and contains non-condensable gases. Therefore, corrections are applied to the Nusselt's formulae [216] to represent the actual rate of condensation and effective heat transfer coefficient.

The pseudo-bond graph model of the condenser is shown in Figure 6.8.

In the pseudo-bond graph model, the storage of mass and energy of the steam is modeled in C_s field element. The governing equations for this element are identical to the governing equations for the boiler, given in Equation6.6. Similarly, the storage of condensate at the bottom of the condenser is modeled by the elements C_h and C_t. This part of the model is identical to the model of the storage tank given in Figure 6.3, because the stored condensate is under-saturated. The actual pressure at the exit of the condenser (bottom) is calculated by adding the differential pressure due to condensate water head and the steam pressure at the 1-junction with bond number 48.

The most complicated part of this model is the representation of the Nusselt's formulae for condensation using the modulated multi-port resistive field element, R_n. Besides a lot of geometrical and physical parameters, the rate of condensation primarily depends upon the temperature of the steam and the condensing surface,

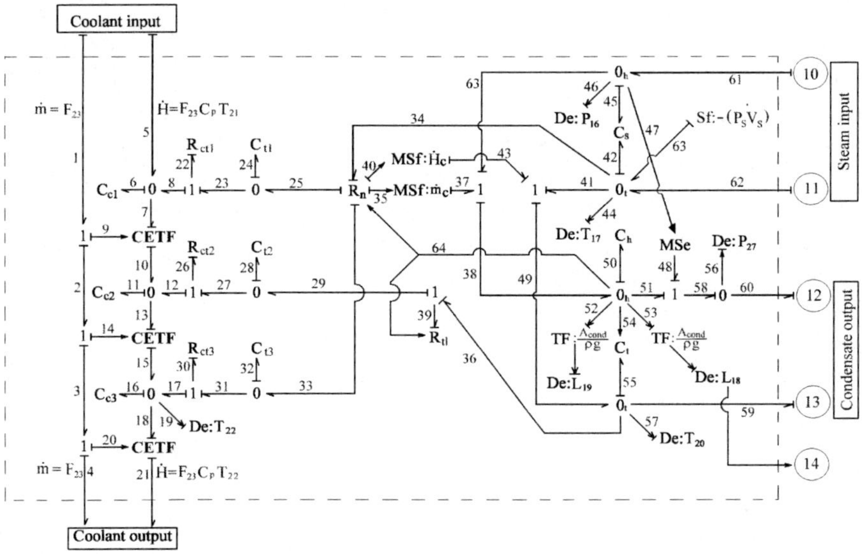

Fig. 6.8 Pseudo-bond graph model of condenser heat-exchanger

i.e. the tubes. Condensate is formed over the vertical tubes in contact with the steam and flows downwards due to gravity. The rate of mass flow of condensate is calculated by this element and represented using a modulated flow source, $MSf : \dot{m}_c$. It is considered that the mass transfer from the steam phase to stored liquid form of the condensate is instantaneous. Condensation occurs by removal of the latent heat of vaporization and subsequent cooling of the flowing condensate deposited over the tubes. This heat is transferred from steam phase to the wall of the tubes through conduction. It is represented in bond numbers 34, 25, and 33 which are connected to the R_n field. The rate of condensation also depends upon the length of the tubes in contact with the steam. This length is dependent on the level of the stored condensate, which is represented as a modulation of the R_n field in bond number 64. It is assumed that the hot condensate which flows downwards transfers thermal energy instantaneously from the steam phase to the stored condensate. This convection energy transfer is modeled using the modulated thermal flow source $MSf : \dot{H}_c$.

The rate of condensation $\dot{m}_c$ and the resulting heat exchange $\dot{Q}$ from steam to the tubes are given by

$$\dot{m}_{ci} = \frac{\pi\, D\, n_t\, g\, \rho_l\, (\rho_l - \rho_v)}{3\mu_l} \left[\frac{4\mu_l\, \lambda_l\, l\, (T_s - T_{ti})}{g\, \rho_l\, (\rho_l - \rho_v)\, (X\, L_v + 0.68\eta c_p (T_s - T_{ti}))} \right]^{\frac{3}{4}},$$

$$\dot{m}_c = \sum_{i=1}^{i=2} \dot{m}_{ci}, \text{ and } \dot{Q} = \sum_{i=1}^{i=2} \dot{m}_{ci}\, (X\, L_v + 0.68\eta\, c_p\, (T_s - T_{ti})), \tag{6.11}$$

where $i = 1,2$ represents the two sections of the tube in contact with the steam, n_t is the number of the U-tubes, ρ_l and ρ_v are respectively the condensate and steam densities given as functions of temperature, μ_l is the dynamic viscosity of the condensate, λ_l represents the heat transfer coefficient of condensate, T_s, T_{t1} and T_{t2} are respectively the saturated temperature of the steam, the temperature of condensing surface of the two sections of the tube in contact with the steam, L_v is the latent heat of vaporization of the steam, c_p stands for the specific heat capacity of the condensate, X is the steam quality, l and D are respectively the length and the diameter of the segment of tubes exposed to the steam. For the condenser in the steam generator installation, the heat transfer from the condensate to the tubes is corrected by a factor of η to account for film rippling and the effects of non-condensable gases [48, 216].

The length of the tube, l, exposed to the steam in each section is calculated as

$$l = l_t - \Phi_{cg}(\Delta P_l), \tag{6.12}$$

where l_t is the total length of each U-tube including the bends, ΔP_l is the differential pressure due to weight of stored condensate (state Q_{50} gives mass) and Φ_{cg} is a non-linear function used to calculate the length of the tube inside the condensate by using geometrical parameters (shape, bends, cross-section, *etc.*) of the condenser and density of the condensate. The simplest approximation is division of the condensate volume by the condenser cross-sectional area excluding the cross-sectional area of all the submerged tubes.

The rate of enthalpy flow of the condensate deposited on the tubes to the condensate storage at the bottom of the condenser is given by

$$\dot{H}_c = \dot{m}_c(X\, L_v + c_p T_s) - \dot{Q}. \tag{6.13}$$

The heat transfer from the stored condensate to the cooling tubes is modeled by R_{tl} element, for which the overall heat transfer coefficient is given by $\lambda_{tl}\Phi_{cg}(\Delta P_l)$, where λ_{tl} is the heat transfer coefficient between the stored condensate and the metallic walls of the tube for unit length of the tube, and Φ_{cg} is the function described in Equation 6.12.

The heat capacity of the coolant is represented by three storage elements, C_{c1}, C_{c2} and C_{c3}, corresponding to Sections 1–3, in the bond graph model of the condenser. The resistive elements R_{ct1}, R_{ct2} and R_{ct3} represent the heat transfer between the flowing coolant and the inner wall of the tubes. The storage elements C_{t1}, C_{t2} and C_{t3} represent the thermal capacity of the metallic tubes of corresponding sections. Since the actual length of these sections is not constant, all the elements described in this paragraph are modulated by the level of the condensate. This modulation (Equation 6.12) is not shown in the bond graph model to preserve the clarity of the drawing.

6.2.6 Bond Graph Model of the Condensate Discharge Valves

The discharge valves are used to control the condensate level inside the condenser. The pseudo-bond graph model of this sub-system is shown in Figure 6.9.

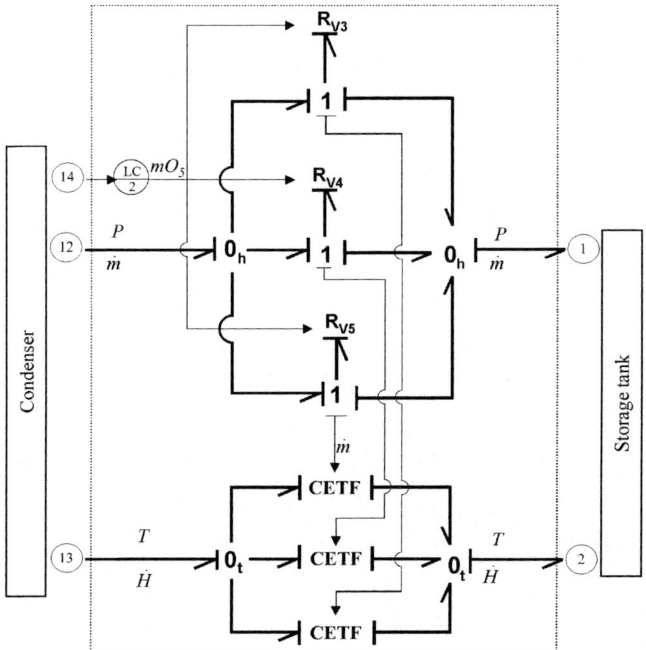

Fig. 6.9 Pseudo-bond graph model of the condensate discharge valves

The valves R_{v3}, R_{v4} and R_{v5} receive the same command, mO_5, given from the level controller. These three valves provide material redundancy and ensure safety of the process. The mass flow of condensate f_i and enthalpy flow $\dot{H}_i$, with $i = 3,4,5$, going through each valve are given by

$$f_i = V_i\_C_d.\sqrt{|\Delta P|}.mO_5,$$
$$\dot{H}_i = f_i.c_p.T_c, \tag{6.14}$$

where ΔP stands for the pressure drop across the valves, $V_i\_C_d$ is the effective discharge coefficient of the i-th valve, and T_c is the temperature of the condensate.

6.3 Model Validation

The global bond graph model of the process shown in Figure 6.10 is validated through comparison between the model outputs (simulation results) and the experimental observations (collected by the sensors). The numerical values of the parameters of the process were obtained through geometrical measurements, experiments and from charts provided in the manufacturers' catalog. The values for the heat transfer correction factor η in the condenser was estimated from experimental data. At $\eta = 1.176$ (about 17.6%), acceptable coherence between the theoretical and experimental results were obtained. This value of η is close to the suggested 20% correction in [216].

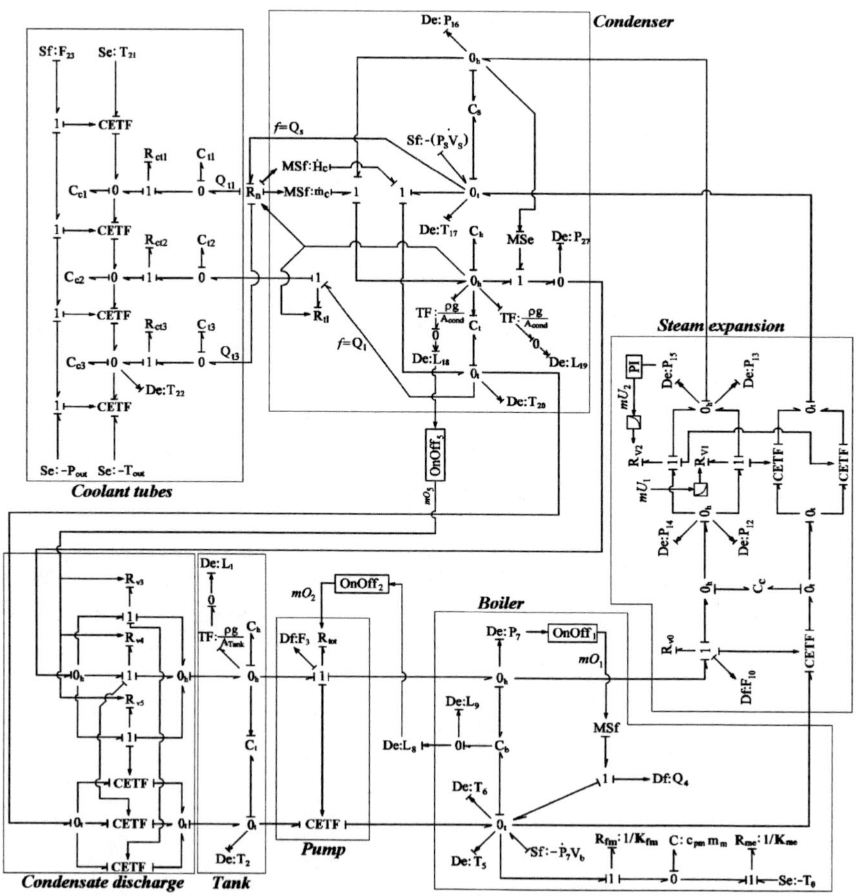

Fig. 6.10 Global bond graph model of the steam generator process

Replicating the behavior of the steam generator process every time is very difficult. This is due to excessive dependence on the environmental temperature variations. The coolant temperature at the inlet of the condenser depends on the outside environment temperature, which also influences the rate of heat loss. In the winter, the condensation and heating rates are higher (more frequency of condensate discharge and heater switching), whereas the opposite holds in the summer. The characteristics of the key equipment used in the steam generator process are given in Table 6.1. The nomenclature and the values for the parameters used in the simulation are given in Section 6.1.2. The thermodynamic functions are used instead of localized values taken from the steam table to allow application in a wide operating range.

Table 6.1 Technical specifications of process components

Boiler	Make: BABCOCK WANSON, Type: REV 60 KW (1993), Rating: 10bars, 89Kg/h, 380V
Valves	Make: SAMSON, Type: 241-1, DN 15 (1995),Rating: 80bars
Condenser	Make: CIAT, Type: UX 168 04 4P CL5 (1993),Rating: 16bars,4bars(coolant),8Ltrs
Pumps	Make: ABB (Asea Brown Boveri), Type: MEUC 90 L 4 (1993),Rating: 4bars(2500kg/h)-10bars(250Kg/h)

The functions $h_{\mathrm{v}}()$ and $h_{\mathrm{l}}()$ return specific enthalpies of steam and water, respectively, as functions of saturation pressure. Likewise, specific volumes of steam and water at saturation, as functions of pressure, are given by functions $\nu_{\mathrm{v}}()$ and $\nu_{\mathrm{l}}()$, respectively. The corresponding saturated steam temperature for a given pressure is determined by the function $Ps2Ts()$ and the density of the steam is given by $\rho_{\mathrm{v}}()$ as a function of pressure. These functions are implemented by using ASME steam tables [167]. The C-language program for the entire table is available as a freeware along with the required DLL (Dynamic Link Library).

The initial conditions used in the simulation are close approximations of their experimental steady state values corresponding to the specified setpoints. As an example, for the boiler operating at an initial pressure P_0 (measured by sensor P_7) and initial liquid volume L_0 (measured by sensors L_8 and L_9), the initial conditions for state variables m and H can be calculated as follows:

$$X_0 = \frac{(V_b - L_0)\ \nu_l(P_0)}{(V_b - L_0)\ \nu_l(P_0) + L_0\ \nu_v(P_0)},$$

$$m = \frac{V_b}{(1 - X_0)\nu_l(P_0) + X_0\ \nu_v(P_0)}, \text{ and } H = m\left((1 - X_0)h_l(P_0) + X_0\ h_v(P_0)\right). \tag{6.15}$$

The simulation results for the pressure (P_7) and water level (L_8) inside the boiler are given in Figure 6.11a. In Figure 6.11b, simulation results for the steam pressure (P_{16}), outlet coolant temperature (T_{22}) and the stored condensate temperature (T_{20}) of the condenser are given. As can be seen from Figure 6.11a, the level in the boiler continues to drop (because of the steam flowing into the condenser). After a long time (depending upon the load), when the level falls below the lower threshold specified by a setpoint, the pumps are switched on to fill the boiler up to the upper threshold specified by the setpoint. It can be seen from Figure 6.11b that the stored condensate initially heats up due to mixing with the high temperature just condensed liquid phase flowing from the U-tubes. The heat transfer from stored condensate to submerged tubes increases with increase in condensate level. Consequently, the stored condensate temperature attends a dynamic steady state.

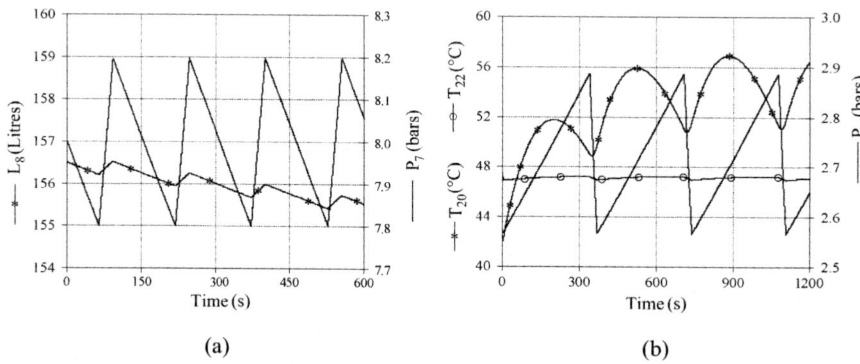

Fig. 6.11 Simulation results for **a** boiler and **b** condenser

The simulation was carried out using initial values corresponding to a sample experimental data obtained from the real process. The data were collected after the process was brought to steady state with regular venting from the boiler and the condenser to remove the non-condensable gases. The simulation result was then compared with that same sample data set. The comparison of experimental and simulation results for the boiler is given in Figure 6.12a. An acceptable agreement between the results for boiler's pressure and the heater output is observed. Note that in the experimental observations, the triggering of the heater (switching on and off) is affected by the measurement noises (of sensor P_7). Furthermore, in the real process, there is always delay between the command and full power output of the heater. The actuator delays and measurement noises have not been considered in the model. This is why, even though the magnitudes of the simulated and experimentally observed responses are close, it is practically impossible to match the phases.

The comparison between the experimental and simulated results for the condenser is given in Figure 6.12b. These results show good match in amplitudes, but

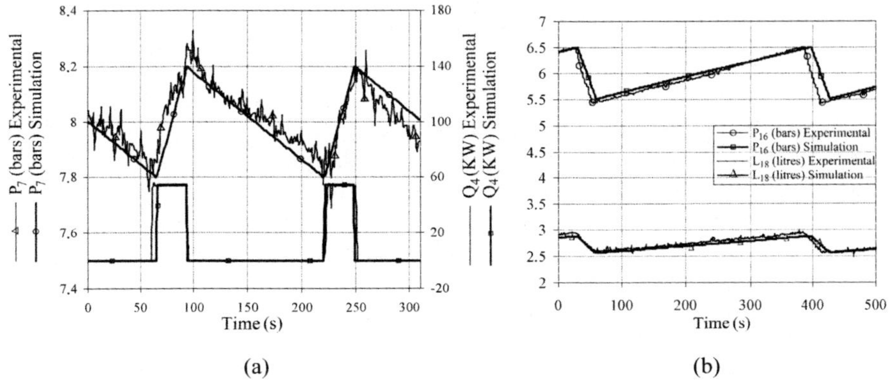

Fig. 6.12 Comparison between the simulated and the experimental results. **a** Boiler. **b** Condenser

slight errors in phases. These drifts are attributed to the measurement noises and actuator delays. Similarly, the experimentally observed values of temperatures in the boiler and the condenser are very close to the simulated results, as shown in Figure 6.13.

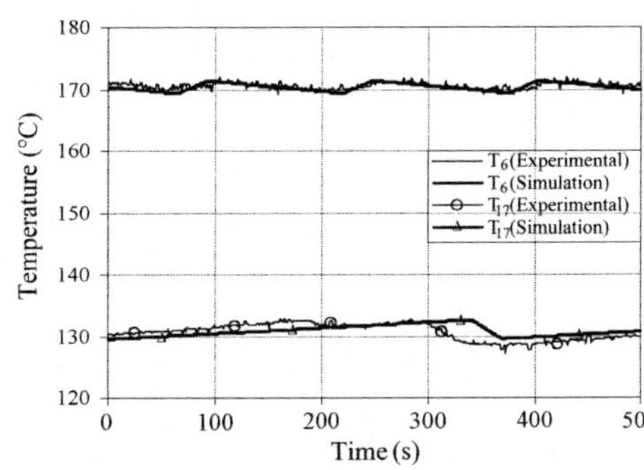

Fig. 6.13 Comparison between the simulated and the experimental results for temperatures in the boiler and the condenser

It was found that the experimentally observed temperature of the liquid in the boiler is approximately 3–4 $°C$ less than the simulated results (the approximate relative error is 1.8 –2.3%). This is because of the assumptions made during the modeling stage where the water and steam have been considered as a homogeneous mixture. This error in the magnitude and the actuator delays are considered in the design of the FDI system. In fact, using residuals computed from comparison of the process data and the output from the behavioral model for FDI does not work well in presence of such response phase drifts. Residuals derived by using derivative forms are robust to phase drifts because they only rely on the local observation. This justifies the use of derivative form of the residuals in monitoring of the steam generator installation.

6.4 Design of the Supervision System

The global model of the steam generator installation is created by linking the sub-models together. The storage elements in the bond graph model are assigned preferred differential causality to derive ARRs. Furthermore, some static constraints between the variables are added to the graph. In the behavioral model, static constraints are either unnecessary or are hidden within the relationships of the elements (*e.g.* linear dependence of rows of a matrix for a field element). The global model in preferred differential causality is shown in Figure 6.14.

The pressure and the temperature of the steam in saturated condition are related by thermodynamic equilibrium conditions given in steam tables. Thus, when one of

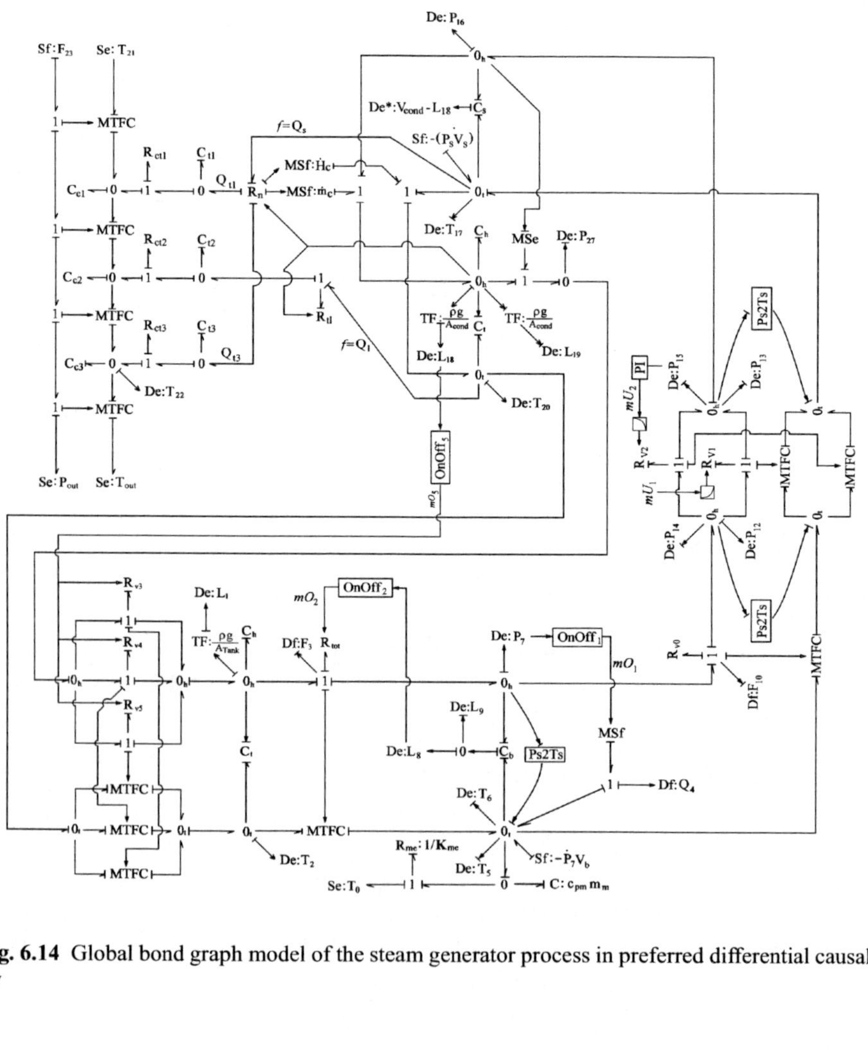

Fig. 6.14 Global bond graph model of the steam generator process in preferred differential causality

them is considered as an independent variable (pressure is considered as independent variable here), the other becomes a dependent variable. In the model, the steam temperature is represented as a function of the steam pressure by using the block Ps2Ts. Similarly, the condenser has a constant volume and measurement of the liquid volume by sensors L_{18} and L_{19} implicitly leads to measurement of the steam volume. Such static constraints are represented as virtual sensors in the model. Simulation studies presented here and also by Åström and Bell [9] of drum boiler dynamics have shown that the dynamic-state metal temperature is close to the saturation steam temperature. Thus, the thermal resistance, R_{fm}, between the boiler and its body is neglected, which eliminates a possible differential-algebraic loop between R_{fm} and R_{me}. This is justified, since $R_{fm} \ll R_{me}$. Note that the non-linear coupling capacitor, C_s, used to avoid algebraic loops in the steam expansion sub-system is not necessary in the preferred differential causality model.

6.4.1 Determination of Hardware Redundancies

When the causalities of some of the sensors are inverted to assign differential causality to storage elements, the causalities of a set of sensors cannot be inverted. Usually, causalities of the sensors which are used to control the process are preferably inverted. The sensors whose causalities cannot be inverted represent direct hardware redundancies [240] in sensor placement. Direct hardware redundancies can be deduced by following the causal paths [226]. Direct hardware redundancies detected in the steam generator installation are given in Table 6.2.

Table 6.2 Direct hardware redundancies of sensors

Redundant sensor	Redundancy with ?
L9	L8
T6	P7
T5	P7
P14	P12
P13	P16
P15	P16
T17	P16
L19	L18
P27	(P16, L18)

In the steam generator, direct redundancies lead to nine static ARRs. These redundancies not only improve the fault isolation algorithm, but also provide the means for system reconfiguration [133, 157, 225, 276] and fault tolerant control [16, 112, 224] of the process in the presence of faults.

6.4.2 Derivation of ARRs

In this section, we discuss only some complex parts of the process. The relations for the C_b field in differential causality, describing the mass and energy storage in the boiler, is given by $[m \;\; H]^T = \Phi_{C_b}(P, T, L)$, where m, H, P, T and L represent the total mass and enthalpy of the mixture, steam pressure, temperature and volume of the liquid phase, respectively. Note that since the steam temperature is a dependent variable, mass and enthalpy flow cannot be calculated from the effort information in the two passive ports. The inverted causality in the active port imparting measured liquid level information is used to calculate those variables as follows:

$$m_l = \frac{L_8}{v_l(P_7)}, \quad m_v = \frac{V_b - L_8}{v_v(P_7)}, \quad X = \frac{m_v}{m_l + m_v},$$
$$m = m_l + m_v, \quad H = m_l.h_l(P_7) + m_l.h_l(P_7), \tag{6.16}$$

where m_l and m_v are, respectively, the masses of liquid and steam phases, and X is the steam quality. The solution given in Equation 6.16 is used to derive residuals for the boiler.

The equations in the model subspace representing the condenser and the cooling circuits cannot be resolved because in the preferred differential causality model, the storage elements C_{c1} and C_{c2} cannot be assigned differential causalities with the given set of sensors. This is exactly what is meant by an under-determined/under-constrained subsystem [240] resulting from our inability to assign proper computational causalities.

However, when a projected subspace is considered whereby the total volume inside the condenser, including both the phases, is taken as the control volume and the net heat transfer to the outside of this control volume in steady state condition is used, a subset of the equations belonging to the subspace can be resolved. The solution can then be obtained as given below:

$$f_{h\_in} - s\left(\frac{V_{Cond} - L_{18}}{v_v(P_{16})}\right) - \dot{m}_c = 0, \tag{6.17}$$

$$\dot{m}_c - \rho.s(L_{18}) - f_{h\_out} = 0, \tag{6.18}$$

$$f_{t\_in} - s\left(\frac{V_{Cond} - L_{18}}{v_v(P_{16})} h_v(P_{16}) + P_{16}.L_{18}\right) - \dot{H}_c - Q_s = 0, \tag{6.19}$$

$$\dot{H}_c - \rho.c_p.s(L_{18}.T_{20}) - Q_l - f_{t\_out} = 0, \tag{6.20}$$

where f_{h_in}, f_{h_out}, f_{t_in}, f_{t_out} are respectively known hydraulic and thermal flows to condenser, $\dot{m}_c$ and $\dot{H}_c$ represent the flow rate of condensation and the enthalpy flow convected by the condensate, the sum of Q_s and Q_l gives the total heat taken by the coolant from the condenser, and s stands for time derivative operator.

Eliminating unknown variables $\dot{m}_c$ and $\dot{H}_c$ from Equations 6.17–6.20 leads to

$$f_{h\_in} - s\left(\frac{V_{Cond} - L_{18}}{v_v(P_{16})}\right) - \rho.s(L_{18}) - f_{h\_out} = 0, \tag{6.21}$$

$$f_{t\_in} - s\left(\frac{V_{Cond} - L_{18}}{v_v(P_{16})} h_v(P_{16}) + P_{16}.L_{18}\right) \\ -\rho.c_p.s(L_{18}.T_{20}) - (Q_s + Q_l) - f_{t\_out} = 0. \tag{6.22}$$

By using steady state assumptions in coolant and U-tubes,

$$Q_s + Q_l = (Q_{t1} + Q_{t3} + Q_l) = F_{23}.c_p.(T_{22} - T_{21}).$$

Thus, Equation 6.22 can be written as

$$f_{t\_in} - s\left(\frac{V_{Cond} - L_{18}}{v_v(P_{16})} h_v(P_{16})\right) - \rho.c_p.s(L_{18}.T_{20}) \\ -F_{23}.c_p.(T_{22} - T_{21}) - f_{t\_out} = 0. \tag{6.23}$$

Note that only two residuals, Equations 6.21 and 6.23, are obtained for the steam-condensate control volume as a subspace containing four constraints given in Equations 6.17–6.20.

The analytical redundancy relations for the other subspaces of the steam generator model are easily derived by using the algorithm and approach given in the previous chapter.

The analytical redundancy relations derived from the bond graph model of the steam generator are given here. The nomenclature of variables used in these equations is given in Section 6.1.2 at the start of this chapter.

1. The storage tank

$$ARR_1: \ (V_{3}\_C_d + V_{4}\_C_d + V_{5}\_C_d)\sqrt{P_{16} + \rho.g.\frac{L_{18}}{A_{Cond}} - P_0 - \rho.g.\frac{L_1}{A_{Tank}}}.mO_5 \\ -F_3 - \rho.s(L_1) = 0, \tag{6.24}$$

$$ARR_2: \ (V_{3}\_C_d + V_{4}\_C_d + V_{5}\_C_d)\sqrt{P_{16} + \rho.g.\frac{L_{18}}{A_{Cond}} - P_0 - \rho.g\frac{L_1}{A_{Tank}}}.mO_5.c_p.T_{20} \\ -T_2.c_p.F_3 - T_2.\rho.s(L_1) - \rho.L_1.s(T_2) = 0. \tag{6.25}$$

2. The water supply system

$$ARR_3: \ F_3 - \frac{(V_{7}\_C_d)^2}{2.K_1}.mO2 + \frac{V_{7}\_C_d}{2.K_1} \times \\ \sqrt{(V_{7}\_C_d)^2 - 4.K_1.\left(K_2 + K_1.(P_7 - P_0 - \rho.g.\frac{L_1}{A_{Tank}}\right)} mO_2 = 0. \tag{6.26}$$

3. The boiler

$$ARR_4 : F_3 - F_{10} - s\left(\frac{(V_{boiler} - L_8)}{\nu_v(P_7)} + \frac{L_8}{\nu_l(P_7)}\right) = 0, \tag{6.27}$$

$$\begin{aligned} ARR_5 : {} & F_3.c_p.T_2 - F_{10}.h_v(P_7) - K_{me}.(\mathbf{Ps2Ts}(P_7) - T_0) + Q_4 \\ & -s\left(\begin{array}{c} \frac{L_8}{\nu_l(P_7)}.h_l(P_7) + \frac{(V_{boiler} - L_8)}{\nu_v(P_7)}.h_v(P_7) \\ -P_7.V_{boiler} + c_{pm}.m_m.\mathbf{Ps2Ts}(P_7) \end{array}\right) = 0. \end{aligned} \tag{6.28}$$

4. The heater

$$ARR_6 : Q_4 - W_p.mO_1 = 0, \tag{6.29}$$

5. The steam expansion system

$$ARR_7 : F_{10} - V_0\_C_d.\sqrt{P_{12}}.\sqrt{P_7 - P_{12}}.\Phi_e(mU_0) = 0, \tag{6.30}$$

$$\begin{aligned} ARR_8 : {} & F_{10} - \sqrt{P_{16}}.\sqrt{P_{12} - P_{16}}.(V_1\_C_d.\Phi_e(mU_1) + \\ & V_2\_C_d.\Phi_e(mU_2)) = 0. \end{aligned} \tag{6.31}$$

6. The condenser

$$\begin{aligned} ARR_9 : {} & [V_1\_C_d.\Phi_e(mU_1) + V_2\_C_d.\Phi_e(mU_2)]\sqrt{P_{16}}.\sqrt{P_{12} - P_{16}} - \\ & \rho.s(L_{18}) - s(\rho_v(P_{16})(V_{Cond} - L_{18})) - (V_3\_C_d + V_4\_C_d + V_5\_C_d). \\ & \sqrt{P_{16} + \rho.g.\frac{L_{18}}{A_{Cond}} - P_0 - \rho.g.\frac{L_1}{A_{Tank}}}.mO_5 = 0, \end{aligned} \tag{6.32}$$

$$\begin{aligned} ARR_{10} : {} & -s\left(\begin{array}{c} (V_{Cond} - L_{18}).\frac{h_v(P_{16})}{\nu_v(P_{16})} + \rho.c_p.T_{20}.L_{18} \\ -P_{16}.V_{Cond} + c_{pCond}.Mt_{Cond}.\mathbf{Ps2Ts}(P_{16}) \end{array}\right) \\ & + h_v(P_{12}).[V_1\_C_d.\Phi_e(mU_1) + V_2\_C_d.\Phi_e(mU_2)].\sqrt{P_{16}}.\sqrt{P_{12} - P_{16}} \\ & - (V_3\_C_d + V_4\_C_d + V_5\_C_d).\sqrt{P_{16} + \rho.g.\frac{L_{18}}{A_{Cond}} - P_0 - \rho.g.\frac{L_1}{A_{Tank}}}. \\ & mO_5.c_p.T_{20} - F_{23}.c_p.(T_{21} - T_{22}) = 0. \end{aligned} \tag{6.33}$$

7. The controllers

$$ARR_{11} : mO_1 - \mathbf{OnOff_1}(P_{7\_ref}, P_7) = 0, \quad (6.34)$$
$$ARR_{12} : mO_2 - \mathbf{OnOff_2}(L_{8\_ref}, L_8) = 0, \quad (6.35)$$
$$ARR_{13} : mU_2 - \mathbf{PI}(P_{15\_ref}, P_{15}) = 0, \quad (6.36)$$
$$ARR_{14} : mO_5 - \mathbf{OnOff_5}(-L_{18\_ref}, -L_{18}) = 0. \quad (6.37)$$

8. The sensors

$$ARR_{15} : T_5 - T_6 = 0, \quad (6.38)$$
$$ARR_{16} : T_6 - \mathbf{Ps2Ts}(P_7) = 0, \quad (6.39)$$
$$ARR_{17} : P_{12} - P_{14} = 0, \quad (6.40)$$
$$ARR_{18} : P_{16} - P_{13} = 0, \quad (6.41)$$
$$ARR_{19} : P_{16} - P_{15} = 0, \quad (6.42)$$
$$ARR_{20} : P_{16} + \rho.g.\frac{L_{18}}{A_{Cond}} - P_{27} = 0, \quad (6.43)$$
$$ARR_{21} : T_{17} - \mathbf{Ps2Ts}(P_{16}) = 0, \quad (6.44)$$
$$ARR_{22} : L_{19} - L_{18} = 0, \quad (6.45)$$
$$ARR_{23} : L_8 - L_9 = 0. \quad (6.46)$$

6.4.3 Practical Fault Signature Matrix and Residual Sensitivity

It is important to distinguish between two different fault signature matrices: a theoretical one which is obtained from ARRs and the one which is really used and implemented to detect and isolate faults. The main difference between these two is that the practical matrix takes into account the relative sensitivity of the residuals to variations of the measurements and the parameters.

In general, the practical fault signature matrix is a refined form of the theoretical one, obtained by taking the actual numerical values of dimensions, environment, operating setpoints, *etc.* into consideration.

Let us consider the case of the residual R_3, deduced from the analytical redundancy relation of the pump (Equation 6.26). It is theoretically influenced by the characteristics of the pump in operation and by the state of immediately preceding and succeeding components obtained through measurements. At the operating point of the steam generator, $P_7 = 8.10^5$ Pa and $\frac{L_1}{A_{Tank}} \leq 1$ m. Thus, the term $\rho.g.\frac{L_1}{A_{Tank}} \leq 10^4$ Pa can be neglected when compared to the value of P_7 in the residual. In fact, the effect of L_1 and A_{Tank} in Equation 6.26 is even less than the effects caused by the measurement inaccuracies of P_7 (the sensor accuracy is $\pm 1.5\%$). Thus in the practical fault signatures matrix, $S_{L_1,3} = S_{Tank,3} = 0$, which means residual R_3 is not sensitive to L_1 sensor fault and tank faults (leakage, *etc.*).

To determine the signature of each component, all the measurements and component parameters are exploited. The practical signature matrix was constructed by using comparative analysis, simulations and offline verifications.

6.4.4 Effect of Hybrid Components

Supervision of hybrid systems is complex due to the different modes in the process. These modes must be identified from the process outputs. Details regarding supervision of hybrid systems can be found in [156]. The steam generator plant contains three hybrid controllers (boiler level, boiler pressure, and condenser level controllers). The output of each controller is either zero or one. Those ARRs which contain conditional participation(s) of actuators modulated by the hybrid controller outputs lead to different structures in different modes. During the operation of the boiler, 2^3 fault signature matrices are possible. The controller outputs unambiguously determine the transition condition from one mode to another. As an example, the signature of P_7 sensor is given in Table 6.3. This signature changes only due to the boiler level controller output mO_2.

Table 6.3 Fault signature of sensor P_7

Condition	r_1	r_2	r_3	r_4	r_5	r_6	r_7	$r_8 \ldots r_{15}$	r_{16}	$r_{17} \ldots r_{23}$
$mO_2 = 0$	0	0	**0**	1	1	0	1	0	1	0
$mO_2 = 1$	0	0	**1**	1	1	0	1	0	1	0

Note that non-conditional sections of the ARRs do not change the fault signatures, even when measurements corresponding to hybrid controller outputs are present in them, because they do not represent any actuator dynamics.

The practical fault signature matrix of the steam generator process, given in Table 6.4, corresponds to the case where all hybrid controller outputs are equal to one. Measurements mU_0, mU_1, mU_2, (controller outputs and user settings) and the soft sensors (mO_1, mO_2 and mO_5) are excluded from the fault signature matrix. In fact, the latter three soft sensors are used to switch between the 2^3 possible hybrid modes. Furthermore, to avoid further hybrid configurations, it is assumed that mU_0 is never

Table 6.4 Practical fault signature matrix of the steam generator process

	D_b	I_b	r_1	r_2	r_3	r_4	r_5	r_6	r_7	r_8	r_9	r_{10}	r_{11}	r_{12}	r_{13}	r_{14}	r_{15}	r_{16}	r_{17}	r_{18}	r_{19}	r_{20}	r_{21}	r_{22}	r_{23}
L_1	1	0	1	1																					
T_2	1	1		1			1																		
F_3	1	1	1	1	1	1	1																		
L_8	1	1				1	1																		1
L_9	1	1																							1
P_7	1	1			1	1	1		1									1							
Q_4	1	1					1	1																	
T_5	1	1															1								
T_6	1	1															1	1							
F_{10}	1	1				1	1		1	1															
P_{12}	1	1							1	1	1	1							1						
P_{13}	1	1																		1					
P_{14}	1	1																	1						
P_{15}	1	1																			1				
P_{16}	1	1	1	1						1	1	1								1	1	1	1		
T_{17}	1	1																					1		
L_{18}	1	1									1	1										1		1	
L_{19}	1	1																						1	
P_{27}	1	1																				1			
T_{20}	1	1		1								1													
Tank	1	0	1	1																					
Pump	1	1			1																				
Boiler	1	1				1	1																		
Condenser	1	1									1	1													
Heater	1	1						1																	
$OnOff_1$	1	1											1												
$OnOff_2$	1	1												1											
$OnOff_5$		1														1									
PI	1	1													1										
V_0	1	1							1																
V_1	1	0								1	1	1													
V_2	1	0								1	1	1													
V_3	1	0	1	1							1	1													
V_4	1	0	1	1							1	1													
V_5	1	0	1	1							1	1													

0, and both mU_1 and mU_2 are not simultaneously closed. In Table 6.4, the terms D_b and I_b appearing in the first and second columns represent the detectability and the isolability.

It is concluded from analyzing this table that the sensor L_1 and the tank have the same fault signature which implies that faults in these two components (L_1 sensor failure, tank leakage or something falling into the tank, *etc.*) are not isolatable. One additional level sensor in this part of the process can be a remedy to this problem. It is further observed that faults in the valves V_1 and V_2 are not isolatable. The isolation of the faults on these valves needs additional number of sensors. In fact, by analyzing the causal structure of the bond graph model, it can be concluded that one flow sensor in one of the branches in the steam expansion system is the minimum requirement for isolability in the concerned sub-space of the model. These decisions are automatically generated by the ModelBuilder software [191]. Similarly, faults in the valves V_3, V_4 and V_5 are also not isolatable and need additional sensors.

6.4.5 Selection of Decision Procedure

Once analytical redundancy relations are obtained and practical fault signature matrices are constructed, the decision procedure is the next important step in the design of a supervision system. The decision procedure is applied to the set of the evaluated residuals at each sampling time and generates the alarms. In this application, we apply a residual pre-processor $\Psi_{\Delta T}$, which takes a moving average of residuals for a duration of ΔT. The threshold is considered by bounding $|\Psi_{\Delta T}(r_i)|$ by a small value ε_i. As explained previously, this non-zero bound ε_i is due to process uncertainties, modeling errors and measurement noises. Then, each element, c_i, of the coherence vector C is obtained by using the following rule:

$$c_i = \begin{cases} 1, \text{ if } |\Psi_{\Delta T}(r_i)| > \varepsilon_i; \\ 0, \text{ otherwise.} \end{cases} \tag{6.47}$$

The chosen value of ε_i must neither be too big (to avoid the missed alarms or non-detection), nor be too small (to avoid the false alarms). In the current application, a moving average of 10 s time window is performed on all the residuals.

The thresholds chosen for scaled residuals of this installation are given in Table 6.5 along with the corresponding scaling factors, ζ. Proper scaling of residuals to obtain uniform residual threshold is discussed in later chapters. Hereafter, the term residual (r_i, $i = 1..23$) in this chapter means its scaled value.

Table 6.5 Experimental residual thresholds after moving averaging over 10 s

r_i	ζ_i	ε_i	r_i	ζ_i	ε_i	r_i	ζ_i	ε_i
r_1	1	0.2	r_9	1	0.025	r_{17}	10^5	0.2
r_2	8.3×10^4	0.2	r_{10}	2.67×10^6	0.0019	r_{18}	10^5	0.2
r_3	0.25	0.4	r_{11}	1	0.5	r_{19}	10^5	0.2
r_4	1	1.0	r_{12}	1	0.5	r_{20}	10^5	0.2
r_5	2.67×10^6	0.1	r_{13}	1	0.5	r_{21}	10	1.0
r_6	5.5×10^4	0.5	r_{14}	1	0.5	r_{22}	1	0.05
r_7	0.022	0.225	r_{15}	10	1.0	r_{23}	1	1.0
r_8	0.022	0.675	r_{16}	10	1.0			

For the steam generator application, the faults can be classified into two categories: abrupt faults and incipient faults.

Detection of abrupt faults is not trivial in the cases where the corresponding parameters of the faulty component are in a time derivative form. Generally, sensor variables appear in single or multiple time derivatives. The detection and isolation of the fault in these situations requires other techniques, like cumulative sum approach [13] or other statistical methods such as pattern recognition. Note that presence of material redundancy in these type of sensors can lead to additional residuals where the sensor variable can be in non-derivative form and thus can improve isolability.

In this work, it is assumed that incipient faults, such as leaks and sediment depositions, are progressive, *i.e.* gradual and irreversible during the period of fault. The degree of the fault and/or its rate of progress have direct influence on the efficacy of detection procedures.

Note that when involved measurements are in derivative form and the rate of the fault is small, the derivatives greatly magnify the effect of noises in the signal in comparison with any real drift in its mean value. Therefore, two-sided cumulative sum ($CuSum_2$) algorithm [13] is used for all residuals containing time derivatives of measurements.

6.5 Online Implementation

The steam generator process (Figure 6.15) is a pilot installation which mimics a reduced scale operation of a part of a power plant. This installation has been specifically designed to test FDI applications. It is possible to simulate most of the faults manually or automatically. A leakage from the condenser, the tank or the boiler can be simulated by opening the valves V_8, V_9 or V_{10}, respectively (refer to the P&ID given in Figure 6.1). A boiler output blockage can be contrived by fully or partially closing the valve V_0. Faults on sensors can be introduced by disconnecting corresponding power supply from switch boards situated in the plant specifically dedicated to this purpose, whereas the controller faults are performed directly from the computer by altering the algorithms. It is also possible to perform faults on the actuators such as the heater or the pumps by manipulating their power supplies.

The sensor readings are calibrated in different units (L_8 for example is calibrated to measure the volume in the boiler in liters, P_7 is calibrated to measure pressure in bars, *etc.*). So, before use in evaluating the residuals, all sensor measurements are scaled to their corresponding SI unit values. Also, all pressure sensors that give gage outputs are converted to absolute values.

6.5.1 Data Acquisition and Toolbox Integration

The basic tasks involved in supervision and process control are detailed in [2]. In ARR based online supervision, the first step involves implementation of the algorithms for the evaluation of the residuals and of the logic of the fault detection and

Fig. 6.15 The steam generator installation

isolation in a software. Panorama [54], which is an easy to use commercial supervision software, is used in this application. It is a client-server based supervision application. The software provides the possibility to assign and modify the process and the alarm variables, to draw the process synopsis, to display the alarms, to formulate and manage the control laws, to maintain the database and to archive and exchange data with other software, *etc.* The steam generator process is linked to the user interface by using a local network bus. The operator can control and exchange the data directly from the computer *via* this bus, the installed hardware, and the internet, by using the corresponding software drivers. Besides Panorama software, Gensym's G2 software [90] has been used to maintain the process database (DB) and the knowledgebase (KB). This KB can be used along with an expert system to provide additional supervision capabilities. One such work is reported in [99].

The scheme of data acquisition is shown in Figure 6.16. The process measurements are acquired at 1 s sampling interval by a program called FCTINTPP (Function to Initialize Panorama Process). This program, implemented in C-language, makes use of the drivers for the data-acquisition cards as well as those of the Panorama. This bridge program is also the ideal place to perform complicated mathematical calculations, such as the evaluation of the residuals. There is a two-way communication between the program interfaces. Four software timers are implemented in FCTINTPP. On receiving the timer messages, FCTINTPP acquires the sensor data from the acquisition cards, retrieves setpoints and commands from Panorama, performs evaluation of the residuals, processes them using decision pro-

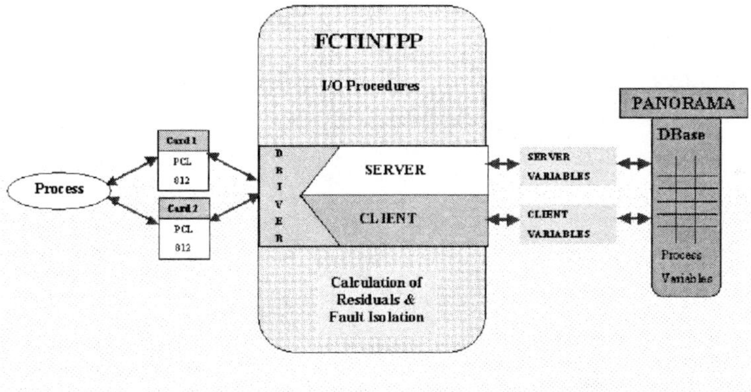

Fig. 6.16 Panorama data acquisition scheme

cedures and finally isolates any fault when present. Alarms generated in FCTINTPP are reported to Panorama where they are displayed to the operator in an appropriate manner.

The sampling period is sufficient for all calculations by using optimized C-language code and it leaves about 80% of processor resources free for use by other programs, such as the network communications, other diagnosis tools, future implementation of complicated and resource intensive algorithms, *etc.*

The results presented in this chapter are the outcome of research within the framework of CHEM project [1]. The steam generator installation served as one of the European test platforms for many different approaches to FDI. The ModelBuilder software discussed in Section 2.7 has been packaged as a toolbox (TB 5.1). The model in TB 5.1 can be refined infinitely (*e.g.* adding new sensors or changing sensor placements, *etc.*) to achieve the desired levels of isolability. It simultaneously provides a behavioral model including possible faults and linked with the residual equations. This behavioral model can then be simulated by using various software [174]. This toolbox plays an important role in the conception of an integrated supervision platform through out its life cycle. One important contribution of ModelBuilder is that it can generate XML (eXtended Markup Language) output of the residual equations and the fault signature matrix for direct transmission to other software in need of these.

Different toolboxes developed in CHEM project provide complementarities in analysis. The communication and interaction architecture between a part of these integrated operations is given in Figure 6.17.

TB 5.1 generates the process model, symbolic expressions for residuals, fault signature matrix and sends them to other interested toolboxes and to the process interface in XML form *via* the CCOM server (an instance of XML Blaster running using Java runtime and communicating over network through remote procedure calls us-

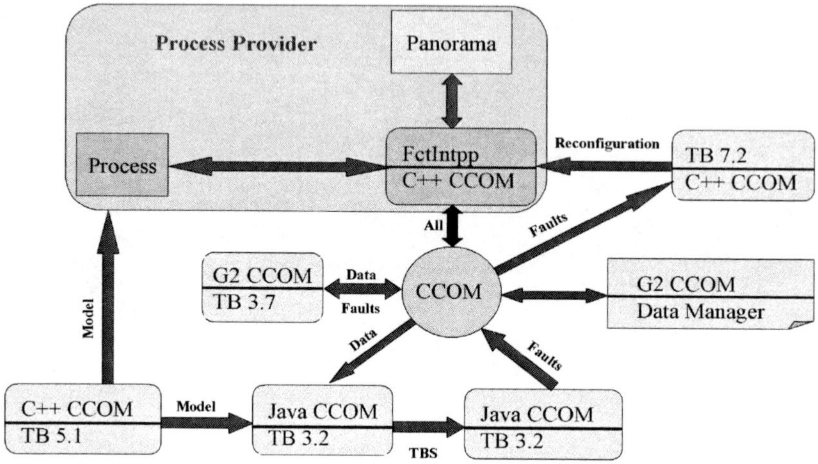

Fig. 6.17 Part of the integrated supervision platform

ing TCP/IP). TCP/IP is the preferred implementation standard for internet based process control, remote monitoring and remote supervision systems [158, 281]. Each of the toolboxes performs a specific task and uses a different methodology from the others. All communication and the data exchange between them is carried out by the CCOM server. Each toolbox can query the data directly from Data Manager (global network database) or latch onto periodic publish messages through CCOM. The standard forms of network communication using subscribe, publish, post, and point to point messaging protocols are implemented. Archived data for a specified duration can be obtained from Data Manager, which is both a DB and KB implemented using G2. Toolboxes can receive filtered and time stamped data indirectly from Data Manager. All toolboxes use XML-Parser Java class to parse and organize received messages to meaningful information and convert them into their corresponding internal data structures.

Data Manager stores time stamped process data (current values, archive, constants, static descriptions, scales and units, *etc.*), faults, alarm states, reconfiguration recommendations, current mode of operation, *etc.* in different storage areas (segments of database). Storage items or pages that contain the process data are created within the storage areas.

TB 5.1 provides the list of variables and equations of the residuals to toolbox (TB 3.2) which then transforms the residuals to polynomial form and generates the temporal band sequences [253] by taking the quantified uncertainties in the sensor measurements and the process parameters into account. Another toolbox, TB 5.2, receives process data from Data Manager *via* CCOM and TB 3.2; evaluates the temporal band sequences interval model and generates a list of probable faults when inconsistency is detected. Simultaneously, toolbox TB 3.7 for intelligent multi-variate statistical process control [201] (iMSPC), which has been precalibrated for the current operating mode and process states, receives process data and on detecting action limit violations on principal components, generates alarms for abnormal process state. TB 3.7 is implemented as a KB using Gensysm's G2. Similar work concerned with the application of MSPC integrated with a KB for real-time batch process supervision, implemented using G2, is presented in [261]. MSPC models based on principal component analysis (PCA) do not require an analytical process model. The statistical model is obtained from the calibrations using the process history DB for a specified operating range. Furthermore, recently MSPC and interval methods have been integrated to achieve good enhancement in diagnosis by using confidence limits on residuals of each variable rather than confidence limits on the overall residual [212].

A top-level toolbox called ReconfAnalyzer (TB 7.2) performs filtering of the corresponding alarms generated from different toolboxes, such as TB 5.2 and TB 3.7. Thereafter, the fault is isolated and a decision support is displayed to the operator in order to perform the appropriate actions and reconfiguration. All reconfigurations by the operator must be reported to TB 7.2 in order to update the current operating mode. TB 7.2 is responsible to alert all other toolboxes and the process interface of any mode changes performed, so that those toolboxes which can adapt to the new environment can reorganize themselves and others who are incapable in the given situation must enter the standby (suspended) state.

The complete integrated platform uses many more toolboxes, *e.g.* TB 3.4 and TB 5.4 for causal graph [105] based FDI, TB 5.3 for Hazard and Operability (HazOp) Studies, TB 3.1 for episode or trend chart generation [7], TB 3.6 for AI based FDI, TB 3.5 and 5.5 for Fuzzy Decision Trees (FDTs) based FDI, *etc.* Discussion on these toolboxes is out of the scope of this book.

Each toolbox has its own method and is implemented using different programming languages and can run in different computers. So, the data exchange with other toolboxes in XML format by using CCOM server requires bridges between the toolbox and the CCOM TCP/IP client component. The bridge depends on the type of programming interface/language and the target platform (Windows, Unix, *etc.*). C++ CCOM, G2 CCOM, Java CCOM bridges used in this integration platform correspond to the programs written in C++ language, G2 interface and Java language, respectively, for Microsoft Windows operating system.

6.5.2 Native Interface

Besides all these toolboxes, the process interface also contains a native implementation to detect and isolate faults. This implementation uses the exact form of the residuals and the set of fault signature matrices (for various hybrid states) presented in Section 6.4.4. Implemented in the FCTINTPP program, it detects and isolates the faults in the process and displays them in the Panorama synopsis window. The output of this implementation provides complementarities with other toolboxes and it has been used as a benchmark to test performance of other toolboxes in CHEM project.

The native interface is shown in Figure 6.18. This interface displays process the state (measurements), residuals, alarms and the corresponding faults. This native interface can be executed in any remote computer. Besides monitoring, these password protected client programs allow specific changes to process setpoints. The results presented in the following sections are obtained by using the native implementation.

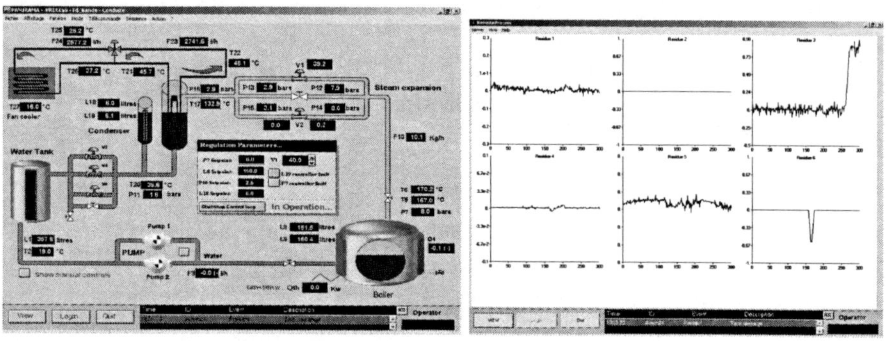

(a) (b)

Fig. 6.18 Supervision interface. **a** Native monitoring. **b** Remote supervision

6.6 Experimental Validation of Fault Scenarios

6.6.1 Process Faults

They are the physical faults that can happen in the steam generator such as wearing of the valves, leakages in the boiler, in the condenser and in the tank, blockages in the pipes and valves, *etc.*

Two separate process faults are treated here. They are leakage of water from the tank and blockage of the valve V_0 at the output of the boiler. The rate of leakage in the tank can be slow or fast. When it is fast (in this case approximately 0.4 L/s discharge in a time window from 250–300 s), the leakage can be detected by a simple comparison between the scaled magnitude of r_1 and its corresponding threshold ($\varepsilon_1 = 0.2$) as shown in Figure 6.19a. During the leakage, the scaled magnitude of r_2 also exceeds the corresponding threshold. All other residuals do not violate their thresholds during the leakage from the tank in absence of any other fault. Thus the large leakage of water from the tank generates a coherence vector $C = [1, 1, 0, ..., 0]$ which has two matches in the fault signatures matrix, given in Table 6.4, corresponding to component 'Tank' and sensor 'L_1'. Thus, even though this fault is always reliably detected; its isolation is not possible by simple approaches. However, a non-deterministic warning like "Tank fault (leakage or dumping) or tank level sensor failure" can be issued by the supervision platform. It can be seen from Figure 6.19a that during operations of the pump ($mO_2 = 1$), the fall in level L_1 is a normal phenomena and it does not affect the magnitude of residuals.

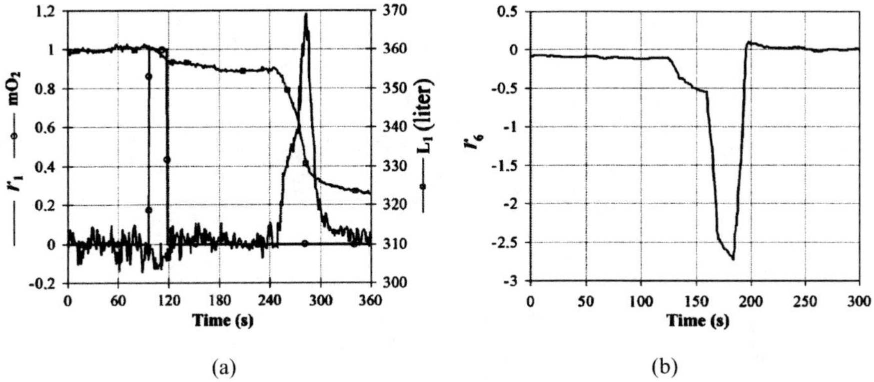

Fig. 6.19 Detection of **a** fast tank leakage and **b** complete boiler output blockage

The same threshold test is applied for the residual r_7 as shown in Figure 6.19b to detect the complete blockage in the valve V_0 during the time window from 120–180 s. This fault, which affects only residual r_7 and generates a unique signature, $C = [0,0,0,0,0,0,0,1,0,..,0]$, is easily detected and isolated.

Detection of very slow rate of water leakage from the tank (approx. 0.03 L/s, in this case, from 180–480 s) may not be possible by a simple threshold test as shown in Figure 6.20a. From the fault signature matrix, the leakage in the tank should result in the discrepancies of the residuals r_1 and r_2, only. In this case, the tank leakage corresponds to a variation of the level of water in the tank given by L_1 sensor measurements which is in a time derivative form and thus the residual deviates from its normal value by a small magnitude (Figure 6.20a). So, to detect this small rate of leakage by a simple redundancy relation, it is necessary to fix a small value of the threshold for residuals r_1 and r_2. However, this makes the detection algorithm very sensitive to these residuals and results in numerous false alarms. That is why the CuSum$_2$ (two-sided cumulative sum) algorithm is useful in detecting such small leakages. In an attempt to achieve robust detection and minimize false alarms due to the influence of unmodeled disturbances, alarm states are declared when the cumulative sum (g_k) of the residual exceeds a two-sided threshold (h) fixed *a priori*. Two cumulative sum (CUSUM) algorithms are used here: the first for detecting an increase in the mean, and the second for detecting a decrease in the mean. The resulting alarm time is given by

$$t_a = \min\left\{k : \left(g_k^+ \geq h\right) \cup \left(g_k^- \geq h\right)\right\}, \tag{6.48}$$

where $g_k^+ = \left(g_{k+1}^+ + y_k - \eta - \sigma\right)^+$ and $g_k^- = \left(g_{k-1}^- - y_k + \eta - \sigma\right)^+$.

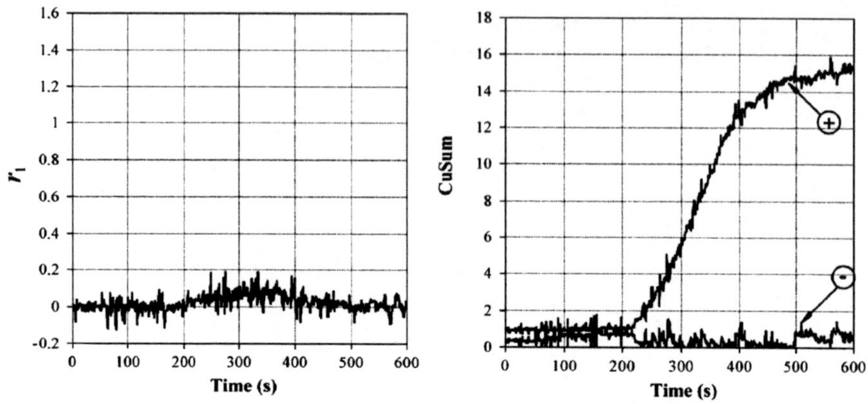

(a) (b)

Fig. 6.20 Detection of slow tank leakage. **a** Moving average. **b** $CuSum_2$

The output from the implementation of the $CuSum_2$ algorithm during small leakage from tank is shown in Figure 6.20b. The $\oplus$ and $\ominus$ symbols in the figures correspond, respectively, to the positive and the negative cumulative sums of the residual r_1 with $\nu = 0.05$. The slow leakage from the tank is detected when the positive side (g^+) of $CuSum_2$ is higher than a threshold ($h_c = 2$) fixed from experience.

In the steam generator application, $CuSum_2$ algorithm is used to construct the elements c_i of the coherence vector C corresponding to all residuals that contain derivatives of measurements. These residuals, their corresponding $CuSum_2$ thresholds (h_{ci} for i-th residual), and the chosen values of ν are given in Table 6.6.

Table 6.6 Thresholds for $CuSum_2$

	r_1	r_2	r_4	r_5	r_9	r_{10}
ν	0.05	0.05	0.1	0.05	0.005	0.001
h_c	2	2	5	1	0.3	0.02

The response to slow rate of leakage from condenser (0.05 L/s) during 150–210 s is shown in Figure 6.21. The positive components of $CuSum_2$ of residuals r_9 and r_{10} exceed their corresponding thresholds and since only these two residuals should be affected by condenser leakage, the fault is uniquely isolated.

Abnormal behavior in the process is detected when any one of the residuals (or its cumulative sum) exceeds the prescribed threshold. When multiple residuals (or their cumulative sums) exceed their corresponding thresholds, it is not guaranteed that they would do so at the same time. All care is taken in the present implementation to adjust the chosen thresholds and other decision procedure parameters to optimize simultaneous (or with a very small delay) detection across the broad range

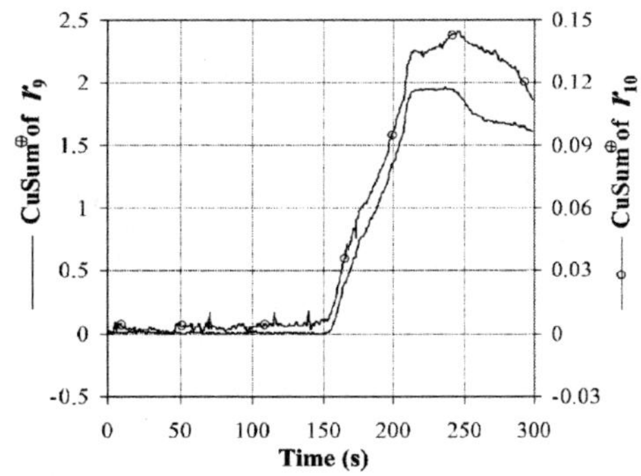

Fig. 6.21 Detection of leakage from the condenser

of residuals. The isolation procedures are activated after 5 s delay from the first instant of detection to give enough time for other residuals (or their cumulative sums) to reflect their consequent dynamics. This delay, fixed through experience on the steam generator process, avoids small periods of false diagnosis.

6.6.2 Sensor Faults

For the steam generator application, all sensor faults can be detected and isolated except the isolation of the fault in the sensor L_1. Note that only abrupt sensor faults are considered in this application. With this assumption, when a sensor is faulty, the changes in the trends and the magnitudes of the corresponding residuals depend on the form of participation of the sensor variable in the equations of those residuals.

Let us consider the case of a fault on P_7 readings. The corresponding fault signature of this component is given in Table 6.3. During the fault, the pump was not running ($mO_2 = 0$) and hence only four residuals (r_4,r_5,r_7 and r_{16}) are sensitive to the fault in the corresponding hybrid state. As explained before, detections of threshold violations in residuals r_4 and r_5 are done using CuSum$_2$, whereas direct thresholds are used to detect discrepancies in r_7 and r_{16}. These results are given in Figure 6.22 in presence of a fault in the pressure sensor P_7 during 90–190 s. The discrepancy is detected in all the four concerned residuals and since the corresponding coherence vector has a unique match in the fault signature matrix (in the given hybrid state), this failure is easily isolated. When the fault appears, one of the components of the CuSum$_2$ of residual r_5 responds to the changes depending on the sense of the fault. In this case, since P_7 sensor measurement suddenly becomes zero from a higher value, the positive component (g^+) of CuSum$_2$ of residual r_5 responds. Similar observations can be made for residual r_4. When the fault disappears, depending on its nature, the opposite component of the CuSum$_2$ (g^- in this case) responds to the sudden change. This implies: there would still be a threshold violation on the opposite component of CuSum$_2$ even after the fault has been repaired and continue to do so for an indefinitely long time or till the entire algorithm is reinitialized explicitly.

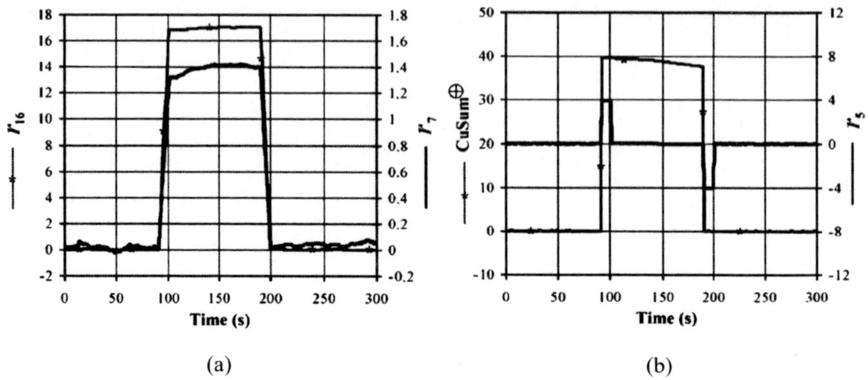

Fig. 6.22 P_7 sensor fault. **a** Normal detection in r_7 and r_{16}. **b** $CuSum_2$ detection in r_5

Likewise, faults in all of the sensors can be detected by using simple threshold tests on the residuals and the two-sided cumulative sums. Except for fault in the sensor L_1, all of the other sensor faults can be isolated in all hybrid configurations. When there is a fault in sensor L_1, the corresponding coherence vector has two identical matches in the fault signature matrix and a non-deterministic warning, as tank fault and/or tank level sensor fault, is made as described in Section 6.6.1.

6.6.3 Actuator Faults

These faults correspond to failures of actuators such as the pumps, the controlled valves and the heater. The failure of the heater is considered here. This can be caused by fuse burn-out due to short circuits, fatigue of the coil, *etc.*

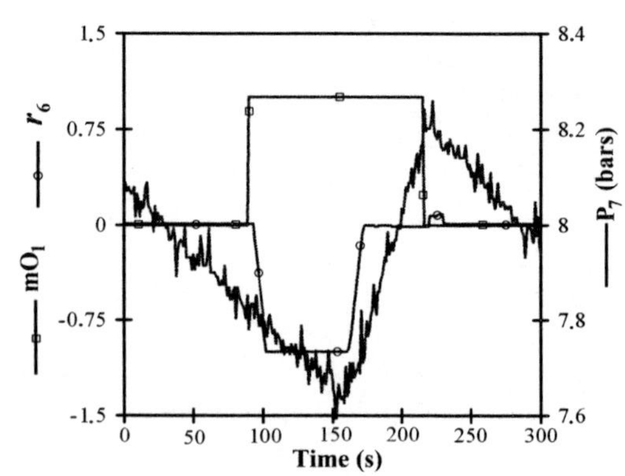

Fig. 6.23 Detection of heater fault

The detection of this fault is shown in Figure 6.23, which corresponds to the deviation of residual r_6 from its normal behavior. It can be seen that during the failure (60–160 s), while heater command mO_1 was non-zero, the steam pressure P_7 continued to drop below 7.8 bars ($P_{7\_ref} = 8 \pm 0.2$ bars). This is because the heater was not working properly during that duration and thus could not provide the power to the thermal resistor. Note that during 60–90 s, while the heater command mO_1 was zero, the fault was not detected.

Note that the actuator may be itself a complex system and it may not be always appropriate to model it through a simple constitutive relation. See Chapter 8 for an application concerning FDI of an actuator.

6.6.4 Controller Faults

A failure in the controller is explained as a discrepancy between its actual output and predicted output given as a function of the setpoint and the measurements involved in the regulation task. Faults in hardware controllers can be simulated by changing the controller parameters (*e.g.* gains, bias, *etc.*). The controller algorithms in the steam generator are implemented in the supervision software Panorama. Thus, controller faults can be easily simulated by deliberately altering the corresponding control algorithms as well as by changing the controller parameters.

As an example, the condenser level controller fault is considered here. The condensate in the condenser is regulated by an On-Off controller according to a level setpoint ($L_{18\_ref} \pm 0.5$ L). When the upper limit of the setpoint is reached the three On-Off valves open and condensate flows to the tank till the level is below the lower value of the setpoint. In the simulated faulty condition, the On-Off level controller does not act even though the level of the condensate inside the condenser has exceeded the prescribed setpoint ($L_{18\_ref} = 6 \pm 0.5$ L). The predicted value of the controller is always calculated according to the non-faulty algorithm and obviously a discrepancy occurs between the measured output mO_5 of the modified controller and the predicted value given by the function $\mathbf{OnOff_5}(-L_{18\_ref}, -L_{18})$ in Equation 6.37.

As shown in Figure 6.24, the fault on the controller is detected when deviation of the residual r_{14} exceeds the already fixed threshold, $\varepsilon_{14} = 0.5$.

The intermittent small jump in Figure 6.24 is due to 1–2 s delay between the command mO_5 sent by the controller and the effective opening or closing of the valves V_3, V_4 and V_5. The false detection caused by this delay (r_{14} becomes ± 1 for the duration of delay) is avoided by 10 s moving average, applied on all the residuals. This reduces the magnitude of residual r_{14}, due to the delay, to ± 0.1 (for 1 s delay). When the delay is more than 5 s, even the moving average exceeds the threshold. Thus, any delay of less than 5 s is considered normal. It is important to note that if the dynamics of the valves are modeled and new residuals accounting for the delays (response time of valves) are introduced, even these small discrepancies can be explained and detected.

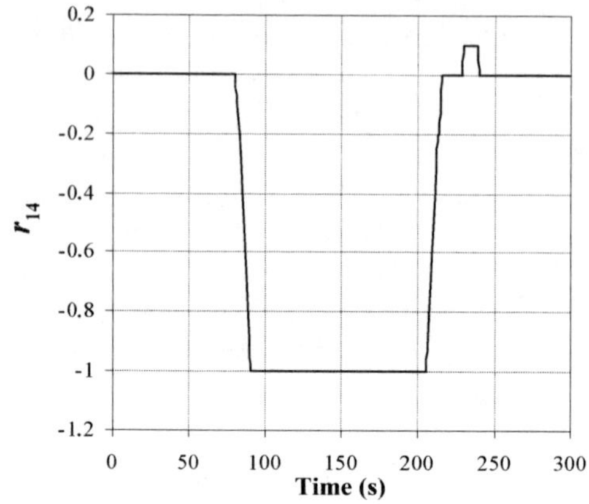

Fig. 6.24 Detection of condenser level controller fault

6.7 Reconfiguration

The reconfiguration strategies [133, 157] are aimed to ensure the continuity of the process operation in the presence of faults. The material redundancy of sensors and some components in the steam generator plant allows such reconfigurations. The fault tolerant pressure control in the boiler is an important task from the point of view of safety. The water in the boiler is heated according to the pressure inside it measured by the sensor P_7. At the saturation point of the steam, there exists a relation, given by the steam tables, between the pressure P_7 and the temperatures of steam and water, which are measured, respectively, by T_6 and T_5. Since these two sensors are only used in ARR_{15} and ARR_{16}, it is easy to isolate their fault. In the case of a single fault when P_7 measurement is wrong, the sensor T_6 is used to control the pressure in the boiler and avoids shutting down of the process. Similarly, if both P_7 and T_6 are faulty, then the pressure in the boiler can still be controlled by the measurements from T_5.

For the steam expansion system, the reconfiguration concerns the control of stem position of the valve V_2 according to pressure P_{15}. So, if P_{15} sensor is faulty, then the control of the valve V_2 can be carried out using either P_{13} or P_{16} because two of these three sensors are redundant.

The level in the condenser is controlled using the sensor L_{18}. The sensor L_{19} can be used in a situation where there is a fault in the sensor L_{18}. The fault in the three discharge valves (V_3, V_4, V_5) between the condenser and the tank cannot be isolated. If a fault is detected in the corresponding sub-space of the structure, the operator may manually ascertain that not all of them are faulty simultaneously and then continue to run the process in a degraded operating mode with the remaining ones. The setpoints for the level and the control laws for these valves can be amended, if deemed necessary.

Similarly, the sensor L_9 measurements can be used in presence of the fault in the L_8 level sensor. The fault in one of the pumps in operation leads to incorporation of the other physically redundant pump to supply water to the boiler.

Different operating modes of the steam generator are possible according to the number of sensors and/or components failed: the normal operating mode which corresponds to the availability of all sensors and components of the plant, different degraded operating modes depending on the faulty sensors and/or components and the stop operating mode. When the instrumentation of the steam generator does not allow to run the process in safe condition, the decision to stop the process for repair and maintenance has to be taken. Assuming a degraded mode exists, there are eight (for three hybrid controllers) different fault signature matrices (of dimension $(35-p_i)\times(23-q_i)$) for each operating mode, as shown in Figure 6.25, where p_i and q_i are two positive integers representing, respectively, the number of components and the number of residuals lost in that mode. In the case of normal operation $p_i = q_i = 0$.

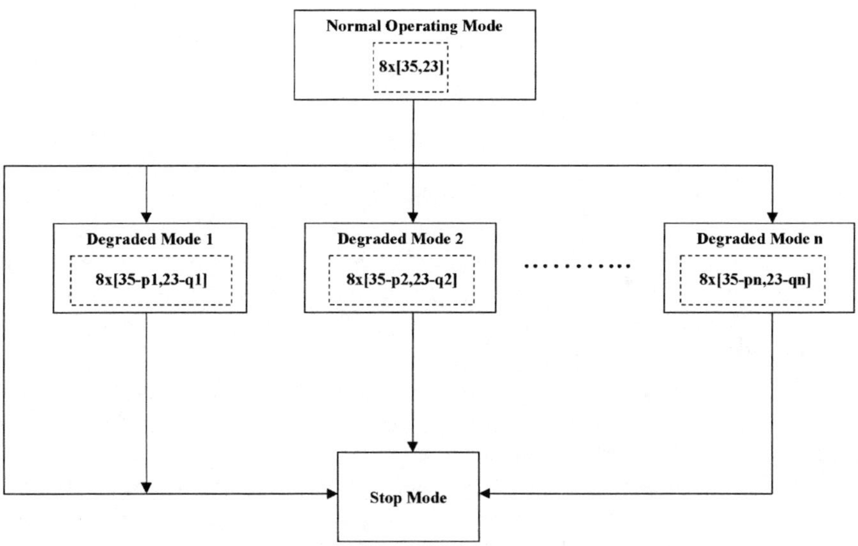

Fig. 6.25 Management of operating modes

The transition from one mode to another depends on the task to be achieved and its corresponding set of instrumentations available. All of these reconfigurations are directly achieved by using the various toolboxes in the supervision system. The current operating mode and equipment availability are stored in DataManager. Readers may see [284] for a hybrid diagnosis scheme developed by using Petri-net and bond graph models.

Here, the case of reconfiguration in presence of P_7 sensor fault is discussed. When sensor T_6 is used instead of sensor P_7, the redundancy relation given by Equation 6.42 between the two sensors is removed from the set of ARR. Every occurrence of P_7 in other relations are replaced by the function relating P_7 to T_6. This function is called **Ts2Ps**(), in this implementation. Thus, any calls to function **Ps2Ts**(P_7) become **Ps2Ts**(**Ts2Ps**(T_6)) $= T_6$. The component P_7 sensor and residual r_{16} are removed ($p_i = q_i = 1$) from the 2^3 new fault signature matrices. This task or complete reconstruction of the equations and fault signature matrix after removing the faulty component from the model are achieved by theModelBuilder (TB 5.1). In the present installation, these equations and the corresponding fault signature matrices are maintained in a database rather than using an online interface with the toolbox TB 5.1. This avoids the delay caused by TB 5.1 during symbolic computations. It is interesting to note that all the isolatable faults in the normal operating mode of the steam generator remain still isolatable after the reconfiguration of heater control using T_6 sensor. However, this is not always the case. In the presence of faults in sensor L_8, the fault signature of the sensor L_9 after the reconfiguration is identical to that for the fault in the boiler. Thus, after such reconfiguration, any fault detected in the residuals containing L_9 leads to complete shut down of the process. There are many such combinations of conflicts arising out of the reconfigurations. These are managed by the toolbox ReconfAnalyzer (TB 7.2) by using the protocols discussed in Section 6.5.1. Detailed reconfiguration of the steam generator process is discussed in Chapter 8.

Upon any reconfiguration, the resulting mode, equipment availability and new process architecture are notified to other toolboxes through CCOM server. Some toolboxes cannot adapt to the new operating mode and are suspended (*e.g.* TB 3.7 using PCA). The new set of equations and fault signature matrices are posted online to other toolboxes which have expressed their interest through direct subscription. As an example, TB 3.2 (based on temporal band sequences) receives the new residual equations and formats them to its compatible polynomial form. It then uses symbolic parsing to decompile the equations and executes evaluation of the temporal bands by using incremental compilation and interpreters. It uses the exact form of the fault signature matrices provided by TB 5.1 and thus detects and isolates most of the faults in diverse operating conditions.

The entire architecture developed in this chapter allows one to supervise and operate the steam generator in normal operation as well in the presence of faults. Finally, it is worthwhile to note that, even though the steam generator is a strongly non-linear hybrid process, ARR based supervision could be successfully applied to it.

Chapter 7
Diagnostic and Bicausal Bond Graphs for FDI

The analytical redundancy relation (ARR) derivation method presented in Chapter 5 follows the causality inversion of detectors (putting them as sources) to derive a closed form ARR expression. In this chapter, we show that closed form expressions for ARRs cannot be derived for all kinds of processes with all kinds of instrumentation because it may not be possible in certain cases to eliminate unknown variables from the model through symbolic algebra. We develop a few substitutions in this chapter which results in a new model structure to directly evaluate the residuals. The modified method leads to the same set of residuals which can be obtained through classical means, if the equations are symbolically resolvable. It also generates only the structurally independent residuals, and thus, reduces the computation time. Furthermore, the developed method leads to a numerical residual generation scheme which can be applied to all situations, irrespective of whether the set of equations can be symbolically resolved, or not.

7.1 Diagnostic Bond Graph

First of all, we define a sensor element, Ds, to measure a signal without qualitatively assigning the nature of signal, *i.e.* effort or flow, to the device. The device measuring the output from a controller is represented as a signal sensor, Ds. When the causality of the signal sensor, Ds, is inverted, we represent it as an element for signal source, called Ss. This element, Ss, has already been proposed in [83] as a source sensor for use in the more general application for system inversion in the design of the control laws. In fact, the models for the FDI design presented in this chapter can be represented together with the behavioral model of the process in a single bond graph model using bicausality [83, 178] notations. However, for simplicity as well as for compatibility with many of the existing software tools, we initially represent both models separately and then follow the bicausality notation.

When preferred differential causality is assigned to the bond graph model using inversion of the sensor causalities (if necessary), the following five compositions are possible:

1. Inverted causality in the effort sensor (De)
2. Inverted causality in the flow sensor (Df)
3. Non-inverted causality in the effort sensor (De)
4. Non-inverted causality in the flow sensor (Df)
5. Inversion of signal sensor, Ds, to signal source, Ss (always inverted)

The causalities of the output ports of sensors, which may be connected to controllers, are not inverted. Sensor output ports are considered in signal domain and we do not extend the notion of causality to that domain.

Let us consider the inverted causality in the effort sensor, De. As proposed in Chapter 5, the flow in the bond connected to this sensor is equated to zero. This element is then represented as an effort source (measurements from real process) and the expression for the source loading (flow variable) is equated to zero. This expression is indeed, a residual (it does not involve any states, because, all storage elements are differentially causalled). We use a virtual flow sensor, which has only computational existence, to measure this output. All virtual sensors measuring residuals are represented with a superscript, '*', *e.g.* Df*. This substitution is shown in Figure 7.1.

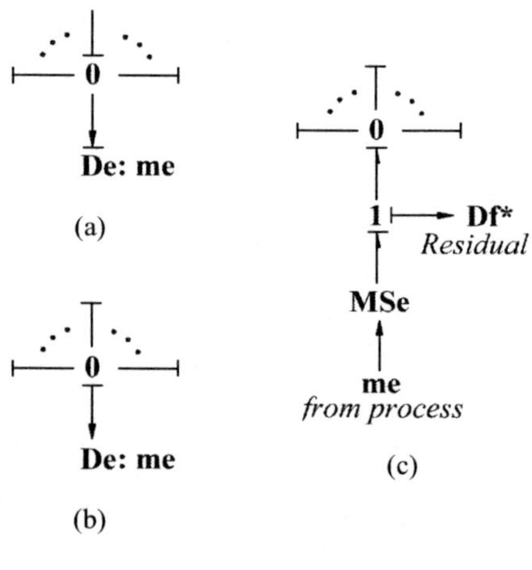

Fig. 7.1 **a** De in behavioural model. **b** Inverted causality in De. **c** Substituted representation for inverted causality in De resulting in the residual sensor Df*

Likewise, the substitution proposed for the inverted causality in the flow sensor is shown in Figure 7.2. Obviously, it is dual (see duality and gyrator equivalence in [173]) of the substitution for the inverted causality in the effort sensor.

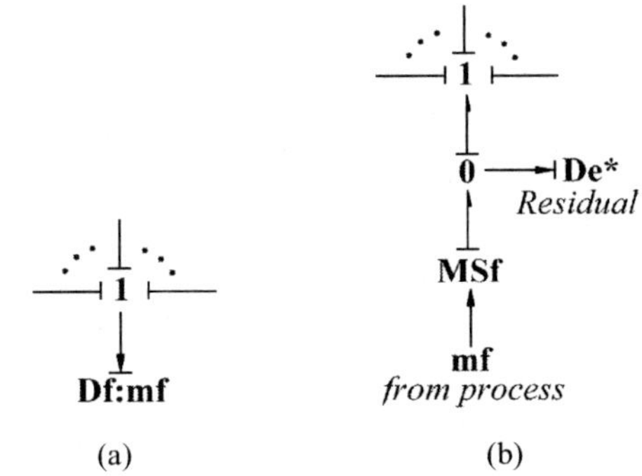

Fig. 7.2 **a** Flow sensor, Df, in inverted causality. **b** Corresponding substitution and residual sensor

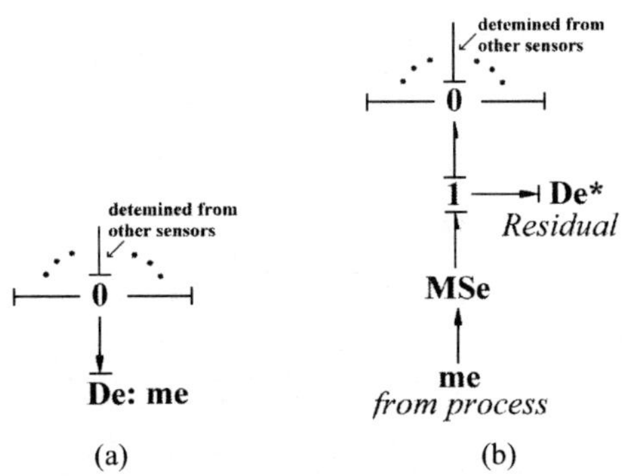

Fig. 7.3 **a** Effort sensor, De, in non-inverted causality. **b** Corresponding substitution

For the effort sensor, whose causality is not inverted (Figure 7.3a), the value measured by the sensor can be determined from the measurements by the other sensors. Thus, the difference between the value obtained from the expression for the effort determined from the other sensors and the one measured by the concerned sensor is zero. This gives rise to a residual, which when represented in a bond graph form, is observed by the virtual sensor shown in Figure 7.3b. The substitution for the flow sensor, whose causality is not inverted, is given by the dual of the substitution in Figure 7.3b.

Note that for those sensors, whose causalities have been inverted, the corresponding residuals are observed in their dual sensors; whereas for those sensors, whose causalities have not been inverted, the corresponding residuals are observed in similar sensors.

In the signal domain, the outputs of the controllers are, generally, measured. In most applications, controller algorithms are implemented in the supervision software. This allows the possibilities for the fault tolerant control (FTC) by changing the control laws to compensate for the errors. The residuals for the controllers are simple comparisons of the measured outputs and the predicted outputs. For this purpose, each signal sensor, Ds, is always converted to a signal source, Ss, and the residual is obtained by simple comparison, as shown in Figure 7.4.

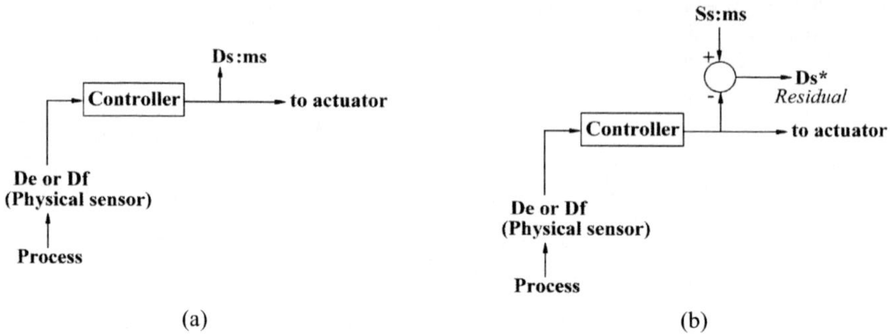

Fig. 7.4 **a** Signal sensor, Ds. **b** Substitution for inverted causality in signal sensor

7.1.1 Derivation of ARR

First of all, all the storage elements in the bond graph model are assigned derivative causality by allowing inversion of detector causalities when necessary. It is evident from the substitutions described in the previous section that there is a virtual residual sensor for each real sensor, irrespective of its type. In [21], Borutzky has developed a similar approach and refers to these virtual residual sensors as residual sinks. Thus, the number of residuals is equal to the number of actual sensors. For a sensor in inverted causality, there exists at least one causal path to it which is different or does not exist when compared to the causal paths to any other sensor in the inverted causality. Note that if any two sensors in inverted causality have identical sets of causal paths, they are mutually redundant. The proof is similar to those given in [61]. Those sensors for which causalities have not been inverted, only the corresponding residual sensors have the causal path from those sensor measurements. As an example, in Figure 7.3b, only the virtual residual sensor, De*, has a causal path from the measured process variable, *me*. This proves that all the virtual residual sensors have distinct signatures, *i.e.* residuals obtained are always structurally independent.

Let us consider the example of a simple two-tank system, shown in Figure 7.5a. A pump feeds flow to the first tank, T_1, at a constant mass flow rate, Q_p. Another tank, T_2, is connected to the first tank, T_1, by a valve, V_b. The flow from the second tank, T_2, is discharged to the environment through an outlet valve, V_o. There are two pressure sensors at the bottom of each tank. It is assumed that all the piping connections are also at the bottom of these tanks. The pseudo bond graph model of the system, in the preferred derivative causality, is given in Figure 7.5b. The pressures and the mass flow rates have been taken as the effort and flow variables, respectively. The causalities of both the sensors are inverted in Figure 7.5b. The value of the C-elements are given by $C_{Ti} = A_i/g$, where $i = 1,2$ enumerates the tanks, A_i is the corresponding cross-sectional area, and g is the gravity.

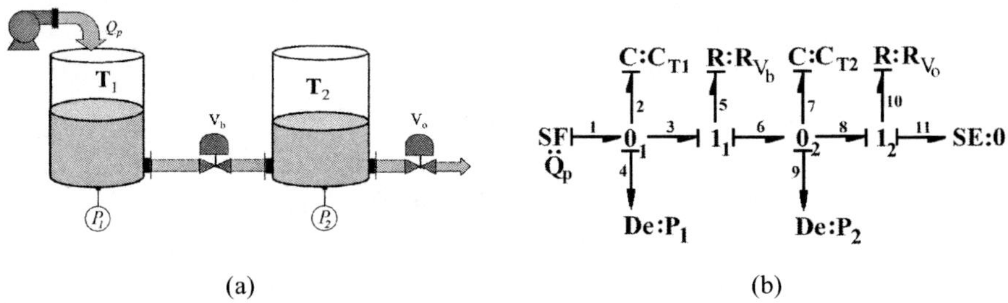

Fig. 7.5 **a** A two-tank system and **b** its bond graph model in preferred differential causality

We leave it to the reader to verify that the following ARRs can be derived from the bond graph model in Figure 7.5b by applying the algorithm given in Chapter 5:

$$ARR_1 = Q_p - \frac{A_1}{g}\frac{dP_1}{dt} - C_{db}.sign(P_1 - P_2)\sqrt{|P_1 - P_2|},$$
$$ARR_2 = C_{db}.sign(P_1 - P_2)\sqrt{|P_1 - P_2|} - \frac{A_2}{g}\frac{dP_2}{dt} - C_{do}.sign(P_2)\sqrt{|P_2|}. \quad (7.1)$$

This method to obtain ARRs cannot be applied where some of the unknown variables cannot be symbolically eliminated. Therefore, we will follow the approach based on sub-graph substitutions as a general solution, disregarding the fact whether unknown variables can/cannot be eliminated through symbolic treatment.

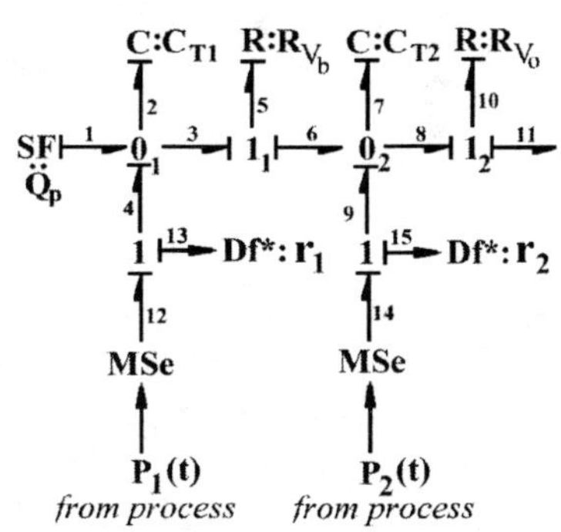

Fig. 7.6 DBG model of the two-tank system

By using the proposed substitutions, the bond graph model in Figure 7.5b can be represented as shown in Figure 7.6. Such a bond graph represents neither the behavioral model nor the inverse model [83, 178]; it has been termed a Diagnostic bond graph (DBG) in [162, 227]. It is easy to verify that when the equations for the two residual sensors are written from this model (*i.e.* expressions for variables f_{13} and f_{15} in Figure 7.6), they exactly match ARR_1 and ARR_2 given in Equation 7.1. Obviously, the fault signature matrix will be same, *i.e.* when the set of equations can be resolved to produce residual's equations (ARR), both the classical method and the method developed here are equivalent.

7.1.2 Example of a Non-resolvable System

Let us consider the controlled three-tank system shown in Figure 7.7, which includes a continuous PI controller and a hybrid On-Off controller. Water is pumped into the first tank, T_1, using a pump from a water source situated at a lower level. An On-Off controller uses the pressure sensor, P_1, output to act on a pump in order to maintain the water level in T_1 within specified lower and upper bounds. The mass flow output from the pump is measured by a flow sensor, F_1, and the output of the On-Off controller is measured by a sensor, u_1. The second tank, T_2, has no sensors to measure its level. The third tank contains a pressure sensor, P_2. The output to the consumer, measured by mass flow sensor, F_2, is controlled at a specified mass flow rate by using a PI controller which acts on the output valve, V_3. Valves V_1 and V_2 are opened at constant stem positions and thus, their coefficients of discharge are constant. Valve, V_0, is not used in the process, but used to simulate leakage from the tank, T_2.

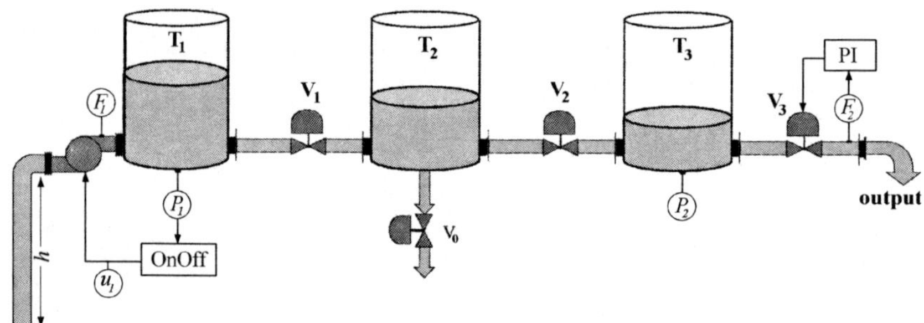

Fig. 7.7 A controlled three-tank system

The pseudo bond graph model for the system in preferred integral causality is given in Figure 7.8. The pump has been modeled by using an R-element. In fact, an ideal pump can be represented as a linear R-element with negative damping (see Chapter 6). In the current model, the pump's characteristic is available in a form that returns mass flow rate as a non-linear function of the opposing pressure head. We assume that this function cannot be inverted, and thus, the R-element representing the pump is always assigned conductive causality.

The bond graph model in preferred derivative causality is given in Figure 7.9, in which two algebraic loops [19, 24, 263, 264] are detected. The first loop (marked by '**A**' in Figure 7.9) is due to the coupling between the R-elements for the pump

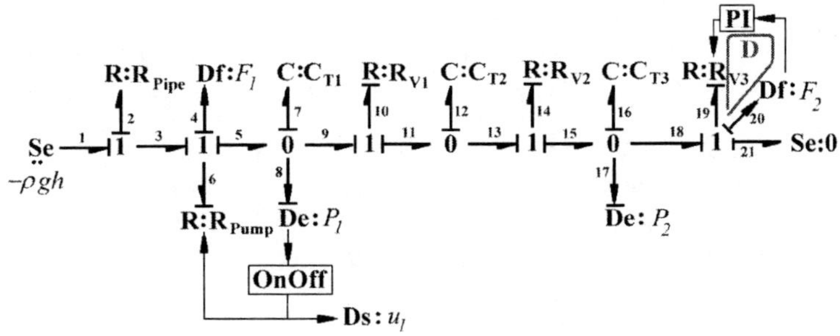

Fig. 7.8 Bond graph model of the three-tank system

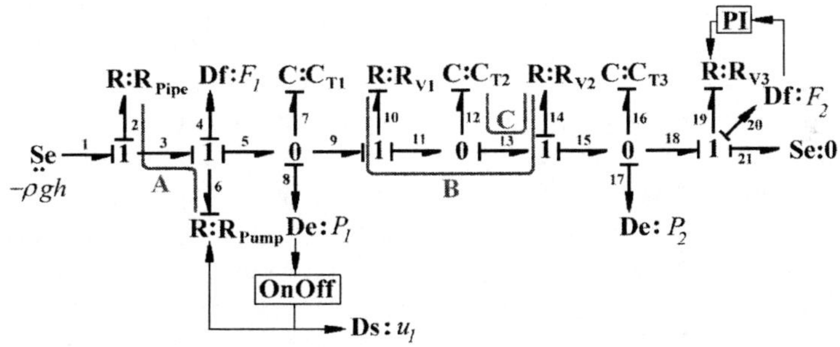

Fig. 7.9 Bond graph model of the three-tank system in preferred differential causality

and the pipe. This loop also exists in the behavioral model in Figure 7.8. In the behavioral model, this loop can be removed by inserting a coupling capacitor or pad [173, 257]. However, this adds an extra state to the model. When this extra state is assigned differential causality, the algebraic loop again resurfaces. One such loop (marked by '**B**' in Figure 7.9) even exists naturally in the model with preferred derivative causality. These types of loops are also called ZCPs or zero-order-causal-paths [43, 263, 264]. A differential algebraic loop also exists in Figure 7.9 (marked by 'C'), which is algebraic coupling between a differentially causalled element and a looping resistance in a ZCP and it cannot be solved symbolically (without introduction of time integrals into the solution) even for linear systems. Note that several numerical solution techniques exist for various kinds of loops; *e.g.* Lagrange multipliers, coupling impedance, break variables, *etc.* [19, 24, 43]. However, for classical FDI purposes, symbolic solutions are desired. We deal with each of these loops separately.

Loop '**A**' exists both in the behavioral model and the model in preferred derivative causality. Let us assume that the pressure of the fluid just before the pump, which is unknown, is given by P_b. The pressure at inlet to the pipe is $P_i = 0$ (reference atmospheric pressure). The pressure at the exit of the pump, P_1, is known. Then the flow through the pump and the pipe, f, is given as

$$f = \Phi_p(P_b - P_1).OnOff(P_1) = C_d.\sqrt{-\rho g h - P_b}, \tag{7.2}$$

where C_d is the coefficient of discharge of the pipe and Φ_p is the non-invertible characteristic function of the pump. There are two equations and two unknowns (f and P_b), which can be solved only in special cases, such as when function Φ_p represents a linear or polynomial relation (see Chapter 6 for model of a pump with a pipe). For FDI purposes, as in the present case, f is measured and hence there are two equations containing a single unknown variable, P_b. This over constraint leads to an ARR, which is written as

$$F_1 - \Phi_p\left(-\frac{F_1^2}{C_d^2} - \rho g h - P_1\right).OnOff(P_1) = 0. \tag{7.3}$$

However, this ARR cannot be derived from the model in Figure 7.9 by using the given computational causality. The loop is resolved in the preferred derivative causality model only due to the presence of the flow sensor.

Let us now consider the differential algebraic loop formed by the paths marked by '**B**' and '**C**' in Figure 7.9, which have arisen due to the preferred differential causality assignment to the element C_{T2}. Here, we only investigate a subset of the equations relating to the flow, f_9, in bond number 9. When this equation cannot be resolved, equations for all other variables that include bond number 9 in their causal path cannot be resolved as well:

$$f_9 = R_{V1}(P_1 - e_{13}), \tag{7.4}$$

$$e_{13} = e_{14} + P_2 = R_{V2}^{-1}(f_{13}) + P_2 = R_{V2}^{-1}\left(f_9 - C_{T2}\frac{de_{13}}{dt}\right) + P_2, \tag{7.5}$$

where R_{V1} represents the characteristics for the valve V_1 given in a similar form as used in Equation 7.1 and R_{v2}^{-1} represents the characteristics for the valve V_2 given in its inverse form. These two equations, Equation 7.4 and Equation 7.5, form the differential algebraic loop between variables f_9 and e_{13}. These equations are no more symbolically solvable. However, a pressure sensor in the tank T_2 or a flow sensor in any one of the valves, V_1 or V_2, can break this loop in the preferred derivative causality model.

The path marked by '**D**' in the behavioral model in Figure 7.8 represents an algebraic loop, which also exists in Figure 7.9. Algebraic loops which arise out of feedback are called control loops. The equation for the flow, f_{19}, in bond number 19 of the model in Figure 7.8 is written as

$$f_{19} = C_{dV3}(x)\sqrt{P_2}, \tag{7.6}$$
$$x = PI(f_{19}),$$

where x is the output from the PI-controller. This implicit form of Equation 7.6 creates a loop in the behavioral model. However, input to a controller from a sensor is treated as a known variable in FDI. Therefore, controllers never lead to loops in the FDI framework.

Therefore, only loops '**B**' and '**C**' remain unsolvable. Adding a sensor to measure flow through valve V_1 (Df at 1-junction connected to R_{V1}) breaks loop '**B**', but loop '**C**' still remains. In contrast, if a sensor is added to measure flow through valve V_2 (Df at 1-junction connected to R_{V2}), loops '**B**' and '**C**' are simultaneously eliminated, but a new loop '**E**' appears between elements C_{T2} and R_{V1}. However, adding an effort sensor (De at 0-junction connected to C_{T2}) eliminates all the loops and leads to a well computable FDI model, for which all ARR can be symbolically derived. A detailed analysis on sensor placement [226] in systems to obtain well computable FDI models is given in later chapters.

Remark 7.1. Sensor failures are very common in harsh industrial environments and the FDI algorithm should continue monitoring the plant even after loss of some sensors. Sensor failures may lead to formation of new loops in the diagnosis model. Usually, multiple diagnostic algorithms are implemented for various fault scenarios and the appropriate one is chosen depending upon the availability of system resources (after fault tolerant control and system reconfiguration) in the current operating mode. Such fault tolerant diagnosis is commonly implemented for distributed control systems and wireless sensor networks.

It is evident from the above discussions that symbolic derivation of all ARRs and the fault signature matrix is not possible for all kinds of models. However, most of the loops can have numerical solutions. In the following, it is shown that DBG based numerical residual evaluation can handle both situations, when the ARRs can and cannot be symbolically obtained.

For the three tank system discussed here, one can obtain three residuals in symbolic form, whereas, five residuals can be obtained numerically. The availability of all the residuals significantly enhances the detection and isolation capabilities of the FDI procedures based on the fault signature matrix. The DBG model for the controlled three-tank system, derived from the model in preferred derivative causality (Figure 7.9), is given in Figure 7.10. The DBG in Figure 7.10 has five external inputs (corresponding to the five sensors) and five outputs (corresponding to the five residuals), which as an input-output relation in block diagram form is given in Figure 7.11.

7.1.3 Fault Signature Matrix from Causal Paths

From the DBG model, the numerical values for the residuals can be obtained. However, for FDI, the fault signature matrix, *S*, should also be available. Because we do not have closed form ARRs, analysis of the causal paths to each residual is used to generate these signatures [240, 251]. As an example, the virtual sensor for residual

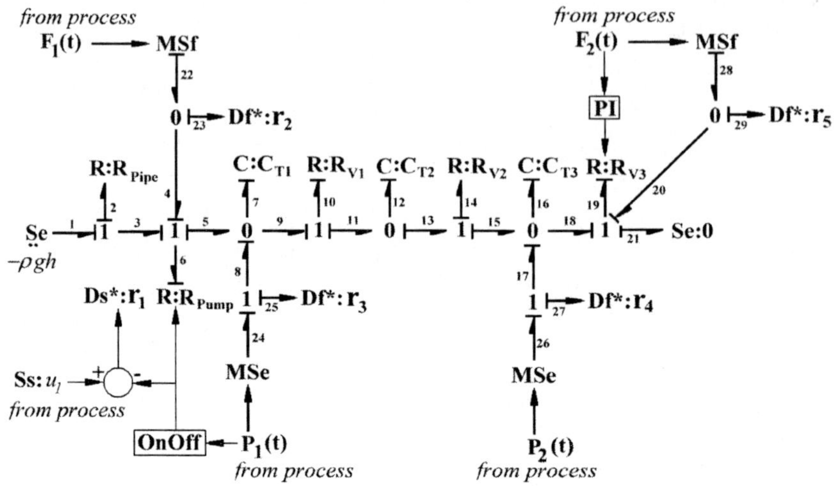

Fig. 7.10 DBG of the three-tank system

Fig. 7.11 Inputs and outputs of a DBG in block diagram form

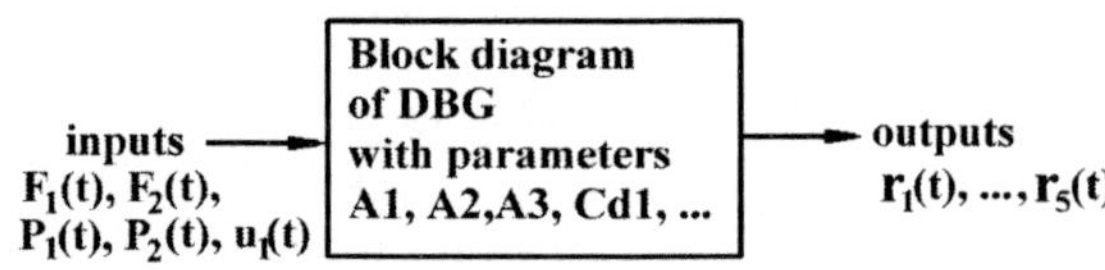

r_2 is represented in the bond number 23 in Figure 7.10. The following are the causal paths to residual r_2:

1. $F_1(t) \rightarrow f_{22} \rightarrow f_{23}$.
2. $-\rho gh \rightarrow e_1 \rightarrow e_3 \rightarrow e_6 \rightarrow R_{pump} \rightarrow f_6 \rightarrow f_4 \rightarrow f_{23}$.
3. $f_2 \rightarrow R_{Pipe} \rightarrow e_2 \rightarrow e_3 \rightarrow e_6 \rightarrow R_{pump} \rightarrow f_6 \rightarrow f_4 \rightarrow f_{23}$.
4. $P_1(t) \rightarrow e_{24} \rightarrow e_8 \rightarrow e_5 \rightarrow e_6 \rightarrow R_{pump} \rightarrow f_6 \rightarrow f_4 \rightarrow f_{23}$.
5. $P_1(t) \rightarrow OnOff \rightarrow R_{pump} \rightarrow f_6 \rightarrow f_4 \rightarrow f_{23}$.

From these causal paths, the elements involved in the residual r_2 are obtained as K_2=[F_1, R_{pump}, R_{Pipe}, P_1, $OnOff$]. This signature can be written in the terms of components as K_2=[F_1, Pump, Pipe, P_1, OnOff].

The fault signature matrix, **S**, for all the devices (sensors, components, controllers) of the three tank system, in Figure 7.7, is given in Table 7.1. From Table 7.1, it can be seen that the fault signatures of the pipe and the pump are identical. Thus, the faults in the pipe and the pump cannot be isolated separately. However, if the pipe and the pump are considered together as a single component, the fault can be isolated. Similarly, due to the differential algebraic loop '**B-C**' in Figure 7.9, the fault signatures of components T_2, V_1 and V_2 are identical, and their faults cannot be isolated. The sensor placement problem to isolate selected faults is discussed later in this chapter.

Table 7.1 Fault signature matrix of the controlled three-tank system

	r_1	r_2	r_3	r_4	r_5	M_b	I_b
u_1	1	0	0	0	0	1	1
F_1	0	1	0	0	0	1	1
P_1	1	1	1	1	0	1	1
P_2	0	0	1	1	1	1	1
F_2	0	0	0	1	1	1	0
Pipe	0	1	1	0	0	1	0
Pump	0	1	1	0	0	1	0
T_1	0	0	1	0	0	1	1
V_1	0	0	1	1	0	1	0
T_2	0	0	1	1	0	1	0
V_2	0	0	1	1	0	1	0
T_3	0	0	0	1	0	1	1
V_3	0	0	0	1	1	1	0
OnOff	1	1	1	0	0	1	1
PI	0	0	0	1	1	1	0

7.2 Simulation and Real Time Implementation of the Residuals

MATLAB®-Simulink® [285] provides a toolbox called 'Real Time Work Shop' for interfacing Simulink® models to real processes. A Simulink® block diagram model representing the DBG of the three tank system is given in Figure 7.12.

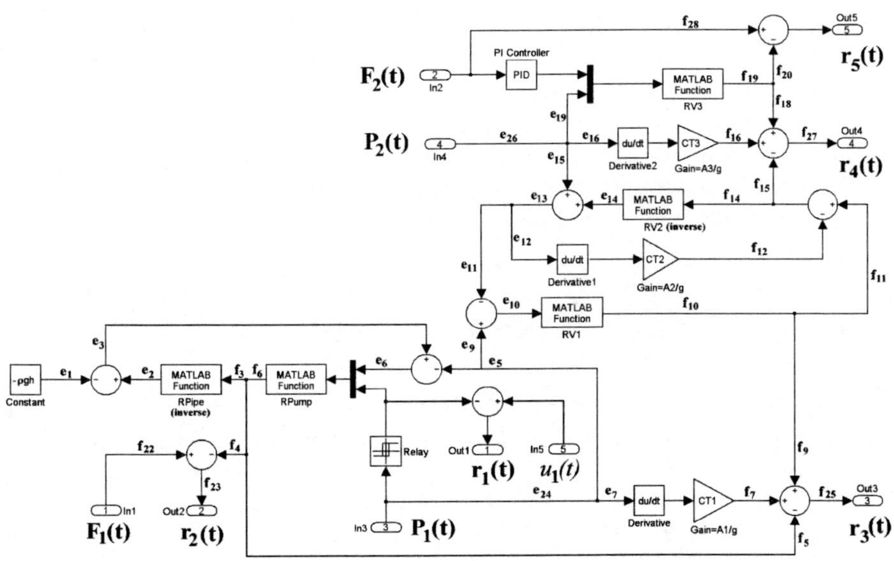

Fig. 7.12 Simulink® model of the DBG of the three-tank system

This model can be represented as a MATLAB®-Simulink® S-function using various bond graph software [36, 100, 101, 174, 262]. Simulink® 'Real Time Workshop' creates optimized C-language code from the block diagram model, which can be interfaced with data acquisition and FDI algorithms.

The theoretical fault signature matrix obtained from the analysis of the causal paths cannot be used directly in real time applications. The theoretical model does not represent the relative sensitivity of residuals to various faults. Analysis of relative effects of different terms of a residual is easy, when it is available in equation form. However, when the residuals are in numerical form, the analysis of such sensitivities for all the residuals can be carried out through simulation or off-line tests. The sensitivity of a residual to different faults is observed to create the corresponding coherence vector. Practical fault signature matrix is then constructed by stacking the coherence vectors. The practical fault signature matrix refines the theoretical fault signature matrix by removing some influences (replacing 1s by 0s).

In most industrial processes, introduction of the process faults to generate the corresponding offline data for testing may be too hazardous, *e.g.* leakage from a storage device containing poisonous gas, blockage of a pipe carrying the coolants, *etc.* Sometimes, introduction of a fault can lead to process instability. Furthermore, there should be a provision in the process to introduce such faults. As an example, in the three tank system described in Figure 7.7, a leakage from tank T_2 can be introduced by opening the valve V_0. However, there is no possibility of introducing leakages from tanks T_1 and T_3.

Therefore, simulation of the process behavior is used to produce a rich database for the offline tests. Almost all the possible faults can be introduced in the simulated model at no risk to the real process and the operators. Furthermore, a fast simulation model reduces the overall time required to construct the process database.

Although simulation models can be used to create the process database containing different failure scenarios, which can, then, be fed into the DBG model to generate the residuals and the corresponding fault signatures, an integrated test model significantly simplifies the overall management. Moreover, the integration methodology presented in the following section can be implemented by using available bond graph modeling software [36, 100, 174].

7.2.1 Integrated System Simulation: Coupling the Models

The basic framework for integration involves coupling of the behavioral and the diagnostic models. This is explained by using the model of the three tank system. The corresponding coupled model is given in Figure 7.13, in which four different zones are separated by dashed lines. Some bond numbers and element nomenclature in the behavioral part and the diagnostic part are the same. This causes no problem as long as these two parts are treated as separate sub-models, which can have their own local nomenclature. The modified behavioral model contains the sources needed to introduce the faults, which are triggered by the user. This triggering mechanism is represented in the 'Events' section of the model, where different types of appearance of the fault (abrupt, progressive, *etc.*) can be implemented.

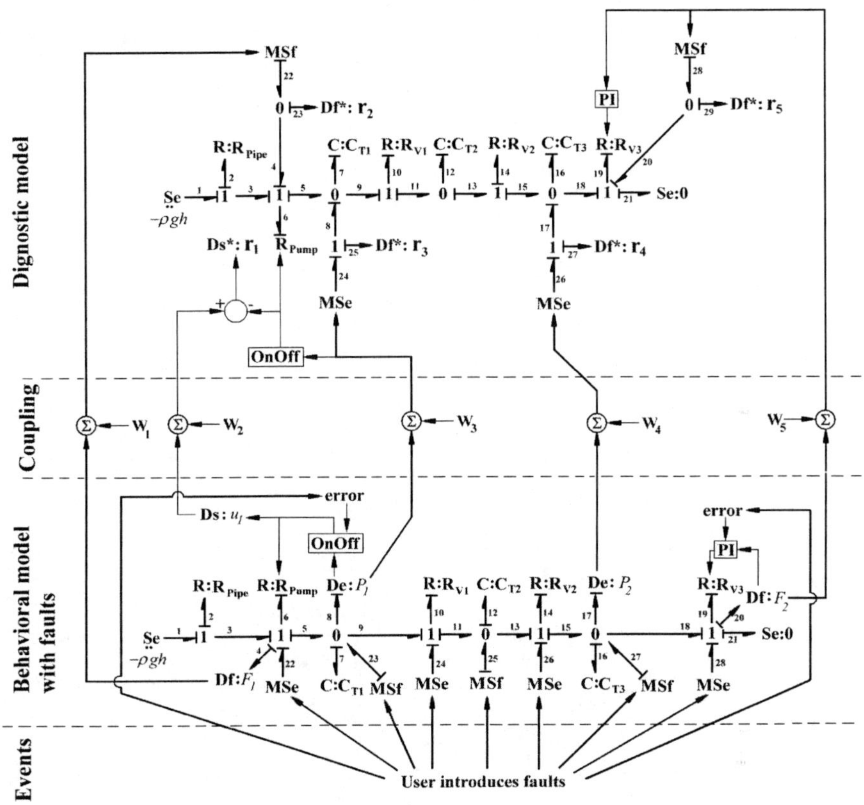

Fig. 7.13 Coupling of behavioural and diagnostic bond graph models

The 'Coupling' section provides the interface between the behavioral sub-model and the diagnostic sub-model. To simulate real instrumentation dynamics and to study the effect of measurement noise, all sensor outputs from the behavioral sub-model are contaminated with noises (W_1 to W_5 in Figure 7.13). The magnitudes and the characteristics of the sensor noises are obtained from the specifications provided by the sensors' manufacturers. Different types of noise sources are available in bond graph modeling software. The 'band-limited white noise' source provided by MATLAB®-Simulink® is one such suitable implementation.

The derivatives in the diagnostic model are calculated in discrete time, while the behavioral model is evaluated in continuous time. This requires a hybrid simulation tool. While MATLAB®-Simulink® is capable of doing so, it is not possible in available bond graph modeling software. Thus, a mechanism to represent the storage elements in derivative causality as a fully integrally causalled sub-model for continuous time domain simulation is developed here.

The electronic circuit for an analog differentiation is given in Figure 7.14a. Bond graph modeling of electronic devices is well detailed in [172, 173]. The bond graph model of the analog differentiation is given in Figure 7.14b, and its corresponding signal flow graph is given in Figure 7.14c. The gain, $G(s)$, between the input and output voltages is determined by using Mason's gain rule.

$$G(s) = \frac{-\mu}{1 + \frac{1}{RCs} + \frac{\mu}{RCs}}. \tag{7.7}$$

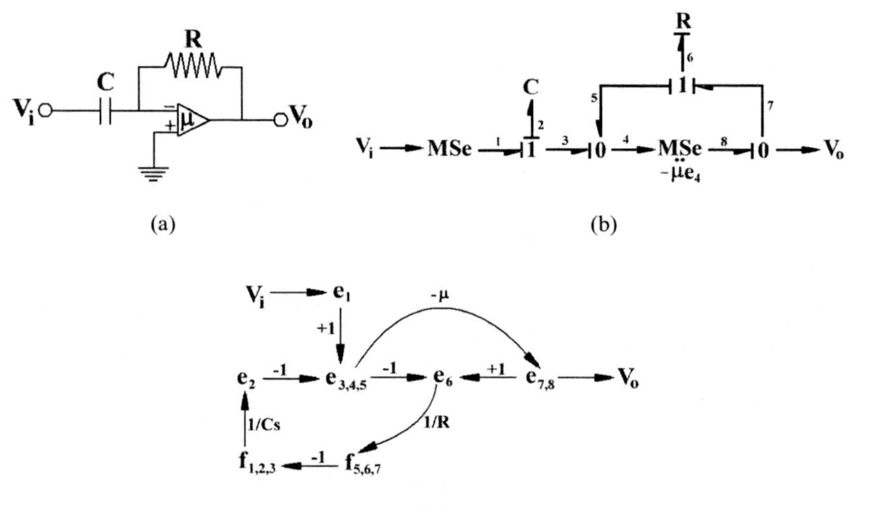

Fig. 7.14 **a** Analog differentiator. **b** Its bond graph model. **c** Signal flow graph

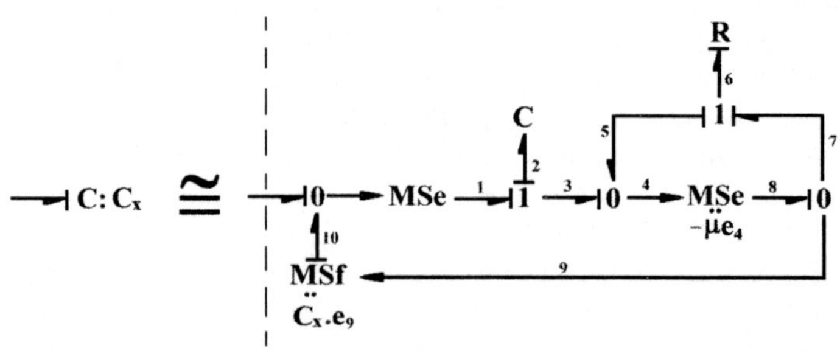

Fig. 7.15 Representation of differentially causalled C-element using analog differentiation

In Equation 7.7, when $\mu \to \infty$, $G(s) \to -RCs$. Taking $R = 1/C$ and $\mu \to \infty$, $G(s) \to -s$, which represents an ideal differentiation. Any differentially causalled storage element can be represented using the model for the differentiator. The passivity of the element is established through a feedback with an appropriate gain, corresponding to the driving point impedance. The representation of a differentially causalled C-element is given in Figure 7.15. Note that the only storage element in model of the differentiator is in integral causality. Thereafter, the resulting global model can be solved by using any continuous time implicit DAE solving algorithm [89, 205].

7.2.2 Simulation Results

The Simulink® model of the three-tank system is given in Figure 7.16.

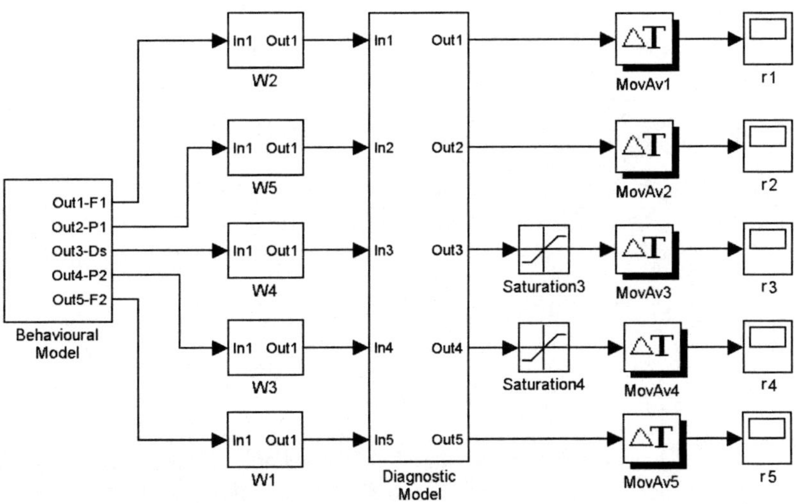

Fig. 7.16 Block diagram of the integrated test model

To simulate reality, noise is added to sensor outputs from the behavioral model before being input to the diagnostic model. The noise magnitude is chosen at 2% of each sensor's output. The block diagram representation of such coupling (W_1 to W_5) is given in Figure 7.17a. Adding a non-zero value to the output produces a mean shift of the signal output and thereby simulates a sensor error.

The block diagram model for the controlled valve is given in Figure 7.17b. Although the output from the PI controller can be any value, the valve can accommo-

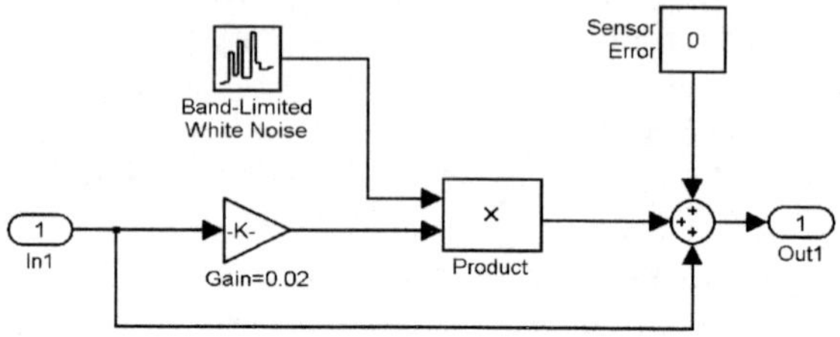

(a)

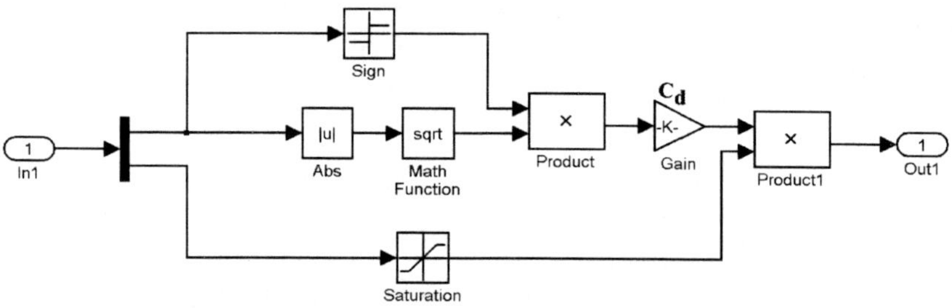

(b)

Fig. 7.17 **a** Block diagram of the sensor data model. **b** Block diagram of the controlled valve

date a command range of 0 to 1 (fully closed to fully open). This is achieved by using a saturation block. It is assumed that the valve has linear characteristics of the form $C_d(x) = C_d.x$.

The characteristic of the pump has been taken according to a simplified model of an impeller pump:

$$Q = \lambda_1.\Delta P + \lambda_2, \tag{7.8}$$

where λ_1 and λ_2 are two parameters and ΔP is the pressure gain across the pump. Obviously, λ_1 has a negative value, which represents energy import.

The integrated model in Figure 7.16 contains two saturation blocks to filter out large instantaneous changes to residuals r_3 and r_4 resulting from the derivatives of the noisy measurements. All residuals are further refined by a moving average [13], which is applied over a 50 s sliding time window. The parameter values used in the simulation are given in Table 7.2. The time response of the behavioral model in normal operation is given in Figure 7.18. The results for the uncontaminated sensor outputs P_1 and P_2, corresponding to the pressure at the bottom of the storage tanks T_1 and T_3, are given along with the unmeasured pressure, P_{mid}, at the bottom of the middle tank, T_2.

The residual responses in the normal operation were used to choose the fixed threshold values given in Table 7.3.

Table 7.2 Parameter values for the three-tank system

Parameter	Description	Value	Unit
A_1	Cross-sectional area of tank T_1	0.1	m^2
A_2	Cross-sectional area of tank T_2	0.05	m^2
A_3	Cross-sectional area of tank T_3	0.04	m^2
C_d	Coefficient of discharge of pipe	0.05	$kg^{\frac{1}{2}} m^{\frac{1}{2}}$
C_{d1}	Coefficient of discharge of valve V_1	0.01	$kg^{\frac{1}{2}} m^{\frac{1}{2}}$
C_{d2}	Coefficient of discharge of valve V_2	0.01	$kg^{\frac{1}{2}} m^{\frac{1}{2}}$
C_{d3}	Coefficient of discharge of valve V_3	0.01	$kg^{\frac{1}{2}} m^{\frac{1}{2}}$
λ_1	Pump parameter	−0.001	m.s
λ_2	Pump parameter	110	kg/s
h	Length of the pipe	10	m
L_s	Level set-point for OnOff controller	1	m
F_s	Flow set-point for PI controller	0.1	kg/s
K_p	Proportional gain of PI controller	1	
K_i	Integral gain of PI controller	0.01	
ρ	Density of the fluid	1000	$kg/\ m^3$
g	Gravity	9.81	m/s^2

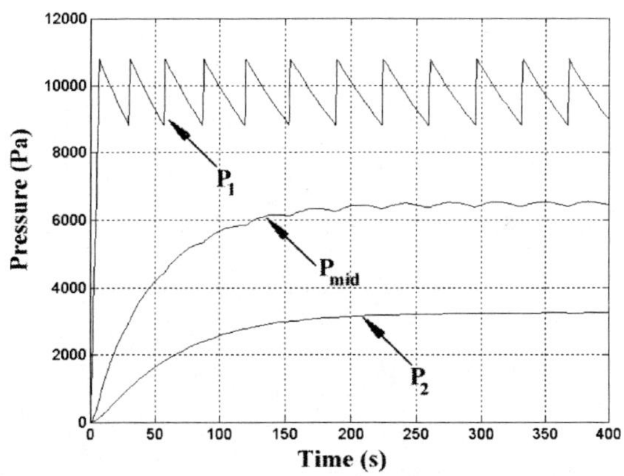

Fig. 7.18 Time response of the three-tank system

Table 7.3 Residual thresholds

ε_1	ε_2	ε_3	ε_4	ε_5
0.02	0.1	0.1	0.1	0.02

Two fault scenarios have been simulated. In the first case, a leakage from the middle tank, T_2, has been simulated during a 400 s time interval, from 600–1000 s. Note that tank, T_2, does not contain any sensors. The response of the residuals, given in Figure 7.19, reveals that residuals r_3 and r_4 are sensitive to this fault. This result

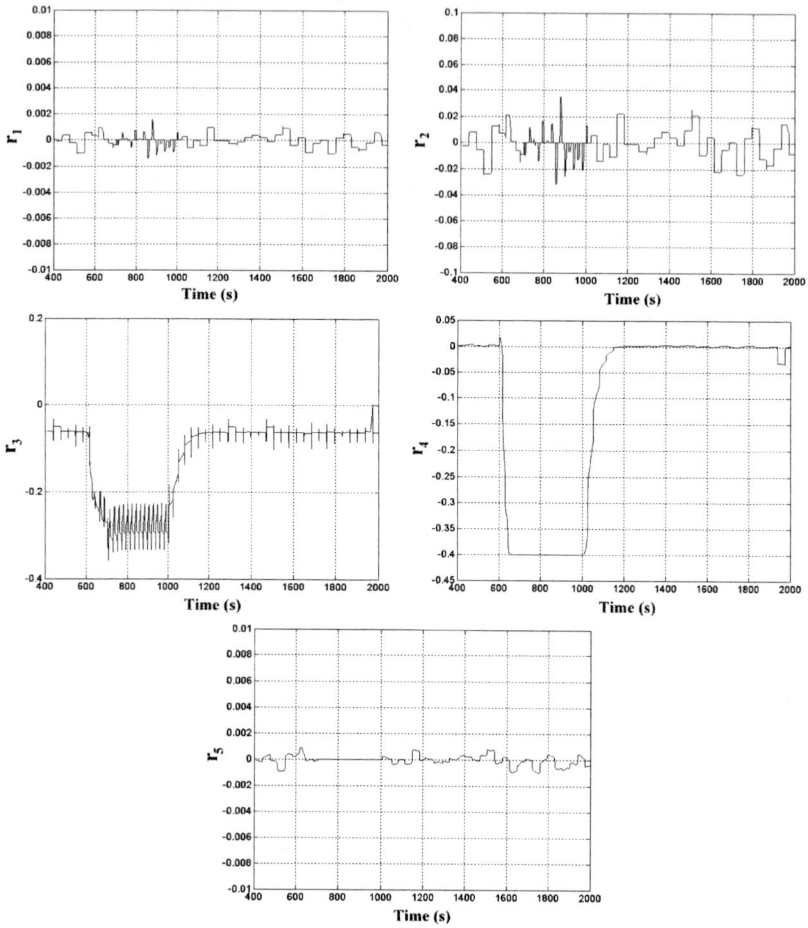

Fig. 7.19 Response of the residuals to leakage in tank T_2

matches the fault signatures given in Table 7.1. From the residual thresholds given in Table 7.3, the resulting coherence vector is $C = [0,0,1,1,0]$. This coherence vector has multiple matches in the signature matrix (Table 7.1) and thus the fault cannot be isolated, but only detected.

In the second fault scenario, the output valve, V_3, has been partially blocked (by 50%) during the time duration of 1200–1600 s. The response of the residuals to the fault is shown in Figure 7.20. The resulting coherence vector, $C = [0,0,0,1,1]$, has multiple matches in the fault signature matrix and hence this fault too can only be detected, but cannot be isolated. For some other type of faults, such as leakage from tanks T_1 or T_3, isolation of the faults can be done directly from the fault signatures given in Table 7.1.

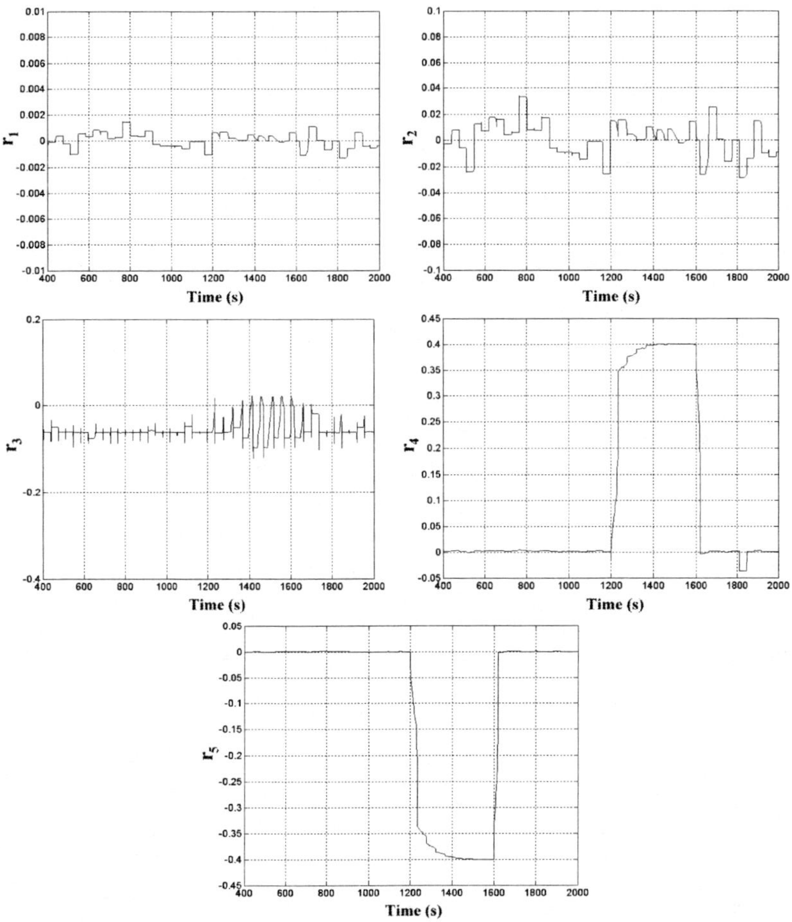

Fig. 7.20 Response of the residuals to blockage of valve V_3

Note that some faults in the three-tank system could be detected from the response of the residuals r_3 and r_4, which would not be otherwise available because the corresponding ARRs cannot be obtained in the symbolic form. Moreover, when some faults are not uniquely isolated due to lack of sufficient number of sensors, they are detected and a fault subspace containing fault candidates is isolated, which in turn helps to carry out quick maintenance work.

7.3 The Initial Conditions Problem

It is not an absolute necessity to invert causalities of all non-hardware redundant sensors to derive ARRs. In fact, such inversions often lead to formation of causal and/or algebraic loops in the model, which need to be avoided. All ARRs for some under determined subspace of the graph, in which a few unknown variables cannot be eliminated due to loops, cannot be obtained. However, the method developed in the previous section would directly obtain some residuals using numerical solutions; thus bypassing the need to derive ARRs. Fault signatures for such systems are usually obtained from causal paths.

The basic requirement for ARR derivation is to decouple the observable subspace of the graph from the unobservable subspace, if any. To do that, a minimum set of sensor causalities must be inverted (however, there is no harm in exceeding this limit as long as it does not lead to formation of irresolvable loops). This means that all causal paths must start from the known outer vertices and terminate at virtual residual sensors, *i.e.* they should not get into a never ending loop somewhere in the observable subspace. For certain systems, when this condition has to be fulfilled, one cannot assign desired causalities in the model, *e.g.* differential causalities to storage elements and preferred causalities to non-linear elements like non-invertible R-fields.

In this section, we consider how to derive ARRs, when some storage elements cannot be assigned preferred derivative causality and because of that the ARRs derived become sensitive to initial conditions.

Consider the spring-mass-damper system and its bond graph model shown in Figures 7.21a,b, respectively.

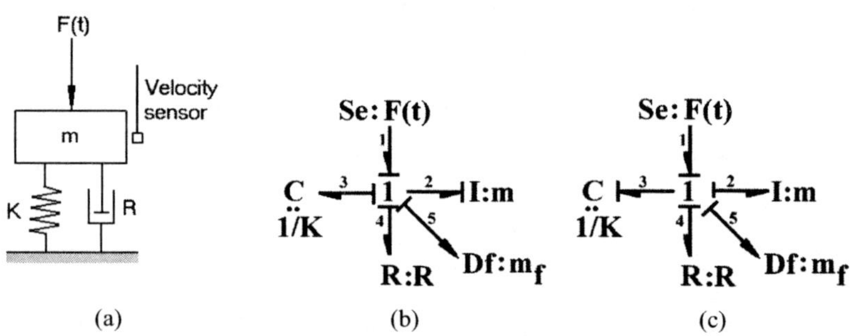

Fig. 7.21 Single-degree-of-freedom system and its BG model

When we assign preferred derivative causality to storage elements, as shown in Figure 7.21c, we find that the sensor causality cannot be inverted. Thus, the sensor does not become the outer vertex of the system. Moreover, the model in Figure 7.21c has a differential loop between the I and C elements, which means its equations cannot be derived. Therefore, let us try to find out the ARRs without inverting the sensor causalities, *i.e.* from the model in Figure 7.21b:

$$e_1 = F(t), \quad e_2 = m.s.f_3, \quad e_4 = R.f_3,$$
$$f_1 = f_3, \quad f_2 = f_3, \quad f_4 = f_3, \quad f_5 \text{ or } m_f = f_3, \quad f_3 = \frac{1}{K}s.e_3,$$
$$e_3 = e_1 - e_2 - e_4, \tag{7.9}$$

where operator s stands for derivative with respect to time. In the above, we have nine equations for eight unknown variables (f_1, f_2, f_3, f_4, e_1, e_2, e_3, e_4). Thus we know *a priori* that there is a redundancy, but how do we eliminate unknown variables, *i.e.* how to assign causalities, to obtain the ARR?

1. Fist of all, when a known quantity is in the left side of the equation, we invert the relation if and only if one unknown variable is in the right hand side and the relation is invertible. So, the equation $m_f = f_3$ becomes $f_3 = m_f$.
2. We now have two equations for the variable f_3, thus upon sequential reduction we obtain the ARR as

$$e_1 = F(t), \quad f_1 = f_2 = f_4 = f_3 = m_f, \quad e_2 = m.s.m_f, \quad e_4 = R.m_f,$$
$$e_3 = F(t) - m.s.m_f - R.m_f,$$
$$f_3 = \frac{1}{K}s.e_3 \Rightarrow m_f = \frac{1}{K}s.\left(F(t) - m.s.m_f - R.m_f\right)$$
$$\text{or} \quad \text{ARR:} \frac{dF(t)}{dt} - m.\frac{d^2(m_f)}{dt^2} - R.\frac{d(m_f)}{dt} - K.m_f = 0. \tag{7.10}$$

Note that the first step is a causality assignment problem and the second step is possible if the set of equations are regular. The DBG essentially ensures that the first step is satisfied, *i.e.* it ensures that sufficient number of known quantities always appear on the right hand side of equations. Let us consider the DBG of the system given in Figure 7.22, in which we find that the C-element cannot be assigned derivative causality; but there are no loops in the model.

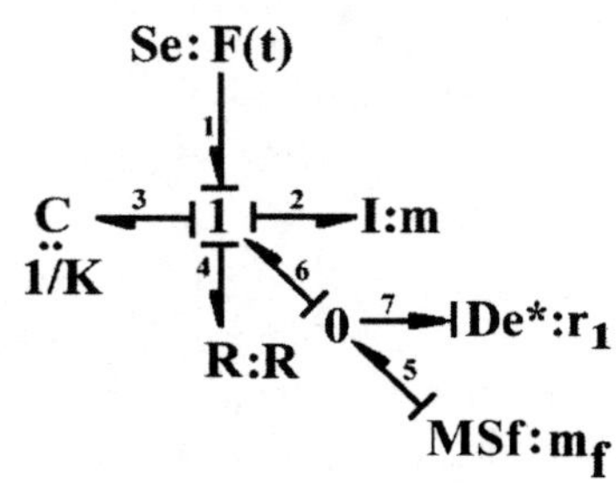

Fig. 7.22 DBG of single-degree-of-freedom system

Deriving the ARR equation ($e_7 = 0$), we obtain

$$\text{ARR:} \quad F(t) - m.\frac{d(m_f)}{dt} - R.m_f - K.\int m_f.dt = 0. \tag{7.11}$$

This ARR is sensitive to initial conditions (due to the integration) and, thus, it cannot be used for real-time process supervision. Therefore, we take a time derivative and obtain

$$\text{ARR:}\ \frac{dF(t)}{dt} - m.\frac{d^2(m_f)}{dt^2} - R.\frac{d(m_f)}{dt} - K.m_f = 0. \tag{7.12}$$

7.3.1 Order of Extra Derivatives

The extra derivative was necessary due to the involvement of terms from an integrally causalled storage element (C_3). To determine how many orders of derivatives should be taken, one may follow this simple procedure:

Check all different causal paths to a residual sensor. Assign an order counter to each causal path. The order of the path $o = l - m$, where l and m are respectively the number of integrally causalled and differentially causalled storage elements in that causal path.

If there are z number of causal paths, to i-th residual sensor, let $n_i = \max o_j$, $j = 1..z$, where o_j is the order associated with j-th causal path. For i-th ARR to be insensitive to initial conditions, $n_i \leq 0$. If $n_i > 0$, then at least n_i number of derivatives are to be applied on that ARR. More derivatives will serve no purpose because they will lead to problems in implementing the decision procedure.

Thus the proper DBG representation of the spring-mass-damper system should be as shown in Figure 7.23.

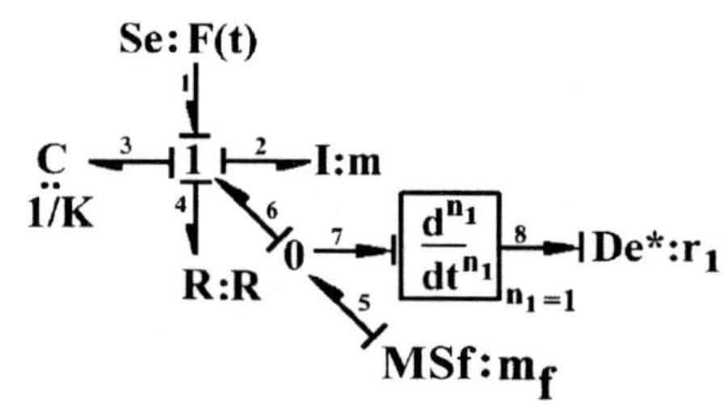

Fig. 7.23 Extended DBG of single-degree-of-freedom system

Let us consider a system shown in Figure 7.24a. In the bond graph model in Figure 7.24b, there is no way to assign preferred derivative causalities to the two storage elements; at the most one storage element can be assigned derivative causality. Therefore, the system is unobservable. Furthermore, the model contains a differential algebraic loop and the sensor is not the outer vertex, which means ARR cannot be derived for the system.

However, when we impose sensor as an outer vertex, both the storage elements acquire integral causality, as shown in Figure 7.25a. Thus by analyzing the causal

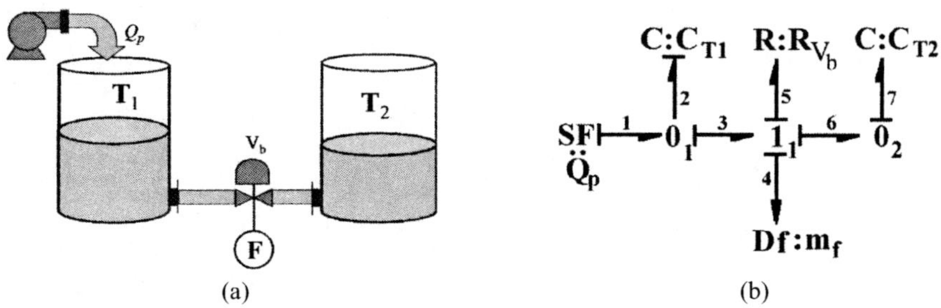

Fig. 7.24 Two-tank system with a flow sensor and its model

paths to the residual sensor in Figure 7.25b, the final version of DBG is obtained in Figure 7.25c.

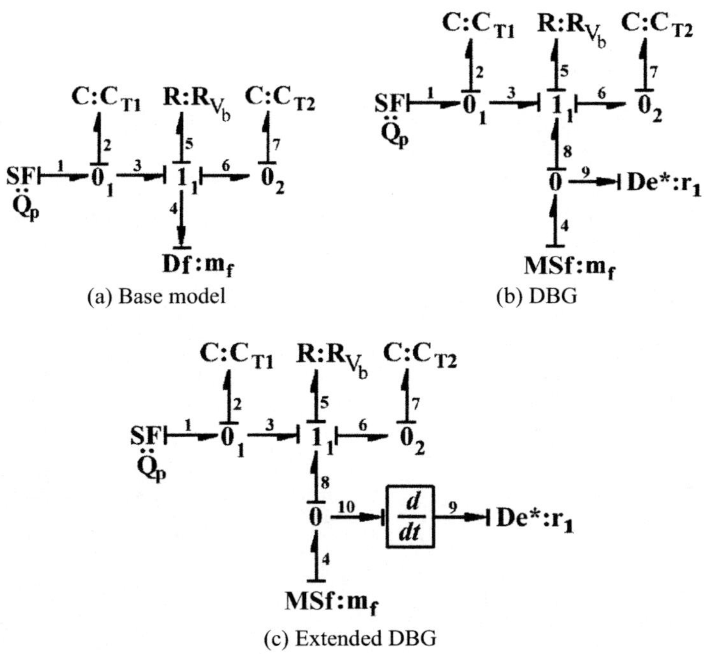

Fig. 7.25 Construction of extended DBG of two-tank system from the base model

The ARR for the system, obtained from Figure 7.25c, is given by

$$\text{ARR:} \quad \frac{1}{C_{T1}} Q_p - \frac{2}{C_d^2} . m_f . \frac{d(m_f)}{dt} - m_f \left(\frac{1}{C_{T1}} + \frac{1}{C_{T2}} \right) = 0. \tag{7.13}$$

The above ARR becomes singular when the valve is closed. Thus by multiplying C_d^2 to both sides of the equation, we obtain

$$\text{ARR:} \quad \frac{C_d^2}{C_{T1}} Q_p - 2.m_f.\frac{d(m_f)}{dt} - m_f.C_d^2 \left(\frac{1}{C_{T1}} + \frac{1}{C_{T2}} \right) = 0, \tag{7.14}$$

which reflects the fact that when $C_d = 0$, for compatibility, there must be $m_f = 0$.

Remark 7.2. In this example, one mode of the system is unobservable and another mode (internal mode, to be precise) is observable. Therefore, one can derive one ARR from the observable subspace of the system. Even if the system is not completely observable, we find that the derived ARR is sensitive to faults in all components and hence the entire system is monitorable. However, faults cannot be isolated from only one ARR containing many fault candidates.

7.3.2 Fault Scenario Simulation

The development of an integrated simulation model for testing residual sensitivity involves coupling of the behavioral and the diagnostic models. The corresponding coupled model to evaluate the ARR generated in Equation 7.14 is given in Figure 7.26, in which dashed lines separate the four different zones and the extra derivative is added before the residual sink.

Now one may ask, how to assign causalities such that the singularity in ARRs (Equation 7.14) is automatically avoided. In this case, it means assigning conductive causality to the R-element for the valve and yet being able to terminate causal paths at virtual residual sensors such that ARRs can be derived. Unfortunately, that is not possible in a classical bond graph model. This is because when causality in a bond is chosen for one power variable, $\chi \in [e, f]$, then the causality of the complimentary variable $\chi' = [e, f] - \chi$ is automatically assigned for that bond. We cannot independently assign these causalities in a bond graph. However, exactly the same condition is necessary in parameter estimation algorithms and also in system inversion. This problem motivates us to use the bicausality notation, which is discussed in the following section.

7.4 Matching Problems in Classical Bond Graph Modeling

In FDI models, the measured quantity is imposed on the system as a pseudo-source, preferred differential causality is used in storage elements and the reactive factors of power from the pseudo-sources are equated to zero. These reactive factors are the ARRs if they are written in symbolic form. In some other notation, sensors are not converted into sources, but retained as they are and instead, the causalities of the sensors are inverted.

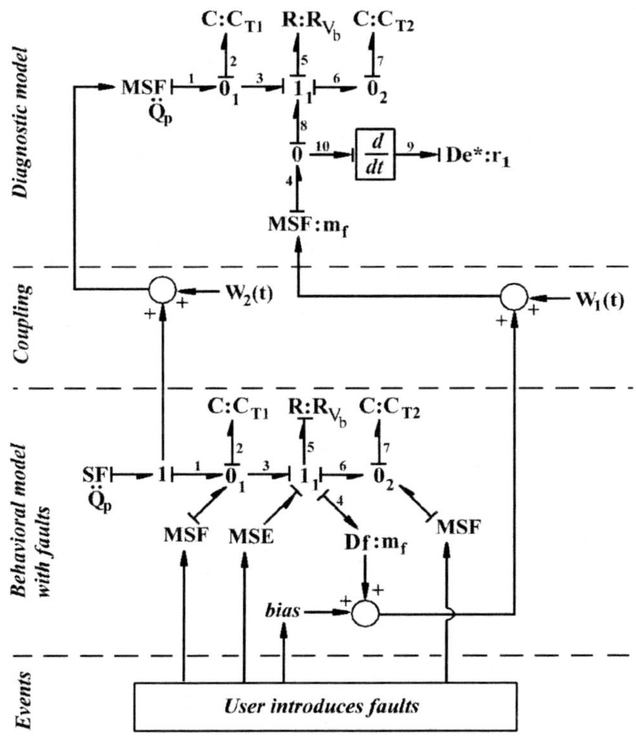

Fig. 7.26 Coupled behavioral and diagnostic bond graph models of the two-tank system

Some forms of causalities are preferred over others, *e.g.*, those forms, which have singularities or multi-valued solution are avoided. For example, consider the force *vs* velocity characteristics for dry-friction as shown in Figure 7.27. This can be modeled with an R-element, but that R-element must be assigned resistive causality. The reason is that a unique force f_v can be obtained for any given velocity v, whereas the reverse is not true. Such restrictions in causality assignment and the requirement of derivative causalities on storage elements correspond to the crossed entries in the incidence matrix (see Chapter 5).

The rules of bond graph modeling are developed from physical viewpoint and therefore they do not allow indiscriminate causality assignment. This restrains bond graph based software from applying efficient causalities to eliminate unknown variables in desired ways so that specific form of ARRs can be obtained.

As an example, consider the bond graph model of a controlled valve given in Figure 7.28a.

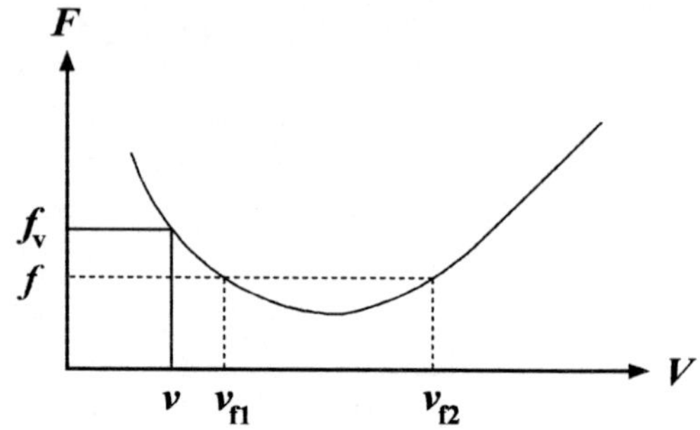

Fig. 7.27 Schematic force-velocity characteristics for dry friction

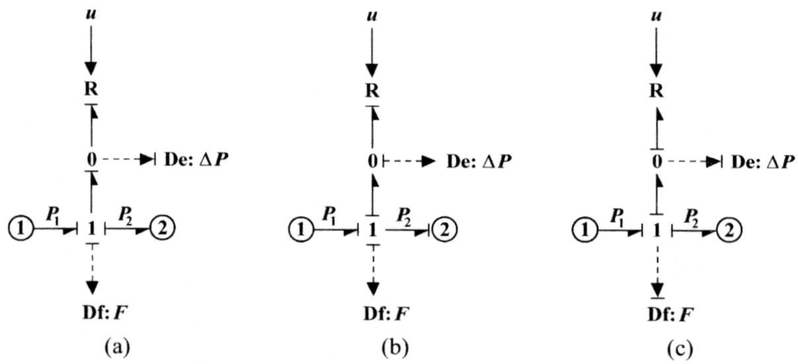

Fig. 7.28 Bond graph model of a controlled valve with different sensor causalities

In the bond graph model, the R-element represents the dissipation offered by the valve, which is modulated by a control signal u. The dotted lines represent possible instrumentations, sensor Df: F to measure flow through the valve and De: ΔP to measure differential pressure across the valve. The R-element has to be assigned conductive causality when De element causality is inverted, *i.e.* ΔP is taken as the starting node (Figure 7.28b), whereas the R-element has to be assigned resistive causality when Df element causality is inverted, *i.e.* F is taken as the starting node (Figure 7.28c).

Assuming that u and ΔP are known, as shown in bond graph and block diagram representations in Figure 7.29a, the flow through the valve can be determined from

$$F = C_d\sqrt{|\Delta P|}.sign(\Delta P).\phi(u). \tag{7.15}$$

In another case, supposing that the flow through the valve (F) and u are known, as shown in bond graph and block diagram representations in Figure 7.29b, the pressure drop across the valve can be determined as

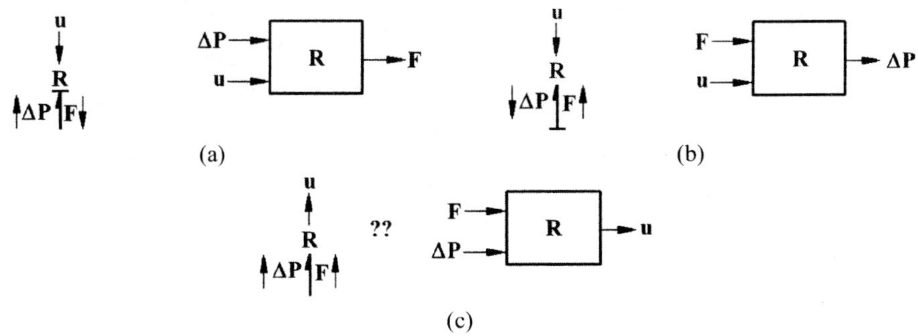

Fig. 7.29 Bond graph causality assignment and corresponding block diagram representation of the controlled valve for determination of flow (**a**), pressure drop (**b**) and control command (**c**)

$$\Delta P = \left(\frac{F}{C_d \phi(u)}\right)^2 sign(F), \tag{7.16}$$

under the condition that $\phi(u) \neq 0$. Therefore, the causal form of R-element given in Figure 7.29a is preferred.

However, consider that we have the values for the flow, F, and pressure drop, ΔP, across the valve. If the flow is non-zero, then we can always uniquely find u, if the function $\phi(\,)$ is single valued and its inverse $\phi^{-1}()$ can be defined, *i.e.*

$$u = \phi^{-1}\left(\frac{F}{C_d\sqrt{|\Delta P|}.\mathrm{sign}(\Delta P)}\right). \tag{7.17}$$

This is represented in block diagram form in Figure 7.29c, but there exists no corresponding valid causality assignment in classical bond graph representation!

We cannot assign causality to the R-element to receive both effort and flow information simultaneously, whereas in an incidence matrix we can simultaneously consider two nodes, F and ΔP, as the starting nodes. This implies that bond graphs can only represent a subset of all possible matching to eliminate unknown variables. Such a need for relaxation of causality rules, especially for non-linear systems, was identified in [119].

As a solution to such matching problems, bicausality notation is used in this chapter. It allows a more relaxed way of computational causality assignment to eliminate unknown variables from the given set of constraints portrayed in a system model.

Automatic selection of counter-oriented information flow path may be bad from the FDI point of view, but it is an important property of bond graph models, which allows power conservation in the junction structure, *i.e.* junction structure neither consumes nor dissipates power.

Therefore, when a modeler constructs a model from one viewpoint, *e.g.* dynamics (effort balance) or kinematics (flow balance), the other set is automatically satisfied by the constraints in bond graph. This is why bond graph modeling is so useful in creating models of physical systems involving passive interactions, especially when different energy domains are interfaced and parts of the model can be modeled easily in some energy domains by using effort equations (dynamics considerations) while other parts belonging to different energy domains may be modeled using flow equations (kinematic considerations).

7.4.1 Notion of Bicausality

The notion of bicausality [83, 176–180] was primarily introduced to study inverse system dynamics with bond graph models. It introduces some additional bond graph elements, among which SS (Source-Sensor), AE (Amplifier of Effort) and AF (Amplifier of Flow) are relevant in the present context. A Sequential Causality Assignment Procedure for Inversion (SCAPI) is applied on regular bond graph models to arrive at a bi-causalled model. Bicausality notation splits the causality assignment for the two factors of power, namely effort and flow. By separating the causal strokes, it allows imposing two complimentary information at one end of a bond. The different ways of information exchange in bi-causalled bonds are shown in Figure 7.30.

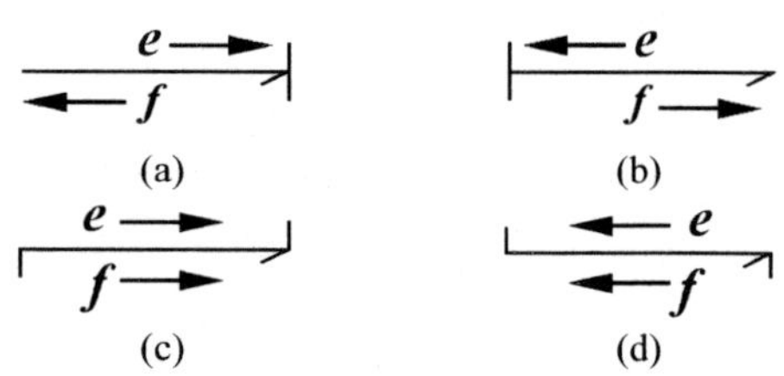

Fig. 7.30 Information exchange in bi-causalled bonds

In Figure 7.30, each bond has two half causal strokes, one for effort and the other for flow. The causal stroke for the flow variable is shown on the side, where the half arrow for power direction is present. On the other side of the power direction, the causality for effort variable is portrayed. When the effort and flow information paths are counter-oriented, as in Figures 7.30a,b, the bond is said to be uni-causalled. Unicausality corresponds to normal causality in bond graph modeling. The cases shown in Figures 7.30c,d correspond to cases, when effort and flow information paths are co-oriented. The rule for bicausal 0 (1)-junction is that only one bond can bring effort (flow) information and other bonds can bring the flow (effort) information. This means that at a bicausal junction, only two bonds can be in bicausality, not more, not less. In other words, at a bicausal junction, there must be one bond bringing in both effort and flow information while there must be another bond taking out both effort and flow information. Tables 7.4 and 7.5, respectively, summarize bicausal configurations for source-sensors, and bicausalled bond graph elements and junctions.

Table 7.4 Source-sensor causality assignment

Causal configuration	Nature of the SS element
SS ——⇀|	Effort source, flow sensor (Se element)
SS |——⇀	Flow source, effort sensor (Sf element)
——⇀| SS ($f=0$)	Zero flow source, effort sensor (De element)
|——⇀ SS ($e=0$)	Zero effort source, flow sensor (Df element)
SS ┌——⇁┘	Flow source, effort source
┌——⇁┘ SS	Flow sensor, effort sensor

Table 7.5 Bicausal configurations for bond graph elements

Bicausal configuration	Assignment statements	Causal configuration	Assignment statements
e, f I:m	$m=\left(\int e.dt\right)/f$	e_2 f_2; e_1 f_1 **0** e_3 f_3	$e_2=e_1$ $e_3=e_1$ $f_3=f_1-f_2$
e, f C: $\frac{1}{C}$	$C=\left(\int f.dt\right)/e$		
e, f R:R	$R=e/f$	e_2 f_2; e_1 f_1 **1** e_3 f_3	$f_2=f_1$ $f_3=f_1$ $e_3=e_1-e_2$
e_1 f_1 $\overset{n}{\text{TF}}$ e_2 f_2	$e_2=e_1/n$ $f_2=n.f_1$		
e_1 f_1 $\overset{n}{\text{TF}}$ e_2 f_2	$e_1=n.e_2$ $f_1=f_2/n$	e_1 f_1 AE:k e_2 f_2	$f_1=0$ $e_2=k.e_1$
e_1 f_1 $\overset{r}{\text{GY}}$ e_2 f_2	$e_2=r.f_1$ $f_2=e_1/r$	e_1 f_1 AE:k e_2 f_2	$f_1=0$ $e_1=e_2/k$
e_1 f_1 $\overset{r}{\text{GY}}$ e_2 f_2	$e_1=r.f_2$ $f_1=e_2/r$	e_1 f_1 AF:k e_2 f_2	$e_1=0$ $f_2=k.f_1$
		e_1 f_1 AF:k e_2 f_2	$e_2=0$ $f_2=k.f_1$

In this book, bicausal notation in bond graph modeling will be used for designing supervision systems [223], *i.e.*, from sensor placement, residual evaluation and FDI to system reconfiguration and fault tolerant control. Bicausal-bond graph (BBG) model based algorithms for sensor placement and fault quantification, by considering single fault hypothesis, are developed in the following sections.

7.4.2 Algorithm for ARR Generation and Construction of FSM

- Each bond connected to a sensor is assigned bicausality so that the direction of both power variables is towards the junction, *i.e.* it imposes both effort and flow information (one of them is zero) on the junction. Obviously, the bond connecting the sensor to the junction becomes the strong bond at that junction. Note that if a bond connected to a sensor cannot be bicausalled, then the corresponding sensor is a direct redundancy.
- The causality from each bicausalled junction is propagated to other elements. Any passive element (I,C and R) which receives information of both the power variables is called a terminating node. Usually non-linear elements, elements with multi-valued relations and storage elements are preferred terminating nodes, because this allows us to write down the ARRs in a flexible way.
- All elements and sensors, which have a causal path to a particular terminating node, are identified.
- All the elements thus identified and the particular terminating node element give the fault signature associated with an ARR, which is derived from the constitutive law of the terminating node.

7.5 Example I: A Two-tank Process

Consider a two-tank process, shown in Figure 7.31 and its bond graph model in Figure 7.32. For the two-tank process, ARR_1 and ARR_2 (corresponding residuals being $r_1$and r_2) can be directly written from the constitutive relation of the pump and the control law for the PI controller, respectively; and then they can be used for construction of a part of the FSM. The bicausal bond graph, shown in Figure 7.33, is used here to analyze the open-loop physical process.

7.5.1 Sensor Placement by Using Bicausal Bond Graphs

In Figure 7.33, the causal paths to the terminating node ($C:C_{\mathrm{T1}}$) are as follows:

$$\begin{aligned} &L_1 \rightarrow e_1 \rightarrow e_2 \rightarrow e_3; R_{V1} \rightarrow f_5 \rightarrow f_4 \rightarrow f_3; \\ &L_2 \rightarrow e_9 \rightarrow e_8 \rightarrow e_6 \rightarrow e_5 \rightarrow R_{V1} \rightarrow f_5 \rightarrow f_4 \rightarrow f_3; Q_{\mathrm{P}} \rightarrow f_{12} \rightarrow f_3. \end{aligned} \tag{7.18}$$

From these causal paths, the components involved in r_3 (corresponding to terminating node, $C:C_{\mathrm{T1}}$) are $K_3 = \left[\mathrm{T}_1\ L_1\ \mathrm{V}_1\ L_2\ Q_{\mathrm{P}}\right]$. The causal paths to the other terminating node ($C:C_{\mathrm{T2}}$) give the components involved in the residual r_4 as $K_4 = \left[\mathrm{T}_2\ L_2\ \mathrm{V}_1\ \mathrm{V}_2\ L_1\right]$.

Fig. 7.31 A controlled two-tank system

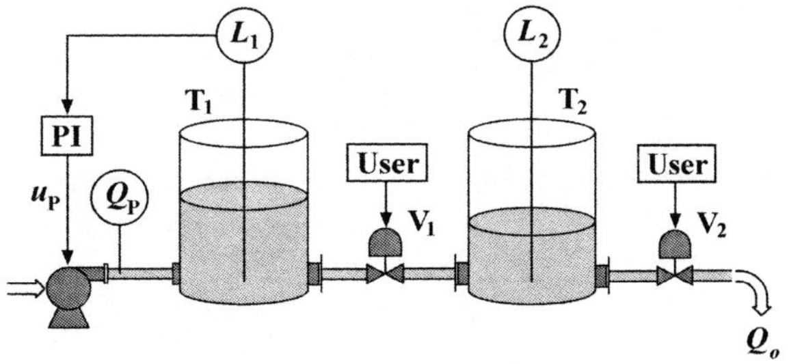

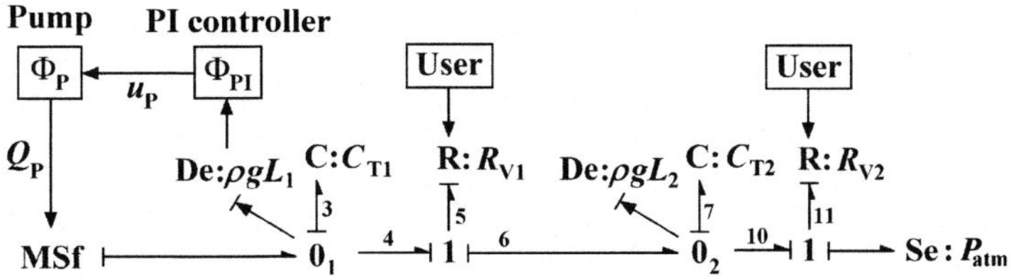

Fig. 7.32 Bond graph model of a controlled two-tank system

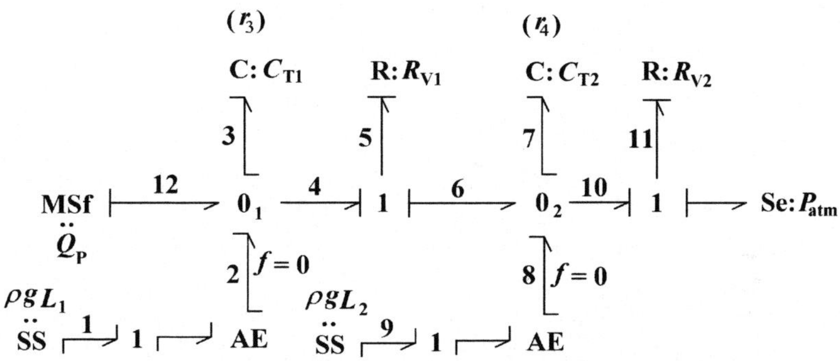

Fig. 7.33 Bicausal bond graph model of a controlled two-tank system

The resulting fault signature matrix (considering no sensor faults as a technological specification) is shown in Table 7.6. In Table 7.6, faults in components T_2 and V_2 are structurally non-isolatable (I_b=0) because they have identical fault signatures. The fault signature matrix can be constructed in many different ways depending upon the choice of bicausality propagation to different terminating elements; nevertheless the same fault isolability results. For example, one may chose R: R_{V1} as a terminating node instead of C: C_{T2} (as in Figure 7.33), as shown in Figure 7.34.

The causal paths to the terminating node ($C : C_{T1}$) are

Table 7.6 FSM obtained by using the first approach

	r_1	r_2	r_3	r_4	I_b
Pump	1	0	0	0	1
PI	0	1	0	0	1
T_1	0	0	1	0	1
T_2	0	0	0	1	0
V_1	0	0	1	1	1
V_2	0	0	0	1	0

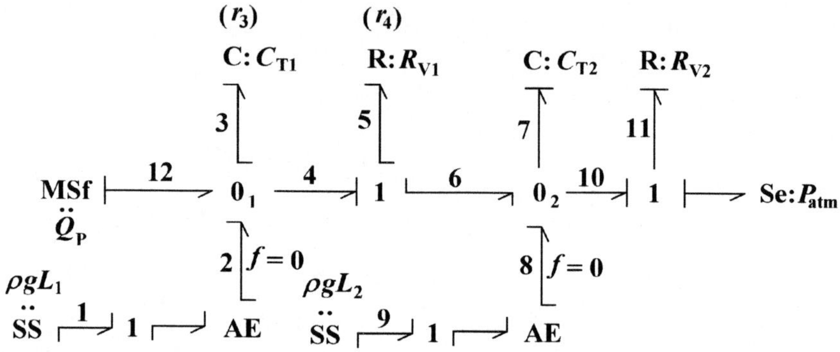

Fig. 7.34 Alternative bicausality assignment to the bond graph model of a controlled two-tank system

$$
\begin{aligned}
&L_1 \rightarrow e_1 \rightarrow e_2 \rightarrow e_3; C_{T2} \rightarrow f_7 \rightarrow f_6 \rightarrow f_4 \rightarrow f_3;\\
&L_2 \rightarrow f_9 \rightarrow f_8 \rightarrow f_6 \rightarrow f_4 \rightarrow f_3;\\
&R_{V2} \rightarrow f_{11} \rightarrow f_{10} \rightarrow f_6 \rightarrow f_4 \rightarrow f_3; Q_P \rightarrow f_{12} \rightarrow f_3.
\end{aligned} \tag{7.19}
$$

Then the components involved in residual r_3 are $K_3 = \left[T_1\ L_1\ T_2\ L_2\ V_2\ Q_P \right]$. Similarly, the causal paths to the terminating node (R: R_{V1}) are

$$
\begin{aligned}
&L_1 \rightarrow e_1 \rightarrow e_2 \rightarrow e_4 \rightarrow e_5; C_{T2} \rightarrow f_7 \rightarrow f_6 \rightarrow f_5;\\
&R_{V2} \rightarrow f_{11} \rightarrow f_{10} \rightarrow f_6 \rightarrow f_5; L_2 \rightarrow e_9 \rightarrow e_8 \rightarrow e_6 \rightarrow e_5.
\end{aligned} \tag{7.20}
$$

Then the components involved in the residual r_4 are $K_4 = \left[\mathrm{V}_1\ L_1\ \mathrm{T}_2\ \mathrm{V}_2\ L_2 \right]$.

The fault signature matrix corresponding to the selected causal structure is given in Table 7.7. Comparing the two matrices (Tables 7.6 and 7.7), one can observe that although the fields under residual columns are different, identical isolability indices are obtained. The non-isolability of faults in T_2 and V_2 (due to the same signature) is quite obvious from physical point of view, as any type of faults (say leakage) in those two components are equivalent, *i.e.* discharging of fluid to the environment. But an additional flow sensor measuring flow through the valve V_2 is sufficient to make all the faults isolatable. This is verified by using the bicausalled bond graph model shown in Figure 7.35.

Table 7.7 FSM obtained by using the second approach

	r_1	r_2	r_3	r_4	I_b
Pump	1	0	0	0	1
PI	0	1	0	0	1
T_1	0	0	1	0	1
T_2	0	0	1	1	0
V_1	0	0	0	1	1
V_2	0	0	1	1	0

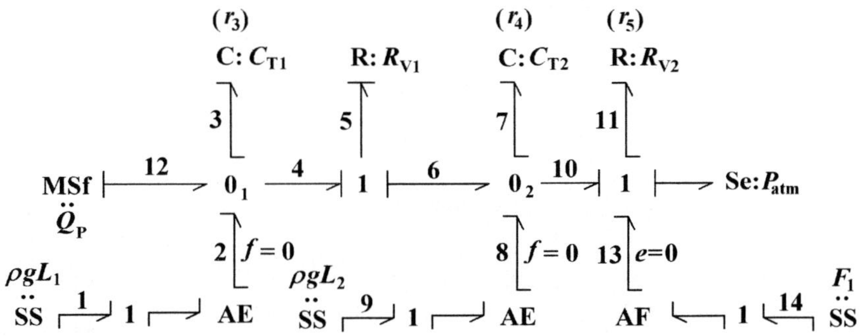

Fig. 7.35 Bicausal bond graph model to verify complete fault isolability

The causal paths to the element C: C_{T1} are identical to those in Figure 7.33. So the components involved in r_3 are $K_3 = \left[\mathrm{T}_1\ L_1\ \mathrm{V}_1\ L_2\ Q_\mathrm{P}\right]$.

The causal paths to the element C: C_{T2} are

$$\begin{aligned} &L_2 \rightarrow e_9 \rightarrow e_8 \rightarrow e_7; \\ &R_{V1} \rightarrow f_5 \rightarrow f_6 \rightarrow f_7; \\ &F_2 \rightarrow f_{14} \rightarrow f_{13} \rightarrow f_{10} \rightarrow f_7. \end{aligned} \tag{7.21}$$

Then the components involved in r_4 are $K_4 = \left[\mathrm{T}_2\ L_2\ \mathrm{V}_1\ F_2\right]$.

The causal paths to the element R: R_{V2} are

$$\begin{aligned} &F_2 \rightarrow f_{14} \rightarrow f_{13} \rightarrow f_{11}; \\ &L_2 \rightarrow e_9 \rightarrow e_8 \rightarrow e_{10} \rightarrow e_{11}. \end{aligned} \tag{7.22}$$

Then the components involved in the residual r_5 are $K_5 = \left[L_2\ \mathrm{V}_2\ F_2\right]$.

Note that, now, there are five residuals corresponding to five sensors. The FSM, shown in Table 7.8, is derived on the basis of the above causal paths and it is found that all component faults are isolatable with the selected instrumentation architecture. Different ways of sensor placement methods based on bond graphs can be consulted in [102, 129, 223].

Table 7.8 FSM corresponding to full fault isolation

	r_1	r_2	r_3	r_4	r_5	I_b
Pump	1	0	0	0	0	1
PI	0	1	0	0	0	1
T_1	0	0	1	0	0	1
T_2	0	0	0	1	0	1
V_1	0	0	1	1	0	1
V_2	0	0	0	0	1	1

7.5.2 Residual Generation: Symbolic Method

A symbolic ARR derivation method, by using bicausality notation, is developed in this section. For symbolic derivation of the ARRs, we have considered the bicausal bond graph model in Figure 7.35.

Since information of both two power variables, f_3 and e_3, are available to the terminating node at the element C: C_{T1},

$$
\begin{aligned}
& e_3 = \rho.g.L_1, \\
& f_3 = f_{12} - f_4 = Q_p - f_5 = Q_p - C_{d1}\sqrt{e_5} = Q_p - C_{d1}\sqrt{\rho.g(L_1 - L_2)}. \quad (7.23)
\end{aligned}
$$

The constitutive relation for the element, C: C_{T1}, in derivative form is given by

$$
\begin{aligned}
& f_3 = \frac{1}{C_{T1}}.\frac{d}{dt}(e_3) = \frac{A_1}{g}.\frac{d}{dt}(\rho.g.L_1) \\
& \Rightarrow Q_p - C_{d1}\sqrt{\rho.g(L_1 - L_2)} - \frac{A_1}{g}\frac{d}{dt}(\rho.g.L_1) = 0 \\
& \Rightarrow \text{ARR}_3 : Q_p - \frac{A_1}{g}\frac{d}{dt}(\rho.g.L_1) - C_{d1}\sqrt{\rho.g(L_1 - L_2)} = 0. \quad (7.24)
\end{aligned}
$$

Similarly, ARR_4 can be derived from constitutive relation of the element, C: C_{T2}:

$$
\begin{aligned}
& f_7 = \frac{1}{C_{T2}}\frac{d}{dt}(e_7) = \frac{A_2}{g}.\frac{d}{dt}(\rho.g.L_2); \\
& f_7 = f_6 - f_{10} = C_{d1}\sqrt{\rho.g(L_1 - L_2)} - C_{d2}\sqrt{\rho.g.L_2} \\
& \Rightarrow \text{ARR}_4 \ : \ C_{d1}\sqrt{\rho.g(L_1 - L_2)} - \frac{A_2}{g}\frac{d}{dt}(\rho.g.L_2) - C_{d2}\sqrt{\rho.g.L_2} = 0. \quad (7.25)
\end{aligned}
$$

ARR_5 is derived from non-linear constitutive relation for the element R: R_{V2}:

$$\begin{aligned} e_{11} &= e_{10} - e_{15} = \rho.g.L_2, \\ f_{11} &= F_2, \text{ and } f_{11} = C_{d2}\sqrt{e_{11}} \\ \Rightarrow \text{ARR}_5 &: F_2 - C_{d2}\sqrt{\rho.g.L_2} = 0. \end{aligned} \tag{7.26}$$

Note that if the last ARR, which represents a deduced redundancy, is not considered, then the ARRs obtained are exactly the same as the ARRs obtained in Chapter 5 from the bipartite graph.

7.5.3 Residual Evaluation and Fault Scenario Simulation

The coupled model for the two-tank process is given in Figure 7.36, in which the *Coupling* section provides the interface between the behavioral and the diagnostic sub-model [227].

The diagnostic model is bicausalled from its input nodes to a set of terminating nodes. The residuals are evaluated in the diagnostic model by using the constitutive relations of the elements at the terminating nodes. This representation allows direct evaluation of residuals and thus bypasses the need to derive ARRs. Therefore, this model can be used to obtain residuals of system models having looping variables.

The two-tank system was simulated with nominal parameter values given in Table 7.9 and a fault in V_1 was triggered at time 505 s by changing the value of C_{d1} from nominal 1.8×10^{-3} $(kg.m)^{1/2}$ to 1.2×10^{-3} $(kg.m)^{1/2}$. As shown in Figure 7.37, residuals r_3 and r_4 deviate from their nominal response; whereas r_5 does not (except for a temporary deviation at the inception of the fault). Note that response of residuals r_1 and r_2, which remain within their thresholds for this fault scenario, are not shown because they are known to be insensitive to process deviations. Hence, the obtained coherence vector is $C = [0,0,1,1,0]$, which has a unique match with the fault signature of component V_1 in the FSM given in Table 7.8, thus isolating the valve V_1 as the faulty component and validating the fault event test simulation.

Table 7.9 Parameters of the two-tank system

Symbol	Description	Value	Unit
C_{d1}	Discharge coefficient of valve V_1	1.8×10^{-3}	$kg^{\frac{1}{2}}m^{\frac{1}{2}}$
C_{d2}	Discharge coefficient of valve V_2	1.5964×10^{-3}	$kg^{\frac{1}{2}}m^{\frac{1}{2}}$
A_i	Floor area of Tank_i $(i = 1,2)$	1.54×10^{-2}	m^2
Q_{Pmax}	Maximum flow rate from pump	1	kg/s
K_p	Proportional gain of PI controller	10^{-4}	Pa^{-1}
K_i	Integral gain of PI controller	10^{-6}	$Pa^{-1}s^{-1}$

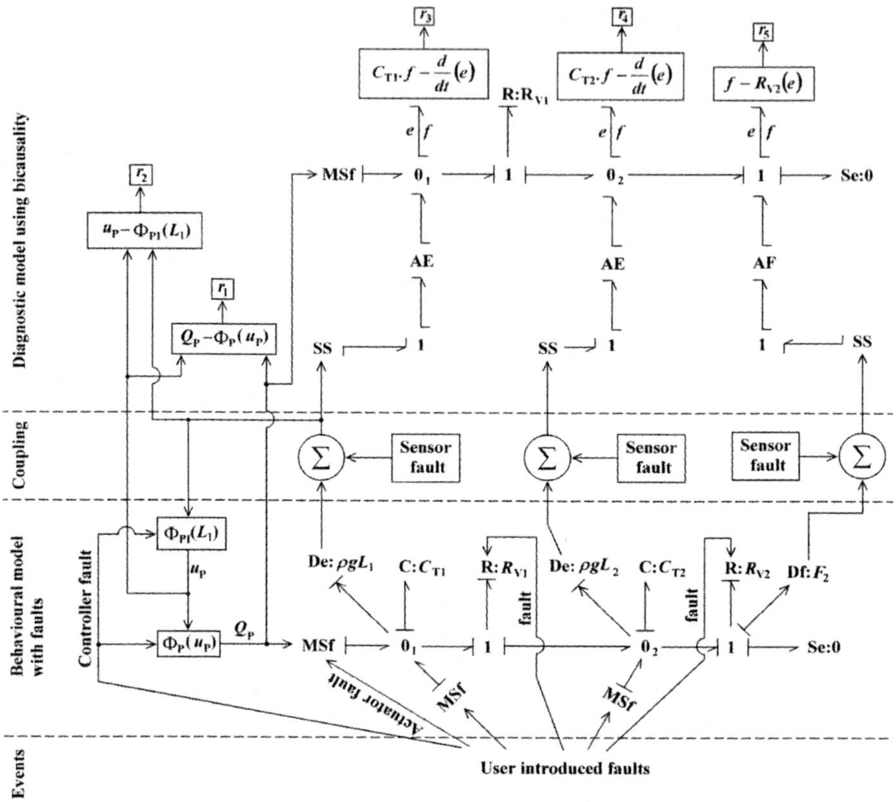

Fig. 7.36 Coupling of behavioural and bicausalled diagnostic bond graph model

7.6 Example II: A Servo-valve Controlled Motor Transmission System

7.6.1 System Description and Bond Graph Model

A servo-valve controlled motor transmission system with pump loading is shown in Figure 7.38. The bond graph model of the system was developed in [57], wherein dynamic behavior was studied through simulation and then it was experimentally validated.

By retaining the assumptions made in [57] and their bond graph model, the FDI scheme for the system has been developed here. For the purpose of FDI analysis, we have modified their bond graph to a diagnostic bond graph by using the notion of bicausality (as shown in Figure 7.39). Note that for the sake of brevity, we have directly used SS elements without additional AE and AF elements, but indicated the corresponding power variable (e or f) as zero depending on the type of detector.

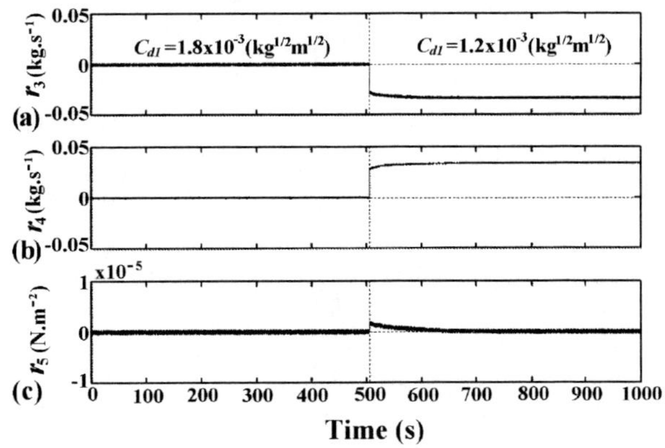

Fig. 7.37 Residual response for two-tank process. **a** Residual-3. **b** Residual-4. **c** Residual-5

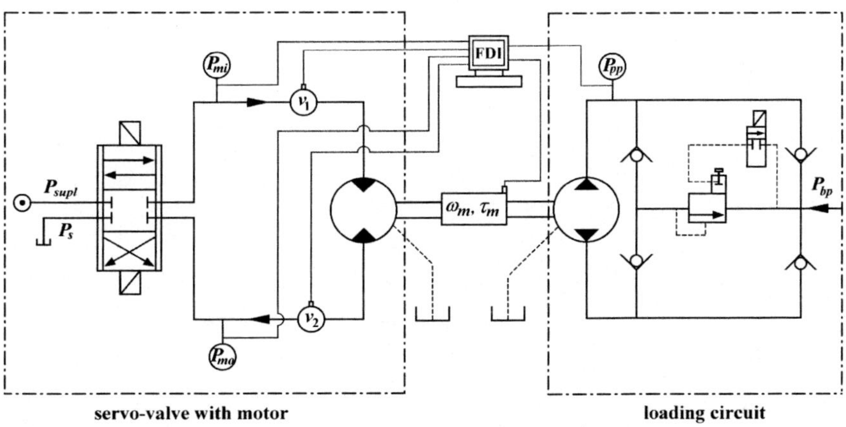

Fig. 7.38 Servo-valve controlled motor transmission with pump loading

The nomenclature of various used variables is given in Table 7.10.

Inlet fluid flow (v_1) from a constant pressure source (P_{supl}) to the motor and also outlet flow (v_2) from the motor to the sump, through the port of the servo-valve, are modulated by the current driving the valve. The modulated flow (v_1 and v_2) through the resistances of the servo-valve ports (R_1 and R_2 in Figure 7.39) are expressed as:

$$v_1 = k_{fi}.x_{sv}.\sqrt{|P_{supl} - P_{mi}|},$$
$$v_2 = k_{fo}.x_{sv}.\sqrt{|P_{mo} - P_s|}. \qquad (7.27)$$

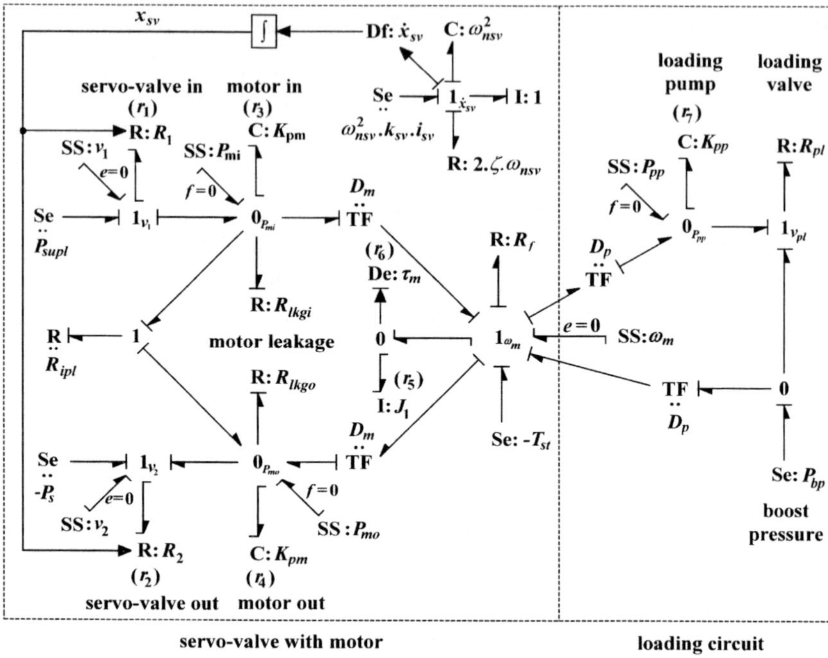

Fig. 7.39 Bicausalled bond graph model of the servo system

The spool displacement (x_{sv}), which is controlled by the current input (i_{sv}) to the servo-valve, is obtained from the solution of a second order differential equation (Equation 7.28) according to the dynamical analysis conducted by Watton [274]:

$$\ddot{x}_{sv} + 2.\zeta.\omega_{nsv}.\dot{x}_{sv} + \omega_{nsv}^2.x_{sv} = \omega_{nsv}^2.k_{sv}.i_{sv}. \tag{7.28}$$

The flow to the pump plenum is proportional to the speed of the motor (ω_m) and the motor can be loaded by controlling the relief valve of the loading circuit. In the present context, FDI analysis is limited to the drive unit. The inlet and outlet pressures of the motor (P_{mi} and P_{mo}), the inlet and outlet flows through the servo-valve (v_1 and v_2), the speed (ω_m) and the torque (τ_m) of the motor, and the pump plenum pressure (P_{pp}) are measured.

7.6.2 ARRs and FSM

Seven structurally independent ARRs (for seven sensors) are obtained from the diagnostic bond graph model as

Table 7.10 Parameters for the servo-valve controlled motor transmission system

Symbol	Description	Value
P_{supl}	Supply pressure to the servo-valve	1.0×10^{7} N.m^{-2}
P_s	Sump pressure	0.0 N.m^{-2}
k_{fi}	Flow coefficient at the servo-valve inlet	5.0×10^{-5} kg$^{-0.5}$.m$^{3.5}$
k_{fo}	Flow coefficient at the servo-valve outlet	5.0×10^{-5} kg$^{-0.5}$.m$^{3.5}$
k_{sv}	Gain of the servo-valve	0.03 m.A^{-1}
ω_{nsv}	Natural frequency of the servo-valve	100 rad.s^{-1}
ζ	Damping coefficient of the servo-valve	0.4
K_{pm}	Effective bulk stiffness of the fluid at the motor plenum	1.0×10^{14} N.m^{-5}
R_{lkgi}	External leakage coefficient of the motor inlet	1.5×10^{12} N.s.m^{-5}
R_{lkgo}	External leakage coefficient of the motor outlet	1.5×10^{12} N.s.m^{-5}
R_{ipl}	Cross-port leakage coefficient of the motor	3.2×10^{12} N.s.m^{-5}
D_m	Volume displacement rate of the motor w.r.t. its shaft rotation	0.68×10^{-6} m^{3}.rad^{-1}
D_p	Volume displacement rate of the pump w.r.t. its shaft rotation	0.68×10^{-6} m^{3}.rad^{-1}
J_1	Motor plus load inertia	0.0014 kg.m^{2}
R_f	Viscous frictional resistance on the motor	0.001146 N.m.s.rad^{-1}
T_{st}	Stiction or coulomb friction torque loss of the motor	2.5 N.m
K_{pp}	Effective bulk stiffness of the fluid at the loading circuit	1.0×10^{14} N.m-^{5}
R_{pl}	Load resistance in the loading circuit	7.5×10^{10} N.s.m^{-5}
P_{bp}	Boost pressure supply	2.0×10^{6} N.m^{-2}

$$
\begin{aligned}
&\mathrm{ARR}_1 : v_1 - k_{fi}.x_{sv}.\sqrt{\left|P_{supl} - P_{mi}\right|} = 0,\\
&\mathrm{ARR}_2 : v_2 - k_{fo}.xsv.\sqrt{\left|P_{mo}\right|} = 0,\\
&\mathrm{ARR}_3 : v_1 - \frac{\dot{P}_{mi}}{K_{pm}} - \frac{P_{mi}}{R_{lkgi}} - D_m.\omega_m - \frac{P_{mi} - P_{mo}}{R_{ipl}} = 0,\\
&\mathrm{ARR}_4 : -v_2 - \frac{\dot{P}_{mo}}{K_{pm}} - \frac{P_{mo}}{R_{lkgo}} + D_m.\omega_m + \frac{P_{mi} - P_{mo}}{R_{ipl}} = 0,\\
&\mathrm{ARR}_5 : D_m.P_{mi} + D_p.P_{bp} - R_f.\omega_m -\\
&\qquad T_{st} - D_m.P_{mo} - D_p.P_{pp} - J_1.\dot{\omega}_m = 0,\\
&\mathrm{ARR}_6 : \tau_m - J_1.\dot{\omega}_m = 0,\\
&\mathrm{ARR}_7 : D_p.\omega_m - \frac{\dot{P}_{pp}}{K_{pp}} - \frac{P_{bp} + P_{pp}}{R_{pl}} = 0.
\end{aligned}
\tag{7.29}
$$

Note that De: τ_m in the diagnostic bond graph model in Figure 7.39 is acting as a functionally redundant sensor because it is measuring an effort which can be deduced from the measured flow ω_m. Therefore, detector De: τ_m cannot accept the bicausality; however the corresponding residual (r_6) is evaluated by subtracting the deduced effort from the measured effort.

The FSM obtained from the ARRs is given in Table 7.11, which reveals that all faults are monitorable ($M_b = 1$) and isolatable ($I_b = 1$) with the given instrumentation architecture.

Table 7.11 FSM for the servo-valve controlled motor transmission system

	r_1	r_2	r_3	r_4	r_5	r_6	r_7	M_b	I_b
k_{fi}	1	0	0	0	0	0	0	1	1
k_{fo}	0	1	0	0	0	0	0	1	1
R_{lkgi}	0	0	1	0	0	0	0	1	1
R_{lkgo}	0	0	0	1	0	0	0	1	1
R_{ipl}	0	0	1	1	0	0	0	1	1
R_f	0	0	0	0	1	0	0	1	1
τ_m	0	0	0	0	0	1	0	1	1
ω_m	0	0	1	1	1	1	0	1	1
R_{pl}	0	0	0	0	0	0	1	1	1

7.6.3 Validation Through Simulation

The nominal parameter values for simulation are given in Table 7.10. The simulations were carried out by using SYMBOLS 2000 software [173, 174]. For simulation, the current input i_{sv} to the servo-valve was chosen as a square wave [57] of step 3–8 mA and a period of 0.5 s, as shown in Figure 7.40a. Gaussian noise amounting to 2% of signal magnitude was superposed on every measured signal to realize noisy sensor outputs. The time responses of motor speed ω_m and pressure P_{mi}, during normal operation, are given in Figure 7.40b.

In the fault scenario simulation, the model was simulated in normal operating mode and then a cross-port leakage fault was introduced after 2 s by changing the value of R_{ipl} from 3.2×10^{12} N.s.m^{-5}(nominal value) to 1×10^{10} N.s.m^{-5}. This resulted in deviations of residuals r_3 and r_4, as shown in Figure 7.41a. All other residuals remained unaffected, thereby leading to a coherence vector $C = [0,0,1,1,0,0,0]$, which has an exact match with the fault signature of R_{ipl} in the FSM given in Table 7.11. Therefore, cross-port leakage fault can be detected and isolated.

Another simulation was carried out by changing the value of k_{fi} from nominal 5.0×10^{-5} kg$^{-0.5}$.m$^{3.5}$ to 1.0×10^{-5} kg$^{-0.5}$.m$^{3.5}$ at a time of 2 s, which corresponds to inlet side blockage of the servo-valve. In this case, the residual, r_1, deviates as shown in Figure 7.41b. So the fault event is isolated because the resulting coherence vector uniquely matches with the fault signature of k_{fi} (Table 7.11).

In [57], it is shown that steady state pressure (P_{mi}) and speed (ω_m) deviate with change in R_{ipl}. However, there may be various reasons behind such deviations (*e.g.* due to change in values of R_{lkgi} and R_{lkgo}). Therefore, faults cannot be isolated

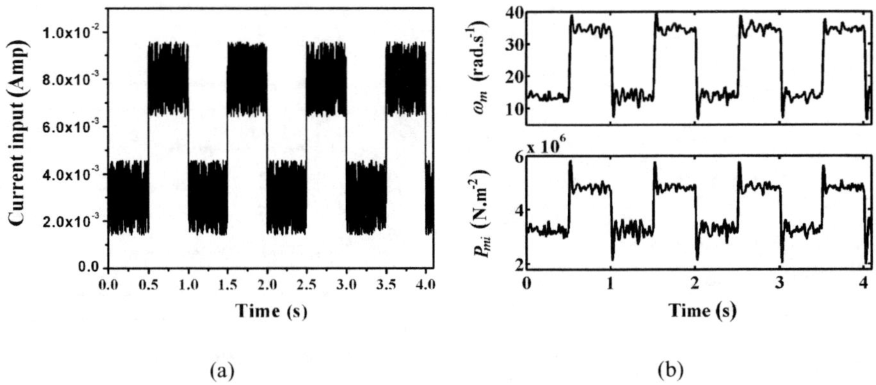

Fig. 7.40 **a** Current input to the servo-valve. **b** Time response of motor speed (w_m) and pressure (P_{mi})

by simple qualitative comparison of outputs. On the contrary, ARRs structurally decouple one fault effect from the other and therefore faults are uniquely isolated from quantitative evaluation of ARRs.

7.7 The Fault Isolation Problem

Fault signatures from structured residuals are diagonal or square type, where only each residual is sensitive to a single fault and insensitive to others. All faults are uniquely isolatable considering both single and/or multiple fault hypothesis with a structured set of residuals. In single fault hypothesis, we assume that only one component may be faulty at a time. Note that if a fault is isolatable considering multiple fault hypotheses, then that fault is also isolatable for single fault case, *i.e.* single fault signature is a subset of multiple fault signature. Therefore, multiple fault hypothesis is a generalized hypothesis and structured set of residuals can isolate a fault generally.

However, structured residuals cannot always be obtained, particularly when the number of fault candidates is more than the number of sensors. In that case, non-square fault signature matrix may result in which some subspaces may be localized where component faults may be isolatable subjected to specific decoupling conditions. The set of structurally independent residuals (*i.e.* a particular set of residuals are sensitive to a particular component) guarantee unique fault isolability only for single fault hypothesis.

This is further explained by using an example fault signature matrix given in Table 7.12, in which θ_i, $i=1$ to 7, are seven independent parameters and r_j, $j=$

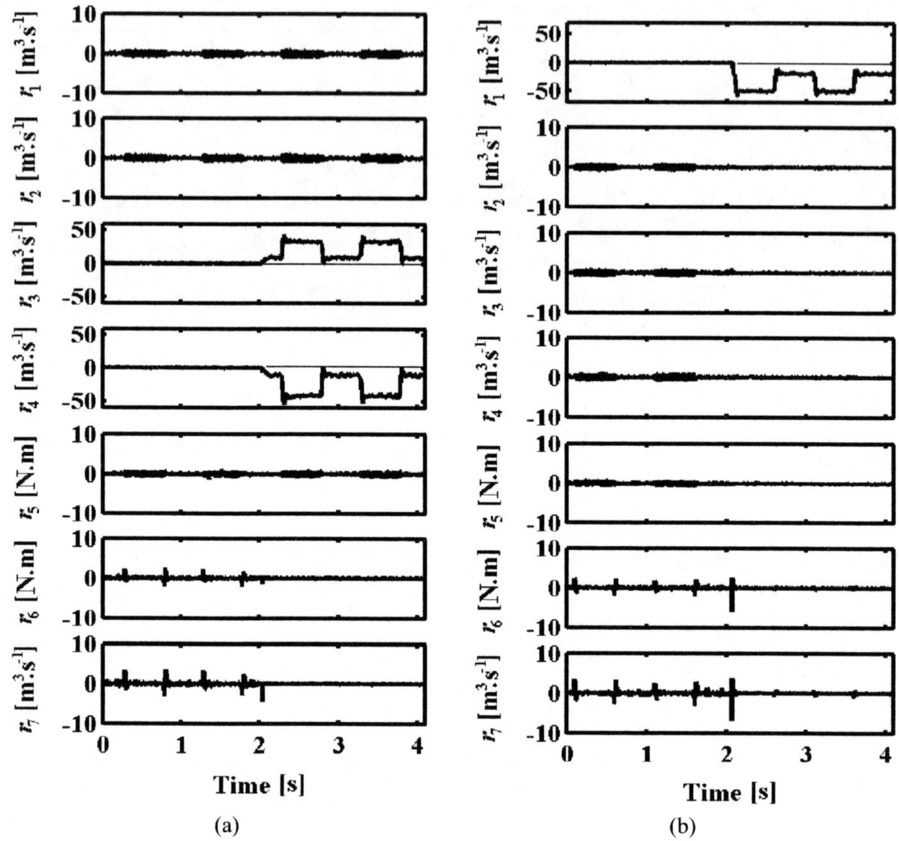

Fig. 7.41 Residual responses of the servo-valve controlled motor transmission system under different types of faults. **a** Cross-port leakage (R_{ipl} is changed). **b** Inlet servo-valve blockage (K_{fi} is changed)

1 to 6, are six residuals. The fault signature matrix shows four groups of faults, S_{P1}, S_{P2}, S_{P3} and S_{P4}. The subspace S_{P1}consists of diagonal or structured set of residuals and therefore, faults affecting this subspace are always isolatable. This structured or diagonal or square part of residuals is marked with a darker shade and the unstructured or non-square part is marked with a lighter shade in Table 7.12.

Consider that only residual r_4 is abnormal, which leads to a coherence vector $C = [0,0,0,1,0,0]$. This coherence vector has a unique match in the fault signature matrix and the component corresponding to parameter θ_4 can be isolated as the fault candidate, under single fault hypothesis.

But in S_{P3}, when r_4 and r_5 both are abnormal, the component corresponding to θ_5 is uniquely isolated as a fault candidate by considering single fault hypothesis, because the set of residuals are structurally independent. However, if multiple fault

Table 7.12 Example fault signature matrix

		r_1	r_2	r_3	r_4	r_5	r_6
θ_1		1					
θ_2	S_{P1}		1				
θ_3				1			
θ_4	S_{P2}				1		
θ_5	S_{P3}				1	1	
θ_6						1	1
θ_7	S_{P4}					1	1

hypothesis is considered, θ_5 is definitely faulty, but one cannot be sure about the state of θ_4.

If r_5 and r_6 both are abnormal, then from the resulting coherence vector $C = [0,0,0,0,1,1]$, a unique fault cannot be isolated, even under single fault hypothesis; only a susceptible subspace S_{P4} containing non-isolatable parameters, θ_6 and θ_7, can be localized.

Suppose that one obtains a coherence vector, which does not match with any of the fault signatures in the FSM, but matches with a linear combination of some fault signatures (*e.g.* $C = [1,0,1,1,1,1]$ in Table 7.12), then it is definitely a multiple fault case.

The foregoing discussion can be summarized as follows:

1. Fault in a parameter $\theta_i \in \mathrm{FS_U}$ with fault signature C_i is isolatable for both single and multiple fault case, if $\forall C_j \in \mathrm{FS_U}$, $j \neq i$; $C_i.\mathrm{AND}.C_j \neq C_i$.
2. In other words, if $\exists\ j \neq i$; $C_i.\mathrm{OR}.C_j = C_i$, then that fault is not isolatable with multiple fault hypotheses.
3. Any non-null coherence vector $C = \sum_{i=1}^{n} \alpha_i.C_i$, $\alpha_i \in [0,1]$, where $\sum$ is an OR operator and more than one $\alpha_i \neq 0$, corresponds to multiple faults.

When a component(s) cannot be uniquely isolated as fault candidate(s), but a susceptible subspace containing more than one component can be located, further refinement through parameter estimation is required for fault isolation. This refinement step is termed the second level isolation procedure. Second and/or further higher level isolation procedures are needed in many practical situations where the system is not sufficiently instrumented. Determination of the initial set of fault candidates is called fault-hypothesis generation. The second level decision procedure is called fault-hypothesis validation, whereby temporal information are used to arrive at a unique fault candidate or at the least, the dimension of the set of initial fault candidates is minimized. These issues are handled in Chapters 9 and 10.

Problems

7.1. The system shown in Figure 7.42 has six state variables. The system is instrumented as shown in the figure (three sensors and a source are the known measurements). Can you derive ARRs for this system? Construct the DBG of the system and identify (1) a closed causal path containing two differentially causalled storage elements and (2) a causal loop containing a differentially causalled storage element and a resistance. Couple the DBG with the corresponding behavior model for online FDI.

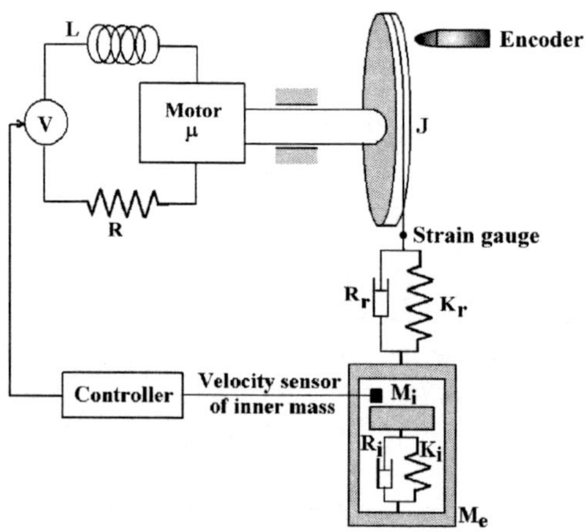

Fig. 7.42 A speed controlled device to lift sensitive equipment

7.2. The actuator of a semi-active vehicle suspension system, with its instrumentation, is shown in Figure 7.43. The input velocity is known and the working fluid is assumed to be compressible (bulk modulus=β). Each check valve offers either a constant finite resistance or infinite resistance, depending upon the pressure across it. Determine the fault signature matrix of the system from causal paths by using bicausality. Couple the bicausalled DBG to the corresponding unicausalled behavior model.

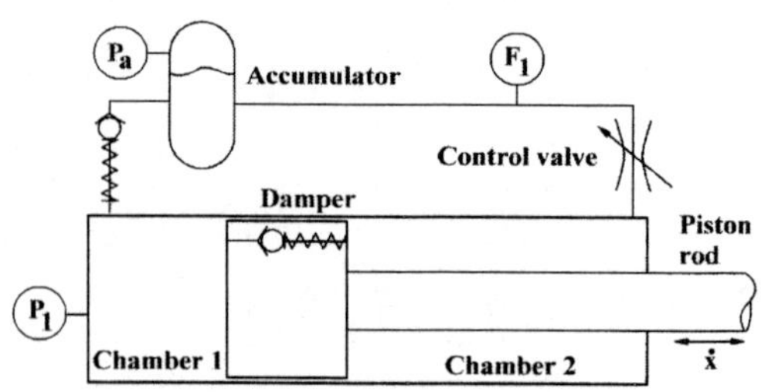

Fig. 7.43 An actuator of a semi-active vehicle suspension system

Chapter 8
Actuator and Sensor Placement for Reconfiguration

8.1 Introduction

Fault Tolerant Control (FTC) is performed through fault accommodation and/or system reconfiguration. In fault accommodation, the objective is to control the system under actual constraints. In system reconfiguration, part of the actual faulty system is replaced by another one [282]. The objective of FTC is to prevent local faults developing into serious failures.

8.1.1 Minimal Sensor and Actuator Placement

The minimum set of actuators and sensors required for operating a system is defined by the controllability and observability conditions, which were discussed in Chapter 3.

FTC requires that faults can be isolated. Reconfiguration can be performed if redundant hardware for the faulty component is available. The objective of actuator and sensor placement scheme for FTC is to maintain the controllability and the observability of the process at all times, *i.e.* even after loss of key hardware. The term key hardware means the set of actuators and sensors needed to operate and monitor the process.

The necessary conditions for functional recoverability [8] from faults is that with the remaining actuators and sensors, the system is functionally controllable and observable, the system can be identified and its parameters can be estimated. This is formally termed as the *model matching technique* [268]. If some key sensors are lost, then their measurements can be reconstructed by using an observer and used to control the system without changing the control law.

8.1.2 Sensor Placement for FDI and FTC

For fault isolation and accommodation, more sensors are to be installed in the process to obtain structured residuals. When a fault is too severe, one may go for reconfiguration by using stand-by devices, called hardware redundancy, to accommodate the fault. Two types of redundancies are possible for actuators and sensors: deduced and direct redundancy. These issues are explained with the help of an academic example in Section 8.4.

8.2 External Model

Automation of large scale complex industrial processes encompasses a broad scope which has to take into account additional functionality such as maintenance, management and supervision facilities. This has led to the concept of distributed intelligent systems where smart sensors and actuators integrate with the information system of the automated process. The local processing powers of smart sensors and actuators increase the quality of the processed information. This additional information extends the range of the possibilities of the global system. In addition to the control possibilities, smart equipment creates a large number of data items which can be communicated to various operators, who manage the process. Furthermore, most complex process engineering systems are hybrid.

The external model, which uses the concept of services offered to the users and organizations based on operating modes, leads to a generic model description, which can be used to specify and verify the functional specifications of smart instruments and hybrid systems [3, 214, 241]. Contrary to the black box models, which carry no information about the component's behavior in different operating situations (whether normal or faulty), external models describe components from the point of view of the user who receives services and can use them for different purposes.

At any time, a system carries out a coherent subset of missions. Each subset is called an User Operating Mode (USOM). The logical conditions to move from one USOM to another are described by a state graph termed *USOM graph management*. This graph defines the logical sequences of operating steps in which some missions start and some others end. The achievement of the mission depends on the services provided by the components of the system. The low level components are directly connected to the physical process, and they provide the elementary services, whereas higher level services and components are realized by proper aggregation of the low level components. In the presence of faults, the alarm decision is taken according to the availability of the elementary system components [51, 52]. The drawback of the external model is that it describes the services and the USOMs in terms of the component's functions, without taking into account the component's physical and dynamic behavior. This leads to some ambiguity. Furthermore, the availability of the resources is not determined by a Fault Detection and Isolation (FDI) procedure.

By associating bond graph models with external models, it becomes possible to obtain the behavioral knowledge of the system's components and to improve their monitoring, and consequently to ensure an improved safety standard. However, in the industrial supervision tasks, human operators consider the running system in the terms of its functions [149, 168]. In order to give the operator, or the industrial user, a clear view of the system's functional organization and to allow the user to estimate the system's ability to achieve its goal, a functional modeling tool based on an external model may be used as the external interface whereas the corresponding bond graph model may be used for the internal representation [189].

8.2.1 External Model in a Bond Graph Sense

From an external point of view, an automation system provides services to users. Users request services for specific purposes, for example: storage, energy transformation, *etc.* The external model describes the device from the point of view of the services, it is able to provide, to external entities (operators, other field instruments, computers, and so on). It introduces the notion of the services, the user operating modes, and the versions of the services.

8.2.2 Services

A service is defined as a procedure whose execution results in the modification of at least one data item in the instrument's database, and/or at least one signal on its output interface. Each service, S_i, is defined by a quintuplet of sets: consumed variables (C_i), produced variables (P_i), conditions of activation (Ca_i), data processing (dp_i) and resources (R_i):

$$S_i = < C_i, P_i, dp_i, R_i, Ca_i > . \tag{8.1}$$

A service is an operational entity corresponding to a physical object of the hardware architecture. The service has a hardware existence, which means that, for a service to be achieved, it needs one or more resources of the process. Besides, a service possesses a functional identity, *i.e.* every service is associated with an activity. Any given service performs only one task at a time. The set of services S, which are offered by an equipment, is finite:

$$S = \{s_i \,|\, i \in [1..n]\}. \tag{8.2}$$

- In the sense of bond graph, the services correspond to some set of bond graph elements such as C for storage component (tank, boiler, *etc.*), TF and GY elements for transformation of energy from one domain to another, R for resistive element (pipe, valve, *etc.*), Se for effort source (pressure pump, voltage generator, *etc.*) and Sf for flow source (flow pump, current generator, *etc.*).

- The services offered by the sensors (to measure) are delivered by effort (De) and flow (Df) detectors. The sensors can of course provide other services, such as keeping memory of minimum and maximum levels, providing alarms, *etc.* In this chapter, only those services that provide a signal as a function of a measured variable are considered.
- The services provided by the equipment of sources of energy (mechanical motor, potential or kinetic energy of fluid and hydraulic pump, *etc.*) are represented by effort (Se) or flow (Sf) sources.
- The services provided by the information system (controllers, positioners, *etc.*) are presented as block diagrams since it is assumed that these systems are not energetic, but only carry the signals. In the bond graph models, these systems are modeled by information bonds represented by full arrows.
- The consumed and produced variables are the power variables (the effort-flow pair).
- The request is represented by an information bond in the bond graph. The services offered by the functional role of the equipment (storage, transport, transformation, *etc.*) are represented by the bond graph elements C, R, TF, *etc.*

Thus, in a bond graph, the services as defined in the external model represent a set of elements necessary for the physical realization of the corresponding service. The execution of these services by a request can be controlled intentionally by an external entity (order coming from an operator or automaton) or by an internal one (from the process/controller). The availability of each service offered by a component is modeled by a Boolean variable associated with an internal or an external request. This Boolean variable is provided by the decision procedure implemented in the FDI system. For example, when a sensor is faulty, the corresponding "measure" service will not be available. A hardware failure implies the unavailability of some services and could bring some missions to a halt. That is where model based FDI allows the analysis of the consequences of a failure on the availability of the services offered by a system.

8.2.3 User Selected Operating Mode (USOM)

As discussed previously, in order to avoid simultaneous attempts at performing incompatible services by the system (initialization and production services, for example), the set of services is organized into coherent subsets, called USOMs. Each USOM contains at least one service, and each service belongs to at least one USOM. At any given time, and according to the technical specifications, a system carries out a coherent subset of missions. Each of these subsets is called an operating mode. Each USOM_i can be associated to a bond graph model noted MBG_i, where i is an index for the operating mode:

$$\text{USOM}_i \rightarrow \text{MBG}_i. \tag{8.3}$$

USOMs are given by a general classification of the operating modes of industrial devices: off-operation, configuration, manual or automatic mode, nominal regime, starting mode, *etc.* An automaton, to change the user operating modes, specifies the USOM (initial USOM, and the others) and the conditions to change from one USOM to another. From this description, a formal specification of the intelligent equipment can be obtained.

- *Versions of services*

According to the resource states decided by the supervision of the process, several versions, V, of the services (nominal and degraded) may be designed. As an example, if the set of services (set of bond graph elements) is $\{S_1, S_2, S_3, .., S_n\}$ and $\{V_1, V_2, V_3, .., V_n\}$ is the corresponding set of versions, then the running operating mode is represented as

$$\text{MBG} = \{S_1(V_1), S_2(V_2), ...S_n(V_n)\}. \tag{8.4}$$

8.2.4 Operating Mode Management

At any given time, the process (or the equipment) runs in a nominal operating mode represented by the corresponding bond graph model.

Let MBG be the set of the bond graph models, and MBG_j be the bond graph model corresponding to the j-th operating mode, and n be the total number of the operating modes:

$$\text{MBG} = \{\text{MBG}_1, \text{MBG}_2, ..\text{MBG}_l, ...\text{MBG}_n\}. \tag{8.5}$$

Let b be the set of transitions:

$$b = \{b_{ij}\}, \tag{8.6}$$

where b_{ij} indicates the required condition to move from MBG_i to MBG_j. The transition condition is represented by the Boolean variable $b_{ij} \in \{0, 1\}$.

It is considered that all mode change conditions (noted C_m) are correctly defined, *i.e.*

$$\forall \text{MBG}_i (i = 1, n) \in \text{MBG}, \exists j \text{ such that } b_{ij} \in C_m \text{ and } b_{ij} \neq 0. \tag{8.7}$$

The running operating mode can be represented as

$$\begin{aligned} \text{MBG}_j &= \{s_{j1}(V_{j1}), s_{j2}(V_{j2}), ...s_{ji}(V_{ji}), ...s_{jm}(V_{jm})\} \\ &= \{S_j\}/s_{ji}(i = 1, k) = \{\{C_{ji}\}, \{R_{ji}\}, \{TF_{ji}\}, \\ &\quad \{GY_{ji}\}, \{Se_{ji}\}, \{Sf_{ji}\}, \{De_{ji}\}, \{Df_{ji}\}\}, \end{aligned} \tag{8.8}$$

where m is the number of services (functions) offered by the system to accomplish the given operating mode mission associated to the MBG_j and k is the number of bond graph elements necessary for the physical performance of the i-th service.

Each USOM (or bond graph model) MBG_j contains at least one service, *i.e.* one bond graph element (otherwise the physical process does not exist).

In fact, the USOM's organization can be described as a deterministic automaton. Let $G = \{\text{MBG}, b, C_m\}$ be the USOM graph. It can be seen as a deterministic transition if $\forall \text{MBG}_i \in \text{MBG}, \exists b_{ij} \in C_m, \exists$ at the most one transition with MBG_i as origin mode and b_{ij} as label. MBG is the set of bond graph models. The sets of the USOMs and the transition conditions can be represented as shown in Figure 8.1.

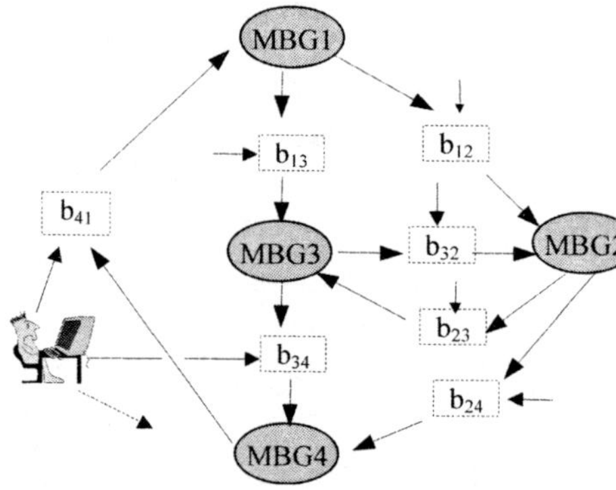

Fig. 8.1 USOM graph in the bond graph sense

The whole set of the available services and its set of associated versions are established for a given USOM by an internal function called "Operating mode management". The request for a change of an operating mode must indicate the destination mode. However, in presence of faults, because of safety, some couples (origin, destination) may not be allowed. As an example, the request "automatic control mode" would be rejected as long as the service "measure" (provided by the sensors) is unavailable.

We illustrate hybrid functional-physical modeling approach by considering a smart actuator as an example in the next section.

8.3 Application to a Smart Pneumatic Valve

A functional model of a general smart instrument is given in Figure 8.2. We consider a specific smart actuator [280] (Figure 8.3) which represents a pneumatic valve chosen as the DAMADICS benchmark [10, 11, 67, 208, 249].

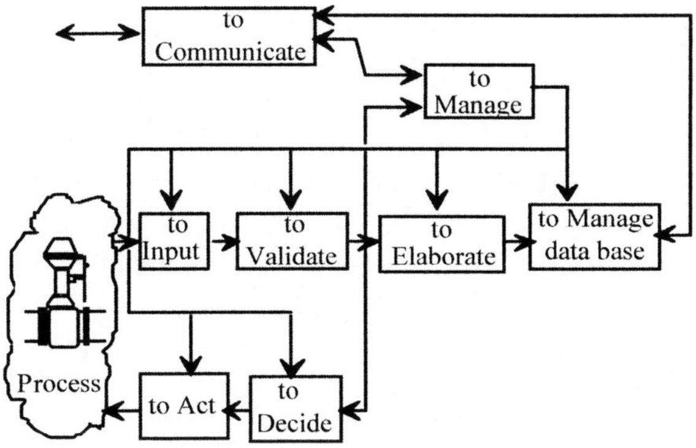

Fig. 8.2 Functional structure of a smart instrument

8.3.1 Description of the System

The actuator consists mainly of a control valve, a servo-motor, and a positioner with a pressurizer.

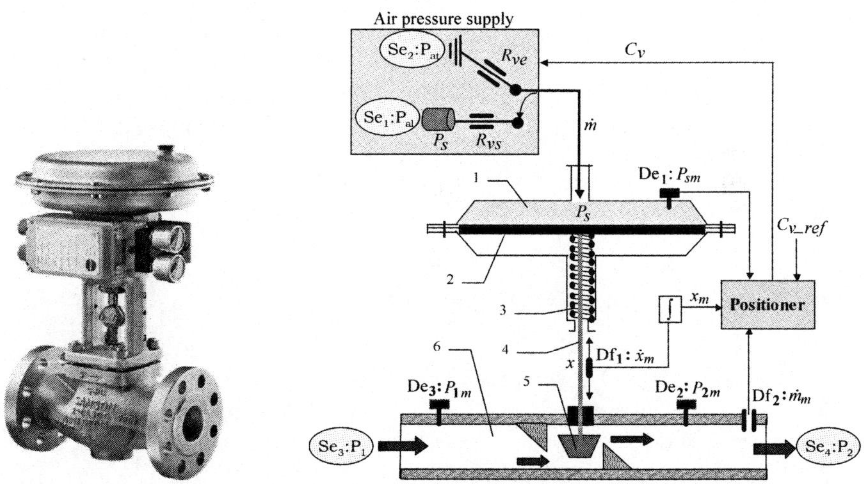

Fig. 8.3 DAMADICS smart actuator (valve)

The control valve (5), which can be a disk, ball, or plug, corrects the flow in the pipe (6) to maintain a desired flow setpoint. The driving force of the fluid is the difference between the upstream pressure ($Se_3 : P_1$) and the downstream pressure

($Se_4 : P_2$). These pressures are measured by the sensors De_3 and De_2. The movement of the valve is accomplished by a servo-motor which consists of a pneumatic chamber (1), a spring (3), a diaphragm (2) and a stem (4). This system is a compressible (air) fluid powered device, in which the fluid stored in (1) acts upon the flexible diaphragm and the spring to provide linear motion of the servomotor stem.

The positioner belongs to the group of microprocessor based smart devices. It is applied to eliminate the control-valve stem mis-positioning produced by the external or internal sources such as friction, pressure unbalance, hydrodynamic forces, *etc.* However, the communication facilities are limited only to two current (4–20 mA) signaling loops. One loop provides a setpoint value, C_{v_ref}, and the second one corresponds to the actual valve stem position. The algorithm is based on PID controller algorithms given by [10]. The positioner output, C_v, acts on the air pressure supply system in order to deliver the desired pressure, P_s. The control signal, C_v, is defined by the valve stem position, x_m, the controlled fluid mass flow, $\dot{m}_m$, the pressure in the pneumatic chamber, P_{sm}, and the reference value, C_{v_ref}. Values for x_m, $\dot{m}_m$, P_{sm}, and C_{v_ref} are provided, respectively, by the sensors Df_1, Df_2, De_1 and the user. The air mass flow delivered by the air supply system depends on the state of the supply (R_{vs}) and the exhaust (R_{ve}) orifices. According to a control signal delivered by the positioner, C_{v_ref}, the pneumatic chamber (1) is exposed to supply pressure, $Se_1 : P_{al}$, or is exposed to exhaust (zero reference pressure), $Se_2 : P_{at}$.

Faults can occur in the process itself, or in the sensors and actuators, for example, a change in the parameters of some component. The FDI specifications are intended to list those faults which have to be considered. In total, 19 faults $\{Flt1...Flt19\}$ are distinguished [137] in the assembly consisting of the control valve, the pneumatic servomotor, and the positioner. These faults are classified into following four groups: control valve faults $\{Flt1...Flt7\}$, pneumatic servo-motor faults $\{Flt8...Flt11\}$, positioner faults $\{Flt12...Flt14\}$ and general/external faults $\{Flt15...Flt19\}$.

8.3.2 Bond Graph Model of the Smart Actuator

The bond graph model in normal operating regime is given in Figure 8.4.

Two forms of energy are modeled here: hydraulic and mechanical. The thermal energy is not modeled, since an isothermal case is assumed. The power variables used for the mechanical energy are the force-velocity $(F,\dot{x})$ pair. The hydraulic (or pneumatic) energy is modeled using pressure-mass flow $(P,\dot{m})$ power variables in a pseudo-bond graph form and pressure-volume flow $(P,\dot{V})$ power variables are used in a true bond graph form.

The inlet air mass flow, $\dot{m}_{16}$, from the supply system to the pneumatic chamber is given by the constitutive equation of modulated resistances, $R : C_{vs}$ and, $R : C_{ve}$. These resistances are modulated by the positioner output signal, Cv. When a fluid is moving through a restrictor, a non-linear relationship exists relating the mass flow rate, $\dot{m}$, to the pressure drop, ΔP, across the restriction and the Boolean variable, b_1.

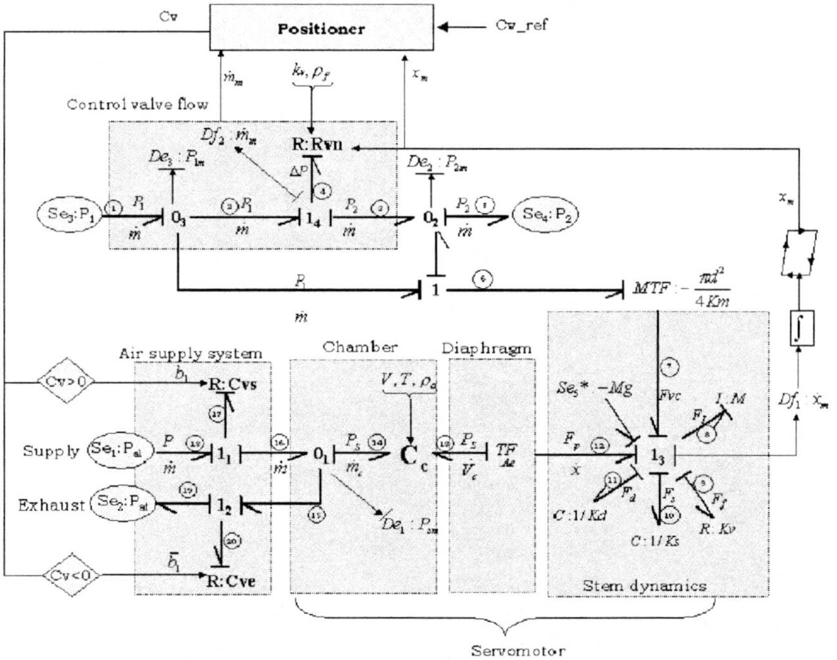

Fig. 8.4 Bond graph model of smart valve in normal operating mode

The compressibility effect and the air mass storage in the pneumatic chamber are modeled by the two-port C-field (C_c). This multiport allows to calculate the pressure which depends on the mass and the volume of air.

$P_1 V_1^n = P_2 V_2^n \Rightarrow \frac{P_1}{P_2} = \left(\frac{V_2}{V_1}\right)^n$ which leads to

$$\dot{P} = \frac{1}{\left[\frac{\rho V}{nP}\right]} \left(\dot{m} - \rho \dot{V}\right), \tag{8.9}$$

where n is the polytropic index, ρ is the fluid density, and V is the volume of the collapsible chamber (which is variable). This C-field has, consequently, one $(P, \dot{m})$ port and one $(P, \dot{V})$ port. This implies that the total energy depends on m and V. The bond graph of this element represents a sophisticated version of models of compressibility effects in air servomechanism, in which the volume is variable. In the bond graph model (Figure 8.4), the bond number 14 models the storage of mass, and the bond number 13 models the storage of volume. The pressure is the same in both these bonds, $(P_{13} = P_{14})$.The constitutive relation for the C_c field is given by

$$\begin{bmatrix} P_{14} \\ P_{13} \end{bmatrix} = \begin{bmatrix} K & -K.\rho_a \\ K & -K.\rho_a \end{bmatrix} \begin{bmatrix} \int \dot{m}.dt \\ \int \dot{V}.dt \end{bmatrix}, \tag{8.10}$$

where ρ_a is the air density and $K = 1/\left[\frac{\rho_a.V}{n.P}\right]$.

In integral causality, the effort variable (pressure) is calculated by the conservation rules at the "0_3" junction and the constitutive relation of the C_c field:

$$e_{14} = \int \frac{n.e_{14}}{\rho_a V}(f_{14} - \rho_a.f_{13})\,dt + e_{14}(0). \tag{8.11}$$

The corresponding block diagram (for $n = 1$, or an isothermal case) is given in Figure 8.5.

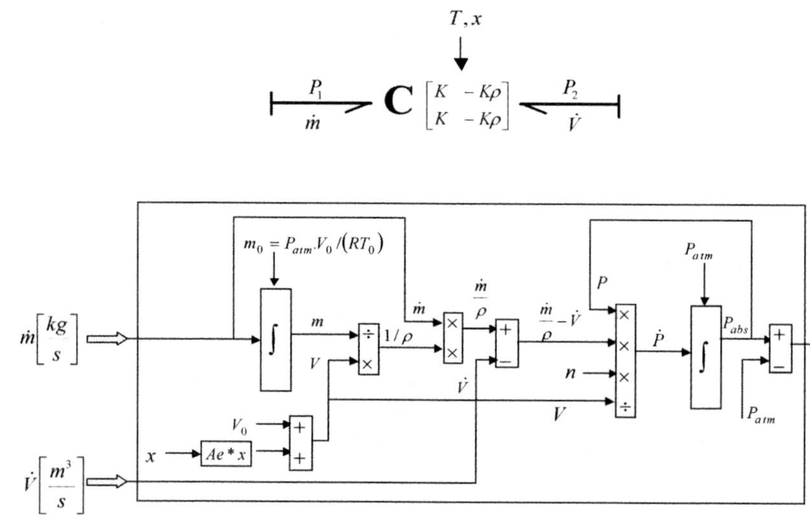

Fig. 8.5 Bond graph and block diagram model of C_c field

The transformation of the pneumatic energy into mechanical energy, which leads to the servo-motor's stem displacement, is modeled by the transformer, TF, whose modulus, A_e, is the cross-sectional area of the diaphragm. The dynamics of the servo-motor's stem is modeled by the conservation of energy represented at the "1_3-junction" linked to the R, C and I elements. $I : M$, K_v, K_s, K_d, $Se_5 : -Mg$, F_{vc} and F_p account, respectively, for the inertia due to the mass of the stem, the force due to the friction [124], the force due to the stiffness of the spring, the force due to the elasticity of the diaphragm, the gravity, the *vena contracta* force (which depends on the pressure of the controlled fluid, diameter of the plug, *etc.*) and the active force. The flow through the control valve is modeled by the modulated R-element, $R : R_{vn}$. The resistance value of $R : R_{vn}$ is the coefficient of hydraulic losses which depends on the stem position, x, which is defined as follows:

$$\begin{aligned}
&\text{Positioner: } Cv = K_p(C_{v_ref} - \tfrac{1}{2} + \tfrac{3}{\pi}\arcsin(\tfrac{x}{0.0381} - \tfrac{1}{2})).\\
&\text{Logic block: } b_1 = \begin{cases} 1, \text{ if } C_v > 0;\\ 0, \text{ otherwise;}\end{cases}\\
&\text{and } \bar{b}_1 = 1 - b_1.
\end{aligned} \tag{8.12}$$

8.3.3 Missions and Versions

Based on the objectives fixed by the technical specifications for the actuator, a non-exhaustive list of missions and associated services, *i.e.* bond graph elements, is defined in Table 8.1.

Table 8.1 List of missions and services for the smart actuator

Number	Missions	BG elements
1	To supply the pneumatic chamber with air pressure	Se_1, C_v, $R{:}C_{vs}$, $R{:}C_{ve}$, Se_2
2	To store the air in the pneumatic chamber	$C{:}C_c$
3	To transform pneumatic energy into mechanical	TF
4	To provide a motion of servo-motor steam	TF, $C{:}1/K_d$, $C{:}1/K_s$, $R{:}K_v$
5	To control the servo-motor steam position	Df_1, Positioner
6	To control the flow of the fluid	Df_2, Positioner
7	To provide a fluid in the pipe	Se_3, $R{:}R_{vn}$
8	To measure the pressure in the pneumatic chamber	De_1
9	To measure the inlet pressure in the pipe	De_3
10	To measure the outlet pressure in the pipe	De_2
11	To measure the servo-motor stem position	Df_1
12	To measure the control valve flow	Df_2
13	To maintain equipment	–

8.3.4 Operating Mode Management of the Smart Actuator

As described previously, the operating mode management, given in Figure 8.6, allows the operator to be informed about the conditions under which the transition from one USOM to another can be performed. In the current approach, and as a complement to the external modeling methodology, each USOM is represented by a bond graph model. It helps the operators to understand the basic physical principles, to visualize the physical phenomena of the process, and to quantify each effect. Furthermore, the bond graph model is used to generate the alarm decisions, and consequently the availability or unavailability of the services (sensor, actuator, or process faults).

The smart actuator's model set includes the following non-exhaustive list of bond graph models.

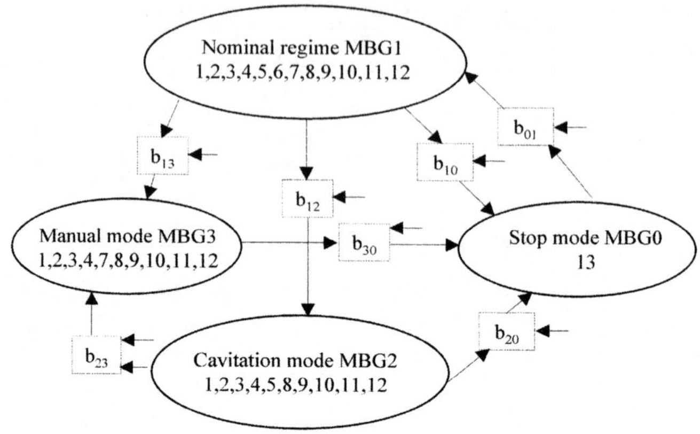

Fig. 8.6 Operating mode management of the smart valve

MBG0, Stop Operating Mode. In this mode, the actuator carries out the mission number 13. Since the process is at rest (no power), the bond graph model is of no interest in this mode.

MBG1, Nominal regime. The missions carried out in the nominal regime are numbers 1–12. The bond graph model corresponding to this regime is given in Figure 8.4.

MBG2, Manual operating mode. In this mode, the smart actuator carries out mission numbers 1–4, and 7–12.

MBG3,Cavitation operating mode. Cavitation occurs if the fluid outlet pressure drops below the vapor pressure of the liquid. As the vapor cavities collapse, noise is generated and damage can occur. Cavitation damage produces a rough, pitted, cinder-like surface. The corresponding bond graph model is given in Figure 8.7. Both flashing and cavitation limit the flow of the liquid through the valve. During flashing and cavitation, bubbles begin to form in the flow stream when the pressure drops below the vapor pressure of the liquid. The bubble formation at the *vena contracta* restricts the amount of liquid that can be forced through the valve. This is commonly known as the choked-flow condition. The flow no longer increases with decreases in downstream pressure. Choked-flow can result in severe damage to the valve, and desired flow requirements may not be reached, which impedes the process. The difference of pressure at the choked flow condition is a function of the flow geometry of the control valve. The experimentally determined coefficient denoted, K_m, is used to define the point of choked flow condition (Figure 8.8). The valve recovery coefficient, K_m, and critical pressure ratio of the liquid, r_c, can be determined from the tables and curves given by the valve's manufacturer [116].

The main missions carried out in this mode are numbers 1–5, and 8–12. Such organization increases the operational safety of the system and forbids access to those USOMs for which the meaningful services cannot be run properly. Consider, for instance, the current USOM: MBG_1 and assume that service 2 (to store the air in

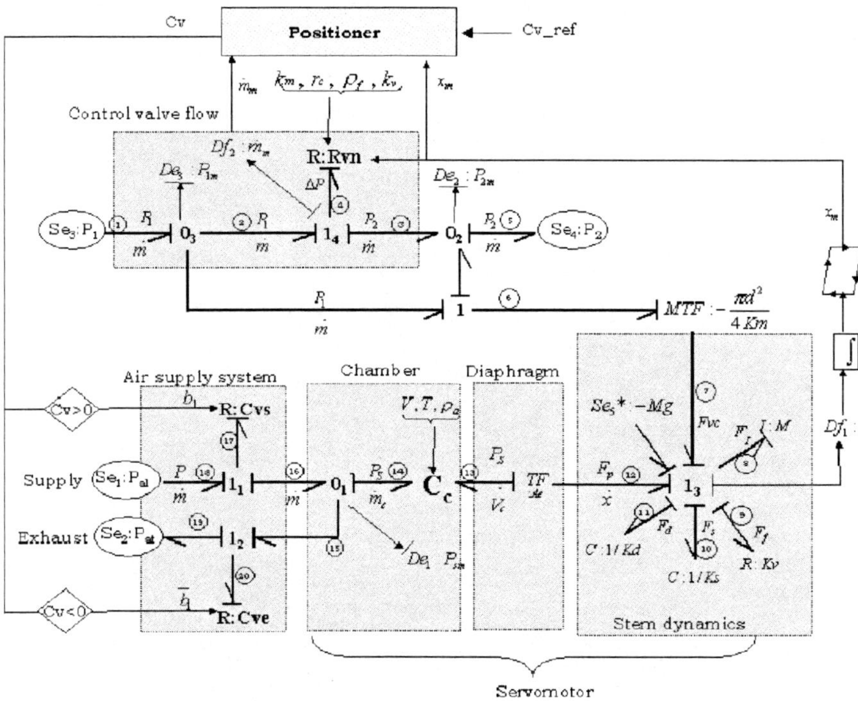

Fig. 8.7 Bond graph model of smart valve in cavitation mode

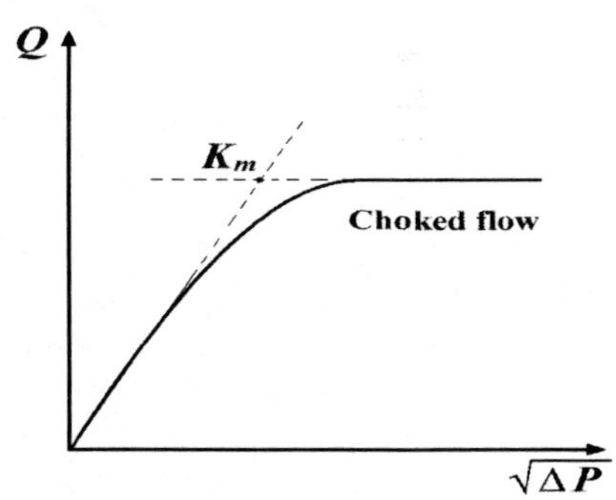

Fig. 8.8 Valve flow characteristics and flow choking

the pneumatic chamber modeled by the two port C_c-field) is not available because of a leakage. Thus, according to Figure 8.6, the transition to USOM: MBG_3 or MBG_2 will be rejected, since service 2 belongs to both these operating modes; only the transition to MBG_0 will be allowed.

The alarm generation in the smart actuator's FDI system is developed in the next section.

8.3.5 Monitoring of the Smart Actuator

By assigning preferred derivative causality to the bond graph model in Figure 8.4, we can derive the following ARRs:

$$\begin{aligned}
\mathbf{ARR}_1 &: \frac{\dot{P}_{sm}.\rho_a(A_e.x_m+V_0)}{n.P_{sm}} + \rho_a.A_e.\dot{x}_m - |C_v|.\sqrt{|Se_1 - P_{sm}|}.b_1 - \\
&\quad |C_v|.\sqrt{|P_{sm} - Se_2|}.\bar{b}_1 = 0, \\
\mathbf{ARR}_2 &: Se_5 - \frac{\pi.d^2}{4.K_m}.(Se_3 - Se_4) - K_v.\dot{x}_m - K_s.x_m + P_{sm}.A_e - K_d.x_m - M.\ddot{x}_m = 0, \\
\mathbf{ARR}_3 &: \frac{\rho_f}{10000}\left(\frac{Df_2}{K_v(x_m)}\right)^2 - Se_3 + Se_4 = 0, \\
\mathbf{ARR}_4 &: Se_3 - De_3 = 0, \\
\mathbf{ARR}_5 &: Se_4 - De_2 = 0.
\end{aligned} \tag{8.13}$$

Note that we have assumed that the measured quantity $\dot{x}_m$ is integrable to obtain x_m. Furthermore, the valve positioner output, C_{vm}, is measured and it results in the additional ARR:

$$\mathbf{ARR}_6 : C_{vm} - K_p\left(C_{v\_ref} - \frac{1}{2} + \frac{3}{\pi}\arcsin\left(\frac{x_m}{0.0381} - \frac{1}{2}\right)\right) = 0, \tag{8.14}$$

where K_p is the gain of the controller and x_m is the valve position.

Because each component involves more than one parameter, it is better in this case to write down the fault signatures in terms of the bond graph elements and parameters rather than the components themselves. Known robust elements, *e.g.* Se_1 and Se_2 for supply and exhaust pressures, are excluded from the FSM. Note that in the FSM, given in Table 8.2, leakage from chamber is related to bond graph element C_c whereas flow cavitation is related to element R_{vn} (function K_v). Based on the binary vector of each component given in Table 8.2, all faults can be detected and isolated except the fault which may affect Df_2 and R_{vn}. If sensor Df_2 is assumed to be robust, then blockage, cavitation or leakage in the valve can be isolated. These fault signatures are used to generate alarms for the operating mode management of the smart actuator, given in Figure 8.6. For example, if the service "To store the air in the pneumatic chamber", associated with the bond graph two-port element C_c (Table 8.1), is not available because of a pneumatic chamber leakage (detected in residual r_1) then the request "nominal regime mode" should be rejected as long as this service (provided by the decision procedure) is not available. Thus, according to Figure 8.6, the transition to USOM: MBG_3 or MBG_2 will be rejected because service-2 is an element of both these operating modes; so only the transition to MBG_0 will be allowed.

Table 8.2 Fault signatures of the smart valve

	r_1	r_2	r_3	r_4	r_5	r_6
Se_3	0	1	1	1	0	0
Se_4	0	1	1	0	1	0
De_1	1	1	0	0	0	0
De_2	0	0	0	0	1	0
De_3	0	0	0	1	0	0
Df_1	1	1	1	0	0	1
Df_2	0	0	1	0	0	0
Leakage (C_c)	1	0	0	0	0	0
Positioner (C_{vm})	0	0	0	0	0	1
TF (A_e)	1	1	0	0	0	0
Spring (K_s)	0	1	0	0	0	0
Blockage/ Cavitation (R_{vn})	0	0	1	0	0	0

8.4 Reconfiguration of a Thermo-fluid System

Let us consider a simple thermofluid process shown in Figure 8.9. Our initial aim is to find if the system is well instrumented; therefore the structure of the open-loop system is studied.

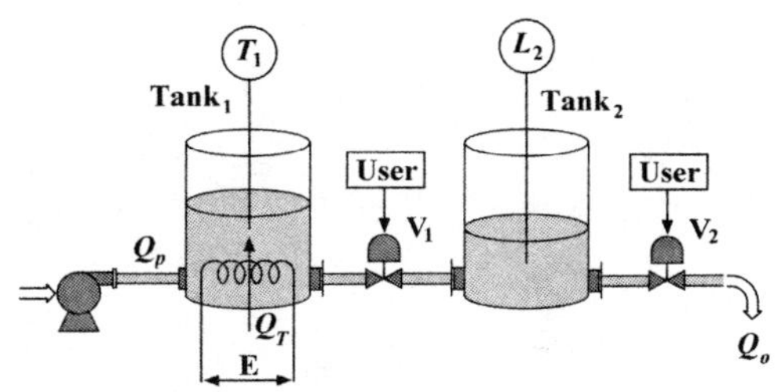

Fig. 8.9 A thermofluid system with arbitrary instrumentation

8.4.1 Minimal Sensor and Actuator Placement

The pseudo BG of the process is shown in Figure 8.10, where CETF represents Coupling Element for ThermoFluids. Two possible causal forms of CETF and the corresponding equations are given in Table 8.3, where c_p is the specific heat capacity of the fluid. Note that the temperature of downstream side (T_2) does not appear in the constitutive relations of CETF. The top part of the BG model (Figure 8.10) concerns hydraulic domain and the bottom part concerns thermal domain, where $C_H = A/g$, $C_T = mc_p$ and A is the cross-sectional area. The states associated with C-elements in the hydraulic and thermal domains are mass (m) and enthalpy (H), respectively.

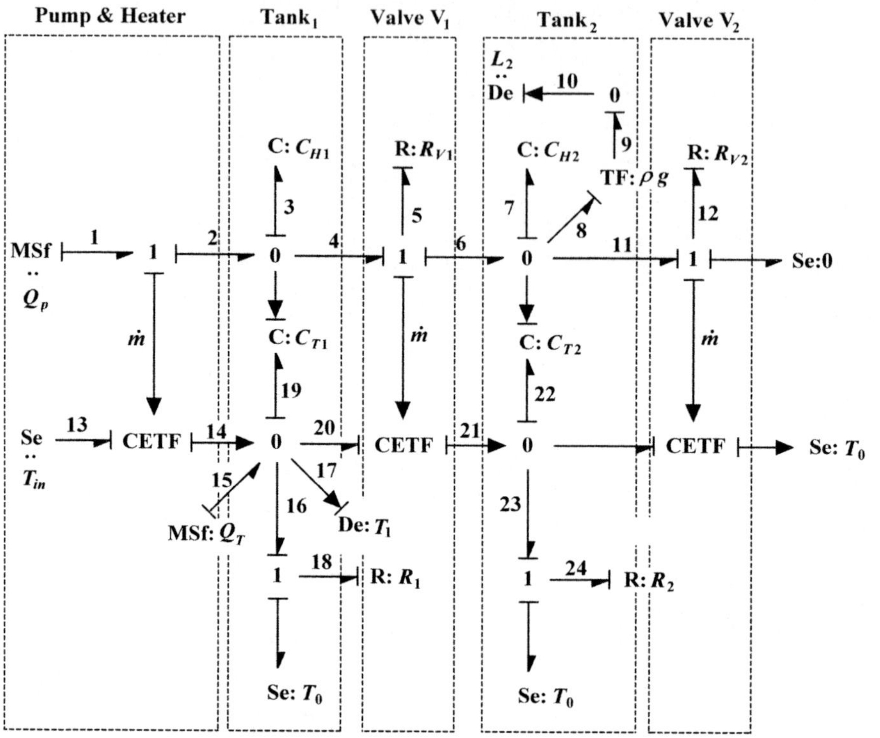

Fig. 8.10 Bond graph model of the thermofluid system

Table 8.3 Causal forms of the CETF element

Causal form of CETF	Equation
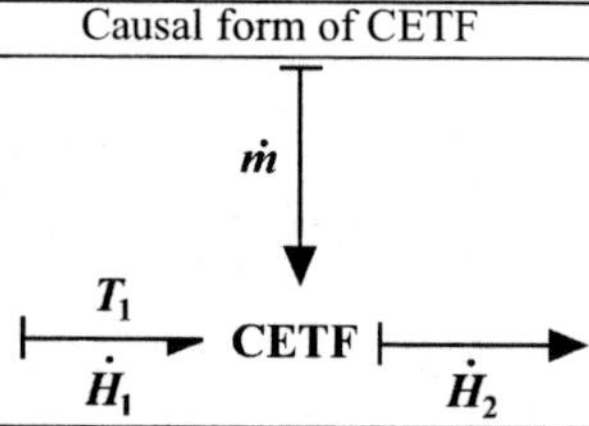	$\dot H_1 = \dot m c_p T_1$ $\dot H_2 = \dot m c_p T_1$
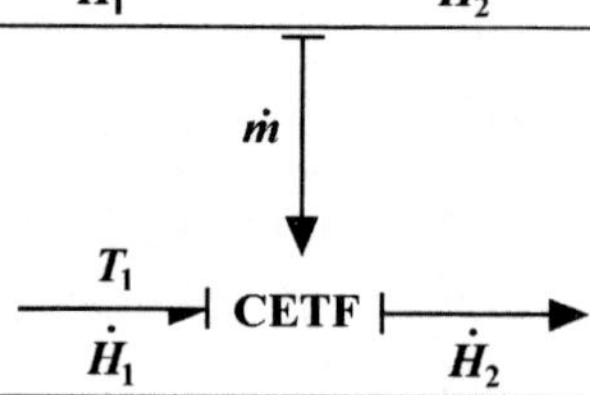	$T_1 = \frac{\dot H_1}{\dot m c_p}$ $\dot H_2 = \dot H_1$

In the process shown in Figure 8.9, the fluid temperature inside Tank_1 is measured by a sensor T_1, while a level sensor L_2 measures the level in Tank_2. In Figure 8.10, element C_{T2} does not have causal path to any detector because effort information cannot pass through bond 21 attached to CETF. Hence the attainability or necessary condition for observability of the system fails for the given sensor configuration.

Let us consider another configuration (Figure 8.11), where the temperature sensor is installed in Tank_2 instead of Tank_1.

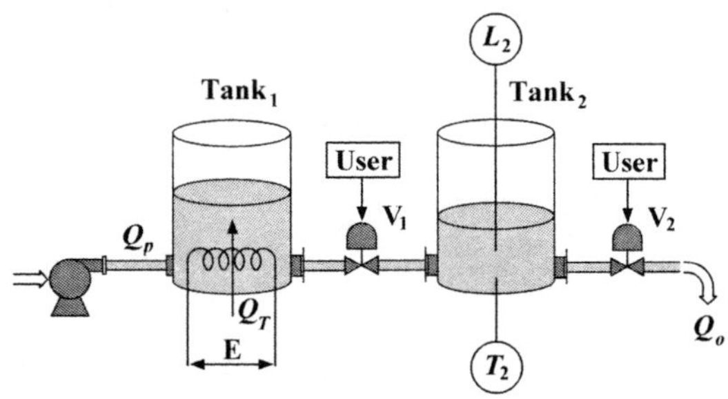

Fig. 8.11 Modified sensor architecture in the two-tank system

Each integrally causalled storage element in the corresponding BG model (Figure 8.12) has a causal path linking it to a detector; thus satisfying the attainability condition for structural observability. The second condition, *i.e.* the sufficient condition for structural observability is tested by using preferred derivative causality, as shown in Figure 8.13. It is seen that both necessary and sufficient conditions for structural observability are satisfied and hence the two sensors in the process constitute minimum sensor architecture with which the system will be observable, which also means that it will be monitorable.

Likewise, it can be verified that the given actuators in the system (pump and the heater) ensure structural controllability of the system and thus they constitute the minimal set of actuators.

For fault isolation and accommodation, more sensors are to be installed in the process to obtain structured residuals. When a fault is too severe, one may go for reconfiguration by using stand-by devices, called hardware redundancy, to accommodate the fault.

Let us now include more sensors and actuators in the process as shown in Figure 8.14, in which a PI controller acts on a pump, two On-Off controllers act on a heater inside Tank_1 and there are some redundant sensors.

On-Off_2 is a stand-by controller meant for FTC, when T_1 becomes faulty. One pump out of the two remains operative and the other is a standby or material redundancy. We assume that valves V_1 and V_2 are never fully closed. It is assumed that controller outputs are known: U_p and U_O, for *PI* and *On-Off* controllers, respectively. Outflows from the pump and heater are also measured (Q_p and Q_T, respectively). The characteristic functions for actuators (pump and heater) and controllers (PI and On-Off) are defined as $\Phi_p, \Phi_h, \Phi_{\text{PI}}, \Phi_{\text{OnOff}}$, respectively.

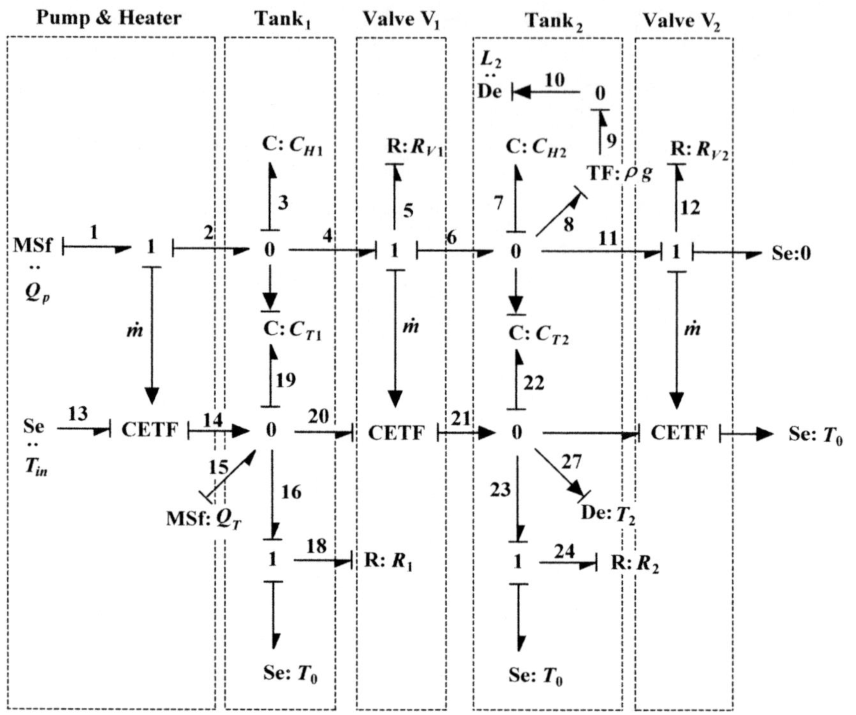

Fig. 8.12 Bond graph model of two-tank system with modified sensor architecture

8.4.2 Determination of Direct and Deduced Redundancies

The BG model of the system in preferred derivative causality (Figure 8.15) is analyzed to determine the redundancies. First of all, essential sensors or base sensors from the view point of FDI are identified (in this case, L_1, L_2, T_1 and T_2). Sensor redundancies are evaluated with respect to the base sensors, *i.e.* causalities of base sensors are inverted. Bond causalities of some sensors must be inverted to assign preferred derivative causality to storage elements (to maintain process observability, *i.e.* minimum sensor set); therefore those sensors are added to the list of base sensors.

There are two occasions when detector causality is not inverted, which lead to two types of redundancies. Where direct causal paths exist from one or more sensors in inverted causality to the redundant sensor, without involving any passive element or two-port element, the resulting redundancy is termed *direct* or hardware redundancy. Sensor L_3 falls in this category. Where causal paths to the sensor in non-inverted causality involve any passive or two-port element, the resulting redundancy is called *deduced* or functional redundancy. Sensor F_1 falls under this category.

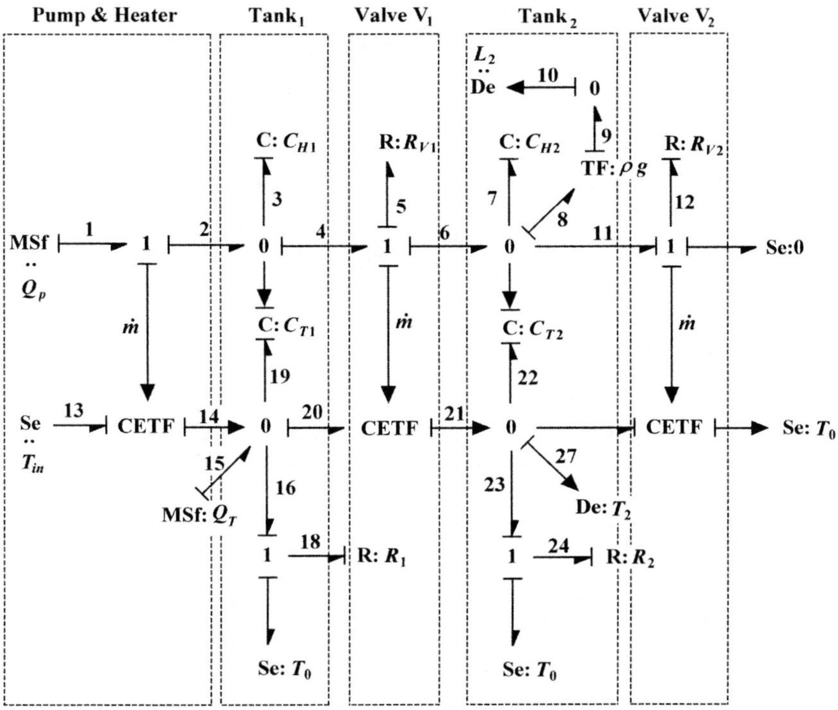

Fig. 8.13 Bond graph model of two-tank system, with modified sensor architecture, in preferred derivative causality

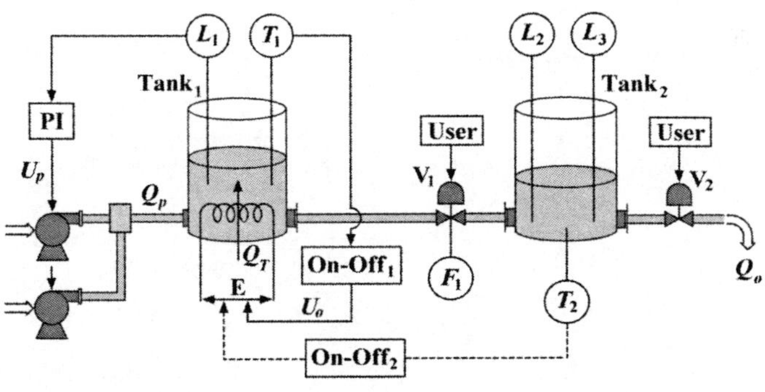

Fig. 8.14 A thermo-fluid system with redundant sensors

8.4.3 Analytical Redundancy Relations and FSM

ARRs for the process are obtained from the BG model in Figure 8.15. Because the process has 10 outputs, 10 ARRs are obtained as

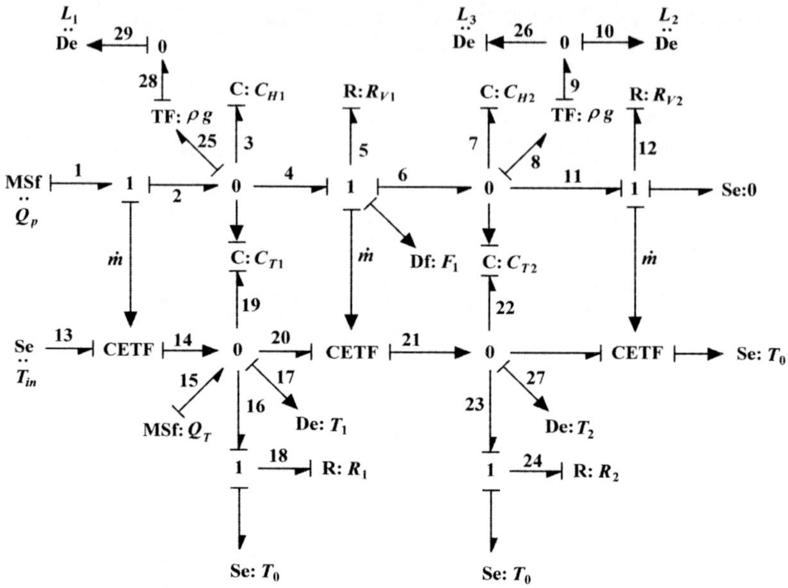

Fig. 8.15 Bond graph model of the thermo-fluid system with redundant sensors

$$
\begin{aligned}
&\mathbf{ARR}_1 : U_p - \Phi_{\mathrm{PI}}(L_1) = 0,\\
&\mathbf{ARR}_2 : Q_p - \Phi_{\mathrm{p}}(U_p) = 0,\\
&\mathbf{ARR}_3 : U_O - \Phi_{\mathrm{OnOff}}(T_1) = 0,\\
&\mathbf{ARR}_4 : Q_T - \Phi_{\mathrm{h}}(U_O) = 0,\\
&\mathbf{ARR}_5 : Q_p - \frac{A_1}{g}.\frac{d}{dt}(\rho.g.L_1) - C_{d1}.\sqrt{\rho.g.(L_1-L_2)} = 0,\\
&\mathbf{ARR}_6 : C_{d1}.\sqrt{\rho.g.(L_1-L_2)} - \frac{A_2}{g}.\frac{d}{dt}(\rho.g.L_2) - C_{d2}.\sqrt{\rho.g.L_2} = 0,\\
&\mathbf{ARR}_7 : L_3 - L_2 = 0,\\
&\mathbf{ARR}_8 : F_1 - C_{d1}.\sqrt{\rho.g.(L_1-L_2)} = 0,\\
&\mathbf{ARR}_9 : Q_T + Q_p.c_p.T_{in} - \rho A_1 c_p.\frac{d}{dt}(L_1 T_1) -\\
&\qquad C_{d1}.\sqrt{\rho.g.(L_1-L_2)}.c_p.T_1 - \frac{T_1 - T_0}{R_1} = 0,\\
&\mathbf{ARR}_{10} : C_{d1}.\sqrt{\rho.g.(L_1-L_2)}.c_p.T_1 - \rho A_2 c_p.\frac{d}{dt}(L_2 T_2) -\\
&\qquad C_{d2}.\sqrt{\rho.g.L_2}.c_p.T_2 - \frac{T_2 - T_0}{R_2} = 0.
\end{aligned}
\tag{8.15}
$$

First of all, the bond graph model is converted into the corresponding DBG form by using the substitutions described in the last chapter. The first four ARRs are obtained for controllers (PI and On-Off) and actuators (pump and heater) by using simple comparison of actual outputs to expected outputs. Then the six remaining ARRs are obtained from the bond graph model given in Figure 8.15. ARRs corresponding to direct and functional redundancies are also included in this set.

The FSM of the process is given in Table 8.4. Note that $Tank_2$ and V_2 are the only components, for which faults cannot be isolated; more sensors are needed to isolate faults in these two components, *e.g.* by placing a flow sensor at the output. The FSM explicitly exposes the direct hardware redundancy because the associated sensor (L_3) appears in only one residual (r_7) which involves some other sensors but no physical process component or parameter. Likewise, functional redundancy in sensor F_1 is also exposed because it appears in only one residual (r_8) which involves at least one physical process component or parameter.

Table 8.4 Fault signatures of the thermo-fluid system

	r_1	r_2	r_3	r_4	r_5	r_6	r_7	r_8	r_9	r_{10}	I_b
PI	1	0	0	0	0	0	0	0	0	0	1
Pump	0	1	0	0	0	0	0	0	0	0	1
On-Off	0	0	1	0	0	0	0	0	0	0	1
Heater	0	0	0	1	0	0	0	0	0	0	1
Tank1	0	0	0	0	1	0	0	0	1	0	1
Tank2	0	0	0	0	0	1	0	0	0	1	0
V_1	0	0	0	0	1	1	0	1	1	1	1
V_2	0	0	0	0	0	1	0	0	0	1	0
L_1	1	0	0	0	1	1	0	1	1	1	1
L_2	0	0	0	0	1	1	1	1	1	1	1
L_3	0	0	0	0	0	0	1	0	0	0	1
F_1	0	0	0	0	0	0	0	1	0	0	1
T_1	0	0	0	0	0	0	0	0	1	1	1
T_2	0	0	0	0	0	0	0	0	0	1	1
Q_p	0	1	0	0	1	0	0	0	1	0	1
U_p	1	1	0	0	0	0	0	0	0	0	1
Q_T	0	0	0	1	0	0	0	0	1	0	1
U_O	0	0	1	1	0	0	0	0	0	0	1

8.4.4 Sensor and Actuator Loss

A process may continue to operate as long as all critical faults can be detected and it remains observable and controllable. However, for continuous monitoring of the process, ARRs and FSM must be modified every time a fault occurs. We consider three categories of sensor failures.

1. Consider failure in a redundant sensor, *e.g.* F_1 or L_3. Redundant sensors appear in only one ARR (a property of redundancy). Therefore, the corresponding ARR is removed from the set of ARRs and the FSM is modified by removing the row and the column corresponding to the sensor and the residual, respectively. There is no need to re-derive ARRs or FSM from the model, but fault isolability is recalculated.
2. Consider fault in a base sensor, say L_2, which has a direct hardware redundancy, with sensor L_3. Component L_3 appears only in one residual, r_7. Then the following procedure applied. (1) ARR_7 is removed from ARRs list and the column corresponding to residual r_7 is removed from FSM, leaving nine ARRs and residuals. (2) In each ARR, variable L_2 is replaced by L_3 and signature (row) of L_2 is copied to L_3. (3) Signature of L_2 is removed from FSM and fault isolability is recalculated.
3. Consider the case, where a failed base sensor is replaced by another sensor, which is a deduced (functional) redundancy. This is not an easy task because it involves symbolic algebra and the structure of the resulting ARRs can change significantly. However, by modifying the BG model and analyzing causal paths, one can easily obtain ARRs and FSM for the process.

When a process component or a base sensor without a redundancy fails, the model must be reconstructed and ARRs and FSM have to be reconstructed. A similar reconfiguration technique is needed for actuator failures.

8.4.5 Automaton Representation of Equipment Availability

The availability of equipment to carry on separate process sub-operations can be represented as separate automatons for process components, controllers, sensors and actuators [249]. Alternatively, it can be represented in a single tree-structure combining all the elements.

Consider I to be the nominal set of actuators of the system and J be one of its subsets of actuators. The set of subsets of I can be represented in an automaton in which the entire set of actuators represents the initial node and other nodes represent a situation of one or more actuator loss. A transition t_{ij} is a Boolean or probabilistic function between a node of level k and the node of level $k+1$, which defines the transition condition from one to another subset of actuators. As an example, let $I = \{a,b,c\}$ be the nominal set of actuators. The set of possible situations is given by: $S = \{\{s(a), s(b), s(c)\}$, such that $s(k) \in \{0 = \text{faulty actuator}, 1 = \text{healthy actuator}\}$ $\forall k \in \{a,b,c\}$. For three initial actuators, the possible situations $\{1,1,1\}$, $\{1,1,0\}$, $\{1,0,1\}$, $\{0,1,1\}$, $\{1,0,0\}$, $\{0,1,0\}$, $\{0,0,1\}$, $\{0,0,0\}$ are respectively represented by the nodes 1, 2, 3, 4, 5, 6, 7 and 8 in the automaton shown in Figure 8.16a.

To distinguish subsets of I with which the system is controllable (*i.e.* the redundant and minimal sets of actuators), they are represented by grey nodes on the

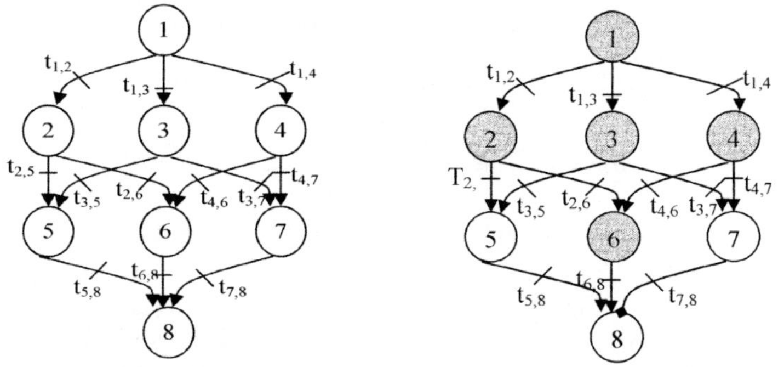

(a) Automaton showing sequences of actuator loss

(b) Automaton showing redundant and minimal actuator sets

Fig. 8.16 Automata of equipment availability

automaton. Assuming that the system is controllable with nodes 1, 2, 3, 4 and 6, the automaton is represented as shown in Figure 8.16b. Nodes 3 and 6 for which the successor nodes are in white, are terminal situations corresponding to minimal sets of actuators.

Similarly, automata are constructed for the sensors and controllers. The system can be operated normally as long as its current state is denoted by grey nodes in all these automata. Upon loss of any hardware, these automata are searched for existence of a solution (transitions to other grey nodes) and if no acceptable solution is found, the process has to be shut down. These automata help in identifying and organizing the situations under which failure of certain kinds of devices is critical.

8.4.5.1 Operating Mode Management

In order to avoid simultaneous attempts at performing incompatible reconfigurations, services offered by the components are organized into coherent subsets, called Operating Modes (*OM*). Each *OM* can be associated with a BG model, as has been shown in Section 8.2.4.

8.4.5.2 Parameter Estimation

Process faults cannot be accommodated as long as the fault is not quantified, *i.e.* fault parameters are not estimated. Real-time parameter estimation is different from partial system identification. However, if the objective is to replace a faulty component by another component available in good condition, *i.e.* perform reconfiguration, then there is no need for parameter estimation. On the other hand, if the faulty component cannot be replaced and one has to perform fault tolerant control then parameter estimation is a necessary precondition. Real-time parameter estimation is discussed in detail in Chapter 10.

8.4.5.3 Actuator Capacity Testing

Once the fault magnitude is estimated, then the next step is to accommodate the fault by suitably changing the control laws. Then process of finding an input sequence to satisfy a constraint given in the form of an output sequence is called system inversion, which is detailed in Chapter 11. The objective is to find an output profile (*e.g.* maximum slew rate), characterized by some parameters and the corresponding input laws such that the operating constraints are satisfied.

8.4.6 Operating Modes of the Thermo-fluid System

For the academic example in Figure 8.14, start, stop and normal operating mode are defined as OM_1, OM_2 and OM_3, respectively. A few other operating modes are discussed in the following.

- Temperature sensor T_1 is faulty and the heater is controlled by another controller, On-Off$_2$, using sensor T_2. We have shown before that the process remains observable after T_1 is lost. This operating mode is called OM_4.

- Level sensor L_1 is faulty in OM_5. However, L_1 can be estimated from ARR$_8$ (Equation 8.15) in terms of L_2 and F_1, as follows:

$$\text{ARR}_8 = F_1 - C_{d1}.\sqrt{\rho.g.(L_1 - L_2)} = 0 \Rightarrow L_1 = L_2 + \frac{F_1^2}{\rho.g.C_{d1}^2}. \tag{8.16}$$

  Alternatively, L_1 can be obtained in terms of L_2 only by solving ARR$_5$. This approach is not preferred because we need a derivative of L_2. Note that if L_1 is faulty and L_2 is unavailable, then system directly enters OM_6.

- Level sensors L_1 and L_2 are both faulty in OM_6. A reconfiguration using sensor L_3 is performed in the similar way as OM_5. Then the following OM sequences result:

$$\begin{aligned} & OM_1 \rightarrow OM_3 \rightarrow OM_5 \rightarrow OM_6 \rightarrow \cdots \rightarrow OM_2, \\ \text{or } & OM_1 \rightarrow OM_3 \rightarrow OM_6 \rightarrow \cdots \rightarrow OM_2. \end{aligned} \tag{8.17}$$

- One pump has failed in OM_8. Since the process consists of a redundant pump (material redundancy), the other pump is used.
- Valve V$_1$ is blocked in OM_{13}. If the objective is to maintain a constant flow rate to the consumer, then the level in Tank$_2$ must be constant, which consequently

requires higher water level in the $Tank_1$. The question now remains, by how much and at what rate we should increase the level setpoint in $Tank_1$? This FTC problem is solved in Chapter 11.

- This case corresponds to multiple non-overlapping faults, *e.g.* failure of T_1 and L_1. Here, both the sensors are independently reconfigured.

Note that in the case of heater failure, leakages from tanks, failure of two pumps, *etc.* for which no reconfiguration or FTC option is available, the process is shut down. Also note that when L_1, L_2 and L_3 are faulty, the observability condition fails and the process is shut down.

The management of these operating modes and transition conditions can be represented in a USOM graph similar to Figure 8.1.

8.5 Application to a Steam Generator Process

The Process and Instrumentation Diagram (P&ID) of a steam generator is given in Chapter 6. The water supply system is made of two pumps (P_1 and P_2) and a pipe. At any given point of time, only one pump can be in operation and the other is redundant.

The BG model of this steam generator process has been developed in Chapter 6. From the BG model in preferred derivative causality, ARR and FSM have been derived and used for FDI in Chapter 6. Our objective is FTC, for which we determine different redundancies, given in Table 8.5, from the BG model. Note that in the saturated regime, steam pressure and temperature are correlated, as given by the steam table or Mollier chart.

Table 8.5 Redundancies in the steam generator process

Device	Redundancy with?	Type	Used in FTC of?
L9	L8	Direct/hardware	Water level in boiler
L19	L18	Direct/hardware	Condensate level
P14	P12	Direct/hardware	Steam expansion system
P13	P16	Direct/hardware	Steam expansion system
P15	P16	Direct/hardware	Steam expansion system
T6	P7	Deduced/functional	Boiler pressure
T5	P7	Deduced/functional	Boiler pressure
T17	P16	Deduced/functional	Steam expansion system
P27	(P16, L18)	Deduced/functional	Condensate discharge
P2	P1	Material	Feed water supply
V1	V2	Direct/hardware	Steam expansion system
V4	V3	Direct/hardware	Condensate discharge
V5	V3	Direct/hardware	Condensate discharge

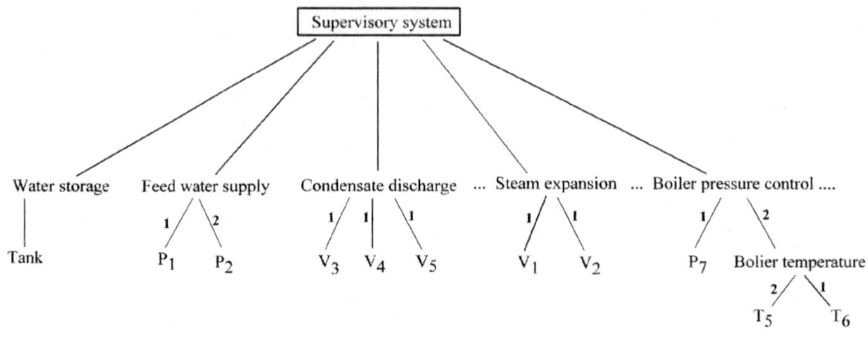

Fig. 8.17 Tree-structured/pyramidal decomposition of basic functionalities of the steam generator with devices needed to achieve them

Reconfiguration is possible only when, for failure of one or more base device(s), there are corresponding redundant devices available in good health. When two or more devices are redundant with a common base device, they are mutually redundant. The coupling between the supervisory layer and the base devices of the steam generator installation is determined from Table 8.5 and it is represented by a tree-structure [225] as given in Figure 8.17.

In Figure 8.17, basic functionalities required for process operation are listed and the associated devices needed to perform those functions are linked in a tree-structure. Where more than one device is available to perform the same task, branches are sequentially numbered indicating a hierarchical preference. Sometimes, equal preference (simultaneous use of devices) is given to all branches. Theoretically, a process can operate normally, as long as at least one device is available for each basic function. When a device fails, the branch associated with it is removed and the system is reconfigured using the next device, according to the defined hierarchy, *e.g.* when P_7 fails, it is reconfigured using T_6 (if available). Note that boiler temperature is not directly linked to the supervisory layer.

8.5.1 Operating Modes of the Steam Generator Process

- Preparation and Start mode (OM_1): close V_0, allow control of boiler level and heater such that steam become saturated. Then open V_0.
- Shut down mode (OM_2): when a fault detected, but cannot be isolated or when an isolated fault cannot be accommodated. Stop heater and pump, allow natural control of steam expansion system and the condenser until boiler pressure falls below a threshold and then shut down the supervision and control systems.
- Normal operation mode (OM_3): after start mode, when all controls are active and there are no faults in any device.
- Degraded operation modes

- Condensate discharge valve failure (OM_4): as long as one of the condensate discharge valves is non-faulty, continue operating the plant.
- Steam expansion system fault (OM_5): alter controller set point. If fault cannot be accommodated within given tolerance limits, enter manual operation mode.
- Partial fault in pump (OM_6): one pump is unavailable and the reconfigured control acts on the second pump, which is not operating with full efficiency.
- Partial heater fault (OM_7): heater is not delivering full power, but is enough to sustain the boiler pressure at given load.

• Reconfigured modes

- Boiler level control (OM_8): Upon failure of L_8, use L_9, if available.
- Boiler pressure control (OM_9)
  · OM_{9a}: upon P_7 failure, use T_6 or T_5.
  · OM_{9b}: upon P_7 and the sensor used in OM_{7a} failure, use the other temperature sensor.
- Condenser level control (OM_{10}): upon failure of L_{18}, use L_{19}, if available.
- Condenser pressure control (OM_{11})
  · OM_{11a}: if P_{15} has failed, use P_{16} for controlling downstream pressure.
  · OM_{11b}: if P_{15} and P_{16} have both failed, use P_{13}.
  · OM_{11c}: if P_{15}, P_{16} and P_{13} have failed, use P_{27} and L_{18} or L_{19}.
  · OM_{11d}: if all pressure sensors in condenser side have failed, use T_{17}.
- Pump failure (OM_{12}): upon failure of P_1, use P_2, if available.

• Fault Accommodation mode (OM_{13}): during steam expansion system fault, estimate the fault, *i.e.* whether a blockage or leakage. If it is leakage, then close V_0 and transit to OM_2 (shut down mode). Otherwise, increase the boiler pressure setpoint and/or decrease the condensate level setpoint (more exposed tubes increase the rate of condensation, which result in pressure drop). The effective coefficient of discharge through V_1 and V_2 is estimated and the desired pressure drop is calculated, which is then used to specify the boiler pressure and condenser level setpoints.
• Critical operation mode (OM_{14}): with the available sensors, faults in all critical components can be detected, but faults in one or more components of the process cannot be isolated.
• Mutually exclusive operating modes (OM_{15}): when FTC or reconfiguration of two or more mutually exclusive control loops is performed simultaneously, *e.g.* $OM_{9a} + OM_{11a}$ (simultaneously P_7 and P_{15} are faulty and they are reconfigured by T_6 and P_{16}, respectively).
• Manual operation mode (OM_{16}): for blockage in steam expansion system, operate V_1 to achieve desired steam flow and pressure drop. If fault cannot be accommodated within given tolerance limits, transit to OM_2.
• Maintenance Mode (OM_{17}): operate full or part of the system to locate faults for repair action. FTC option is usually turned off, but continuous monitoring is desired for safety of maintenance workers.

8.5.2 Experimental Results

We consider a fault in the most critical sensor in the installation, *i.e.* the boiler pressure sensor. Note that in nominal operation, boiler pressure is controlled between 7.8 and 8.2 bars through an On-Off controller. We consider a system without FTC implementation, for which the steam pressure (P_7) and temperature sensor (T_6) readings are plotted in Figures 8.18 and 8.19, respectively. It is seen from these results that the boiler temperature drops during P_7 failure (from 300–450 s), because the heater is automatically switched off as a safety measure.

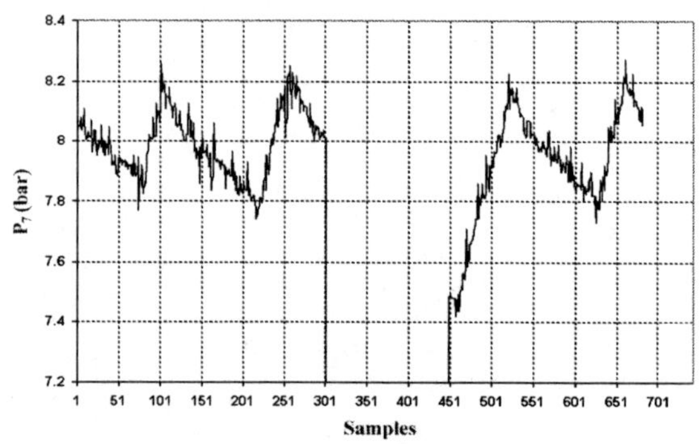

Fig. 8.18 Boiler pressure sensor reading

Fig. 8.19 Boiler temperature sensor reading

Fault accommodation is implemented by transiting to OM_{9a}, *i.e.* controlling the heater by using temperature sensor (T_6) and setting a new setpoint for the On-Off controller to maintain the steam temperature between 169°C and 172°C, which corresponds to a steam pressure range between 7.7 and 8.4 bars. Measured steam pressure and temperature, when the fault tolerant control was activated in the supervision platform, are given in Figures 8.20 and 8.21, respectively.

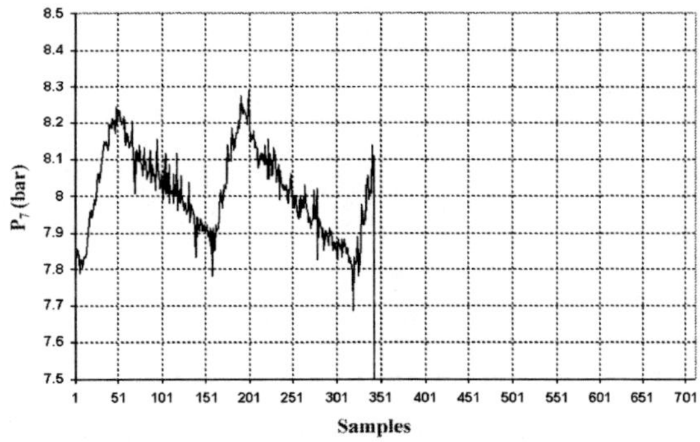

Fig. 8.20 Boiler pressure sensor readings showing permanent fault

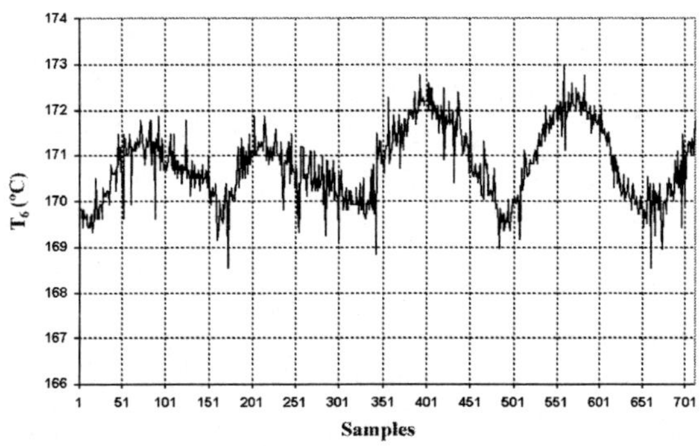

Fig. 8.21 Boiler temperature sensor readings with reconfiguration

Problems

8.1. A four tank system is shown in Figure 8.22. Devise the minimum sensor architecture (key sensors) needed to monitor this plant. If faults in Tank1, Tank2, the two pumps and the two valves are to be monitored, design a set of redundant sensors so that all considered faults would be monitorable even after loss of some of the key sensors. Express the hardware availability architecture in the form of an automaton. Gravity head must be included in the model. Use activated bonds if necessary.

If all the pipes are connected to the bottom of the tanks, how would the key sensor architecture change?

8.2. The schematic diagram of a penicillin fermentation process is shown in Figure 8.23. Acid or base is added to the fermenter to maintain a constant pH within the fermenter. The substrate is periodically removed and replenished. Moreover, the temperature is controlled through a heat exchanger. Air is supplied from the bottom of the tank; however it may be neglected in the mass and energy balance equations.

Create an external model of the system by associating various services (and their versions) with devices in different operating modes. Relate the services to various elements in the bond graph model of the process. By assuming that the pressures and

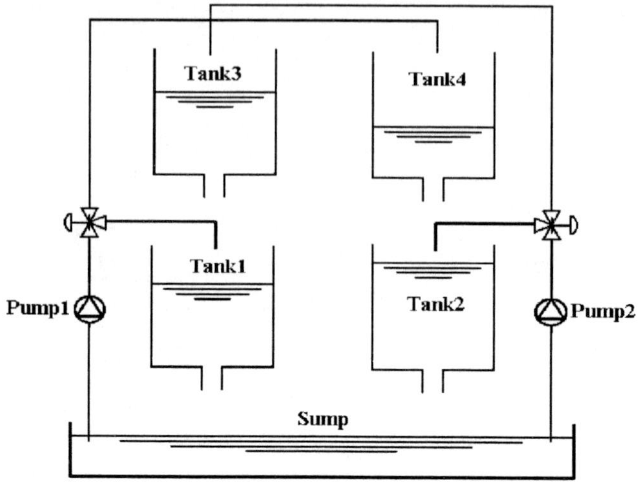

Fig. 8.22 A four-tank system

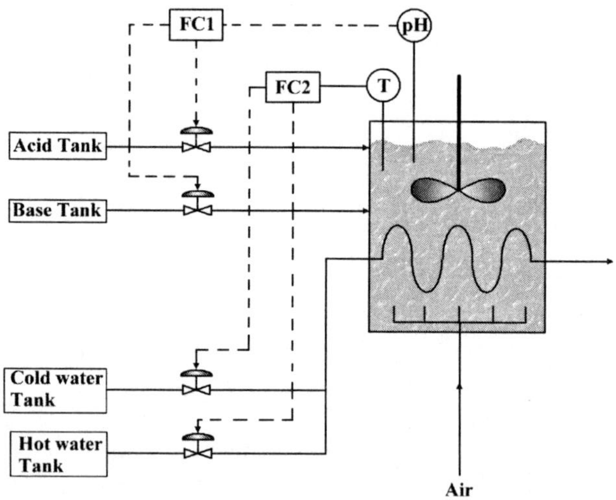

Fig. 8.23 A penicillin fermentation process

the temperatures in all external supply units (acid, base, cold and hot water tanks) are known, design sufficient hardware redundancies such that the process can be operated under loss of key hardware (those needed to maintain pH and temperature). List different possible operating modes of the process.

8.3. A continuous stirred tank reactor (CSTR) is shown in Figure 8.24. This process is similar to the one in the previous problem. An exothermic reaction takes place inside the CSTR in which the rate of heat liberation (or the rate of reaction) is assumed to be a function of the temperature of the water and the substrate mixture inside the CSTR and the time elapsed since the last substrate replenishment. It is assumed that the outflow from the CSTR is filtered so that neither consumed nor pure substrate leaves the container. The temperatures of water supply units are known. The inlet steam temperature and the steam flow rate are also measured. These measurements along with those shown in Figure 8.24 are sufficient to control the process. By assuming these to be the key sensors, what additional sensors are required to isolate the faults in the process?

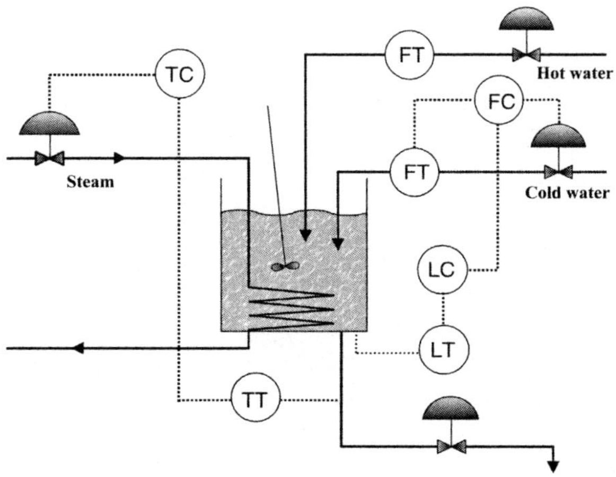

Fig. 8.24 A continuous stirred tank reactor

Discuss the reconfiguration scheme required in the process for a blockage in the hot water supply line. What modifications are required in the process such that it can be reconfigured to accommodate blockage faults in the steam supply and cold water supply lines? How would the modified architecture relate to the one in the previous problem?

8.4. A computer controlled hydraulic actuator is shown in Figure 8.25. Create a bond graph model of the actuator by considering compressible fluid flow and port leakages (cross-port and direct). Use bicausality to determine the set of sensors needed to isolate the faults in the actuator, motor-pump (pressurizer) supply unit and the proportional valve.

List the set of elementary services (missions) provided by various components and relate them to elementary bond graph elements in a table. Determine the various operating modes of the system.

Fig. 8.25 An active hydraulic actuator

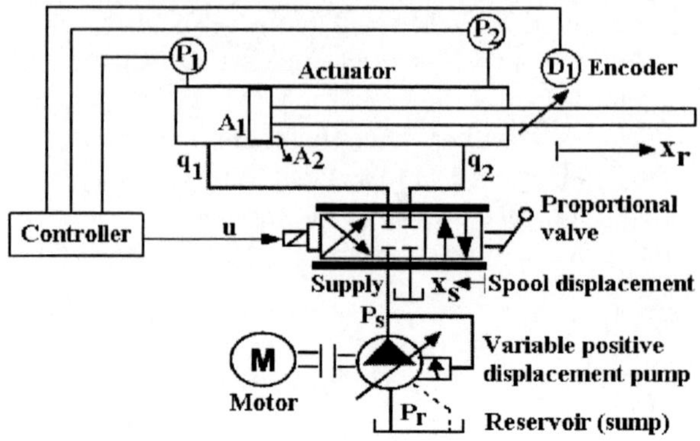

Chapter 9
Isolation of Structurally Non-isolatable Faults

9.1 Introduction

If the number of fault parameters is more than the number of residuals or residuals are unstructured then some faults cannot be isolated even under single-fault assumption. In this chapter, a second level decision procedure or fault hypothesis testing for fault isolation is developed. Both qualitative and quantitative trend analysis techniques are applied for fault disambiguation.

The model based fault detection using analytical redundancy is extended in this chapter with a multi-tier fault isolation/localization technique, which uses single hypothesized fault parameter estimation from residuals. Unlike observer based and parity relation based robust residual generation schemes for complete decoupling of the effects of uncertainties in the model, which are robust but applicable to a select class of non-linear models, the model uncertainties are propagated to the residual and an adaptive threshold is used with residuals in intervals for alarm filtering. Following this so-called *passive approach*, the moving mean value of the threshold is used to represent the actual residual trajectory and this value is then used with the ARR expression to calculate values for each hypothesized fault parameter, while assuming the rest of the fault candidates are non-faulty.

In this chapter, we assume that:

1. The system structure and nominal parameter values are known, the system is state-observable; *i.e.* state variables can be estimated for the non-faulty system.
2. At a time, only single independent parameter of the system may be faulty, *i.e.* single-fault-hypothesis.
3. Sensors and actuators are considered non-faulty.
4. For a given process input and output vector, a unique set of parameter values exists.
5. The process has a monostable steady-state [4].

The monostable characteristics is defined as follows.

Consider a controlled finite dimensional system under suitable regularity assumptions:

$$\dot{x}(t) = f(x(t), u(t)), \qquad (9.1)$$
$$y(t) = g(x(t), u(t)).$$

Now assume that the following property holds: for each constant input $u = a$, there is a unique steady state x_a (*i.e.* $\dot{x}(t)|_{u=a} = 0$). The output $y(t)$ will also have a unique solution at $u = a$. That means, $f(x,a) = 0$ has the unique solution at $x = x_a$ and every solution of the system $\dot{x} = f(x,a)$ converges to this state x_a. In the terminology of [4], we say that such a system has a monostable steady-state step response characteristics.

9.2 Residuals and Robustness

Robust decision procedures minimize misdetection and false alarms by treating the residual noises. There are various methods to generate robust residuals, which are insensitive to modeling uncertainties and sensor noises. One of the approaches to robustness, known as *active approach*, is based on generating residuals that are insensitive to uncertainty, but sensitive to faults. Different methods to achieve active robustness have been developed: unknown input observers, robust parity equations, *etc.* which were discussed in Chapter 5. Perfect decoupling of residuals from uncertainties arising out of parameters is limited by the number of measurement signals and their locations [49, 93]. It is possible to perform a perfect decoupling if there are few disturbances or if the modeling errors concern a restricted number of parameters [93] which may be treated as structured uncertainties [234]. In contrast, optimization techniques are used to find a partial decoupling. An alternative approach to achieve robustness, called as passive, tries to accomplish robustness in the decision-making stage [207]. In a *passive approach*, the effect of the parameter uncertainty is propagated to the residuals and then an adaptive threshold is used to envelop these residuals to achieve robustness [109].

Following an active approach in a linearized model, modeling uncertainties due to the linear approximation, the error in the estimate of the model parameters and sensor noises, around a given setpoint, can be accounted for in different ways: using unknown inputs to represent model uncertainties [50], representing higher order terms as unstructured uncertainties, using a set of models corresponding to different parameter settings [131], and using frequency domain models (transfer functions) with parameters in intervals [103] *etc.* In a passive approach, the sensor noise effect is amplified further because derivatives of measurements influence the residuals. Thus, simple comparison of the residual to a fixed-threshold is impractical in noisy environment because the effect of a fault may be masked by the noise. This calls for the use of statistical change detection algorithms [13, 16].

Therefore, the problem of robustness in FDI with respect to modeling uncertainties and noise effects encompasses a very broad area: see applications to polynomial non-linear systems (in the parity space framework) [93, 237], bilinear systems [131] and control affine non-linear systems (using an observer based approach) [74, 78, 104]. When the system under consideration is subject to unknown disturbances or unknown inputs, their effect has to be decoupled from the residuals to avoid false alarms. When uncertainties associated with parameters and noises in sensors and actuators can be statistically quantified, then under certain conditions, their effects can be analytically decoupled from the residuals using robust model-based approaches [50, 198]. Such analytical techniques or active approaches can be considered as a preprocessing stage, which lead to robust residuals. The existing methods for decoupling unknown inputs and disturbances can handle systems with polynomial non-linearity. For general class of non-linearity and unclassified disturbances, raw residuals are not robust (*e.g.* non-zero mean). Thus, statistical measures of residual are obtained from experimental or simulation data and then those measures are used to adjust the residual signal. This phase, followed in passive approaches, may be considered as a *residual post-processing*.

By assuming that the uncertainty caused by modeling error depends on the operating conditions, this uncertainty can be modeled either theoretically or empirically, and accordingly the residual thresholds can be varied. These adaptive threshold techniques or passive approaches were first proposed in [197]. The passive approach does not attempt to decouple perfectly the effect of uncertainty in the residual. On the contrary, it is based on propagating the parameter uncertainty to the residual, and then bounding the uncertainty in the residual by using an interval. As long as the residual value remains within this interval, it is assumed that there is no fault in the process, because the change in residual value can be either due to the parameter uncertainty or due to a fault [64, 65].

The passive approach has some advantages over the active approach: robustness in the fault detection can be achieved without using any approximation to simplify the parameter representation, even when there are many uncertain parameters in the model. One drawback of this approach is that weak faults, which produce a small residual deviation remaining within the interval bounds set corresponding to parameter uncertainty, may not be detected. An alternative to this passive approach is the possibilistic approach which is based on fuzzy theory [70, 206, 237]. This approach uses a set of interval models, which are parameterized by a probability distribution, and it is essentially a compromise between the interval and the probabilistic approaches [28].

From the foregoing discussions, we conclude that robustness in fault detection can be achieved in the residual generation or in the decision-making stage. While residual preprocessing in *active approach* is analytical, residual post processing in *passive approach* is generally numerical. The former is easy to implement and robust, but handles limited types of system non-linearity. The latter is not constrained by the type of non-linearity, but is less robust in the sense that there could be some misdetection of small magnitude faults. Further post processing steps such as cumulative sums may be used to detect small magnitude faults, but then likelihood of false alarms increases. Therefore some kind of trade-off is made depending on the specific system and other available complimentary diagnosis tools.

The threshold may be adaptive to changes in time, t (as parameters may vary with time, *e.g.* due to wear and tear), input function (u) and one or more factors, which are sensitive to changes in external environment condition, *e.g.* temperature, pressure *etc.* [103, 208]. Let us assume that the process is non-stationary and the residual cannot be approximated by a Gaussian distribution. Then a general form for adaptive threshold [230] is expressed with separation of input and temporal dependency of variables and it is given as

$$\delta_{th}(u_k, t_k) = \psi(t_k).\eta(u_k) \pm z.\phi(t_k).\sigma(u_k), \tag{9.2}$$

where subscript k stands for k-th sample, $\psi(t_k)$ and $\phi(t_k)$ are time varying coefficients modifying the mean and the standard deviation, respectively, u_k is the input, t_k is the time, and z is a coefficient related to the confidence level [273]. The mean and the standard derivations of the residual, $\eta(u_k)$ and $\sigma(u_k)$, respectively, are expressed as

$$\begin{aligned} \eta(u_k) &= (\delta_{\eta i} + u_k.\lambda_{\eta i}), \\ \sigma(u_k) &= (\delta_{\sigma i} + u_k.\lambda_{\sigma i}). \end{aligned} \tag{9.3}$$

Here, $\lambda_{\eta i}$ and $\lambda_{\sigma i}$ are assigned small values to make residuals oversensitive to faults and unknown inputs.

When the effects of time, environmental pressure and temperature *etc.* on the residuals are negligible, $\psi(t_k) = 1$ and $\phi(t_k) = 1$. Then the simplified form of the time-invariant-threshold for i-th residual is given by

$$\delta_i(u) = \mu_i(u) \pm \varepsilon_i(u), \tag{9.4}$$

where

$$\begin{aligned} \mu_i(u_k) &= \eta(u_k) = (\delta_{\eta i} + u_k.\lambda_{\eta i}), \\ \varepsilon_i(u_k) &= z.\sigma(u_k) = z.(\delta_{\sigma i} + u_k.\lambda_{\sigma i}), \end{aligned}$$

i.e. the mean and standard deviation are taken as a first order approximation in terms of the input. The actual order of approximation needed is usually determined from the test data.

9.3 Localization of Fault Subspace

Let n be the number of sensors, *i.e.* there are n ARRs, and let r_n be the set of residuals. Let $S = (\theta_1\ \theta_2\ \ldots\ \theta_m)$ be the parameter space and there be Z dependent parameters represented by a set S_d. Then the independent parameter space will be

$S_i = S - S_d$. An example may be an electrical coil, in which two parameters for inductance and resistance are correlated and only one is chosen as the independent parameter. Then, in the fault signature matrix, let the fault in $l\,(l \leq m - Z)$ parameters be isolatable using the available instrumentation. Then let S_P (dimension $m - Z - l$) be a non-empty parameter sub-space containing the non-isolatable parameters. Furthermore, let n' $(n' \leq n)$ residuals, $r'_n\,(r'_n \subset r_n)$, which are diagnosed abnormal by the decision procedure are such that they lead to multiple matches in the fault signatures matrix. Then, $\bar{S}_P \subset S_P$ is the susceptible parameter sub-space or the set of fault candidates (let m' be their number). We estimate each parameter $\theta_i \in \bar{S}_P (i = 1, ..., m')$ assuming the rest of the parameters $(j = 1, \cdots, m'$and $i \neq j)$ are nominal. This leads to m' number of models, which must be simultaneously solved. There is no change to the topology of the bond graph model. Essentially each fault model is the same bond graph with different parameter values.

The above definitions are further explained by using an example fault signature matrix given in Table 9.1, where θ_i, $i = 1$ to 6, are independent parameters and r_j, $j = 1$ to 3, are the residuals. The fault signature matrix (Table 9.1) shows two groups of faults, S_{P1} and S_{P2}, respectively, which are isolatable from each other, but individual faults within each group are not isolatable from the other faults in the group.

Table 9.1 Example fault signature matrix

				r_1	r_2	r_3	M_b	I_b
θ_1	S			1	0	0	1	1
θ_2	S			0	1	0	1	1
θ_3	S	S_P	S_{P1}	0	0	1	1	0
θ_4	S	S_P	S_{P1}	0	0	1	1	0
θ_5	S	S_P	S_{P2}	0	1	1	1	0
θ_6	S	S_P	S_{P2}	0	1	1	1	0

When residual r_1 is abnormal, it leads to a coherence vector $C = [1 \;\; 0 \;\; 0]$, which has a unique match in the fault signatures matrix and the component corresponding to parameter θ_1 is isolated as the fault candidate. Similar localization of θ_2 is possible, when only residual r_2 is abnormal. All other component faults are not isolatable and they constitute the set $S_P = S_{P1} \cup S_{P2}$.When only residual r_3 is abnormal, $\bar{S}_P = S_{P1}$, and when both residuals r_2 and r_3 are abnormal, $\bar{S}_P = S_{P2}$. Under single fault hypothesis applied to the given fault signature matrix, the initial fault isolation step gives a fault candidate list

$$\Im_F \in [\theta_1 \quad \theta_2 \quad (\theta_3\,\theta_4) \quad (\theta_5\,\theta_6)],$$

where only in the case of the last two elements, with more than one dimension, do we need further refinement through parameter estimation. This refinement step is termed the second level isolation procedure. Second and/or further higher level isolation procedures are needed in many practical situations where the system is not sufficiently instrumented.

9.4 Methodology for Single Fault Isolation

The methodology for single fault isolation is divided into two parts: (1) parameter estimation and (2) hypothesis testing by using a multi block model.

9.4.1 Parameter Estimation

For any fault in a component, the corresponding robust residuals in which the faulty parameter is involved are theoretically non-zero. But if those robust residuals are equated to zero, the faulty value of the parameter can be estimated with sufficient accuracy, assuming that values of the rest of the parameters are nominal and the faulty parameter does not change during estimation.

Let there be a fault sub-space $\bar{S}_P$ containing m' number of independent parameters $\theta_1 \cdots \theta_{m'}$. Assume that only one of them is faulty, say θ_i (the nominal parameter value of that faulty component), which is involved in

$$\text{ARR} : f(U,\theta,Y) = f\left(U,\dot{U},\cdots,(\theta_1\ \theta_2\cdots\ \theta_i\cdots\theta_{m'})^T,Y,\dot{Y},\cdots\right) = 0, \tag{9.5}$$

where $U = [u_1\ u_2\ u_3\cdots]^T$ is the input vector, $\theta = [\theta_1\ \theta_2\cdots\ \theta_{m'}]^T$ is the nominal parameter vector and $Y = [y_1\ y_2\ y_3\ \cdots]^T$ is the output vector. During a fault, if the residual corresponding to the ARR given in Equation 9.5 is evaluated with nominal parameter values, then it deviates from its corresponding threshold. Then the value of the parameter associated with fault hypothesis $\bar{\theta}_i$ can be estimated, either algebraically or numerically, from the relation

$$f\left(U,\ \dot{U},\ \cdots,\ (\theta_1\ \theta_2\ \cdots\ \bar{\theta}_i\ \cdots\ \theta_{m'})^T,\ Y,\ \dot{Y},\ \cdots\right) = 0, \tag{9.6}$$

assuming that rest of the parameters are normal.

However, due to uncertainties in the model, we have considered that $f(U,\theta,Y)$ is bounded by an interval $\delta(U) = \mu(U) \pm \varepsilon(U)$, where a static shift $\mu(U)$ describes the mean path of adaptive threshold for each residual. Therefore, the estimation of parameter is obtained by equating the non-robust residual to the contemporary mean of the threshold $\mu(U)$, *i.e.* by solving θ_i from

$$f(U,\theta,Y) = \mu(U). \tag{9.7}$$

The threshold standard deviation $\varepsilon(U)$ is not used in the estimation, because we assume that the uncertainties are unstructured and uncorrelated, which means that they cannot be independently associated with any particular parameter. Therefore, uncertainty in estimated parameter is bounded and is directly related to the uncertainty associated with the residual thresholds. One limitation of this estimation methodology is that the estimation is possible only when residuals are persistently abnormal.

This method of limited parameter estimation is simpler as compared to other estimation techniques (*e.g.* recursive least squares, Frisch Scheme, multiple inference, Takagi-Sugeno multi-models, Fuzzy Cluster models and non-linear regression analysis) presented in [237], parity-space based fault estimation techniques for deterministic systems and Kalman filter based fault estimation techniques for stochastic systems presented in [16, 237]. In contrast, the simpler fault estimation method presented here follows an algebraic approach and the actual fault is not estimated in a single pass. The consequent trend obtained from process dynamics and those predicted from hypothesized faults are used in a second-level decision procedure.

9.4.2 Parallel Simulation of Bank of Fault Models

Once the set of estimated parameter values $\theta_1 \cdots \theta_{m'}$ are obtained corresponding to m' possible single-faults, the list of fault candidates may be reduced to a smaller set by using some predefined rules, which define plausible bounds for each parameter. For example, the discharge coefficient of a valve cannot be negative; a pump can work in a fixed direction, *etc.* Only those estimated parameter values, which satisfy their feasibility domain, say a number $k \leq m'$, are considered as relevant fault candidates $\theta_1 \cdots \theta_k$. For isolating the fault, k number of fault models $\mathrm{FM}_i, i = 1 \cdots k$ (bank of fault models) are simulated, where each fault model $\mathrm{FM}_i \, (i \in [1 \cdots k])$ uses the parameter set $(\theta_1, \theta_2, \cdots \bar{\theta}_i, \cdots, \theta_k)$. If a fault has actually occurred in the i-th parameter, then FM_i should replicate the faulty plant behavior and the output response (Y^*) from it should match best with that (Y) from the plant. This response matching norm can be used to isolate the component corresponding to the i-th parameter as the actual fault candidate. The parallel models are activated by a special switching mechanism when abnormality in residual(s) is detected by the first phase decision procedure. The identification procedure thus described is treated as a second phase refining decision procedure and is shown schematically in Figure 9.1.

The actual fault candidate can be localized, if (1) the transient response corresponding to a fault is unique from the response corresponding to all other $k-1$ faults, both qualitatively and quantitatively, and (2) the transient response is closest to the actual response of the plant. If more than one transient response from fault models are equally close to the plant response, the fault cannot be isolated, but the list of fault candidates is reduced. Fewer fault candidates help in quickly carrying out maintenance work.

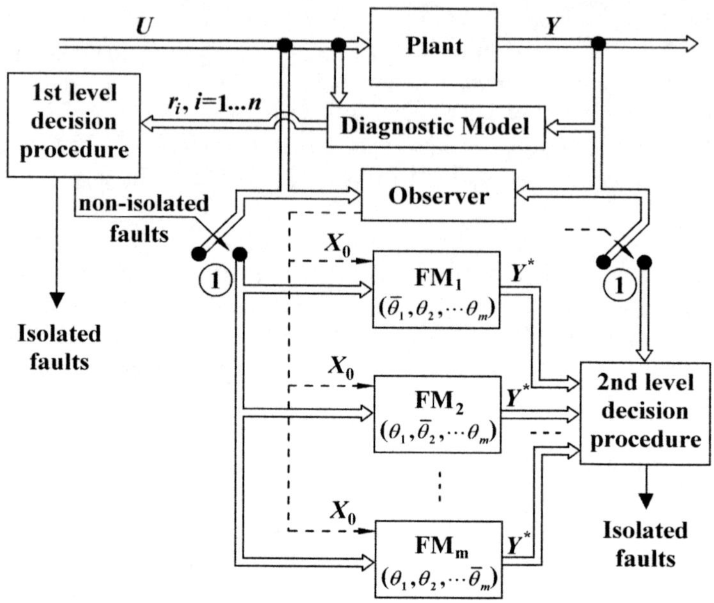

Fig. 9.1 Schematic diagram for isolation of structurally non-isolatable fault through bank of models approach

Before solution of fault models corresponding to different fault hypotheses, values of state variables in non-faulty state are estimated from an observer and copied as initial values to each of the m number of fault models. The observer may be based on an open-loop observer, a Kalman filter or an extended Kalman filter [237], depending on the application. In case of a closed-loop observer, faults in the system influence the observer, which has been used extensively to develop observer based FDI schemes in Chapter 5. However, in our case, the initial values chosen to run fault models are taken corresponding to a time at which the process was working without any fault. In other words, we are using estimates of state variables corresponding to the last known non-faulty state, *i.e.* when the residual was within the described threshold. However, note that a database of process and observer history is required for this purpose. Therefore the kind of observer used is unimportant as long as the response time of the observer is sufficiently fast. However, there will be a time-shift between the response of the fault models and the actual process response due to the time delay suffered in detection of the fault, copying of the initial states from the observer *etc.* This aspect is discussed in the example application.

9.5 Application to a Controlled Two-tank System

A controlled two-tank system is shown in Figure 9.2. The process consists of two tanks T_1 and T_2 connected by a pipe with a valve. A sensor (P_2) measures the pressure in tank T_2. Tank T_1 is fed by a pump and the water level in the tank T_2 is controlled by a PI level controller acting on the pump, which provides inlet flow Q_P. The water flows from the tank T_1 to the tank T_2 through a manually operated valve V_1.

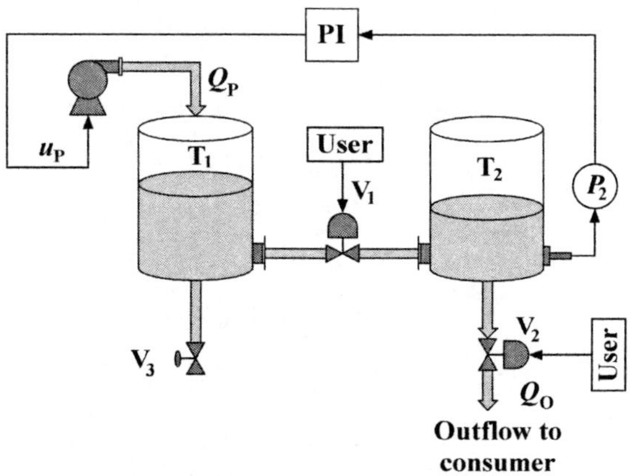

Fig. 9.2 Schematic diagram of a two-tank system

Table 9.2 Model parameters of two-tank system

Symbol	Description	Value
K_p	Proportional gain of controller	10^{-4} m.s
K_I	Integral gain of controller	5×10^{-6} m
S_{Pt}	Setpoint of the controller	5000 $N.m^{-2}$
f_{max}	Maximum outflow from pump	10 $kg.s^{-1}$
A_i	Cross-section area of tanks T_i (i = 1, 2)	1.54×10^{-2} m^2
C_{d1}	Discharge co-efficient of valve V_1	1.593×10^{-2} $kg^{\frac{1}{2}}m^{\frac{1}{2}}$
C_{d2}	Discharge co-efficient of valve V_2	1.596×10^{-2} $kg^{\frac{1}{2}}m^{\frac{1}{2}}$
P_{atm}	Atmospheric pressure (gage)	0 $N.m^{-2}$

To satisfy observability conditions, we assume that the valves V_1 and V_2 are never fully closed. The quantity of water outflow Q_O to the consumer is manually controlled by the valve V_2. The valve V_3 can be used to simulate leakage from tank T_1. Leakage from tank T_2 cannot be isolated as there is no difference between leakage from valve V_2 and that from tank T_2.

We assume that the output of all controllers and sources are measured. The model variables and process parameters are given in Table 9.2. The bond graph models of the system with integral causality, and using bicausality in measurement ports and preferred differential causality for derivation of ARRs, are shown in Figure 9.3a,b, respectively. Instead of using AE and AF elements, the activations are directly shown near the bonds ($f = 0$) in Figure 9.3b.

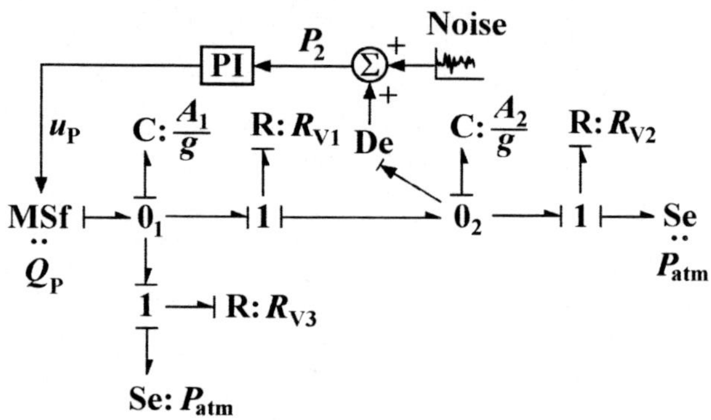

(a)

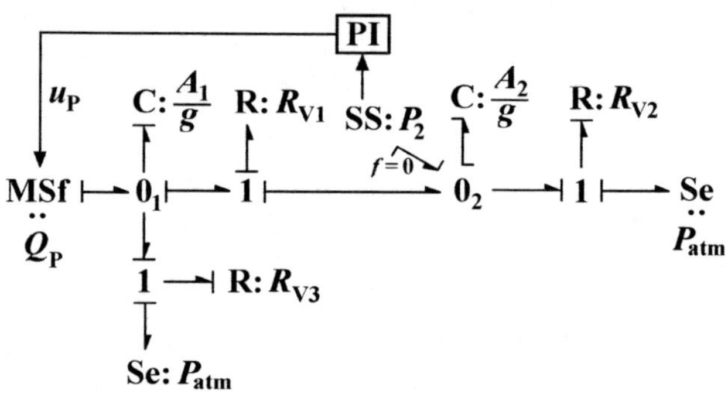

(b)

Fig. 9.3 **a** Behavioral bond graph model of two-tank system. **b** Bicausalled bond graph model of two-tank system in preferred differential causality

9.5.1 ARRs and FSM

There are three sensors, *i.e.* u_P, Q_P and P_2; so three structurally independent ARRs can be derived as follows:

$$\text{ARR}_1 : u_P - \Phi_{\text{PI}}(P_2) = 0,$$
$$\text{ARR}_2 : Q_P - \Phi_{\text{P}}(u_P) = 0,$$
$$\text{ARR}_3 : Q_P - \frac{A_2}{g}.\frac{d}{dt}(P_2) - C_{d2}\sqrt{P_2 - P_{atm}}$$
$$-C_{d3}\sqrt{P_2 + \left(\frac{\frac{A_2}{g}.\frac{d}{dt}(P_2) + C_{d2}\sqrt{P_2 - P_{atm}}}{C_{d1}}\right)^2 - P_{atm}}$$
$$-\frac{A_1}{g}.\frac{d}{dt}\left[P_2 + \left(\frac{\frac{A_2}{g}.\frac{d}{dt}(P_2) + C_{d2}\sqrt{P_2 - P_{atm}}}{C_{d1}}\right)^2\right] = 0. \quad (9.8)$$

The first two ARRs (ARR_1 and ARR_2) are obtained directly from control laws of the PI controller and the pump, whereas ARR_3 is derived from the bicausalled bond graph model given in Figure 9.3b. In ARR_3, variable C_{d3}, which is discharge coefficient of valve V_3, appears in the third term. Since nominal value of C_{d3} is zero, it does not affect the residual during normal operation of the process. Inclusion of this term in ARR_3 allows estimation of the degree of leakage from tank T_1 during faulty operation.

Note that the behavior of the regulator (PI controller) in the control-loop tends to hide the fault effects. For example, consider that a leakage through valve V_2 produces higher output flow. Therefore, the control loop tries to accommodate it by forcing more flow through the pump to tank T_1, which consequently brings the level in tank T_2 to normal operating point, as defined by the associated set-point. From observation of the measurement signal only (level in tank T_2), the fault will be weakly detectable in the transient regime. Note that level in tank T_1 has increased, but it is not observed. On the contrary, the residual is persistently sensitive to the fault, because the increased input pump flow rate is involved in the ARR_3. Therefore, diagnosis based on residuals evaluated from ARRs remains sensitive to faults hidden by the action of the control-loop components.

The fault signature matrix is derived on the basis of the three residuals and is given in Table 9.3. Model Builder software [191] can also be used for this purpose. Note that leakage from tank T_2 is not separately considered in Table 9.3. From a physical viewpoint, it is not possible to distinguish between flow through valve V_2 and leakage from the tank T_2. Flow through valve V_2 and leakage from tank T_2, both discharge water to the environment against same pressure difference. Therefore, a leakage through valve V_2 is equivalent to leakage from tank T_2. Considering all these, C_{d2} represents total discharge coefficient for the out-flow from the tank T_2, whether through valve V_2 or through leakage. When the estimated value of C_{d2} is less than nominal, there is a blockage in the valve V_2, whereas when the estimated value of C_{d2} is above nominal, there is a leakage from tank T_2, either through valve V_2 or at some other location.

Table 9.3 Fault signature matrix of two-tank system

		r_1	r_2	r_3	$\mathrm{M_b}$	$\mathrm{I_b}$
S	Pump	0	1	0	1	1
S	PI	1	0	0	1	1
S; $S_\mathrm{P} = \bar{S}_\mathrm{P}$	$C_{\mathrm{d}1}$	0	0	1	1	0
S; $S_\mathrm{P} = \bar{S}_\mathrm{P}$	$C_{\mathrm{d}2}$	0	0	1	1	0
S; $S_\mathrm{P} = \bar{S}_\mathrm{P}$	$C_{\mathrm{d}3}$	0	0	1	1	0

Fault signatures in Table 9.3 indicate that faults in three components (C_{d1}, C_{d2}, C_{d3}; *i.e.* component wise fault in V_1, V_2 and/or T_2, and T_1) have identical signature and hence they are not isolatable with the given instrumentation in the system.

For simulation, Gaussian noise amounting to 2% of the simulated sensor outputs is superposed to realize noisy sensor outputs. Furthermore, to realize parameter uncertainty, we have used parameter values given in Table 9.3 to evaluate the residuals, but used 5% random error in coefficient of discharges and 2% random error in areas to generate a set of data to simulate the process behavior. For the threshold selector of r_3, we have chosen $\delta_{\eta 3} = 0.0026$, $\lambda_{\eta 3} = 0.042$, $\delta_{\sigma 3} = 0.007$ and $\lambda_{\sigma 3} = 0.035$. We have chosen 97% confidence level in this work, for which $z = 2.17$.

Moving average of each sensor data is then taken over a fixed-width sliding time window, corresponding to Δt number of samples, resulting in $l - \Delta t$ number of averaged points from l number of sensor data. Then episode derivative is calculated by taking two averaged data points at a certain gap, say m points. This yields $(l - \Delta t)/m$ number of episode derivative values. This scheme of filtering noisy sensor data to obtain episode derivatives is shown in Figure 9.4. In the present application, $l = 4020$ number of simulated data are taken from sensor P_2 and for moving averaging, Δt is considered as 20 samples. Then to obtain episode derivative, m is considered as 10. The number of episode derivative points thus generated is 400. The averaged values of sensor output and their episode derivatives are used to calculate residuals and estimate parameters for fault isolation. Note that selection of values for l, Δt and m is dependent on the specific signal and the dynamics of the concerned application.

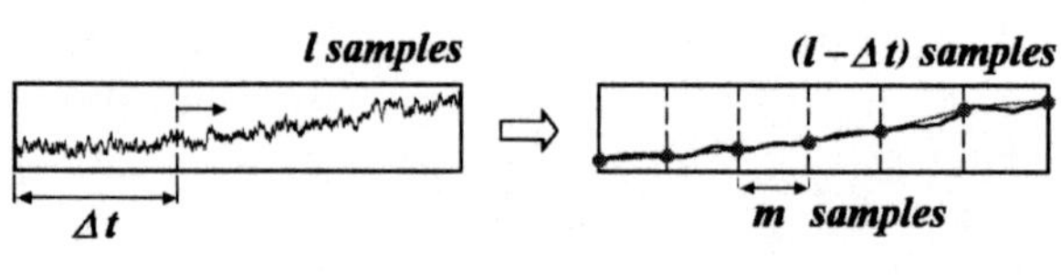

Fig. 9.4 Scheme of sensor noise filtering for obtaining episode derivatives

9.5.2 Parameter Estimation

In this example, r_3 is sensitive to an abrupt fault in valve V_1, but the residual response does not persistently show the abnormality. As can be seen from the expression for ARR_3 given in Equation 9.8, the parameter C_{d1} appears inside derivatives and residual response remains normal for the whole time span except during the transients after introduction of the fault in valve V_1. The last term of ARR_3 vanishes in steady state, *i.e.* when P_2 quickly becomes constant due to action of the PI controller. So the residual sensitivity to the fault in steady state operation is due the third term, which affects the residual only when C_{d3} is non-zero (a multiple fault situation). Corresponding to an abrupt single-fault in V_1, a temporary change in r_3 is observed as shown in Figure 9.5. Thus, the fault in V_1 is weakly detected. The danger with this is that this weak detection might be overlooked as a false alarm by the Decision Support System (DSS).

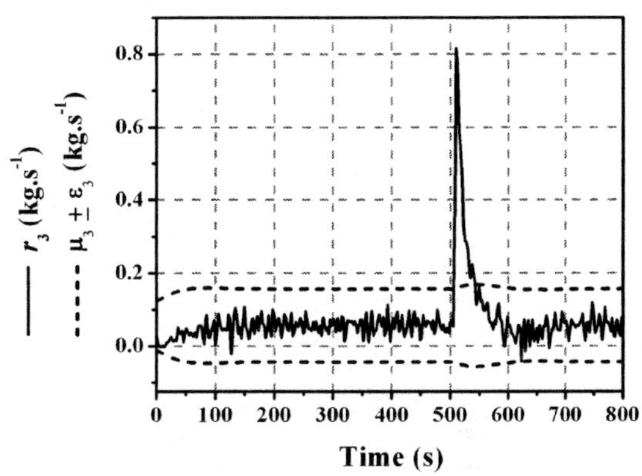

Fig. 9.5 Residual response under V_1 blockage

We consider the other case, *i.e.* when r_3 is persistently abnormal, and r_1 and r_2 are normal. This condition is satisfied when fault occurs either in the valve V_2 and/or T_2, or in valve V_3. In these situations, variables C_{d2} and C_{d3} can be estimated through closed-form solutions obtained from the expression for ARR_3. The estimated discharge coefficient ($\bar{C}_{d2}$) of V_2 (including T_2 leakage) is the positive real root of the quadratic equation

$$A.\bar{C}_{d2}^2 + B.\bar{C}_{d2} + C = 0, \tag{9.9}$$

where

$$A = \frac{1}{C_{d1}^2}\frac{dP_2}{dt},$$

$$B = \frac{2\sqrt{P_2 - P_{atm}}}{C_{d1}^2}.\frac{A_2}{g}.\frac{d^2P_2}{dt^2} + \frac{1}{C_{d1}^2.\sqrt{P_2 - P_{atm}}}\left(\frac{dP_2}{dt}\right)^2.\frac{A_2}{g} + \frac{g}{A_1}.\sqrt{P_2 - P_{atm}},$$

$$C = \frac{dP_2}{dt} + \frac{2}{C_{d1}^2} \cdot \left(\frac{A_2}{g}\right)^2 \cdot \frac{dP_2}{dt} \cdot \frac{d^2P_2}{dt^2} + \frac{A_2}{A_1} \cdot \frac{dP_2}{dt} + \frac{g}{A_1} \cdot (Q_P - \mu_3).$$

In Equation 9.9, C_{d1} and C_{d3} are considered as nominal values. Because the nominal value of C_{d3} is zero, it does not appear in Equation 9.9. Similarly, in the case of leakage from tank T_1, the estimated discharge coefficient of valve V_3 is given by

$$\bar{C}_{d3} = \left(Q_P - \mu_3 - \frac{A_2}{g} \cdot \frac{d}{dt}(P_2) - C_{d2}\sqrt{P_2 - P_{atm}} - \frac{A_1}{g} \cdot \frac{d}{dt}(x)\right) \frac{1}{\sqrt{x - P_{atm}}}, \quad (9.10)$$

where,

$$x = P_2 + \left(\left(\frac{A_2}{g} \cdot \frac{dP_2}{dt} + C_{d2}\sqrt{P_2 - P_{atm}}\right) \frac{1}{C_{d1}}\right)^2,$$

and all other parameter values (C_{d1} and C_{d2}) are taken as nominal.

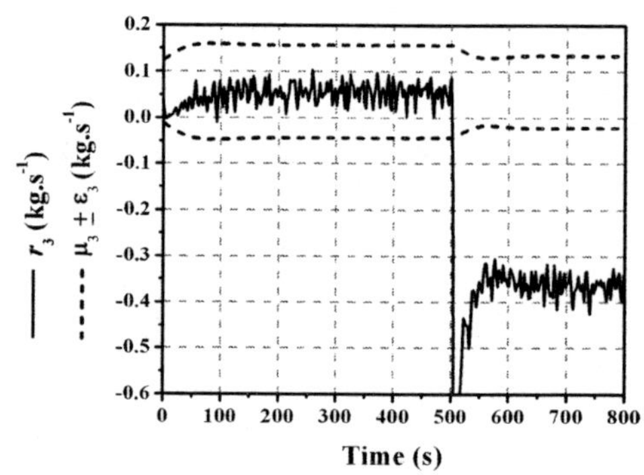

Fig. 9.6 Residual response under V_2 blockage

This estimation procedure is validated by simulating the bond graph model in Figure 9.3a in normal operating mode and with faults in valve V_2 or tank T_2 (C_{d2}), and tank T_1 (C_{d3}), one at a time introduced in the plant.

It is observed from the simulation results in Figure 9.6 that r_3 is abnormal after 500 s, when blockage event was introduced in the valve V_2 by reducing the coefficient of discharge (C_{d2}) from nominal value $1.596\times10^{-2}\sqrt{\text{kg.m}}$ to 1.0×10^{-2} $\sqrt{\text{kg.m}}$. The estimated value of C_{d2}, given by $\bar{C}_{d2}$ in Figure 9.7, closely tracks the actual one both before and after the inception of fault.

By assuming a tank leakage, the value of C_{d3}, given by $\bar{C}_{d3}$, which can result in the same values of residual is also estimated. These estimated values are shown in Figure 9.7 and it can be noticed that $\bar{C}_{d3}$ goes negative after the fault inception at 500 s, which is not a realistic phenomenon. So the fault in V_2 is directly isolatable in this case.

In other simulations, leakages in V_2 and tank T_1 are introduced at 500 s, respectively. The estimation is possible because r_3 is persistently abnormal after fault for both cases as shown in Figures 9.8 and 9.9.

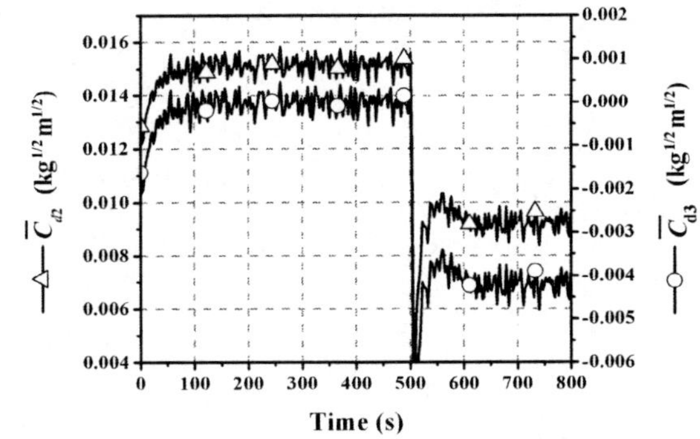

Fig. 9.7 Parameter estimation under V_2 blockage

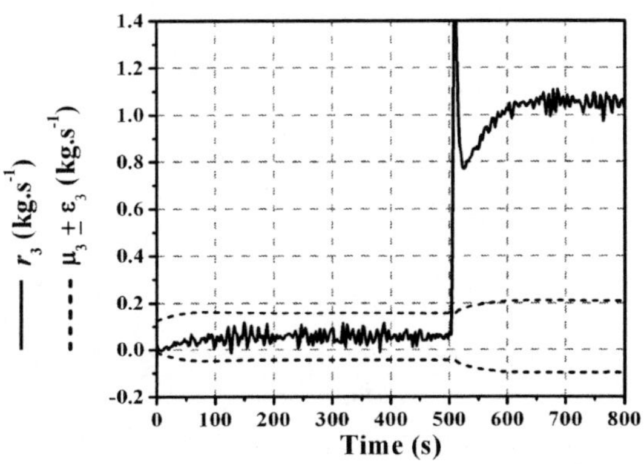

Fig. 9.8 Residual response under V_2 or T_2 leakage

The leakage in V_2 (or from T_2) is simulated by changing the value of C_{d2} from $1.596\times10^{-2}\sqrt{\text{kg.m}}$ (nominal) to 3.0×10^{-2} $\sqrt{\text{kg.m}}$ (faulty). The average estimated $\bar{C}_{d2}$ is about 2.95×10^{-2} $\sqrt{\text{kg.m}}$ after fault (as shown in Figure 9.10), which is sufficiently close to the actual value. The value of $\bar{C}_{d3}$, which can result in the same value of r_3 is also estimated and in this case it is realistic (about 9.97×10^{-3} $\sqrt{\text{kg.m}}$); so fault cannot be localized at this stage. Similar problem is encountered for tank T_1 leakage, where two estimations provide realistic values, as shown in Figure 9.11.

9.5.3 Improvement of Isolability Using Bank of Fault Models

Up to this stage, one can obtain the estimation of faulty parameters based on the response of residuals but cannot conclude which one is the correct estimation or, in other words, the fault cannot be localized. The parallel simulation (bank of fault models) method is followed here as a solution for fault isolation. Since there are two non-isolatable parameters (C_{d2} and C_{d3}) in this example, two models are required for isolating the faults. The first model, devoted to isolate the fault in valve V_2 (or leakage from tank T_2), uses the estimated coefficient of discharge of that valve

Fig. 9.9 Residual response under T_1 leakage

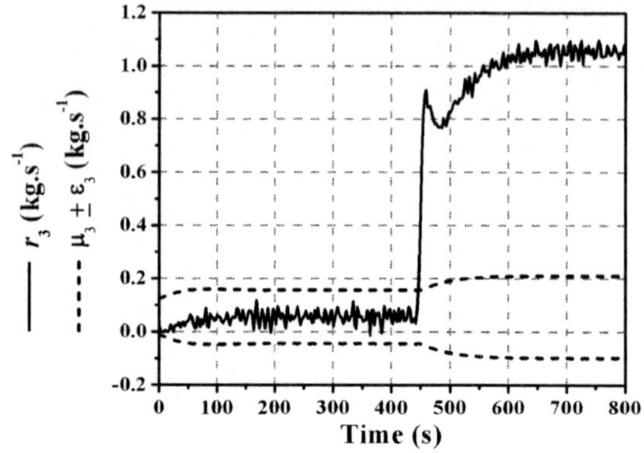

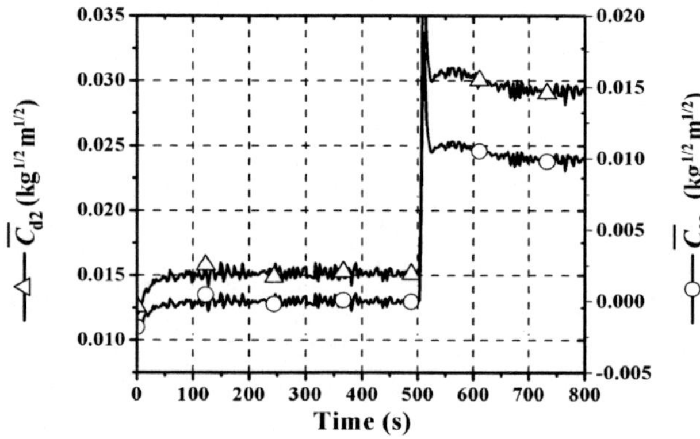

Fig. 9.10 Parameter estimation under V_2 leakage

($\bar{C}_{d2}$), while all other parameters in the model have their nominal values. On the other hand, the second model, devoted to isolate the fault in valve V_3, uses the estimated coefficient of discharge of some hole in tank T_1 (*i.e.* that of valve V_3 or $\bar{C}_{d3}$); while all other parameters in the model have their nominal values.

When any fault occurs in valve V_2, only the first model replicates the faulty plant behavior and the predicted output from the first model matches best with the measured output from the plant. The same logic is also applicable for the second model, *i.e.* when the valve V_3 faulty, the second model replicates the faulty plant behavior. The two parallel models are activated by a special switching mechanism when abnormality in residual r_3 is detected by the first phase decision procedure.

The implementation scheme of the developed architecture is shown in Figure 9.12 by using two switched bond graph models. The two fault models are identified as:

- FMV_2: the fault model using total estimated discharge coefficient from T_2

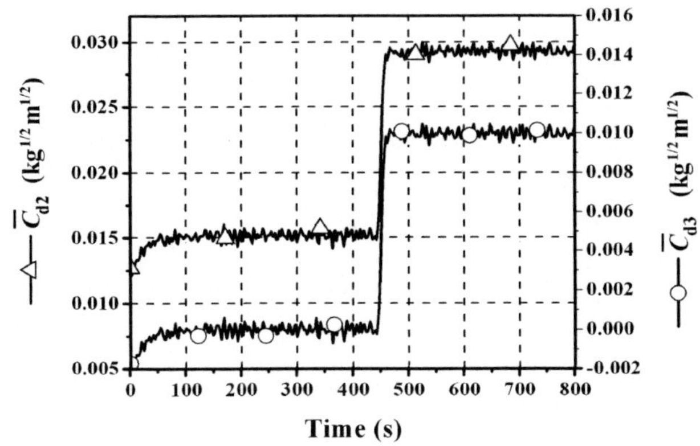

Fig. 9.11 Parameter estimation under T_1 leakage

- FMT_1: the fault model using estimated discharge coefficient of leakage from T_1 (*i.e.* discharge coefficient of the valve V_3)

9.5.4 Validation Through Simulation

The initial conditions for the switching models are determined from the sensors installed in the plant. Note that the process under consideration has all its eigenvalues as large real negative numbers, when its model is linearized around any operating point in its response trajectory. Therefore, we have used an open-loop observer. In the presence of a fault in the process, the process and the observer outputs are different. However, we are using estimates of state variables corresponding to the last known non-faulty state, *i.e.* when the residual was within the prescribed threshold.

The model in Figure 9.12 was simulated in normal operating mode and with leakage fault in valve V_2 (C_{d2}), and tank T_1 (C_{d3}), one at a time introduced in the plant. Let us consider the leakage in V_2 first, which is introduced in the plant at 490 s by changing the value of C_{d2} to $3.0\times10^{-2}\sqrt{\text{kg.m}}$ from its nominal value. The fault is detected at 500 s when r_3 crossed the threshold (Figure 9.8). The estimated values ($\bar{C}_{d2}$ and $\bar{C}_{d3}$) reached steady state at about 800 s, as shown in Figure 9.10. Thereafter, proper initial conditions are assigned to the fault models and they are simulated. Note that the sample number corresponding to the point, at which initial conditions for observers are obtained, is recorded and it is used later to extract process history for comparison with fault models. Sensor responses from the plant as well as from the two fault models are plotted in Figure 9.13. In Figure 9.13, time zones **A**, **B**, **C** and **D** correspond to different operating conditions listed below:

A. Normal operation and observer is tracking the plant

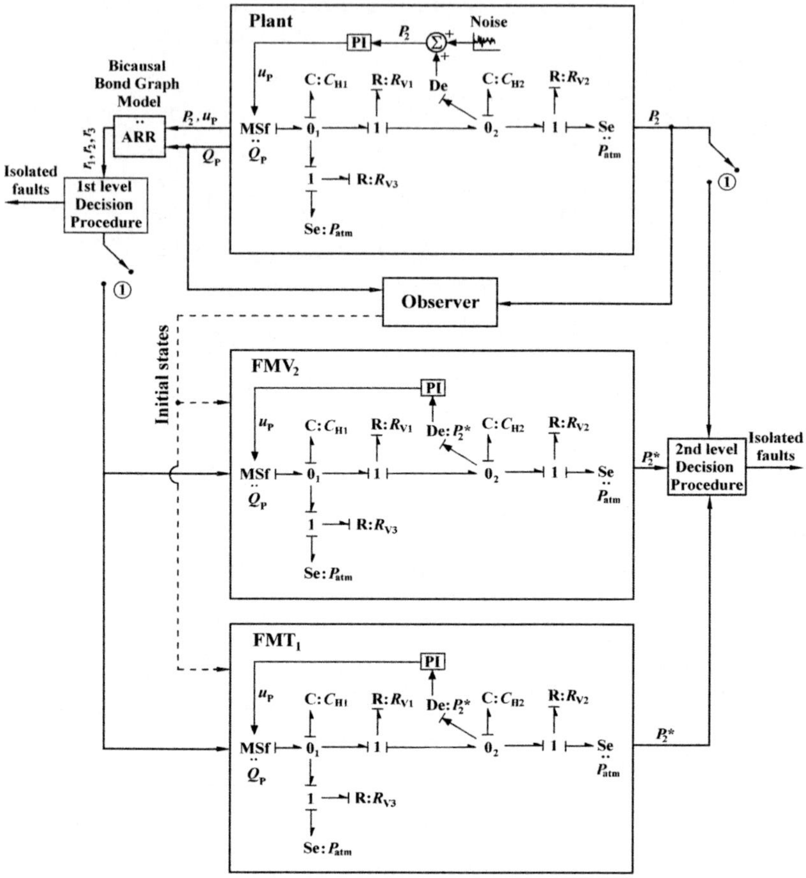

Fig. 9.12 Multi-block switching bond graph model of two-tank system

B. Fault detection, initial state recording from observer just before fault inception and parameter estimation
C. Parallel running of fault models (using estimated parameters) and fault isolation
D. Correct estimation of faulty parameter replaced in observer, which again starts tracking the faulty plant

The response pattern of the fault model that matches best with the plant response is identified and the component corresponding to the estimated parameter used in the model is concluded to be faulty. But no such matching pattern is clearly visible from Figure 9.13. The trend matching procedure can be related to the intervention analysis, provided in [28] for time series modeling, and to further works based on outliers in identification [134].

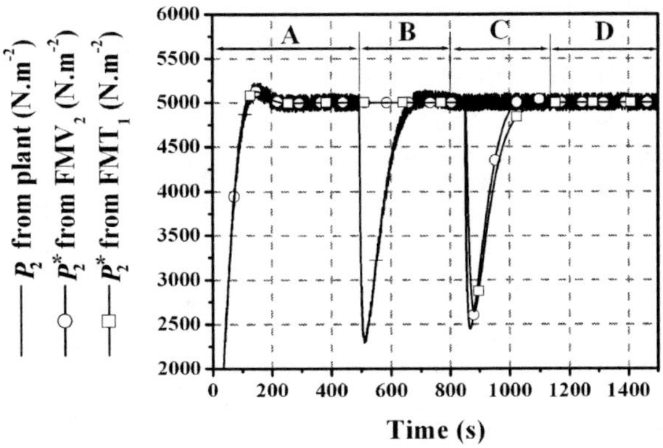

Fig. 9.13 Scheme of fault isolation: comparison of responses from plant and fault models

9.5.5 Qualitative Trend Analysis

We follow a method for pattern recognition by using a Qualitative Trend Analysis (QTA) [58, 248], *i.e.* qualitative representation framework developed in [117]. The set of primitives used in the framework to capture the basic events is shown in Figure 9.14. The steps in the hierarchical feature-extraction process are: the assignment of the primitives to each time interval, the combination of primitives and episodes generation, the sequencing of the episodes with the corresponding interval length to develop the trend, and finally the generation of an overall structure or profile. The response patterns are studied in a small observation time window (800–1050 s) and are shown in Figures 9.15–9.17. It is observed from those figures that the response trend and profile from FMV$_2$ match best with that of plant, so a leakage through valve V$_2$ (or from tank T$_2$) can be isolated as the real fault.

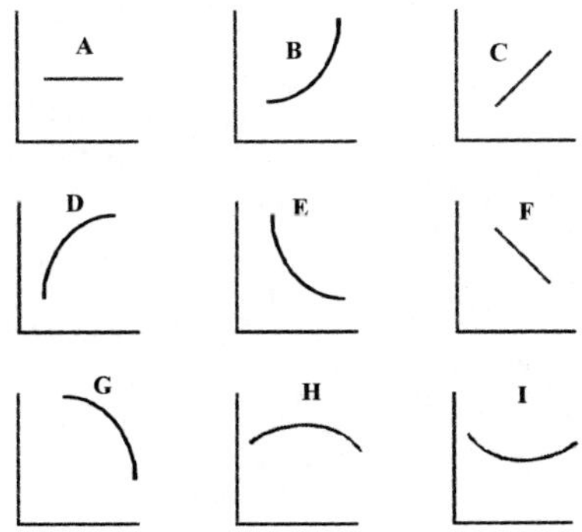

Fig. 9.14 Set of primitives to be used for qualitative trend generation

In simulation of another fault scenario, *i.e.* leakage from tank T$_1$, the estimated discharge coefficients of V$_2$ and V$_3$ ($\bar{C}_{d2}$ and $\bar{C}_{d3}$, respectively) are set in the fault models after 800 s (when those estimations have reached a steady state value in

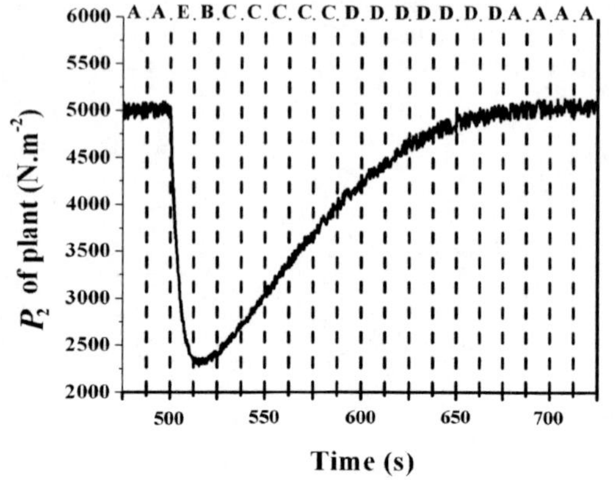

Fig. 9.15 Sensor response from plant. Trend: (A 2) (E 1) (B 1) (C 5) (D 7) (A 4). Profile: (5000 A2 5000) (5000 E1 2380) (2380 B1 2400) (2400 C5 4000) (4000 D7 4950) (4950 A4 5050)

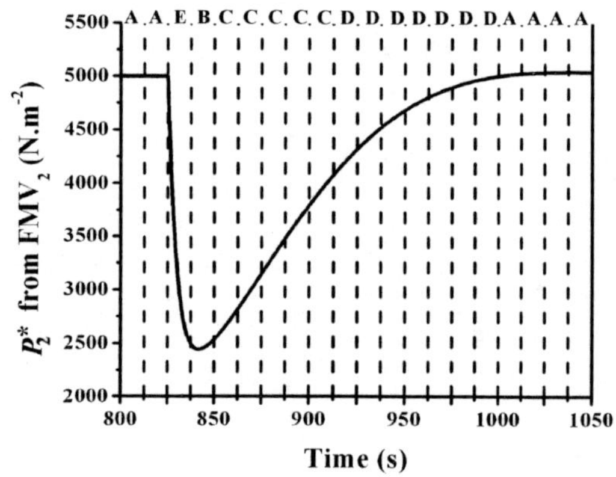

Fig. 9.16 Sensor response from FMV_2. Trend: (A 2) (E 1) (B 1) (C 5) (D 7) (A 4). Profile: (5000 A2 5000) (5000 E1 2500) (2500 B1 2525) (2525 C5 4075) (4075 D7 5000) (5000 A4 5045)

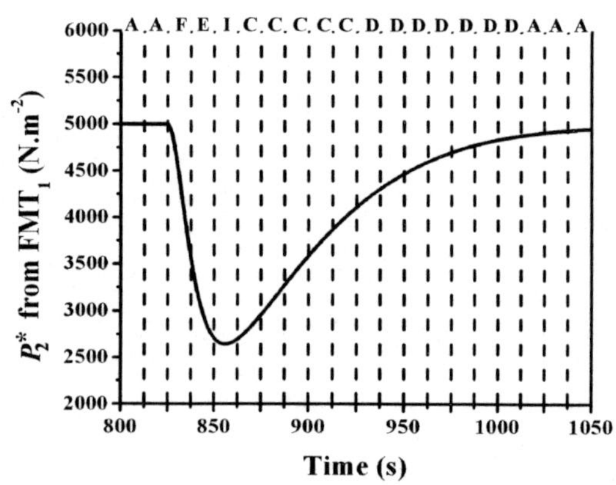

Fig. 9.17 Sensor response from FMT_1. Trend: (A 2) (F 1) (E 1) (I 1) (C 5) (D 7) (A 3). Profile: (5000 A2 5000) (5000 F1 3600) (3600 E1 2700) (2700 I1 2680) (2680 C5 4100) (4100 D7 4855) (4855 A3 4940)

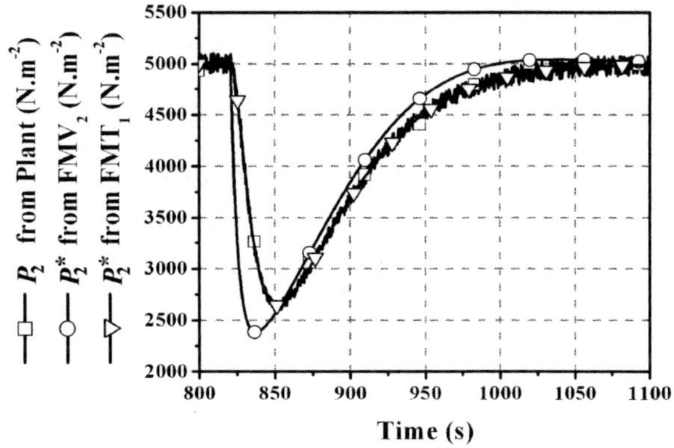

Fig. 9.18 Comparison of responses from plant and fault models

Figure 9.11). The initial state values are copied from the observer to the fault models before initiating the simulations. Sensor responses from the plant as well as from two fault models are compared in a small observation window (800–1100 s), which is shown in Figure 9.18. Note that the starting point of response from the plant has been shifted from 500–800 s to show the comparison in a common time scale. It is observed from Figure 9.18 that the plant response pattern closely matches with the predicted response from the fault model FMT_1; so a leakage from tank T_1 is localized as the actual fault.

Remark 9.1. By using the methodology developed in this chapter, the isolation capability of the FDI system is improved when the existing instrumentation is insufficient to localize the faults. However, the fault parameter estimation is done after the fault and a time delay is to be provided before comparing the selected plant outputs with the output from the fault models. While the fault detection is almost instantaneous, its isolation is delayed.

> The results from qualitative trend analysis (QTA) change with the change in the interval size. The interval size is generally chosen from experience, *i.e.* from known response time of the process variables. This is a subjective judgment and may lead to misdiagnosis.

Further note that the developed methodology has limited scope, because it is only applicable for isolation of single abrupt faults. A general approach to handle different types of faults is developed in the next chapter.

Problems

9.1. A spring-mass-damper system is shown in Figure 9.19. Variables m_f and m_e represent the measured variables obtained from two sensors (Df and De, respectively) for the velocity (flow) and the spring force (effort), respectively. Create a bicausalled bond graph model and show that, under single fault hypothesis, the fault in spring (K) can be isolated because of the uniqueness of its fault signature, but the fault in mass (M) and damper (R) are not isolatable.

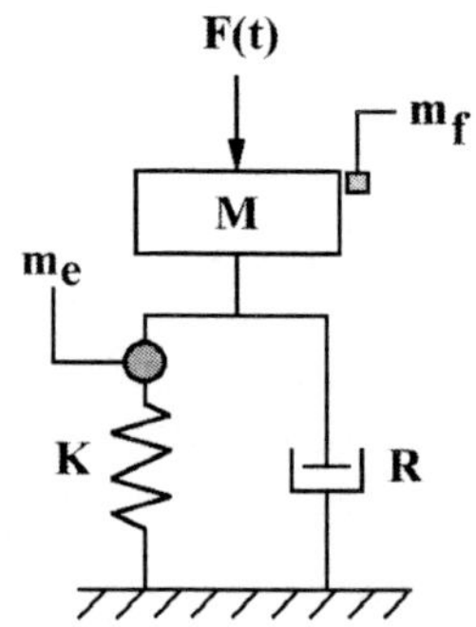

Fig. 9.19 An instrumented spring-mass-damper system in which faults cannot be isolated

The nominal values of parameters are $M = 5$ kg, $K = 100$ kN/m, $R = 1$ kN.s/m. The sampling time is 1 ms.

Because there are two non-isolatable fault parameters in this problem, two fault models are required for second phase fault isolation. The first model, devoted to isolate the fault in mass M, will use the estimated mass (M_{est}) while all other parameters in the model have their nominal values. On the other hand, the second model, devoted to isolate the fault in damper R, will use the estimated damping resistance (R_{est}) while all other parameters in the model have their nominal values. Schematically show the resulting FDI architecture. The parameter estimation may be done from the ARRs or from the bicausalled model as follows:

$$M_{est} = \frac{1}{\dot{m}_f}\left(F(t) - R_{nom} m_f - m_e\right), \tag{9.11}$$

$$R_{est} = \frac{1}{m_f}\left(F(t) - M_{nom} \dot{m}_f - m_e\right). \tag{9.12}$$

Use a sinusoidal excitation, say $F(t) = \sin(50t)$, and simulate for 0.5 s duration with a fault introduced at 0.25 s by changing R value to 100 N.s/m (from nominal 1000N.s/m). Use Equations 9.11–9.12 and show that while steady estimates are obtained for both the parameters in the non-faulty regime (0–0.25 s), in the faulty case (0.25–0.5 s), steady estimate is obtained for the value of R_{est} whereas there is no steady estimate for M_{est}. Verify this through simulation and discuss how the steadiness of estimate can be used for fault isolation.

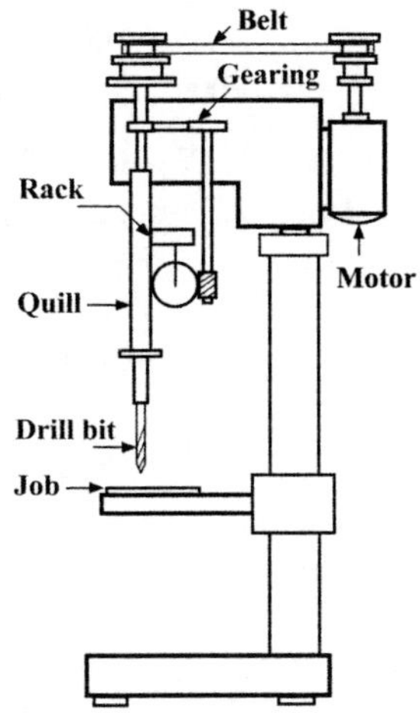

Fig. 9.20 Schematic drawing of a belt-driven drilling machine

Perform another fault scenario simulation for 0.5 s duration with a fault introduced at 0.25 s by changing M value to 25 kg (from nominal 5 kg). Show that a similar approach based on steadiness of estimated variables (R_{est} and M_{est}), as a second level decision procedure or fault detection filter, can be used for fault isolation.

9.2. A belt driven single spindle drilling machine with DC power supply and power feed is shown in Figure 9.20. The feed rate is assumed to be constant. If the drill encounters a hard spot, the thrust or end load on the drill builds up enormously and may lead to drill bit breakage. On the other hand, if the drill encounters a soft spot (*i.e.* impurity layer of low hardness material, blow holes, *etc.*), the end load decreases. We need to monitor both the cases along with the phenomena of belt slippage and motor fault in an FDI scheme. A motor current sensor is installed as the only sensor to take output from the plant.

The DBG model of the process, developed with suitable assumptions, is given in Figure 9.21. The nomenclature of various variables used in the model is given in Table 9.4.

The load torque, M, and the thrust or end load, P_x, are obtained by the following empirical relations:

$$M = C_M.D^{1+x_1}.s_0^{y_1}, \qquad (9.13)$$
$$P_x = C_p.D^{x_2}.s_0^{y_2},$$

where M is in kg.mm, P_x is in kg, D is the diameter in mm, s_0 is the feed rate in mm/revolution. Parameters C_M, C_p, x_1, x_2, y_1 and y_2 are material constants given in Table 9.5, where BHN stands for Brinell Hardness Number. The net load torque can be written as

$$\tau_d = M + P_x.g_r, \qquad (9.14)$$

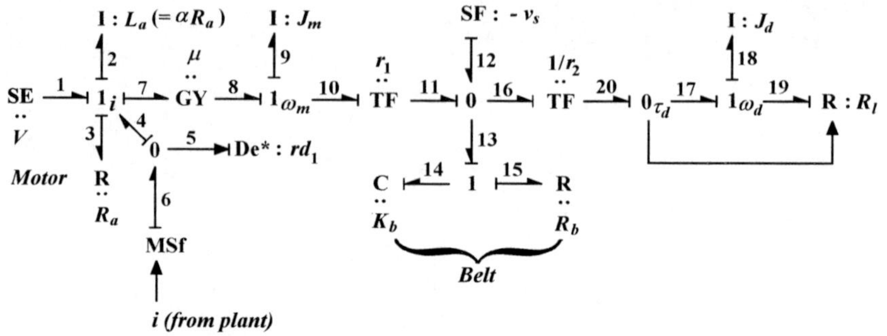

Fig. 9.21 DBG model of the drilling machine

Table 9.4 Parameters of the drilling machine

Symbol	Description	Nominal values
V	Supply voltage	180V
R_a	Armature resistance of DC motor	2.5Ω
L_a	Armature inductance of DC motor	0.07 H
α	Proportionality constant (L_a/R_a)	0.028H/Ω
μ	Equivalent gyrator modulus for motor	1.614 N.m/A
J_m	Rotary inertia of the motor side pulley	0.0025 kg.m^2
J_d	Rotary inertia of the drill side pulley	0.005 kg.m^2
r_1	Radius of motor side pulley	0.12 m
r_2	Radius of drill side pulley	0.3 m
K_b	Stiffness of the belt	2×10^6 N/m
R_b	Damping of the belt	2 N.s/m
g_r	Equivalent gearing ratio	2.069×10^{-5} rev.s/rad.mm
v_s	Slip velocity	0 m/s

where g_r represents the total equivalent gearing ratio.

Table 9.5 Empirical coefficients

Material	C_M	C_p	x_1	x_2	y_1	y_2
Steel, σ_u=75 kg/mm^2	34	85	1.9	1.0	0.8	0.7
Alloy steel, σ_u=115 kg/mm^2	75	270	2.1	1.15	0.76	0.77
Cast iron, BHN=190	23	60	1.9	1.0	0.8	0.8
Cast iron, BHN=150	20	52	1.9	1.0	0.8	0.8
Bronze	12	31	1.9	1.0	0.8	0.8

The entire material load can be represented as a non-linear R-element (R_l) in the behavioral bond graph model. In the inverted causality model, the load from the material side is given by assigning conductive causality to the load resistance, as shown in Figure 9.21. The net load torque is a function of the feed. As feed (s_0) is directly related to rotational speed (ω_d) of the drill spindle pulley, the net load torque, τ_d, is also a function of ω_d. If y_1 and y_2 are of the same value in Equation 9.13, as in the case when cast iron or bronze is used as the working material, ω_d can have a closed form algebraic solution in terms of τ_d. But in the general case, as for steel (σ_u = 75 kg/mm^2) considered for the present problem, $y_1 \neq y_2$. In such cases, ω_d can be obtained numerically from speed-torque characteristics. In the present problem, a third order polynomial approximation within a range of values of rotational speed may be used to obtain ω_d as a function of τ_d, as

$$\omega_d = -3.6 \times 10^{-7}\tau_d^3 + 3.27 \times 10^{-4}\tau_d^2 + 0.124\tau_d - 1.0232677. \qquad (9.15)$$

Equation 9.15 is the constitutive reation for the load resistance in the DBG model in Figure 9.21. In the behavior model, the load resistance is in resistive causality and its constitutive relation is given by Equation 9.14 with $s_0 = \omega_d \cdot g_r$.

Show that with the single sensor, it is not possible to isolate faults if parameters R_a, v_s and ω_d deviate from their nominal values. To arrive at this result, use either causal path analysis on the DBG to construct the FSM or show that bicausality cannot be assigned such that all three corresponding terminating nodes can be simultaneously bicausalled.

Introduce a fault in the behavioral model by changing the resistance (R_a) from nominal value of 2.5 Ω to a faulty value of 5 Ω after 1 s of simulation. Construct the following three fault models:

- FMI. Fault Model using estimated Impedance (R_a, and consequently, L_a)
- FMEL. Fault Model using estimated External Load from job material
- FMBS. Fault Model using estimated Belt Slip

Show that the output pattern from the first fault model matches with the actual plant response (measured armature current) and thus the fault can be isolated.

Note: temporary large values of derivatives of some signals may be cropped (by using Simulink®'s saturation block) to obtain better results.

9.3. A spray polymer coating process is shown in Figure 9.22. It is assumed that all material properties (densities, specific heats, *etc.*) are known. The monomer and activator chambers may be considered as constant pressure and temperature sources in the model.

The dominant modes of heat transfer are convection and radiation. The heat transfer coefficients depend on the rate of spraying and the thickness of the deposited layer. Note that the heat transfer through the substrate and along the length of the coating may be neglected. The coating quality (uniformity of thickness) may be compromised for a variety of reasons: nozzle blocking, improper flow rate, belt slip, non-smooth drive, bad drying, *etc.* If all heat transfer coefficients are known and single fault hypothesis is assumed, then determine the faults in the process which cannot be isolated. Schematically show the fault models required to isolate those faults.

9.4. A metal inert gas (MIG) welding process is shown in Figure 9.23. The inert gas (argon) shields the weld pool from oxidation. Welding is one of the important

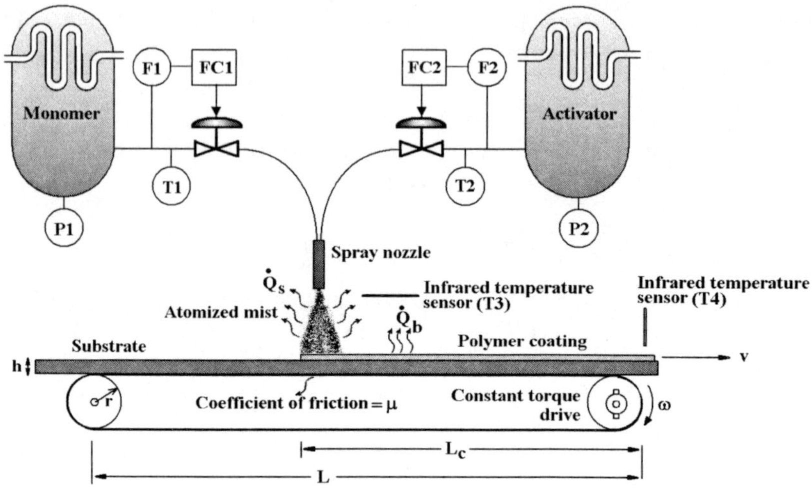

Fig. 9.22 A polymer coating process

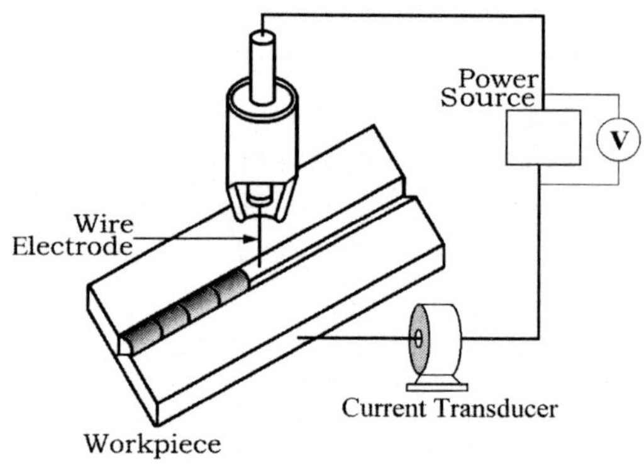

Fig. 9.23 A MIG welding process

sub-processes in most manufacturing processes. As a result of this, there is a need for different technologies to predict precisely the weld quality with respect to the different welding operating conditions. Weld quality can be measured directly or indirectly. Direct methods are visual inspection and vision sensing of the weld puddle. Indirect methods are arc sensing, infrared sensing, radiographic sensing, inductive sensing, arc sound sensing, ultrasonic sensing and acoustic emission sensing.

In the considered process, the welding voltage, current, wire (electrode) feed-rate, table feed-rate and temperatures at a few selected points are measured. In this case, a part of the work-piece melts and the base metal fuses with the electrode flux material. Moreover, there is significant heat transfer through the work-piece.

Discuss how the fault isolation problem in the MIG welding process is related to or different from that in the polymer coating process. Highlight the need for simulation (or prediction, *e.g.* by an ANN) based FDI in this problem.

Chapter 10
Multiple Fault Isolation Through Parameter Estimation

10.1 Introduction

For unambiguous fault isolation, one needs to use a lot of sensors in the process such that structured residuals can be generated. When residuals are unstructured, one may follow estimation based FDI approaches wherein the temporal behavior of the process (from measurements) is used to estimate the process parameters and then these process parameters are compared to their nominal values for fault isolation [17, 113, 114, 236, 237]. Furthermore, even when one or more faults in process parameters are directly isolated, it is required to estimate the degree of those faults to take appropriate fault accommodation measures, called fault tolerant control.

In this chapter, we assume that during nominal operation of the process, all the parameter values are properly estimated and a well developed model, which replicates the process behavior, exists. We deal with process faults, which may occur during subsequent operation of the process.

The standard recursive least squares optimization technique attempts to estimate the parameter values, which give minimum output error [44, 244, 283]. The objective function is defined over a given range of time, T, as

$$J = \frac{1}{2}\int_{0}^{T} (y(t,\theta) - \bar{y}(t,\bar{\theta}))^2.dt, \tag{10.1}$$

where t is the time variable, θ is the vector of actual parameter values, $\bar{\theta}$ is the vector of the estimated parameter values, $y(t,\theta)$ is the process output vector and $\bar{y}(t,\bar{\theta})$ is the output from a simulation model using estimated parameter values. The search space can be constrained if limits for some parameter values are known and they are imposed as appropriate penalty functions. When parameters of a system change due to some fault, without changing the structure of the system (*i.e.* the model remains same), then the estimation procedure applied over a time window (0 to T) identifies the new set of parameter values. If different weights are to be assigned to minimize different output errors $e_i(t,\theta) = y_i(t,\theta) - \bar{y}_i(t,\bar{\theta})$, $i = 1..n_y$, then the objective function can be rewritten as

$$J = \frac{1}{2}\int_0^T (y(t,\theta) - \bar{y}(t,\bar{\theta}))^T.\mathbf{Q}(t).(y(t,\theta) - \bar{y}(t,\bar{\theta})).dt, \qquad (10.2)$$

where $\mathbf{Q}(t) \in \mathbb{R}^{n_y \times n_y}$ is a time varying positive semi-definite weighting (cost) matrix.

The major limitation of this algorithm is that the initial conditions in the process are unknown and they too have to be estimated along with the parameters. Furthermore, the new parameter vector is assumed to be constant over the optimization time window and the optimization procedure requires too many simulations to be carried out, which is time consuming and thus unsuitable for real-time FDI applications.

Clearly, assumption of parameters being constant over a given time window can be applied to identify single or multiple abrupt faults occurring at the same time, but this approach is not appropriate to handle cases related to intermittent or progressive faults. Of course, slowly varying parameters corresponding to slow progressive faults can be handled, if the chosen time window is sufficiently small with respect to the rate of fault progression and convergence can be achieved with fewer number of data points.

Two important assumptions made in this chapter are: (1) for a given process input and output vector, a unique set of parameter exists; and (2) sensors, actuators and controllers are robust, *i.e.*, non-faulty. A passive FDI approach is followed in this work by using an adaptive threshold selector to counter static and time-varying drift associated with model uncertainties.

This chapter concerns the case, where the fault signature matrix is non-diagonal, and some faults cannot be isolated under assumption of multiple-faults. In the parameter subspace corresponding to diagonal residuals, multiple faults are easily isolated and the associated parameters are estimated by using purely algebraic relations (see Chapter 9). Thus, we can decouple the unstructured residual set and the parameters corresponding to them. Only those fault candidates which have influence on the residuals in the unstructured residuals need to be estimated. Furthermore, parameter estimation is essential for fault accommodation through FTC, because the quantitative information is required to enable the implementation of required operational changes (See Chapter 11).

10.1.1 Adaptive Thresholds for Robust Diagnosis

A simple adaptive threshold has been discussed in the previous chapter (Section 9.2). However, in most cases, it is not possible to separate temporal and input dependency. In this section, we consider an example system and develop an adaptive threshold for it by considering various uncertainties in the model.

Robustness analysis is done with the objective of designing control laws guaranteeing the stability for parameter variations inside a known variation interval. Two types of models for robustness study of linear systems are used. The first is the canonical form of uncertain system state equation given as

$$x = [\mathbf{A}_n + \Delta\mathbf{A}]x + [\mathbf{B}_n + \Delta\mathbf{B}]u,$$
$$y = [\mathbf{C}_n + \Delta\mathbf{C}]x + [\mathbf{D}_n + \Delta\mathbf{D}], \tag{10.3}$$

where $\mathbf{A}_n, \mathbf{B}_n, \mathbf{C}_n$ and $\mathbf{D}_n$ are the state-space matrices using nominal parameter values; and $\Delta\mathbf{A}, \Delta\mathbf{B}, \Delta\mathbf{C}$ and $\Delta\mathbf{D}$ are the deviations in those matrices dues to uncertainties in parameter values.

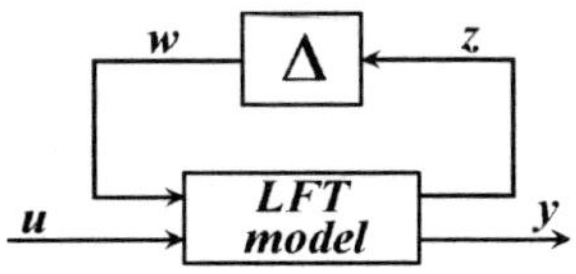

Fig. 10.1 Standard interconnection or internal feedback loop model

The second model is referred to as the standard interconnection model or internal feedback loop. In this form, the parameter uncertainties are separated from the nominal model and represented as feedback loops of internal variables (see Figure 10.1) as follows:

$$\begin{aligned} x &= \mathbf{A}_n x + \mathbf{B}_1 w + \mathbf{B}_{2n} u, \\ z &= \mathbf{C}_1 x + \mathbf{D}_{11} w + \mathbf{D}_{12} u, \\ y &= \mathbf{C}_{2n} x + \mathbf{D}_{21} w + \mathbf{D}_{22n} u, \\ w &= \Delta.z, \end{aligned} \tag{10.4}$$

where $\mathbf{B}_{2n} = \mathbf{B}_n, \mathbf{C}_{2n} = \mathbf{C}_n, \mathbf{D}_{22n} = \mathbf{D}_n$, z is an auxiliary output (internal variable), w is an auxiliary (internal) input and Δ is a diagonal matrix containing relative uncertainties of parameters. This model is the basic form used in μ-analysis and μ-synthesis which use linear fractional transformation (LFT). Various matrices used in Equation10.4 must satisfy some specific properties [234].

The bond graph representation of Equation 10.4 may be achieved by defining the variations as multiplicative errors [234]. Consequently, uncertain parts appear as modulated sources (internal feedback loop) on the nominal bond graph. Another way of representing the same set of equations is an incremental bond graph [25]. Both these approaches are shown to be same in [26].

Consider a two tank system shown in Figure 10.2a and its bond graph model in Figure 10.2b. The process has two pressure sensors which measure the fluid pressure at the bottom of the two tanks. Let us further consider that the constitutive relations of R-elements in the model are linear.

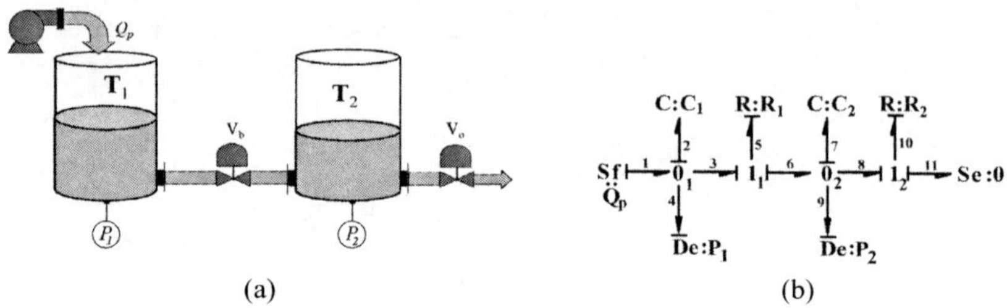

Fig. 10.2 A linear two-tank system and its bond graph model

The diagnostic bond graph (DBG) of the system is shown in Figure 10.3, where storage elements are in differential causality.

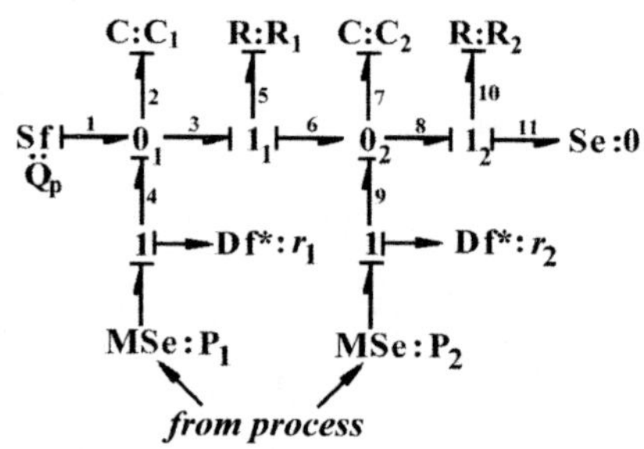

Fig. 10.3 Diagnostic bond graph of the two-tank process

First of all, we consider certain uncertainties in the parameter values.

R-elements: because the actual value of the R-element is never known, it can be written as $R_n \pm \Delta R = R_n(1+\delta_R)$, where R_n is the average estimated (nominal) value and $\pm\Delta R = \pm\delta_R R_n$ is the uncertainty in estimation. When the R-element is in conductive causality, its constitutive relation may be written as

$$\begin{aligned} f &= \frac{1}{R_n \pm \Delta R} e = \frac{1}{R_n}\left(1 \pm \delta_{1/R}\right) e \\ &= \frac{e}{R_n} \pm w_{1/R}, \end{aligned} \tag{10.5}$$

where $\delta_{1/R}/R_n$ is the uncertainty in estimating the conductance $(1/R_n)$ and $w_{1/R}$ may be considered as a disturbance. Such a way of separating uncertainties is termed Linear Fractional Transformation (LFT). The bond graph representation the R-element in conductive causality with the uncertain parameter value is then given in Figure 10.4.

LFT forms for other elements including two-ports (TF and GY elements) can be likewise defined (see [65]). Then the DBG in Figure 10.3 can be drawn in LFT form as given in Figure 10.5.

Now one may derive the ARRs from the DBG model in LFT form as follows:

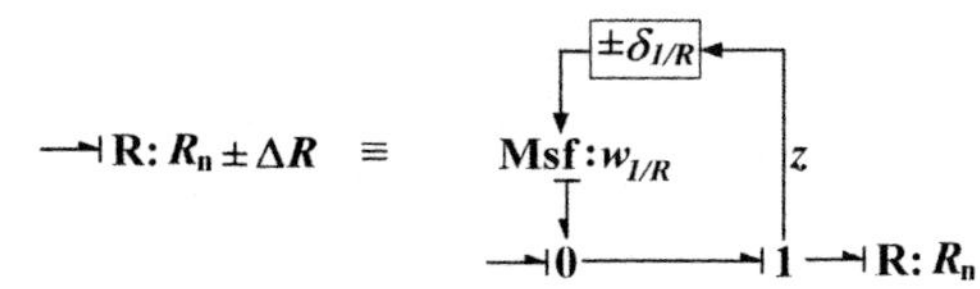

Fig. 10.4 R-element with conductive causality transformed to LFT form

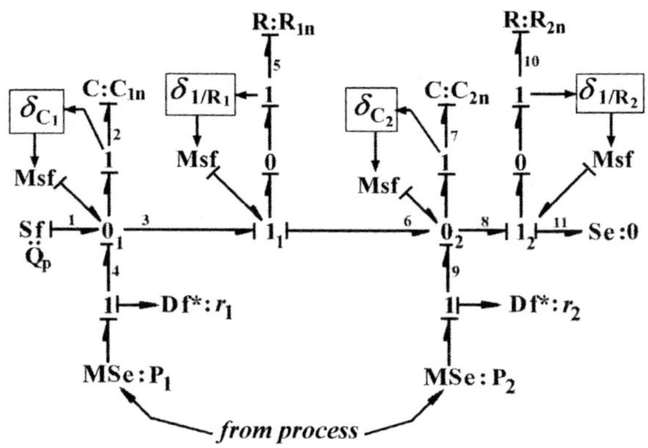

Fig. 10.5 Bond graph model of two-tank system in LFT form

$$
\begin{aligned}
ARR_1 &: Q_p - C_1\frac{dP_1}{dt} - \frac{P_1-P_2}{R_1} \pm \delta_{C_1} C_1 \frac{dP_1}{dt} \pm \delta_{1/R_1}\frac{P_1-P_2}{R_1} \\
&= ARR_{1n} \pm a_1, \\
ARR_2 &: \frac{P_1-P_2}{R_1} - C_2\frac{dP_2}{dt} - \frac{P_2}{R_2} \pm \delta_{1/R_1}\frac{P_1-P_2}{R_1} \pm \delta_{C_2} C_2 \frac{dP_2}{dt} \pm \delta_{1/R_2}\frac{P_2}{R_2} \\
&= ARR_{2n} \pm a_2.
\end{aligned} \tag{10.6}
$$

The ARRs derived above have been separated into their nominal part and uncertain parts (a_1 and a_2). Thus the corresponding residuals are $r_1 = r_{1n} \pm a_1$ and $r_2 = r_{2n} \pm a_2$. Note that the uncertainties in various parameters are uncorrelated. This means there is as much likelihood of them canceling out each other as the likelihood of them adding up. Therefore, we will take most conservative estimate and write the uncertain terms as follows:

$$
\begin{aligned}
a_1 &= \left|\delta_{C_1} C_1 \frac{dP_1}{dt}\right| + \left|\delta_{1/R_1}\frac{P_1-P_2}{R_1}\right|, \\
a_2 &= \left|\delta_{1/R_1}\frac{P_1-P_2}{R_1}\right| + \left|\delta_{C_2} C_2 \frac{dP_2}{dt}\right| + \left|\delta_{1/R_2}\frac{P_2}{R_2}\right|.
\end{aligned} \tag{10.7}
$$

Let us assume that the pump gives a pulsating flow (due to action of some controller) and the pressures in the two tanks are known for certain time duration. The threshold generator must use the contemporary inputs and a nominal model with uncertainties [69]. A block diagram representation of the threshold generator of the two-tank system is shown in Figure 10.6.

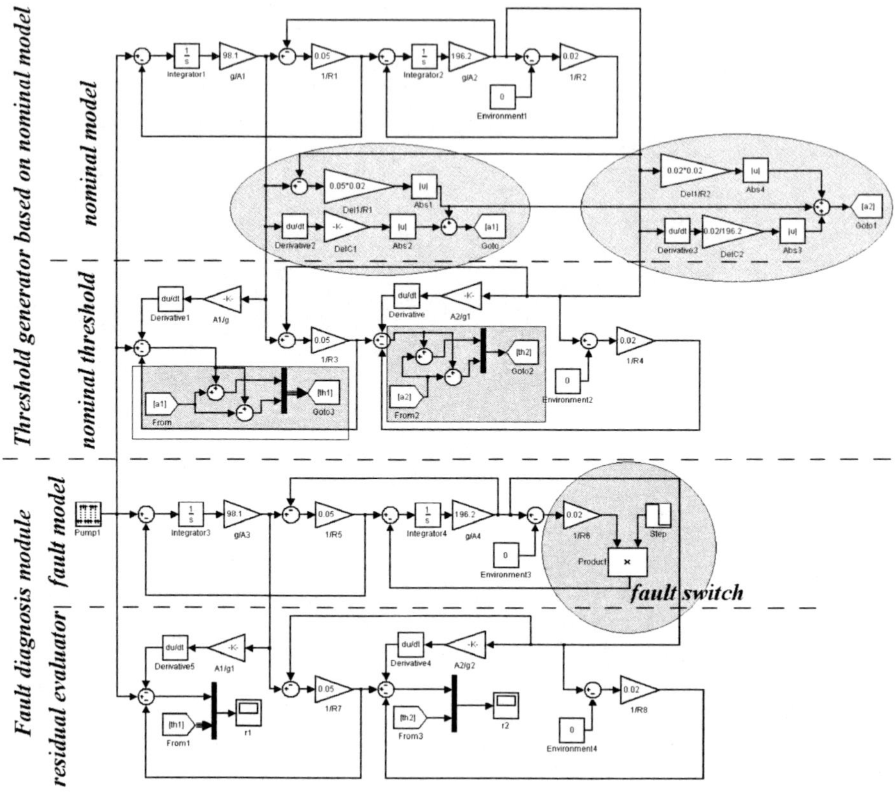

Fig. 10.6 Threshold generator and residual evaluation

In Figure 10.6, the top part uses a nominal model (block diagram form of BG in Figure 10.2b) and a diagnostic model (block diagram form of DBG with LFT in Figure 10.5) to generate the residual thresholds.

Because the system is observable, it is possible to execute the nominal behavior model with estimated initial states. Note that the nominal model must be executed with actual plant inputs. If the process is controlled then controllers should be taken out of the nominal model and actual controller outputs must be used.

In Figure 10.6, the regions shown inside ovals correspond to evaluation of values a_1 and a_2. These two values are used in rectangular regions in the figure to calculate the thresholds. The bottom part of the model in Figure 10.6 is the usual FDI framework where there is a provision to introduce a fault in the valve V_2 (region marked with a circle). The effective value of parameter $1/R_2$ is modified by multiplying it with a step function which is initially 1 and changes to 0.8 after 5 s; *i.e.* 20% valve leakage fault is introduced after 5 s. The bottom most part of the figure is the residual evaluator, where actual residual values are plotted along with their nominal thresholds (note the use of thick vertical lines which represent signal muxing blocks).

The model is simulated with a fixed step size of 0.02 s for duration of 10 s with both state variables initialized to zero. The pump, as an input source, gives a pulse train of amplitude 10 with pulse width of 25 samples, pulse period of 50 samples and 0.02 s sampling time. The same input is given to all sub-models. The nominal parameter values are chosen $C_{1n} = A_{1n}/g$ (with $A_{1n} = 0.1$), $R_{1n} = 20$, $C_{2n} = A_{2n}/g$ (with $A_{2n} = 0.05$) and $R_{2n} = 50$, in consistent units. The uncertainty in each variable, with 99% confidence, is considered to be $\pm 2\%$, *i.e.* $\delta_{C_1} = \delta_{1/R_1} = \delta_{C_2} = \delta_{1/R_2} = 0.02$.

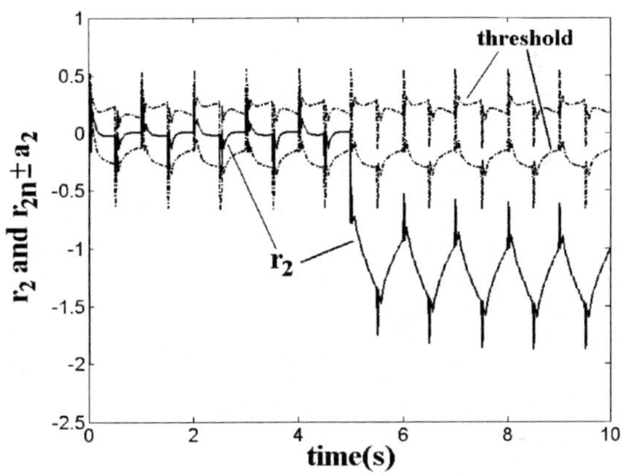

Fig. 10.7 Response of residual r_2 and its adaptive threshold

Looking at Equation 10.6, residual r_2 is sensitive to fault in parameter R_2. The residual response obtained from simulation is shown in Figure 10.7. This result shows that the residual remains within its adaptive threshold (dotted lines) during normal operation and deviates from it under a fault. Note that the residual has sharp peaks (due to the nature of the input and use of derivatives) and yet the adaptive thresholds [62, 213, 252] envelop the residual properly.

Likewise, measurement uncertainties (static drifts) can be included in uncertain part of the residuals to modify the thresholds [230]. However, measurement noise has to be accounted for differently as discussed in Chapter 9.

10.1.2 Localization of Fault Subspace

Consider an example FSM given in Table 10.1, in which θ_i, $i = 1..6$, are six independent parameters, and r_j, $j = 1..3$, are three residuals. When residual r_1 is abnormal, it leads to a coherence vector $C = [1,0,0]$, which has a unique match in the FSM and the component corresponding to parameter θ_1 in fault subspace S_{P1} is isolated as the fault candidate. Residual r_1 is sensitive to a single parameter fault and hence it constitutes the structured part of the FSM. Residuals r_2 and r_3, along with parameters θ_2 to θ_6, constitute the unstructured part of the FSM. Note that in the unstructured part of the FSM, each residual is sensitive to more than one fault.

Table 10.1 Example fault signature matrix

		r_1	r_2	r_3
θ_1	S_{P1}	1	0	0
θ_2	S_{P2}	0	1	0
θ_3	S_{P3} (with θ_4)	0	0	1
θ_4	$S_{P2,3}$ (θ_2 to θ_6)	0	0	1
θ_5		0	1	1
θ_6		0	1	1

The unstructured part of FSM can be further sub-classified into different fault sub-spaces having common fault signatures, *i.e.* binary OR operations with other fault signatures does not change some fault signatures. In other words, some fault signatures may result from one or more faults appearing simultaneously. For example, when only residual r_2 is abnormal, the coherence vector is $[1,0]$ (considering only the unstructured residuals and its FSM) and localization of θ_2, in fault subspace S_{P2}, as the fault candidate is possible. Similarly, if only residual r_3 is abnormal, then the fault could be due to change in values of θ_3 or θ_4 or both (fault subspace S_{P3}). If both residuals r_2 and r_3 are abnormal, then the resulting coherence vector $[1,1]$ could be due to change in values of a single parameter (θ_5 or θ_6) or due to simultaneous faults in more than one parameters (fault subspace $S_{P2,3}$), *e.g.* θ_5 and θ_6, θ_2 and (θ_3 and /or θ_4), *etc.* When a fault subspace containing more than one parameter is identified as the fault candidate, one needs to estimate all the parameters in that fault subspace to isolate which parameters in it are actually faulty and by what magnitude?

10.2 Fault Isolation by Parameter Estimation

Each residual must stay within its threshold during normal process operation. When parameters of the system change from their nominal value by sufficient magnitude (*i.e.* more than the uncertainties used to define residual thresholds) then only one can detect faults. During faulty operation of the process, residuals are outside their prescribed thresholds. If we can determine such a set of parameter values which when used with the available measurements from the process (which already has fault(s)) would give rise to residuals bounded by their thresholds, then the new set of parameter values can be considered to be the new nominal values of the parameters in the process [97, 144, 222]. If some of the new nominal values are sufficiently different from their corresponding old nominal values then they can be isolated as the actual faults.

The above-mentioned fault isolation problem though parameter estimation is simplified if the number of parameters to be estimated is small. In this chapter, we will estimate the parameters corresponding to fault candidates (obtained from analysis of the FSM) by minimizing residuals. Optimization over a very small time window is performed to over-constrain the solution and in cases where the number of unknowns exceeds the number of available algebraic relations. Moreover, uncertainties associated with parameters, sensor bias and noise are to be taken care of during formulation of the cost function.

Let a fault subspace S_p, containing m parameters and n' residuals ($n' \leq n$), define the fault hypothesis, *i.e.* the set of fault candidates. Then each parameter $\theta_i \in S_P\,(i = 1, \cdots, m)$ is estimated by using least squares optimization of n' ARRs, with the assumption of the identifiability condition, *i.e.* for a given input and output vector, there exists a set of unique process parameters. The optimization problem can be stated as

$$\text{Minimization of } \; J\big|_{t=t_k} = \frac{1}{2} \sum_{j=k-q}^{k} \left[\sum_{i=1}^{n'} \bar{r}_i^2 \big|_{t=t_j} \right] + \Theta_P, \tag{10.8}$$

where $\bar{r}_i\big|_{t=t_k}$ represents i-th abnormal residual at k-th sample, which has been already normalized to account for process uncertainties, $q \geq 0$ is number of additional consecutive samples taken for optimization and Θ_P is the penalty function. Residuals are normalized as follows:

$$\bar{r}_i\big|_{t=t_k} = q_i \frac{r_i\big|_{t=t_k} - \psi_i(t_k).\eta_i(u_k)}{z_i.\phi_i(t_k).\sigma_i(u_k)}, \tag{10.9}$$

where q_i is a weight (normally $q_i = 1$) and other variables are the residual properties used in defining its adaptive threshold (see Chapter 9).

Additional consecutive samples are needed when $m > n'$. One can consider more samples to over constrain the solution, provided that parameters are assumed to remain constant during that period. For example, let $S_P = S_{P2,3}$ as shown in Table 10.1; then the number of parameters to be estimated is five (θ_2, θ_3, θ_4, θ_5, θ_6), and two residuals (r_2 and r_3) are sensitive to faults in these parameters. Therefore, at least three more independent equations ($q = 2$ gives four more equations) are to be obtained by using temporal information or time evolution of the residuals. Note that when the sampling interval is very small, consecutive samples may not yield true information of temporal behavior; in other words the optimization problem stated in Equation 10.8 becomes ill-conditioned. Therefore, depending on the system's response characteristics and sampling interval, additional samples taken to formulate the cost function may be taken at larger intervals. Moreover, temporal information ($q \geq 0$) is useful only when the system operates in the transient regime or the input is continuously varying with time.

Penalties may be added to the objective function, J in Equation 10.8 to constrain the parameter values within known physical limits. Such problems may otherwise be solved through various constraint optimization algorithms.

Each estimated parameter value has certain degree of uncertainty associated with it. These arise from the uncertainties in measurements, which were used to estimate the parameters. Moreover, external disturbances and uncertainties in the process structure (unmodeled dynamics) *etc.* also influence the uncertainty of the estimated parameter. Even the initial known parameter values are uncertain.

Consider the schematic representation shown in Figure 10.8 where a is the initial estimate of a parameter. A probability density function describes the spread around point a as its mean value, where the actual parameter value may lie. This spread is restricted by choosing a certain confidence value, *e.g.* 99% of the total area under the normalized probability distribution function.

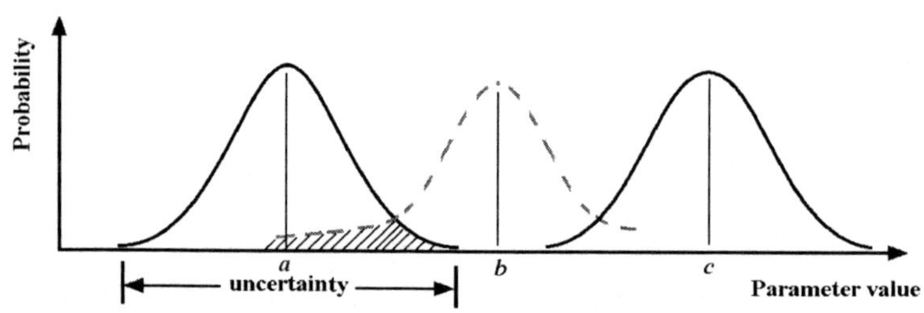

Fig. 10.8 Uncertainty in parameter estimation

Let us assume that at some point of time, the estimated value of the same parameter becomes b. However, the value b itself is uncertain and its spread intersects some region of the spread around value a. Existence of this common region (shaded area in Figure 10.8) implies that if the actual parameter value lies in this region, there is a likelihood of it being estimated either as a or as b or some other value in its neighborhood. Therefore, when the estimated parameter value is b, even though it is different from a, the concerned parameter (or component) cannot be called faulty [107]. When there is an intersecting region as shown in Figure 10.8 and its area falls below a certain predefined threshold then one may issue a warning, but an alarm should not be raised. Only when the estimated parameter value is sufficiently different from the original estimate, say c, can one safely diagnose that the value of the parameter has changed and the corresponding component has a fault.

10.3 Example I: A Linear Two-tank System

Consider a two tank system shown in Figure 10.2a and its bond graph model in Figure 10.2b. The process has two pressure sensors which measure the fluid pressure at the bottom of the two tanks. Initially, we take up a simplified problem and consider that the constitutive relations of R-elements in the model are linear. The objective is to determine all system parameters (C_1, R_1, C_2 and R_2) and initial conditions of state variables from the available measurements.

This problem is classically referred to as partial system identification [87, 143], *i.e.* where the model structure is already known and only parameters and initial conditions are to be determined. The input to the system is known and therefore only the physical process (*i.e.* the open-loop system) will be considered.

Further assume that both the tanks were initially empty (*i.e.* initial conditions of states are known). Because we know the structure of the system, the objective is to find only the parameters of a known model such that the output of the model is sufficiently close to that of the actual process.

A Simulink® block diagram model of the two tank system ('Tank2.mdl') is shown in Figure 10.9 in which parameters are passed a vector p. The model is simulated with a fixed step size of 0.02 s for duration of 10 s with both state variables initialized to zero. The pump, as an input source, gives a pulse train of amplitude 10 with pulse width of 25 samples, pulse period of 50 samples and 0.02 s sampling time. The nominal parameter values are chosen as $\mathrm{p} = [98.1, 0.05, 196.2, 0.02]$ where the elements of p respectively correspond to $1/C_1 = g/A_1$ (with $A_1 = 0.1$), $1/R_1$ (with $R_1 = 20$), $1/C_2 = g/A_2$ (with $A_2 = 0.05$) and $1/R_2$ (with $R_2 = 50$), in consistent units. The simulated output of the model (Figure 10.10) with these parameter values, $y_{ref}(t)$, is considered as the output from the actual process. From here onwards, we will presume that the parameters values used in the model to get these outputs are unknown to us.

10.3.1 Output Error Minimization

Let us now consider some guess parameter values in vector p and run the model. Let its output be $y(t)$, which will be different from $y_{ref}(t)$. We can write a least-square-error minimization program to modify the entries in vector p such that $y(t) \rightarrow y_{ref}(t)$ with sufficient accuracy. This program listing in MATLAB® is given below in two

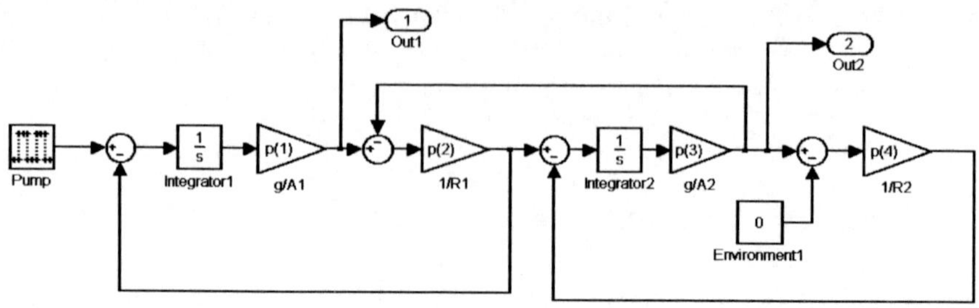

Fig. 10.9 Simulink® model of the linear two-tank system in file 'Tank2.mdl'

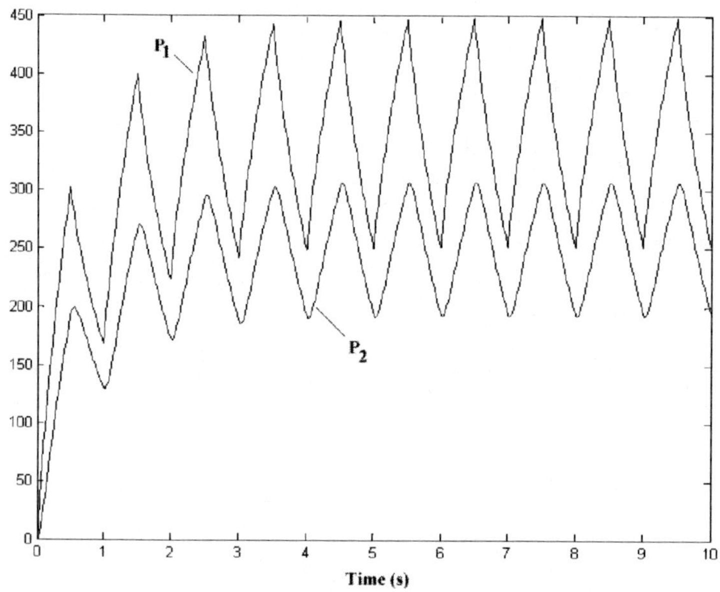

Fig. 10.10 Response of the linear two-tank system

parts. The first part defines the objective function to be optimized as follows in a file named 'Tank2fun.m':

```
function f=Tank2fun(b)
global p;
p=b;
myopts = simset('Solver','ode4','FixedStep',0.02);
[t,x,y] = sim('Tank2', 10, myopts);
global yref;
n=size(y); m = n(1);
z=0;
```

```
for i=1:m
z=z+0.5*(y(i,1)-yref(i,1))^2+0.5*(y(i,2)-yref(i,2))^2;
end;
f=z;
global prec;
prec=[prec;p];
```

The guess parameters are passed to the function as the argument `b`. The parameter vector p is then assigned and then command "`sim('Tank2', 10, myopts);`" executes a simulation of the two-tank system model in file 'Tank2.mdl' with the parameter vector as `p`. Note that variable `p` must be global so that Simulink® can access the variable. The results (`t`, `x` and `y`, for time, states and outputs, respectively) are stored in the returned variables. Note that the simulation options are passed to the solver in the variable `myopts`, which specify that the solver should output data at fixed intervals of 0.02 s. This interval must match the sampling time at which data is acquired from the real system.

The objective function is calculated by using available process data (in global variable `yref`) and the model output. The result, `f`, is returned by the function `Tank2fun`. The last line of the program creates a stack (array) of guess values chosen during optimization procedure in a global variable so that they can be plotted later. This is unnecessary in a practical application because only the final values are important. The main program code is given below:

```
global yref;
global p;
global prec;
prec=[];
%generate process data
A1=0.1; A2=0.05; R1=20;R2=50;
p=[9.81/A1,1/R1,9.81/A2,1/R2];
myopts = simset('Solver', 'ode4', 'FixedStep',0.02);
[t,x,y] = sim('Tank2', 10, myopts);
yref=y;
plot(t,y)
%process data is now available
% run optimization loop with initial guess values
b=[9.81/A1*0.5,1/R1/2,9.81/A2*2,1/R2*3];
options=optimset('MaxIter',1000000,'MaxFunEvals',...
                 1000000,'TolFun',1e-4);
b=fminsearch(@Tank2fun,b,options);
% plot the results
subplot(4,1,1); plot(prec(:,1))
subplot(4,1,2); plot(prec(:,2))
subplot(4,1,3); plot(prec(:,3))
subplot(4,1,4); plot(prec(:,4))
```

In the above program, the remaining text in a line following ‘%’ marks is commented and three or more points at the end of a line indicate continuation. The initial part of the program concerns generation of plant data. Because we do not have the real process at hand, we simulate our model with some parameter values to generate the output `yref`. The objective is to find out those parameter values, which were used to obtain `yref`.

The next part of the program assumes some initial guess parameter values. We have intentionally chosen them to be far away from the actual (presumed unknown) values. The optimization is done by calling the MATLAB® function `fminsearch` which is a simplex optimization process based on Nelder-Mead's algorithm. The arguments of `fminsearch` are a pointer to the function containing the objective function, initial guess values and a set of optimization options.

The estimated parameter values at each iteration step are shown in Figure 10.11, where the results converge to some fixed values. The last part of Figure 10.11 is zoomed and shown in Figure 10.12 which reveals the convergence of the estimated parameter values to those values which were used to generate the process data.

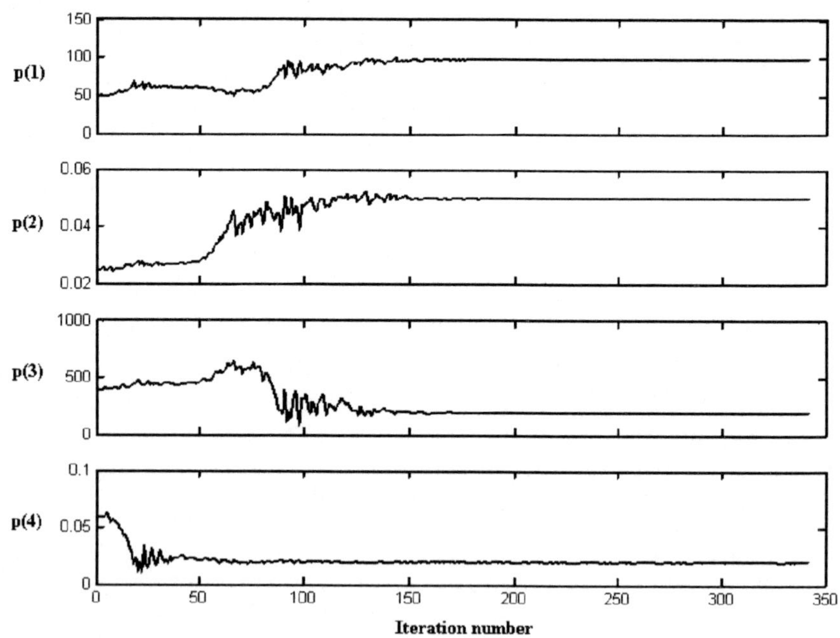

Fig. 10.11 Estimated parameters of the linear two-tank system

The developed method may be used for real-time process supervision, but it has two major problems:

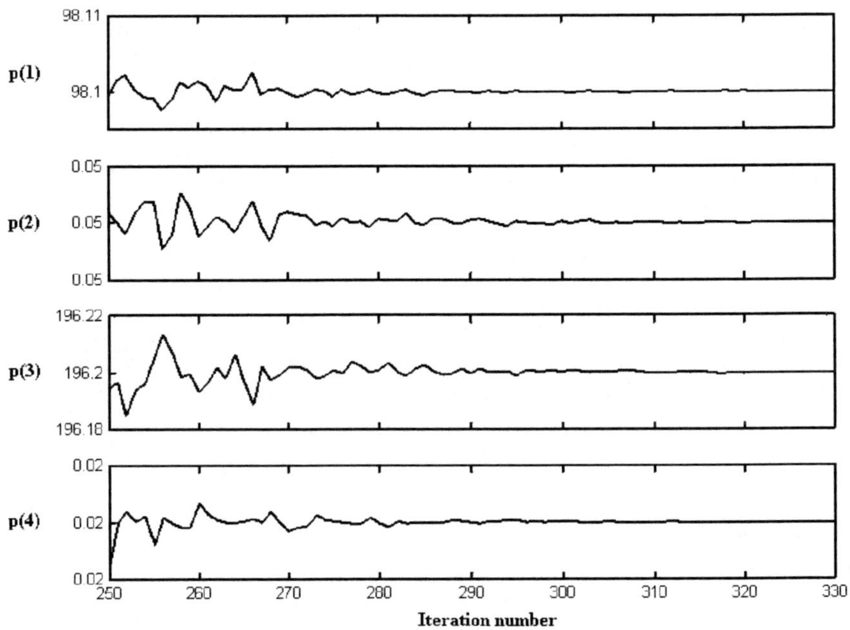

Fig. 10.12 Last few steps of the estimation showing convergence

1. The above listed estimation program was executed in MATLAB® version 6.5 running on an Intel® Pentium-4® 3GHz machine. The optimization loop converged in 11 s. This much time may be very large for real-time supervision.
2. Initial conditions were assumed to be known, which is acceptable when parameters are estimated offline. However, in an online application, we have to consider them as unknowns and then we can specify the initial conditions of two states (`1/s` blocks) in the Simulink® model ('`Tank2.mdl`') as `p(5)` and `p(6)`. The number of unknowns to be estimated then becomes 6. We leave it to the reader to verify that all six unknowns cannot be estimated properly from two known measurements and that the optimization procedure does not converge to fixed values.

10.3.2 Optimization of Least Squares of ARRs

If we assume that fault effects do not cancel each other then, for fault detection, we have to concentrate only on estimating parameter values. Therefore, we not only need to decouple initial conditions from the parameter estimation problem but also need to speed up the estimation process.

Let us now consider optimizing the least-square of residuals instead of the least-square-error in measurements. In this case, we simply have to change the objective function. The main program remains almost the same.

The ARRs for the considered process can be derived and shown to be as follows:

$$ARR_1 : Q_p - C_1\frac{dP_1}{dt} - \frac{P_1 - P_2}{R_1} = 0,$$
$$ARR_2 : \frac{P_1 - P_2}{R_1} - C_2\frac{dP_2}{dt} - \frac{P_2}{R_2} = 0. \tag{10.10}$$

To evaluate these ARRs, the pump output must be known, which indeed is true because sources are considered to be known variables in FDI. Therefore, we modify the block diagram and insert the third output as shown in Figure 10.13.

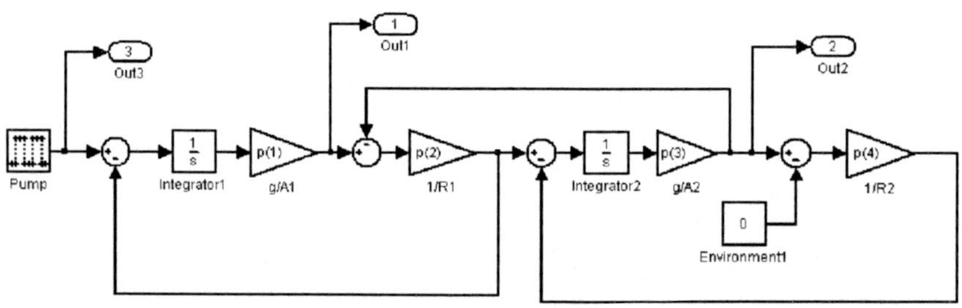

Fig. 10.13 Simulink® block diagram of two-tank system with export of the input variable

Then modified objective function code, with target value of each residual as zero, may be written as follows:

```
function f=Tank2fun(b)
global yref;
z=0;
n=size(yref); m=n(1);
Ts=0.02; % sampling time
for i=2:m
    r1= yref(i,3)-1/b(1)*(yref(i,1)...
       -yref(i-1,1))/Ts-b(2)*(yref(i,1)-yref(i,2));
    r2= b(2)*(yref(i,1)-yref(i,2))-1/b(3)...
       *(yref(i,2)-yref(i-1,2))/Ts-b(4)*yref(i,2);
    z=z+0.5*(r1^2+r2^2);
end
f=z;
global prec;
prec=[prec;b];
```

In the above program, two residuals (variables `r1` and `r2`) are formulated by using available process data and guess parameters. Then the objective function is the sum of the square of residuals over the entire time range.

The main program may as well be modified as follows:

```
global yref;
global p;
global prec;
prec=[];
A1=0.1; A2=0.05; R1=20;R2=50; QT1=1; QT2=0.5;
p=[9.81/A1,1/R1,9.81/A2,1/R2,QT1,QT2];
myopts = simset('Solver', 'ode4', 'FixedStep',0.02);
[t,x,y]= sim('Tank2', 10, myopts);
yref=y;
plot(t,y);
b=[9.81/A1*0.5,1/R1/2,9.81/A2*2,1/R2*3];
options=optimset('MaxIter',1000000,'MaxFunEvals',...
                  1000000,'TolFun',1e-4);
b=fminsearch(@Tank2fun,b,options);
subplot(4,1,1); plot(prec(:,1))
subplot(4,1,2); plot(prec(:,2))
subplot(4,1,3); plot(prec(:,3))
subplot(4,1,4); plot(prec(:,4))
```

Note that the last two parameters in vector p (*i.e.* `QT1` and `QT2`) are passed to the Simulink® model as initial conditions to states (mass of liquid in tank1 and in tank2, respectively). They may be assigned any arbitrary positive values.

The simulation results (process input and outputs) are shown in Figure 10.14 and the estimated parameter values are shown in Figure 10.15. This estimation procedure took less than 1 s to converge. However, the final estimated values (excluding initial conditions which cannot be estimated from optimization of residuals) are $\mathrm{p} = [107, 0.0489, 201, 0.0195]$ which are marginally different from the actual values, *i.e.* $\mathrm{p} = [98.1, 0.05, 196.2, 0.02]$.

We can make three important observations from these results:

1. Optimization of residuals is several times faster than optimization of output error.
2. Initial conditions are decoupled from parameter estimation problem. One can estimate parameters of the system even when the plant response is obtained with any unknown initial conditions. Therefore, this method can be implemented for real-time supervision.
3. The accuracy of parameter estimation is not as good as that obtained from an integral model based approach. However, there will always be some uncertainties in parameter estimation (nobody knows actual parameter values with absolute accuracy) and most often process noise effects will lead to some more domain of uncertainty. These issues have already been discussed in Section 10.2 (see Figure 10.8).

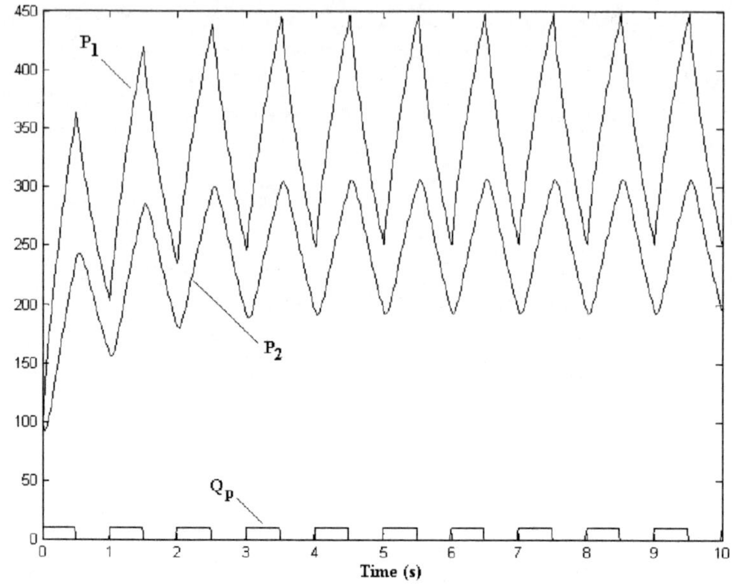

Fig. 10.14 Input and output variables of the two-tank process

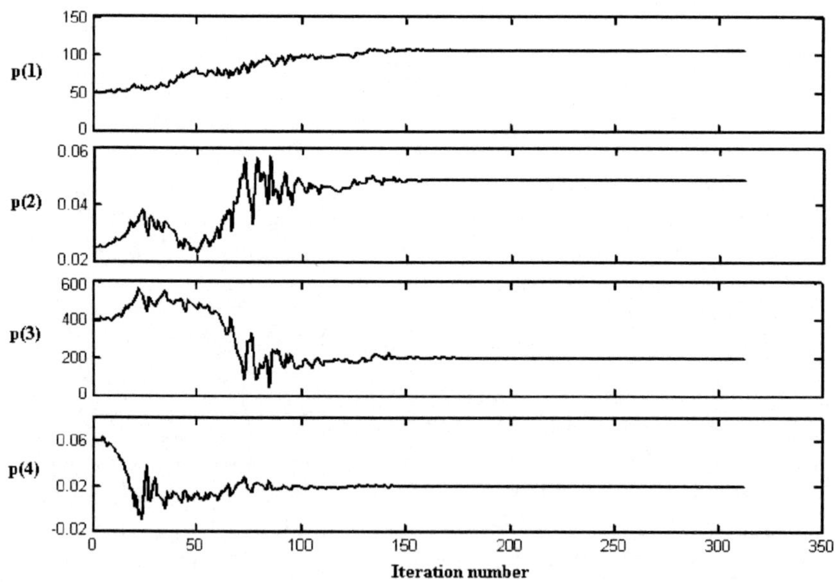

Fig. 10.15 Parameter values estimated by minimizing residuals

10.3.3 Optimization by Using Diagnostic Bond Graph

In the considered problem, ARRs could be derived and thus residual expressions could be written directly in the formulation of the objective function. If residuals have to be computed implicitly, then one has to use a DBG and convert it to a residual generator. Although closed form ARRs could be derived for the considered system, we will show that the same results can be obtained by following the more general approach of implicit residual generation.

The implicit residual generator of the two-tank system in DBG form (Figure 10.3) is derived from Figure 10.2b and then it is converted to Simulink® block diagram form as shown in Figure 10.16.

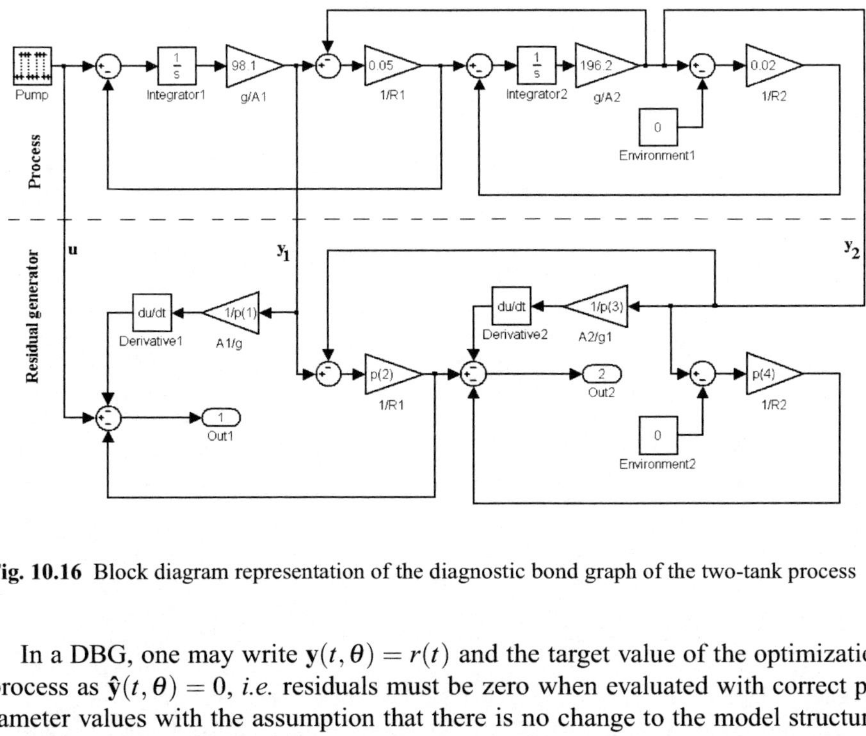

Fig. 10.16 Block diagram representation of the diagnostic bond graph of the two-tank process

In a DBG, one may write $\mathbf{y}(t,\boldsymbol{\theta}) = r(t)$ and the target value of the optimization process as $\hat{\mathbf{y}}(t,\boldsymbol{\theta}) = 0$, *i.e.* residuals must be zero when evaluated with correct parameter values with the assumption that there is no change to the model structure, *i.e.* the constraints are static. In discrete-time, residual evaluation from ARRs uses derivatives calculated by using the data in the previous sample. The obtained derivative will not be a good approximation if the sampling time is large. Therefore, at k-th sampling instance, one may calculate the residual for the just previous step, *i.e.* $(k-1)$-th step. Then the residuals of the two tank system (see Equation 10.10) may be evaluated as

$$r_1(k-1) = Q_p(k-1) - C_1\frac{P_1(k) - P_1(k-1)}{T} - \frac{P_1(k-1) - P_2(k-1)}{R_1},$$

$$r_2(k-1) = \frac{P_1(k-1) - P_2(k-1)}{R_1} - C_2\frac{P_2(k) - P_2(k-1)}{T} - \frac{P_2(k-1)}{R_2} \quad (10.11)$$

where T is the sampling time and k is the sample number. Note that residual calculation is delayed by one step. The discrete-time residual evaluator obtained from the DBG of the two-tank system is shown in Figure 10.17, where blocks having gain $(z-1)/0.02z$ are used to calculate discrete-time derivatives with sampling period of 0.02 s.

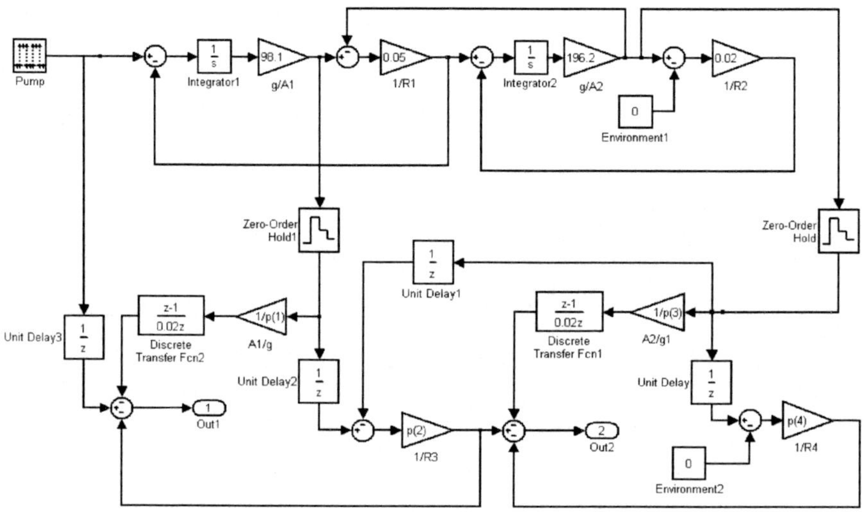

Fig. 10.17 Block diagram representation of the diagnostic bond graph of the two-tank process with residual evaluation in discrete time

The top part in Figure 10.16 is the actual process and the bottom part is the diagnostic (residual generator) part. The objective is to determine parameters in the diagnostic part such that square of the residuals are minimized. Any arbitrary initial conditions may be assigned to the two integration blocks in the model. The output of the model in Figure 10.16 (two residuals) may be used to define the objective function as given in the code below:

```
function f=Tank2fun(b)
myopts = simset('Solver', 'ode4', 'FixedStep',0.02);
global p;
p=b;
[t,x,y]= sim('Tank2', 10, myopts);
ystar=y;
z=0;
```

```
n=size(ystar); m=n(1);
for i=1:m
   z=z+0.5*(ystar(i,1)^2+ystar(i,2)^2);
end
f=z;
global prec;
prec=[prec;p];
```

The main program may be accordingly written as follows:

```
global p;
global prec;
prec=[];
A1=0.1; A2=0.05;
R1=20;R2=50;
b=[9.81/A1*0.5,1/R1/2,9.81/A2*2,1/R2*3];
options=optimset('MaxIter',1000000,'MaxFunEvals',...
                 1000000,'TolFun',1e-4);
b=fminsearch(@Tank2fun,b,options); %estimates
subplot(4,1,1); plot(prec(:,1))
subplot(4,1,2); plot(prec(:,2))
subplot(4,1,3); plot(prec(:,3))
subplot(4,1,4); plot(prec(:,4))
```

The parameter estimation from this approach with any arbitrary initial conditions to the plant states gives exactly the same results as those obtained earlier. Moreover, the program takes less than 5 s to converge, wherein the majority of the time is spent in solving the behavioral model in the top part of Figure 10.16. In the actual case, there is no need to solve the behavioral model because the actual process data would be directly available. If the time spent to solve the behavioral model, 380 times, is taken out then the actual convergence time of the purely optimization problem is less than 1 s. Thus the DBG representation satisfies all the necessary conditions (*i.e.* speed, solvability and initial condition decoupling) for real-time parameter estimation based FDI.

10.4 Example II: A Refrigerator Subsystem

A thermo-fluid process considered here is composed of four main components: a compressor, a condenser, capillary tubes and the evaporator, as shown in Figure 10.18.

A mathematical model of the process, excluding capillary tube and evaporator unit is derived in [270] with the following assumptions:

1. The refrigerant entering the compressor-shell is single-phase.

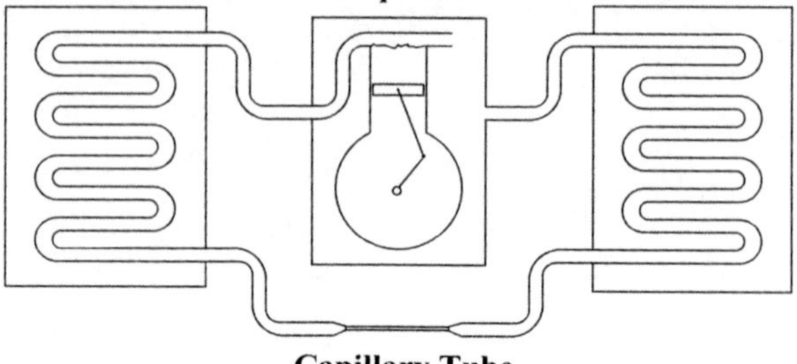

Fig. 10.18 Thermo-fluid process (refrigerator) schematic

2. The motor inefficiencies (*e.g.* friction and electric motor losses) are transformed into heat.
3. The thermal effects of the lubricating oil may be lumped with the compressor-shell.
4. The average mass flow rate is uniform due to constant compressor speed.
5. The condenser heat transfer process is in superheat phase.

Table 10.2 Parameter values for the refrigerator subsystem model

Variable	Description	Value
$\dot{W}_{in}$	Compressor power	140 W
η_t	Thermal efficiency of compressor	0.75
h_{cm}	Convective heat transfer coefficient of motor	8.5 W/(m^2.K)
h_{cs}^{in}	Convective heat transfer coefficient of inside of shell	4.0 W/(m^2.K)
h_{cs}^{out}	Convective heat transfer coefficient of outside of shell	2.0 W/(m^2.K)
h_{cc}	Convective heat transfer coefficient of condenser	100 W/(m^2.K)
$\dot{m}$	Mass flow rate	0.005 kg/s
M_m	Mass of compressor motor	10 kg
M_f	Mass of fluid (refrigerant)	4.0 kg
M_c	Mass of condenser	4.0 kg
M^*	Equivalent heat capacity of shell and lubricating oil $M^* = M_s.c_s + M_{oil}.c_{oil}$	452 J/K
T_{fi}	Inlet fluid temperature	310 K
a_m	Surface area of compressor motor	0.076 m^2
a_s^{in}	Surface area of inside of compressor-shell	0.097 m^2
a_s^{out}	Surface area of outside of compressor-shell	0.04 m^2
a_c	Surface area of superheat condenser	0.114 m^2
c_m	Specific heat of compressor motor	452 J/(kg.K)
c_{vf}	Specific heat of fluid at constant volume	1814 J/(kg.K)
c_{vc}	Specific heat of compressor metal	452 J/(kg.K)
c_{pf}	Specific heat of fluid at constant pressure	2177 J/(kg.K)
r_m	Radius of compressor motor	0.06349 m
r_s	Radius of compressor-shell	0.07937 m
ε_m	Emissivity of compressor motor	0.64
ε_s	Emissivity of compressor-shell	0.64
σ	Stefan-Boltzmann constant	5.67×10^{-8}W/(m^{2}K^4)
T_∞	Ambient temperature	300 K

The dynamics of these components are derived from the first law of thermodynamics and heat transfer principles applied to the Control Volumes (CV) shown in Figure 10.19. We have considered the system as shown in Figure 10.19 for bond graph modeling, simulation and FDI analysis. The nomenclature and the parameter values used in the model are given in Table 10.2.

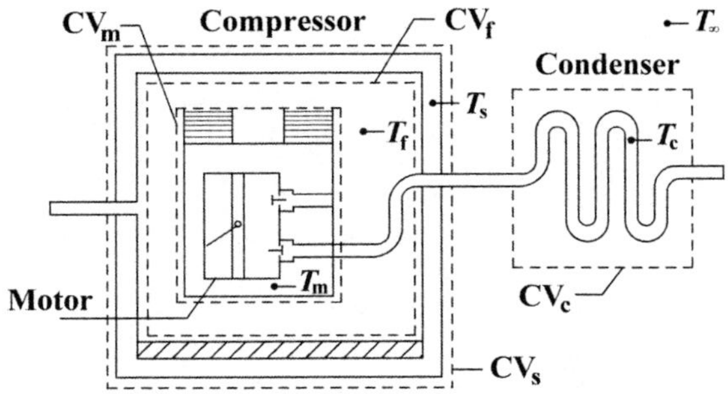

Fig. 10.19 Compressor and superheat condenser schematic

10.4.1 Bond Graph Model and the ARRs

To derive the ARRs, the pseudo-bond graph model of the process is assigned preferred differential causality with bicausality at sensor ports, as shown in Figure 10.20. In the pseudo bond graph representation, temperature and heat flow rate are taken as the generalized effort and the generalized flow variables, respectively, to model free convection and radiation heat transfer, whereas temperature and enthalpy flow are, respectively, taken as the generalized effort and flow variables to model forced convection. It is assumed that the refrigerant flow rate ($\dot{m}$) and its temperature at inlet (T_{fi}) are known, which are shown as sources in the bond graph model.

Four temperature sensors (thermocouples) T_m, T_f, T_s and T_c are installed in the system to measure temperatures of the motor, the refrigerant, the shell and the superheat-condenser, respectively. These temperature sensors are represented by SS elements in the bond graph model. The Coupling Element for Thermo-Fluids (CETF) is used to model enthalpy flow due to forced convection.

The compressor motor releases some energy in the form of thermal radiation which is absorbed by the compressor-shell. According to Stefan-Boltzmann Law, if a hot object at temperature T_h radiates energy to its cooler surroundings at temperature T_c, the net radiation loss rate takes the form

$$\varepsilon.\sigma.a\left(T_h^4 - T_c^4\right), \tag{10.12}$$

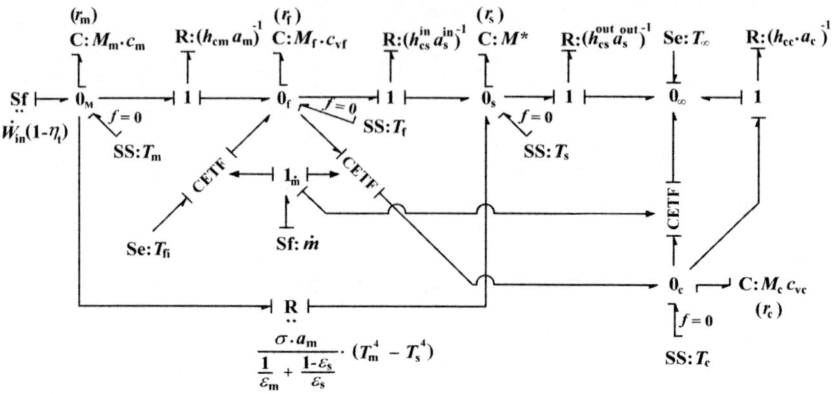

Fig. 10.20 Bond graph model of thermo-fluid process with preferred derivative and preferred inverted detector causality

where ε is the emissivity of the object, σ is the Stefan-Boltzmann constant, and a is the surface area. We can model the heat transfer due to radiation by using a two-port non-linear R-field, shown in Figure 10.20. The non-linear constitutive relation of the R-field (see Figure 10.20) is given as

$$\dot{Q} = \varepsilon.\sigma.a_m \left(T_m^4 - T_s^4\right), \tag{10.13}$$

where

$$\varepsilon = \frac{1}{\frac{1}{\varepsilon_m} + \frac{1-\varepsilon_s}{\varepsilon_s}} \tag{10.14}$$

is the equivalent emissivity of the compressor motor (ε_m) and the shell (ε_s).

Analysis of the causal paths on the differentially causalled model reveals that the system is structurally observable with four sensors. Corresponding to the four sensors, four ARRs are derived as follows:

$$\begin{aligned}
\text{ARR}_m &: \dot{W}_{in}.\left(1-\eta_t\right) - \frac{\sigma.a_m}{\frac{1}{\varepsilon_m} + \frac{1-\varepsilon_s}{\varepsilon_s}\left(\frac{r_m}{r_s}\right)^2}.\left(T_m^4 - T_s^4\right) + \\
&\quad h_{cm}.a_m.\left(T_m - T_f\right) - M_m.c_m.\dot{T}_m = 0, \\
\text{ARR}_f &: \dot{m}.c_{pf}.\left(T_{fi} - T_f\right) + h_{cm}.a_m.\left(T_m - T_f\right) + \\
&\quad h_{cs}^{in}.a_s^{in}.\left(T_s - T_f\right) - M_f.c_{vf}.\dot{T}_f = 0, \\
\text{ARR}_s &: \frac{\sigma.a_m}{\frac{1}{\varepsilon_m} + \frac{1-\varepsilon_s}{\varepsilon_s}\left(\frac{r_m}{r_s}\right)^2}.\left(T_m^4 - T_s^4\right) - h_{cs}^{out}.a_s^{out}.\left(T_s - T_\infty\right) - \\
&\quad h_{cs}^{in}.a_s^{in}.\left(T_s - T_f\right) - M^*.\dot{T}_s^m = 0, \\
\text{ARR}_c &: \dot{m}.c_{pf}.\left(T_f - T_c\right) - M_c.c_{vc}\dot{T}_c - h_{cc}a_c\left(T_c - T_\infty\right) = 0.
\end{aligned} \tag{10.15}$$

Evaluations of ARR$_m$, ARR$_f$, ARR$_s$ and ARR$_c$ result in residuals r_m, r_f, r_s and r_c, respectively. The FSM shown in Table 10.3 is obtained from the ARRs by considering η_t, $\dot{m}$, h_{cs}^{out} and h_{cc} as the parameters of key components. The FSM has a structured part and an unstructured part, which are marked in Table 10.3. In the structured part, faults in η_t and $h_{cs}{}^{out}$ are isolatable under both single and multiple fault hypotheses. In the unstructured part, if only residual r_c is abnormal, then fault in h_{cc} is isolatable under both single and multiple fault hypotheses. However, when both r_c and r_f are abnormal, then $\dot{m}$ is a definite fault candidate, but it is not possible to qualitatively determine whether h_{cc} is faulty or not. Therefore, parameter estimation is needed in this case for fault disambiguation.

In the present application, FDI is not the only issue, but the focus is on parameter estimation to obtain quantitative information and nature of failure: progressive, abrupt or any other kind.

Table 10.3 FSM of the refrigerator subsystem

	Residuals			
Variables	r_m	r_s	r_c	r_f
η_t	1	0	0	0
h_{cs}^{out}	0	1	0	0
h_{cc}	0	0	1	0
$\dot{m}$	0	0	1	1

In the next section we consider simultaneous deviations in all the parameters and continuously estimate parameters values for all fault candidates.

10.4.2 Fault Isolation Through Parameter Estimation

The integrally causalled bond graph model (with 2% Gaussian noise added to sensor outputs) of the compressor-condenser sub-system has been simulated under normal operating mode by using nominal parameter values given in Table 10.2. During simulation we have considered 5% deviation of nominal parameter values to realize parametric uncertainties. This means, simulations are carried out using a data set, which marginally deviates from the parameter values used to evaluate the residuals. It is assumed that the parametric uncertainties are random and uncorrelated.

During normal operation, there are normal process degradations leading to increase in thermocouple temperatures. We have simulated this phenomenon by using linear reduction in parameter values (η_t, $\dot{m}$, h_{cs}^{out} and h_{cc}) at a very low decay rate, progressively with time. Assuming ambient temperature to be constant, this change is evident on a large time frame, as shown in Figure 10.21.

Consequently, the residuals diverge over time, as shown in Figure 10.22, and, therefore, their corresponding upper and lower thresholds should be modified to

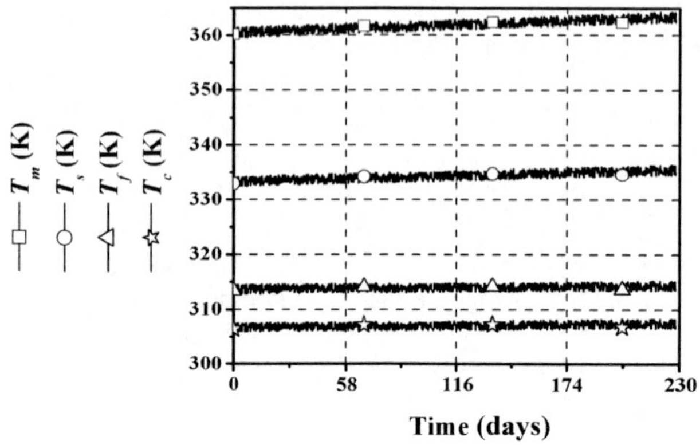

Fig. 10.21 Temperature responses from sensors during no fault

envelop the residuals with the desired confidence limit at all times during normal process operation. A general form for adaptive threshold is given in Chapter 9, which is also applicable for this chapter. The variables, $\lambda_{\eta i}$,$\lambda_{\sigma i}$ *etc.*(see Section 9.2) can be assigned small values to make residuals oversensitive to faults and unknown inputs.

Table 10.4 Adaptive threshold parameters

Threshold parameters/ functions	Residuals			
	r_m	r_s	r_f	r_c
Ψ_t	$1-1.8\times10^{-8}t$	$1-9.4\times10^{-8}t$	$1-1.26\times10^{-7}t$	$1-2.8\times10^{-6}t$
δ_{ni}	-5.418 W	-0.15 W	-2.0 W	-0.163 W
λ_{ni}	0.0	0.0	0.0	0.0
z	2.17	2.17	2.17	2.17
ϕ_t	1.0	1.0	1.0	1.0
$\delta_{\sigma i}$	0.05 W	0.05 W	2. 4 W	2. 4 W
$\lambda_{\sigma i}$	0.0	0.0	0.0	0.0

The values of threshold parameters considered for this example are given in Table 10.4. In practice, experimental data available in the form of a process history database are used to determine the fixed threshold parameters or time varying threshold functions. Residuals are normalized by using the thresholds given in Table 10.4 and the normalization function given in Equation 10.9 so that during normal operation they are bounded within ±1.

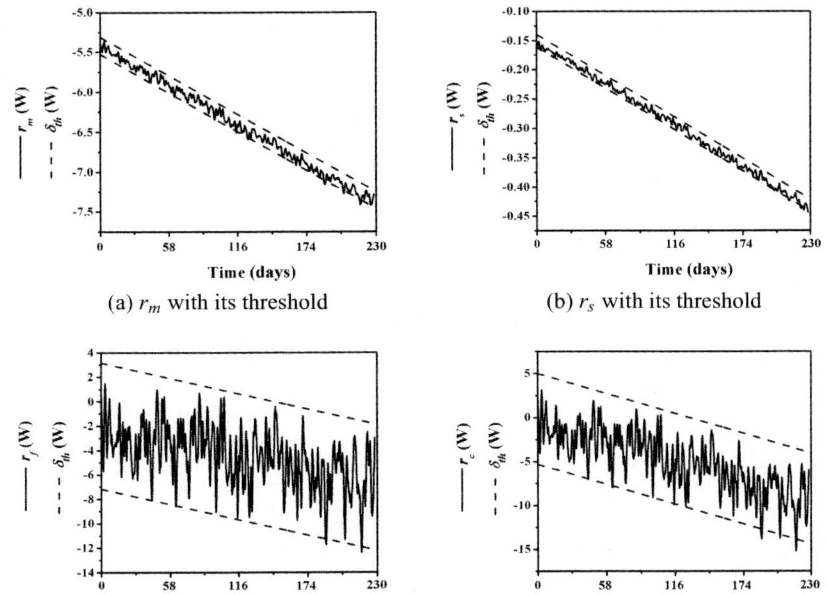

(a) r_m with its threshold (b) r_s with its threshold

(c) r_f with its threshold (d) r_c with its threshold

Fig. 10.22 Selection of adaptive threshold parameters on un-normalized fault-free residuals

Abrupt, progressive and intermittent (hunting) failures are considered in the following fault scenarios:.

- Condenser fan motor failure: condenser fan motor failure results in loss of forced convection, which is simulated by reducing h_{cc} abruptly from nominal 100 W/(m^2.K) to a smaller value of 50 W/(m^2.K) at $t = 2500$ s.
- Outside covering of compressor-shell: the parameter h_{cs}^{out} is considered to drift progressively with time and settle at half of its nominal value. In the simulation, we have introduced the fault at $t = 4000$ s by assuming a function $h_{cs}^{out}(t) = h_{cs}^{out}|_{t=0}\left(1-\frac{1}{2}\left(1-e^{-r(t-t_f)}\right)u_s(t_f)\right)$, where $r = 0.1$ s^{-1}, $t_f = 4000$s and u_s is the unit step or Heaviside function. Note that for the chosen value of decay rate (r), this fault can be considered to be almost abrupt.

- Compressor piston leakage: this fault is also of progressive type and it is simulated by reducing $\dot{m}$ to half of its value starting at t =2500 s with a small decay rate, $r = 5 \times 10^{-4}$ s^{-1}.
- Compressor motor hunting failure: hunting behavior of compressor motor is an indicator of an incipient fault. To simulate this behavior, a square wave function has been used to decrease the value of η_t abruptly to 0.4 and sustain that value for 1200 s, then restore the value of η_t back to its normal value of 0.75 and maintain

it for 1200 s, and thereafter the cycle repeats. This type of hunting in motor is an indication of major failure of the motor in near future. This fault is introduced at $t = 4000$ s.

The sampling time taken in the simulation is 60 s. The variations of temperatures with time, both before and after the faults, are shown in Figure 10.23.

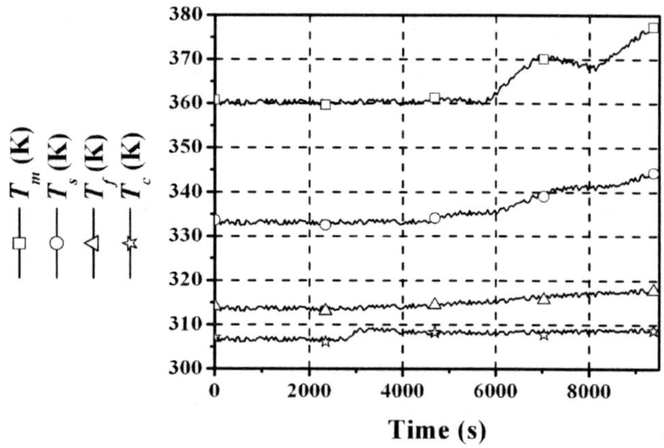

Fig. 10.23 Responses of temperatures from plant model simulation before and after faults

The noisy sensor data are processed before they are used for residual evaluation and parameter estimation. A MATLAB® function, filtfilt, is used for zero-phase digital filtering by processing the input data in both the forward and reverse directions [182]. After filtering in the forward direction, it reverses the filtered sequence and runs it back through the filter. The resulting sequence has precisely zero-phase distortion and doubles the filter order. In addition to the forward-reverse filtering, it attempts to minimize start-up transients by adjusting initial conditions to match the DC component of the signal and by prepending several filter lengths of a flipped, reflected copy of the input signal. Discrete derivatives are obtained numerically from two successive filtered data points. The derivatives of the filtered outputs are again filtered with filtfilt function. We have taken 5 and 10 data point history to filter the sensor signals and derivatives, respectively.

Normalized residuals, shown in Figure 10.24, are processed further for feature extraction in second phase by using cumulative sum algorithm and alarm state is declared on the basis of threshold crossing by the cumulative sum (Figure 10.25). The common threshold for normalized residuals is shown by the shaded area in Figure 10.24. All the residuals and their cumulative sums drift from nominal zero value at different times, upon introduction of different faults.

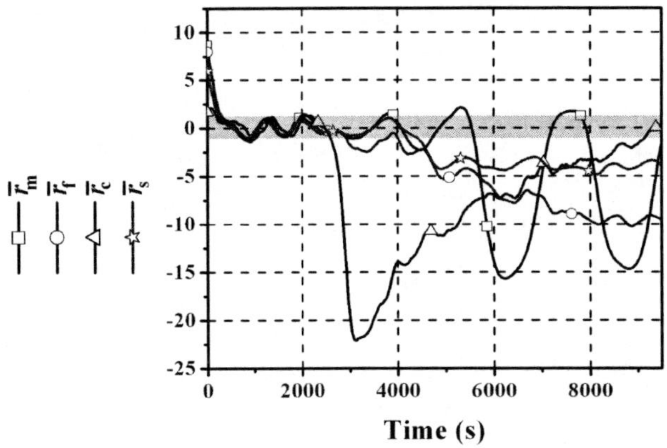

Fig. 10.24 Residual responses both before and after the faults

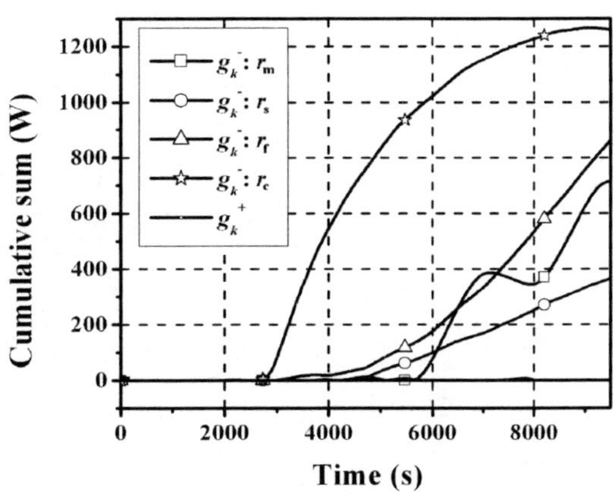

Fig. 10.25 Cumulative sums of residuals

Note that residual r_c drifts at the fault inception, but then starts returning back to the nominal regime (or inside the threshold) as if the fault has been repaired, which is misleading. This is because we have simultaneously simulated condenser fan motor failure (reducing h_{cc}) and compressor piston leakage (reducing $\dot{m}$) both of which appear in the residual r_c. This is the classic case of some fault effects being masked by other fault effects. So to resolve this type of ambiguities resulting from coupled dynamics in a multiple fault scenario, each parameter needs to be estimated. The optimization algorithm has been applied here to estimate the parameters. The estimated parameter values, taking $q = 0$ (see Equation 10.8), are plotted in Figure 10.26 and it is seen that the estimated parameter values approximately track the actual parameter values both before and after the fault. This validates the fault events during this time span.

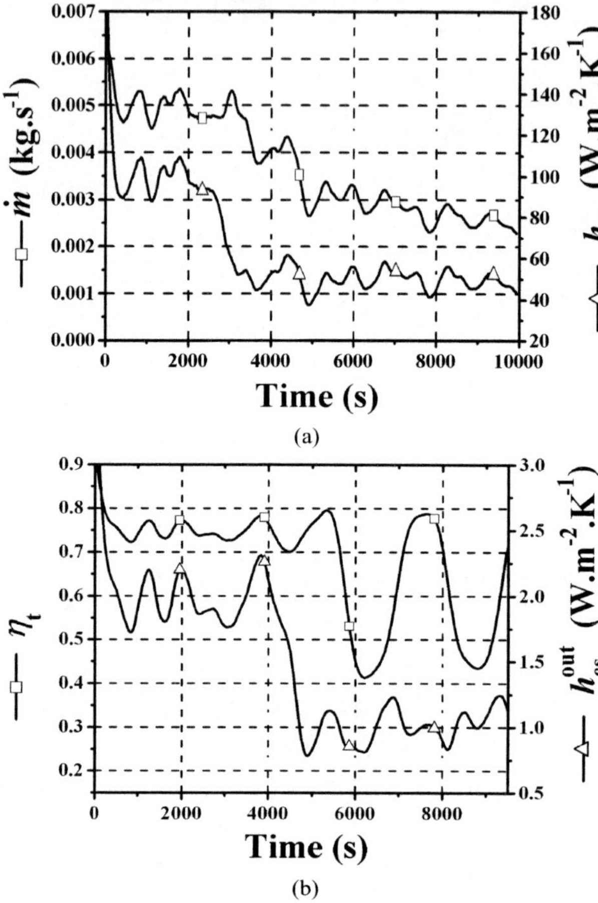

Fig. 10.26 Estimated parameters of the refrigerator subsystem. **a** Estimated $\dot{m}$ and h_{cc}. **b** Estimated η_t and h_{cs}^{out}

10.5 Example III: A Non-linear Two-tank System

10.5.1 The System and Its Bond Graph Model

A controlled two-tank system is shown in Figure 10.27. The model variables and process parameters along with their nomenclatures are given in Table 10.5. The process consists of two tanks T_1 and T_2, which are connected by a pipe with a valve. To satisfy the observability conditions, we assume that the valves V_1 and V_2 are never fully closed. The valve V_3 is used to simulate leakage in tank T_1.

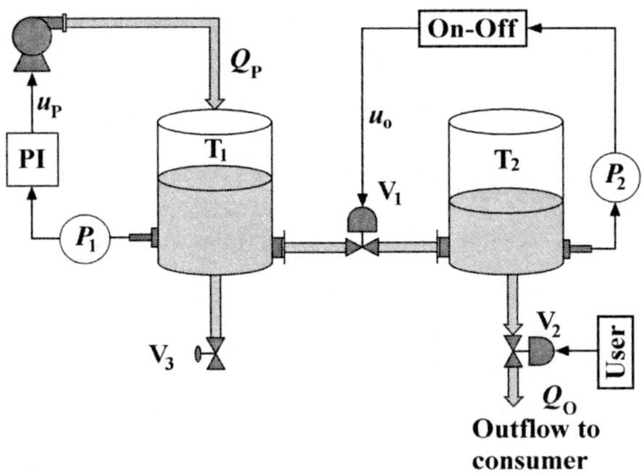

Fig. 10.27 Process and Instrumentation (P&I) diagram of the two-tank system

Table 10.5 Parameters of the two-tank system

Symbol	Description	Value
K_p	Proportional gain of PI controller	1×10^{-4} m.s
K_I	Integral gain of PI controller	1×10^{-6} m
S_{Pt}	Setpoint of the PI controller	5000 N.m^{-2}
P_{max}	Upper limit of On-Off controller	2000 N.m^{-2}
P_{min}	Lower limit of On-Off controller	1000 N.m^{-2}
f_{max}	Maximum outflow from pump	10 kg.s^{-1}
A_i	Cross-sectional area of tanks T$_i$ ($i = 1, 2$)	1.54×10^{-2} m^2
C_{d1}	Discharge coefficient of valve V_1	2.5×10^{-4} kg$^{\frac{1}{2}}$m$^{\frac{1}{2}}$
C_{d2}	Discharge coefficient of valve V_2	2×10^{-4} kg$^{\frac{1}{2}}$m$^{\frac{1}{2}}$
C_{d3}	Discharge coefficient of valve V_3	0.0 kg$^{\frac{1}{2}}$m$^{\frac{1}{2}}$
P_{atm}	Atmospheric pressure (gage)	0.0 N.m^{-2}

The tank T_1 is filled by a pump and the inlet flow (Q_P) to the tank T_1 is controlled by a PI controller, which is actuated by a pressure sensor (P_1) installed at the bottom of the tank T_1. The control command u_P from the PI controller actuates the pump flow in order to maintain the pressure in tank T_1 according to a setpoint S_{Pt}. The pump characteristics and the PI controller function are represented by $\Phi_P(u_P)$ and $\Phi_{PI}(P_1)$, respectively. The PI control law and the pump flow saturation law have appeared earlier in Chapter 5. The water level in tank T_2 is controlled by an On-Off controller (Φ_O), which is actuated by another pressure sensor, P_2, installed at the bottom of the tank T_2. The control command, u_o, from the On-Off controller monitors flow through valve V_1 to keep the pressure at the bottom of tank T_2 in the range P_{min} to P_{max}. The control law is given as

$$u_o = \begin{cases} 0 & \text{if } P_2 > P_{max} \\ 1 & \text{if } P_2 < P_{min} \end{cases} = \Phi_O(P_2)\,. \tag{10.16}$$

The bond graph models of the system in integral causality and bicausality (for derivation of ARRs) are shown in Figures 10.28 and 10.29, respectively.

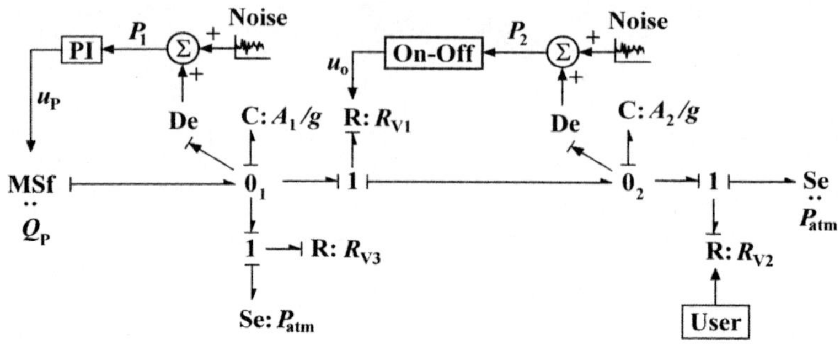

Fig. 10.28 Bond graph model of the two-tank system in integral causality

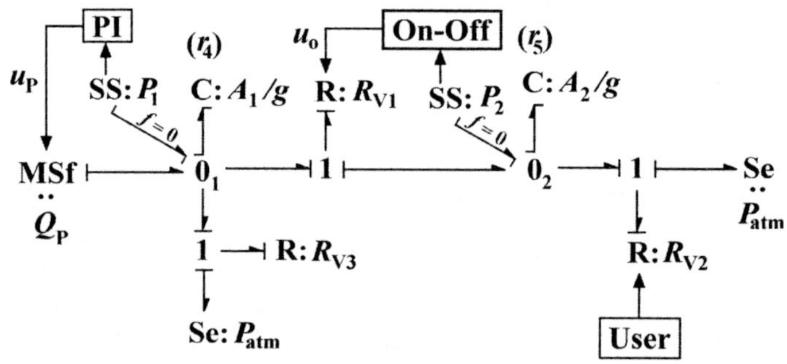

Fig. 10.29 Bicausal bond graph model of the two-tank system

10.5.2 Residual Generation and Fault Detection

There are two sensors, two measured signals from controllers (outputs from PI and On-Off controllers) and an internal source (pump) in the process. Therefore, five ARRs are obtained from the bicausal bond graph model given in Figure 10.29 as

$$
\begin{aligned}
&\mathrm{ARR}_1(\text{Pump}): Q_P - \Phi_P(P_1) = 0,\\
&\mathrm{ARR}_2(\text{PI controller}): u_P - \Phi_{PI}(P_1) = 0,\\
&\mathrm{ARR}_3(\text{On-Off controller}): u_o - \Phi_O(P_2) = 0,\\
&\mathrm{ARR}_4: Q_P - \frac{A_1}{g}.\dot{P}_1 - \left(C_{d1}\sqrt{|P_1 - P_2|}\right).u_o - C_{d3}\sqrt{|P_1|} = 0,\\
&\mathrm{ARR}_5: \left(C_{d1}\sqrt{|P_1 - P_2|}\right).u_o - \frac{A_2}{g}.\dot{P}_2 - C_{d2}\sqrt{|P_2|} = 0,
\end{aligned}
\tag{10.17}
$$

whose evaluations yield five residuals r_i $(i = 1...5)$.

Considering only process faults, *i.e.* based on the assumptions that sensors cannot be faulty, the potential faults are considered only in the controllers, pump and the valves: valve V_1 (C_{d1}), valve V_2 (C_{d2}) and valve V_3 (C_{d3}). The resulting FSM is given in Table 10.6.

Table 10.6 FSM of the two-tank system

		r_1	r_2	r_3	r_4	r_5
Structured	Pump	1	0	0	0	0
	PI	0	1	0	0	0
	OnOff	0	0	1	0	0
Unstructured	C_{d1}	0	0	0	1	1
	C_{d2}	0	0	0	0	1
	C_{d3}	0	0	0	1	0

As explained in the previous chapters, it is not possible to distinguish between leakage through valve V_2 and leakage from tank T_2, because functionally both of them discharge water to the environment and that too at the same pressure difference. Considering all these, C_{d2} represents the total discharge coefficient from tank T_2, whether through the designated valve or through leakage. When the estimated value of C_{d2} is less than nominal, then there is a blockage in valve V_2; whereas when the estimated value of C_{d2} is above nominal, then there is a leakage from tank T_2, either through valve V_2 or at some other location.

From the unstructured part of the FSM (Table 10.6), different fault scenarios corresponding to different coherence vectors (last two items) can be derived as shown in Table 10.7.

Table 10.7 Multiple fault scenarios for the two-tank system

Case	Coherence vector (r_3 r_4)	Fault candidates
A	(0 0)	No fault
B	(1 0)	C_{d3}
C	(0 1)	C_{d2}
D	(1 1)	$C_{d1}/(C_{d1}\&C_{d2})/(C_{d1}\&C_{d3})/(C_{d2}\&C_{d3})/\ (C_{d1},C_{d2}\&C_{d3})$

10.5.3 Fault Isolation Through Parameter Estimation

In case D, to refine the diagnosis further, all the discharge coefficients have to be estimated simultaneously by using the optimization methodology developed earlier.

This may be treated as a second level decision procedure. Here, we have simulated a type D fault signature [0, 0, 0, 1, 1] by simultaneously reducing the value of C_{d1} (blockage in V_1) and increasing the value of C_{d3} (leakage from T_1), while the value of C_{d2} remains unaltered. Since the number of unknowns exceeds the number of residuals, one additional temporal information ($q = 1$) is used to define the objective function specified for parameter estimation.

The integrally causalled bond graph model of the system, shown in Figure 10.28, is simulated and uncorrelated Gaussian noises amounting to 2% of the simulated sensor outputs are superposed to realize real sensor outputs (P_1 and P_2). These noisy sensor data are filtered out using the MATLAB® function, filtfilt. To filter out the noise from sensor signals and their derivatives, 20 data points are considered in the filtfilt function. The blockage fault in valve V_1 is simulated by changing C_{d1} from its nominal value of $2.5\times10^{-4}\sqrt{\text{kg.m}}$ to $2\times10^{-4}\sqrt{\text{kg.m}}$ and the leakage fault in tank T_1 is introduced by changing the value of C_{d3} from its nominal zero value to $1\times10^{-4}\sqrt{\text{kg.m}}$. Both the faults are simultaneously introduced at 840 s. The time responses of sensor signals (P_1, P_2 and Q_P) from the plant model both before and after the faults are shown in Figure 10.30.

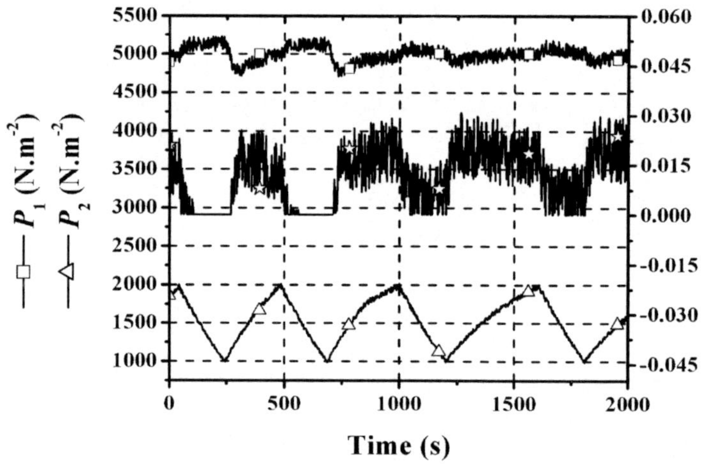

Fig. 10.30 Sensor responses from plant model both before and after the faults

The filtered sensor data are used for the evaluation of the residuals and their cumulative sums. Note that the behavior of the regulator (PI and On-Off controller) in the control-loop tends to hide the fault effects. From observation of the measurement signals only, the faults will be weakly detectable in the transient regime. In contrast, the residual is persistently sensitive to the fault. Therefore, diagnosis based on residuals remains sensitive to faults hidden by the action of the control-loop components. The residual responses (r_4 and r_5) are plotted in Figure 10.31.

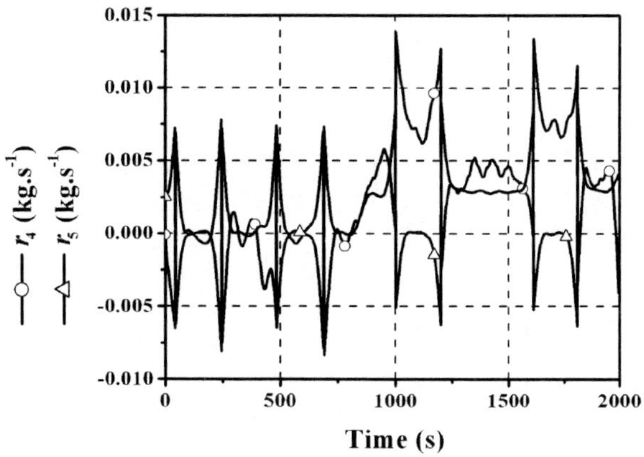

Fig. 10.31 Residual responses both before and after the faults

It is difficult to find thresholds for these residuals, because their values are intermittently too high because of the action of hybrid On-Off controller. Moreover, the residual responses do not improve sufficiently even after backward moving averaging over a 50 sample window, as shown in Figure 10.32. Therefore, decision procedures based on simple thresholds would lead to intermittent false alarms and would not be robust. On the other hand, the two-sided cumulative sum (Figure 10.33) offers a clearer picture about the occurrences of the faults and hence thresholds on cumulative sums are used here for fault detection. Two cumulative sum (CUSUM) algorithms have been discussed in Chapter 6. In this case, the elements of the coherence vector, C, are determined from the two-sided CUSUM.

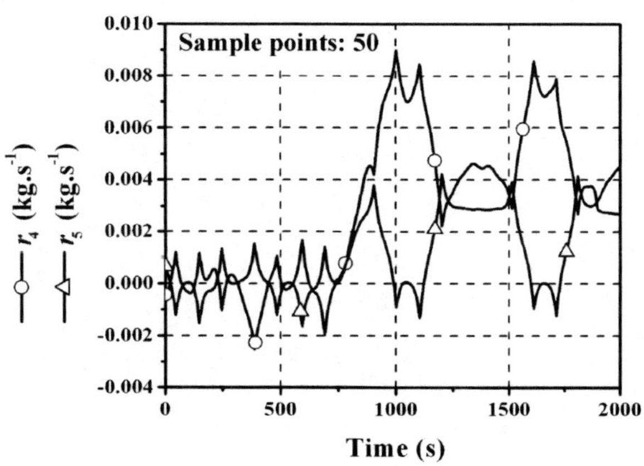

Fig. 10.32 Moving averages of residuals

Note that a complex adaptive threshold (see Section 10.1.1) can be defined to envelop the residuals. If we consider u_o (On-Off controller signal) as one of the inputs then we can use it to modulate a simple adaptive threshold developed in Chapter 9 by separating the upper and lower bounds of the threshold of each residual as shown in Figure 10.34. The threshold defined in Figure 10.34 is sensitive to hybrid transitions, *i.e.* it adds static shifts just before and after the time instants corresponding to $du_0/dt \neq 0$. However, this scheme of adaptive thresholding would require an *a priori* knowledge of when du_0/dt will be non-zero and hence it is difficult to implement. On the other hand, the adaptive threshold developed in Section 10.1.1 is more accurate and easier to implement.

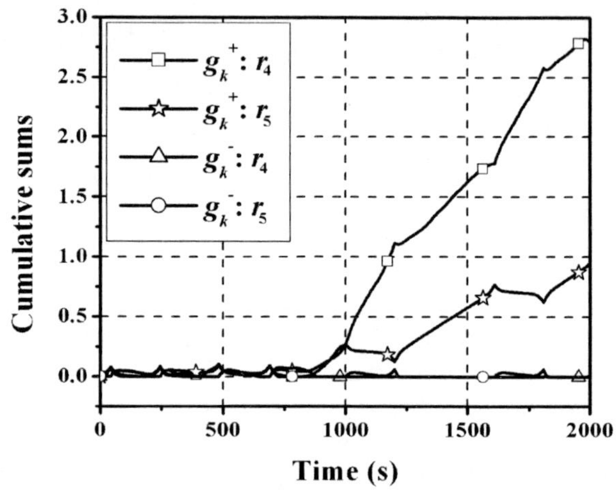

Fig. 10.33 Cumulative sums of residuals

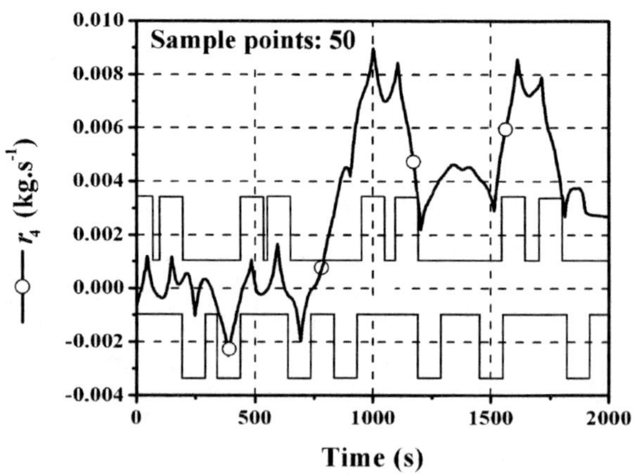

Fig. 10.34 Adaptive threshold for a two-tank system residual

If u_o (On-Off controller signal) is zero then the terms containing C_{d1} in ARR_4 and ARR_5, given in Equation 10.17, vanish as those terms are multiplied with u_o. Therefore C_{d1} cannot be estimated during the Off period of the controller, but C_{d2} and C_{d3} are estimated during that duration by solving out simple algebraic equations obtained after eliminating the terms containing C_{d1} from ARR_4 and ARR_5

(see Equation 10.17). The estimations of the three discharge coefficients are shown in Figure 10.35 and Figure 10.36, which validates the fault scenario simulation. The durations for which valve V_1 remained closed are indicated by the shaded portions in Figures 10.35 and 10.36.

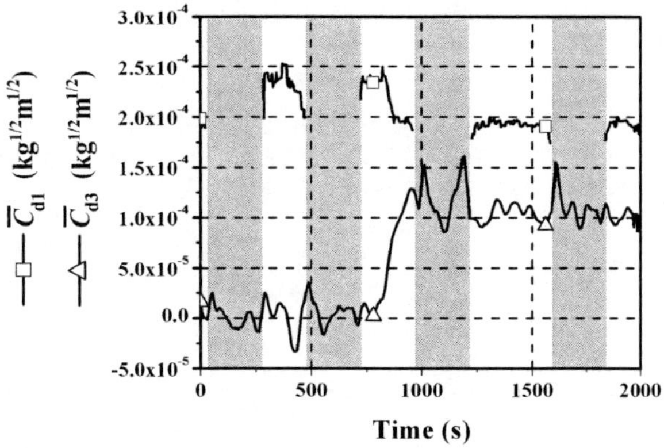

Fig. 10.35 Responses of estimated C_{d1} and C_{d3}

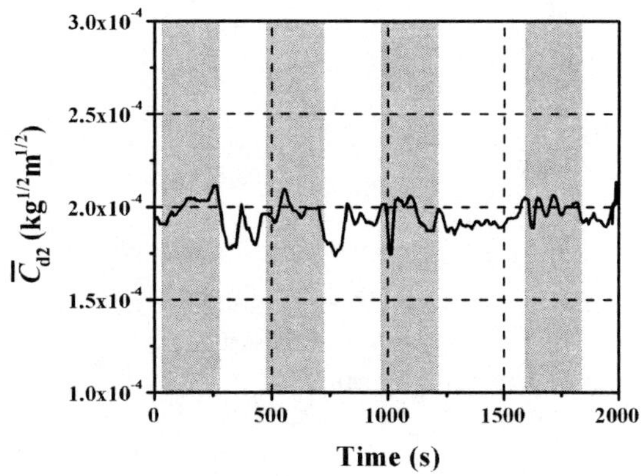

Fig. 10.36 Response of estimated C_{d2}

10.6 Optimization by Using Residual Sensitivity

All optimization problems stated previously in this section were solved through Nelder-Mead's simplex algorithm. Each parameter in the fault subspace, *i.e.* $\theta_i \in S_p$ $(i = 1, ..., m)$ was estimated by using least squares optimization of n residuals with the optimization problem formulated as $\min_{\theta} J(\theta)$,where

$$J(\theta) = \frac{1}{2} \sum_{j=k-q}^{k} \bar{\mathbf{r}}^T(t_j).\mathbf{Q}.\bar{\mathbf{r}}(t_j), \tag{10.18}$$

is the objective function, $\theta \in \mathbb{R}^m$, $\mathbf{Q} \in \mathbb{R}^{n \times n}$ is a positive semi-definite weighting function, $\bar{\mathbf{r}}(t_j) \in \mathbb{R}^n$is the normalized residual vector at j-th sample, k is the present sample number and $q \geq 0$ is the number of past samples considered. All residuals in are normalized as given in Equation 10.9.

A gradient search optimization algorithm offers further improvement in parameter estimation by reducing the number of iterations. This approach uses partial derivatives of the objective function with respect to parameters, called sensitivities. Considering the objective function

$$J(\theta) = \frac{1}{2} \int_0^T (\mathbf{y}(t,\theta) - \hat{\mathbf{y}}(t))^T \mathbf{Q}(t)(\mathbf{y}(t,\theta) - \hat{\mathbf{y}}(t))dt, \tag{10.19}$$

where $\hat{\mathbf{y}}(t)$ is the target and $\mathbf{y}(t,\theta)$ is the output from a simulation model, we have

$$\frac{\partial J(\theta)}{\partial \theta_i} = \int_0^T (\mathbf{y}(t,\theta) - \hat{\mathbf{y}}(t))^T \mathbf{Q}(t) \frac{\partial \mathbf{y}(t,\theta)}{\partial \theta_i} dt. \tag{10.20}$$

The second derivative of $J(\theta)$ can be written by using Gauss-Newton approximation as

$$\frac{\partial^2 J(\theta)}{\partial \theta_i^2} \simeq \int_0^T \left(\frac{\partial \mathbf{y}(t,\theta)}{\partial \theta_i} \right)^T \mathbf{Q}(t) \left(\frac{\partial \mathbf{y}(t,\theta)}{\partial \theta_i} \right) dt, \tag{10.21}$$

where $\dfrac{\partial \mathbf{y}(t,\theta)}{\partial \theta_i}$ is the vector of output sensitivities with respect to parameter θ_i.

Gradients of the objective function defined in Equation 10.18 may be derived in an analogous manner. Because all residuals are normalized, one may consider $\mathbf{Q}$ to be an identity matrix and then

$$\frac{\partial J(\theta)}{\partial \theta_i} = \sum_{j=k-q}^{k} \sum_{i=1}^{n} \bar{r}_i(t_j). \frac{\partial \bar{r}_i(t_j)}{\partial \theta_i}, \tag{10.22}$$

and

$$\frac{\partial^2 J(\theta)}{\partial \theta_i^2} = \sum_{j=k-q}^{k} \sum_{i=1}^{n} \left(\frac{\partial \bar{r}_i(t_j)}{\partial \theta_i} \right)^2, \tag{10.23}$$

where $\dfrac{\partial \bar{r}_i(t_j)}{\partial \theta_i}$ is a residual sensitivity with respect to parameter θ_i. Note that residual sensitivities are not only used for efficient optimization but their numerical values are also used to develop practical fault signature matrix [163], which is used in the actual decision support system. If ARRs are available in closed symbolic form, the sensitivity relations can be derived as

$$\frac{\partial r_i(t_j)}{\partial \theta_i} = Eval\left(\frac{\partial \mathrm{ARR}_i}{\partial \theta_i}\right)\Bigg|_{t=t_j}, \tag{10.24}$$

where $\frac{\partial \mathrm{ARR}_i}{\partial \theta_i}$ is a symbolic relation called residual sensitivity function.

10.6.1 Gauss-Newton Optimization

The optimization based on Gauss-Newton approach follows the following algorithm:

1. Compute $J(\theta)$, $\frac{\partial J(\theta)}{\partial \theta}$ (a vector) and $\frac{\partial^2 J(\theta)}{\partial \theta^2}$ (another vector) at the given guess values θ.
2. Find a step change $\Delta\theta$ such that $\frac{\partial^2 J(\theta)}{\partial \theta^2}\Delta\theta = \frac{\partial J(\theta)}{\partial \theta}$.
3. Update θ in the direction of steepest gradient, *i.e.* $\theta := \theta - \Delta\theta$.
4. Evaluate $J(\theta)$. If the new value of objective function is lower than its previous value then proceed to step (1); else use a bisection technique by choosing $\Delta\theta := \Delta\theta/2$ and repeat steps (3) and (4).
5. The algorithm terminates when the step size or the value of the objective function is less than some predefined threshold.

Step (2) is normally solved through a singular value decomposition-based pseudo-inverse. Note that if the different terms in the ARRs contain only a single variable, then $\frac{\partial^2 J(\theta)}{\partial \theta_i \partial \theta_j} = 0 \forall i \neq j$. Then step (2) of the algorithm may be written as the solution of $\mathbf{H}\Delta\theta = \frac{\partial J(\theta)}{\partial \theta}$, where $\mathbf{H}$ is the Hessian matrix given by

$$\mathbf{H} = \begin{bmatrix} \frac{\partial^2 J(\theta)}{\partial \theta_1^2} & 0 & \cdots & \\ 0 & \frac{\partial^2 J(\theta)}{\partial \theta_2^2} & & \vdots \\ \vdots & & \ddots & \\ & & & \frac{\partial^2 J(\theta)}{\partial \theta_m^2} \end{bmatrix}. \tag{10.25}$$

10.6.2 Example

We consider the same system used in Section 10.3 as a parameter estimation problem. We consider that there is no fault in the process. The objective is to use the entire data set and estimate the parameters of the system. To compute better approximation of derivatives, we use the residual evaluation in the form used in Equation 10.11. The measured process outputs are assumed to be independent variables.

From Equation 10.11, the residual sensitivities with respect to the parameter vector, *i.e.* $\theta = [1/C_1\ 1/R_1\ 1/C_2\ 1/R_2]$, are

$$\begin{aligned}
\frac{\partial r_1(k-1)}{\partial \theta_1} &= \frac{P_1(k)-P_1(k-1)}{T\theta_1^2}, \frac{\partial^2 J(\theta)}{\partial \theta_1^2} \\
\frac{\partial r_1(k-1)}{\partial \theta_2} &= -\left(P_1(k-1)-P_2(k-1)\right), \\
\frac{\partial r_1(k-1)}{\partial \theta_3} &= 0, \\
\frac{\partial r_1(k-1)}{\partial \theta_4} &= 0, \\
\frac{\partial r_2(k-1)}{\partial \theta_1} &= 0, \\
\frac{\partial r_2(k-1)}{\partial \theta_2} &= P_1(k-1)-P_2(k-1), \\
\frac{\partial r_2(k-1)}{\partial \theta_3} &= \frac{P_2(k)-P_2(k-1)}{T\theta_3^2}, \\
\frac{\partial r_2(k-1)}{\partial \theta_4} &= -P_2(k-1).
\end{aligned} \tag{10.26}$$

Then the sensitivities of the objective function with respect to parameters are given as

$$\begin{aligned}
l_1 &= \frac{\partial J}{\partial \theta_1} = \sum_{k=2}^{N} r_1(k-1)\frac{\partial r_1(k-1)}{\partial \theta_1}, \\
l_2 &= \frac{\partial J}{\partial \theta_2} = \sum_{k=2}^{N} r_1(k-1)\frac{\partial r_1(k-1)}{\partial \theta_2} + r_2(k-1)\frac{\partial r_2(k-1)}{\partial \theta_2}, \\
l_3 &= \frac{\partial J}{\partial \theta_3} = \sum_{k=2}^{N} r_2(k-1)\frac{\partial r_2(k-1)}{\partial \theta_3}, \\
l_4 &= \frac{\partial J}{\partial \theta_4} = \sum_{k=2}^{N} r_2(k-1)\frac{\partial r_2(k-1)}{\partial \theta_4},
\end{aligned} \tag{10.27}$$

where N is the total number of samples.

The second derivatives of the objective function are approximated as

$$h_1 = \frac{\partial^2 J}{\partial \theta_1^2} = \sum_{k=2}^{N} \left(\frac{\partial r_1(k-1)}{\partial \theta_1} \right)^2,$$
$$h_2 = \frac{\partial^2 J}{\partial \theta_2^2} = \sum_{k=2}^{N} \left(\frac{\partial r_1(k-1)}{\partial \theta_2} \right)^2 + \left(\frac{\partial r_2(k-1)}{\partial \theta_2} \right)^2,$$
$$h_3 = \frac{\partial^2 J}{\partial \theta_3^2} = \sum_{k=2}^{N} \left(\frac{\partial r_2(k-1)}{\partial \theta_3} \right)^2,$$
$$h_4 = \frac{\partial^2 J}{\partial \theta_4^2} = \sum_{k=2}^{N} \left(\frac{\partial r_2(k-1)}{\partial \theta_4} \right)^2, \quad (10.28)$$

which are the terms in the main diagonal of the Hessian matrix.

The objective function is then coded in MATLAB® as follows:

```
function [f,l,h]=GaussNewtonObj(b)
global yref;
z=0;
n=size(yref); m=n(1);
Ts=0.02; % sampling time
l(1)=0; l(2)=0; l(3)=0; l(4)=0;
h(1)=0; h(2)=0; h(3)=0; h(4)=0;
for i=2:m
  r1=yref(i-1,3)-1/b(1)*(yref(i,1)-yref(i-1,1))/Ts...
     -b(2)*(yref(i-1,1)-yref(i-1,2));
  r2=b(2)*(yref(i-1,1)-yref(i-1,2))-...
     1/b(3)*(yref(i,2)-yref(i-1,2))/Ts-b(4)*yref(i-1,2);
  z=z+0.5*(r1^2+r2^2);
  l(1)=l(1)+r1*(yref(i,1)-yref(i-1,1))/(Ts*b(1)^2);
  l(2)=l(2)-r1*(yref(i-1,1)-yref(i-1,2))...
                +r2*(yref(i-1,1)-yref(i-1,2));
  l(3)=l(3)+r2*(yref(i,2)-yref(i-1,2))/(Ts*b(3)^2);
  l(4)=l(4)-r2*yref(i-1,2);
  h(1)=h(1)+((yref(i,1)-yref(i-1,1))/(Ts*b(1)^2))^2;
  h(2)=h(2)+(-(yref(i-1,1)-yref(i-1,2)))^2+...
        ((yref(i-1,1)-yref(i-1,2)))^2;
  h(3)=h(3)+((yref(i,2)-yref(i-1,2))/(Ts*b(3)^2))^2;
  h(4)=h(4)+(-yref(i-1,2))^2;
end
f=z;
global prec;
prec=[prec;b];
```

The main optimization loop can be coded as given below:

```
global yref;
global p;
global prec;
prec=[];
A1=0.1; A2=0.05;R1=20;R2=50; QT1=1; QT2=0.5;
p=[9.81/A1,1/R1,9.81/A2,1/R2,QT1,QT2];
myopts = simset('Solver', 'ode4', 'FixedStep',0.02);
[t,x,y] = sim('Tank2', 10, myopts);
yref=y;plot(t,y);
b=[9.81/A1*0.5,1/R1/2,9.81/A2*2,1/R2*3];
step=[0,0,0,0];fac=0.5;
[f,l,h]=GaussNewtonObj(b);
H=diag(h,0);
step=(pinv(H)*l')';
bs=b;
while (fac>1e-4 & f>1e-4)
    bs=bs-step*fac;
    [fs,l,h]=GaussNewtonObj(bs);
    if (fs>=f)% bisection
        bs=bs+step*fac;fac=fac/2;
    else
       f=fs;H=diag(h,0);step=(pinv(H)*l')'; fac=0.5;
   end
end
subplot(4,1,1); plot(prec(:,1))
subplot(4,1,2); plot(prec(:,2))
subplot(4,1,3); plot(prec(:,3))
subplot(4,1,4); plot(prec(:,4))
```

In the code written above, pinv function is used to avoid singularity problems by taking pseudo-inverse of the Hessian matrix. The results shown in Figure 10.37 reveal convergence to the same values as in Figure 10.15, but in a smaller number of iterations. If the tolerances are further relaxed, then the Gauss-Newton optimization yields an almost accurate result within 20 iterations.

10.7 Sensitivity Bond Graphs

In the last section we used residual sensitivities to implement Gauss-Newton optimization. The implementation was straightforward due to availability of ARRs in symbolic form. However, the sensitivity relations were derived manually.

A bond graph based approach, called sensitivity bond graphs, was developed in [84, 88] to estimate parameters of the system by using a recursive least squares algorithm. Following standard gradient search principles, partial derivatives of the

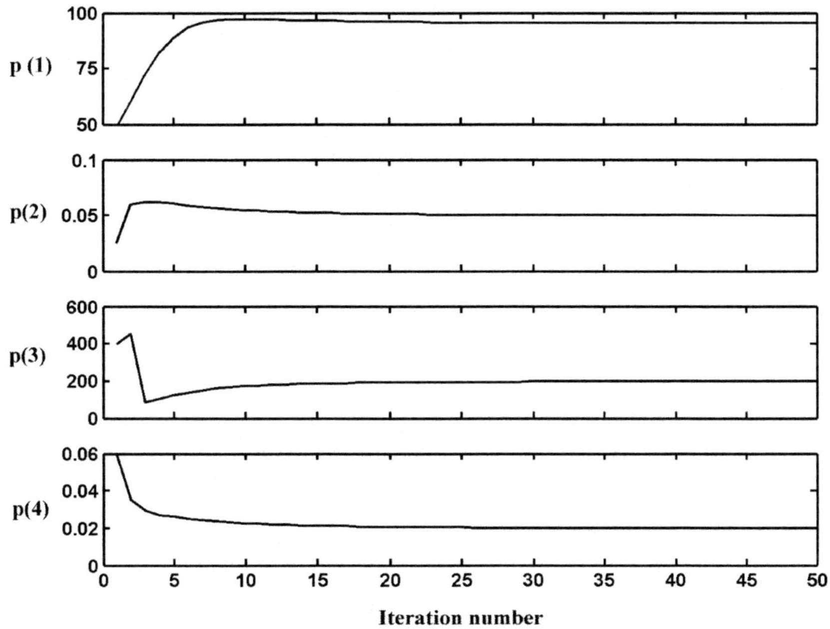

Fig. 10.37 Variation of parameter values during the omptimization loop

cost function with respect to parameters were used to derive a set of equations, which were then represented in a sensitivity bond graph form. The sensitivity matrix [75, 259] is then used to improve the performance of estimation procedure, *i.e.* for quicker convergence. It was shown that sensitivity bond graphs can be obtained by direct manipulation of the behavioral bond graph model of the process.

Sensitivity bond graphs offer a direct way to obtain gradient information in symbolic form such that sensitivity functions can be easily evaluated. Note that incremental bond graph representation developed in [25] offers another solution to generation of sensitivity functions.

Usual sensitivity bond graph gives sensitivities of model's outputs to parameters in the model. Because we require sensitivities of residuals to parameters, we have to use a DBG. Moreover, in many cases, ARRs cannot be symbolically derived and a DBG can be used to obtain numerical values of sensitivities [222].

10.7.1 Diagnostic Sensitivity Bond Graphs

Sensitivity bond graph theory has been presented for general non-linear functional form of constitutive relations in [84] and integrally causalled linear and nonlinear C-elements are treated in the examples. In a DBG, storage elements are differentially causalled and the corresponding constitutive relations of sensitivity storage-elements are written accordingly. Consider a differentially causalled linear C-element connected to i-th bond, which has power variables e_i and f_i related by the following constitutive relation:

$$f_i - C\dot{e}_i = 0, \tag{10.29}$$

where C is a parameter and superposed dot denotes time derivative. Partial derivative of Equation 10.29 with respect to a parameter θ gives

$$\frac{\partial f_i}{\partial \theta} - C\frac{\partial \dot{e}_i}{\partial \theta} - \frac{\partial C}{\partial \theta}\dot{e}_i = 0. \tag{10.30}$$

Assuming that all parameters are uncorrelated,

$$\begin{aligned} \frac{\partial f_i}{\partial \theta} - C\frac{\partial \dot{e}_i}{\partial \theta} - \dot{e}_i = \frac{\partial f_i}{\partial \theta} - C\frac{\partial \dot{e}_i}{\partial \theta} - \frac{1}{C}f_i = 0, \quad \text{for } \theta = C; \\ \frac{\partial f_i}{\partial \theta} - C\frac{\partial \dot{e}_i}{\partial \theta} = 0, \forall \theta \neq C. \end{aligned} \tag{10.31}$$

Equation 10.31 defines the constitutive relation for a sensitivity bond graph element called **sC**. Note that the first constitutive relation in Equation 10.31 resembles the constitutive relation in Equation 10.29, if $\frac{\partial e_i}{\partial \theta}$ and $\frac{\partial f_i}{\partial \theta}$ are taken as the power variables and $\frac{1}{C}f_i$ is taken as a modulating term. This modulating term appears only in the sensitivity model corresponding to parameter C.

Consider several sensitivity bond graph models corresponding to sensitivities with respect to different parameters. In the sensitivity model corresponding to parameter C, the first constitutive relation in Equation 10.31 is used whereas in other sensitivity layers, the second constitutive relation in Equation 10.31 is used. A part bond graph containing a C-element and its corresponding sensitivity bond graph representations in scalar and vector forms are shown in Figure 10.38, in which bonds with a circle over them are vector bonds. The power variables in vector bonds are $[e, \partial e/\partial \theta_1, \partial e/\partial \theta_2, ..]^T$ and $[f, \partial f/\partial \theta_1, \partial f/\partial \theta_2, ..]^T$; and the constitutive relation of the vector **sC** element is given by Equations 10.29 and 10.31 taken together.

Similarly, consider a non-linear R-element connected to j-th bond, which represents the dissipation offered by a hydraulic valve. Its constitutive relation may be written as

$$\begin{aligned} f_j - C_d\sqrt{e_j} = 0, \\ e_j > 0, f_j \geq 0, \end{aligned} \tag{10.32}$$

where $C_d \geq 0$ is the discharge coefficient. Partial derivative of Equation10.32 with respect to a parameter θ gives

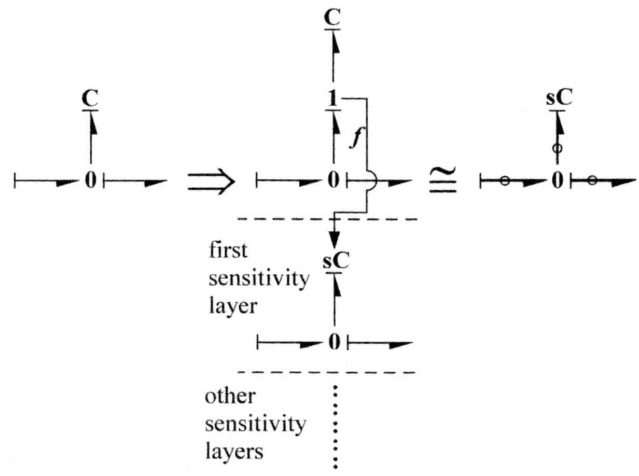

Fig. 10.38 Sensitivity BG of differentially causalled C-element

$$\frac{\partial f_j}{\partial \theta} - \frac{C_d}{2\sqrt{e_j}}\frac{\partial e_j}{\partial \theta} - \sqrt{e_j} = 0, \quad \text{for } \theta = C_d;$$
$$\frac{\partial f_j}{\partial \theta} - \frac{C_d}{2\sqrt{e_j}}\frac{\partial e_j}{\partial \theta} = 0, \forall \theta \neq C_d. \tag{10.33}$$

Equation 10.33 defines the constitutive relations for a non-linear **sR** element corresponding to the hydraulic valve. However, according to Equation 10.33, $\sqrt{e_j}$ appears as modulating signal in all sensitivity models.

For the optimization process, sources are not associated with any variable parameter and hence their sensitivities are assigned zero value. The sources in the context of a DBG mean the inputs and outputs of the actual process.

10.7.2 Example of the Use of Sensitivity Bond Graphs for FDI

We consider the controlled two-tank system in Section 10.5 (Figure 10.27). The bond graph model of the process is already developed in Figure 10.28. We have five residuals (see Equation 10.17) out of which residuals r_4 and r_5 can be used to formulate the objective function given in Equation 10.18 such that process parameters can be estimated. Therefore, we will concentrate on the physical part of the process (*i.e.* exclude all control loops) and formulate the DBG as shown in Figure 10.39.

There are five physical parameters in the model corresponding to two C-elements and three R-elements. Two sensitivity bond graph layers of the DBG in Figure 10.39 are given in Figures 10.40 and 10.41, where the modulating signals f_{T1}, e_{V1}, e_{V2} and

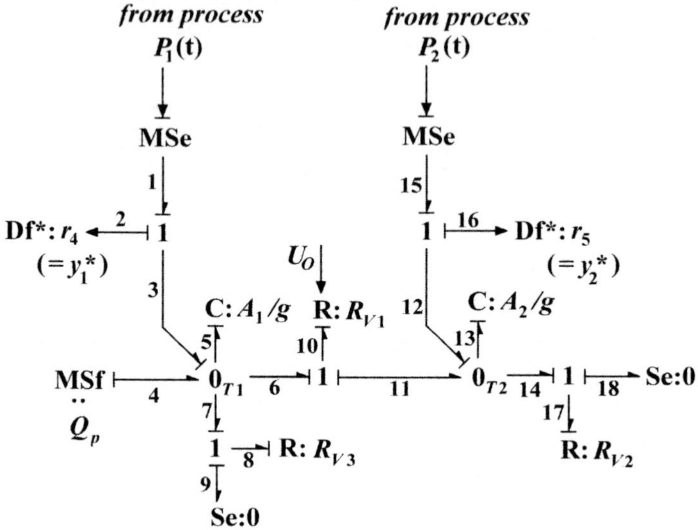

Fig. 10.39 Diagnostic bond graph of the controlled two-tank process

e_{V3} are, respectively, the flow in the bond attached to C: $C_{T1} = A_1/g$, and efforts in the bonds attached to R: R_{V1}, R: R_{V2} and R: R_{V3} (pressure drop across valves).

The sensitivity layers for parameters $C_{T1} = A_1/g$ (Figure 10.40) and $C_{T2} = A_2/g$ are not used in this example because the tank cross-sections are assumed to be constants. Three sensitivity layers for parameters C_{d1}, C_{d2} and C_{d3}, which are identical in structure but use different constitutive relations (Equation 10.33), are used in this example. These layers and the DBG in Figure 10.39 may be merged into a single vector bond graph.

The nominal parameter values of the system under consideration are given in Table 10.5. The process response in Figure 10.30 shows the fault-free and faulty operation of the process, where the fault (as described in Section 10.5) was introduced at $t = 1000$s by changing the of parameters to $C_{d1} = C_{d2} = 2 \times 10^{-4}\sqrt{\text{kg.m}}$ and $C_{d3} = 1 \times 10^{-4}\sqrt{\text{kg.m}}$. During faulty operation of the process (*e.g.* at $t = 1500$s, when $U_O \neq 0$) without considering process noises, the three valve coefficients were estimated by using Nelder-Mead's simplex algorithm over the data in the last four samples and it was found that the optimization procedure converged after 62 iterations (see Figure 10.42a) with arbitrarily selected guess values and a maximum tolerance limit prescribed as $J(\theta) \leq 10^{-16}$. On the other hand, the Gauss-Newton algorithm implemented using the residual sensitivities from the bond graph model converged after merely seven iterations (see Figure 10.42b) with the same tolerance limit and same initial guess values as used in the previous case.

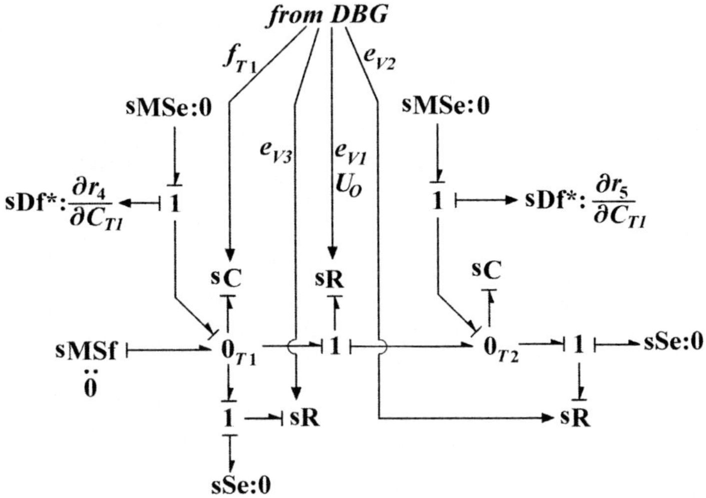

Fig. 10.40 Sensitivity bond graph layer of the controlled two-tank process for parameter C_{T1}

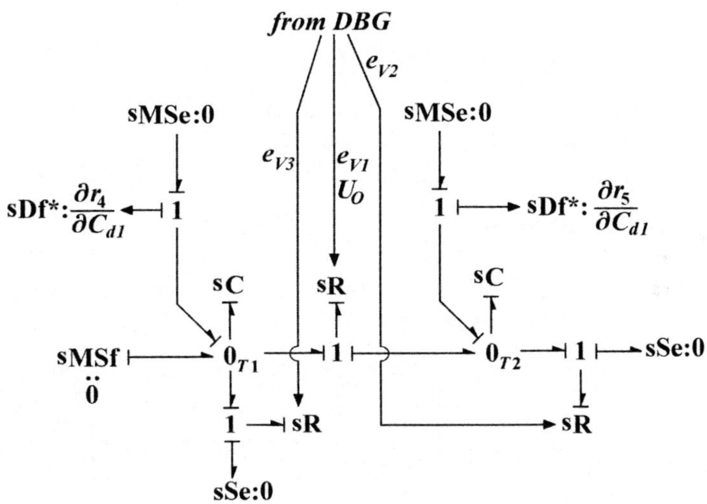

Fig. 10.41 Sensitivity bond graph layer of the controlled two-tank process for parameter C_{d1}

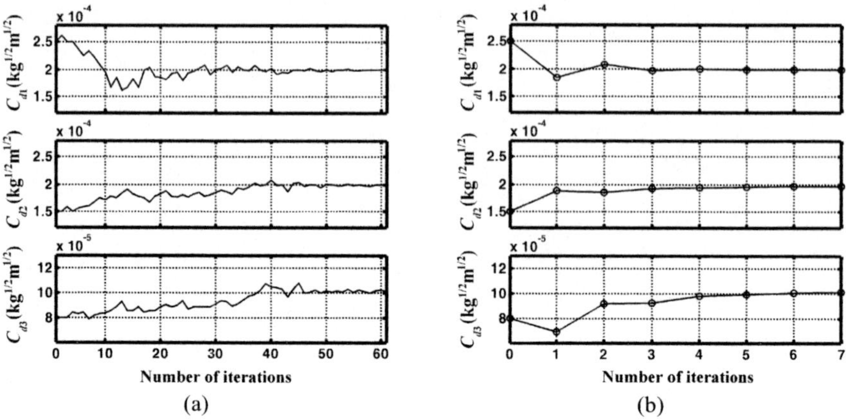

Fig. 10.42 Change in parameter values during optimization using **a** Nelder-Mead's and **b** Gauss-Newton algorithms

In practical problems, one has to consider process noises and then it is not possible to estimate parameters with sufficient accuracy, such as prescribing a strict convergence criterion as $J(\theta) \leq 10^{-16}$.

For FDI applications, rough estimates of parameter values are sufficient and thus the tolerance limit can be significantly relaxed.

When Gauss-Newton optimization procedure was applied at every sample point ($t = 0$ to 2000) in Figure 10.31 (same as the residuals obtained from DBG) to continuously estimate the discharge coefficients of all three valves, the same results as shown in Figures 10.35 and 10.36 could be obtained in very short time.

Remark 10.1. Optimization of least square of residuals provides temporal evolutions of rough estimates of the required parameters, which are useful in preliminary fault isolation and then for implementation of fault accommodation schemes, finer estimates of parameters may be obtained by using more temporal information, *e.g.* by finer sampling.

Problems

10.1. Construct a diagnostic bond graph model of the system shown in Figure 10.43 in LFT form. The LFTs of transformer and gyrator elements should be considered carefully.

Derive the ARRs and separate the uncertain parts which will be necessary to define an adaptive threshold.

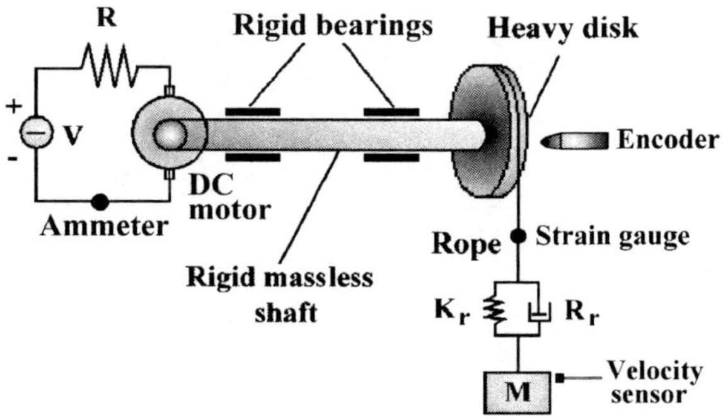

Fig. 10.43 An electro-mechanical system with its instrumentation

10.2. A two-tank system is shown in Figure 10.44. It is assumed that the pump flow and the flow through all the valves are measured. It is further assumed that the process is in unsteady state. The conical tanks are modeled as non-linear C-elements and the valves are modeled as non-linear R-elements. Construct an integrally causalled sensitivity bond graph model of the process such that five parameters, namely cone angle α, two discharge coefficients and initial levels in the two tanks can be estimated. Identify the difficulties faced in constructing a diagnostic sensitivity bond graph model of the system such that all parameters excluding the initial levels in the two tanks can be estimated and suggest ways to get around it.

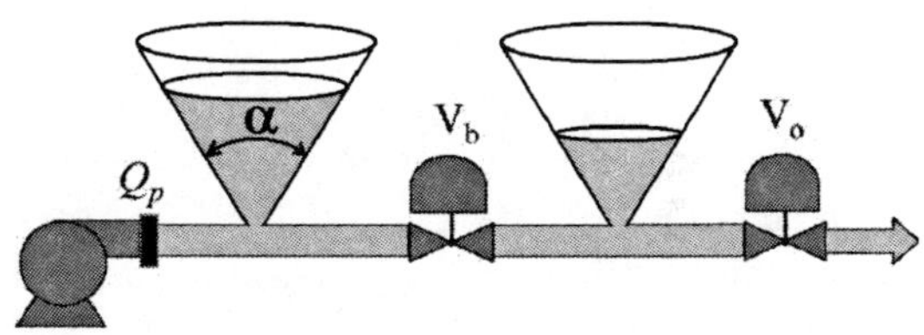

Fig. 10.44 A non-linear two-tank system with conical tanks

Should an integrally causalled model be used for FDI in this case?

10.3. Construct a diagnostic bond graph model of the thermofluid system shown in Figure 10.45 in LFT form. Consider uncertainties in three parameters, namely the tank cross-section, discharge coefficient of the valve and the specific heat of the fluid. Define the adaptive threshold for the ARRs. Discuss the problems encountered in handling the thermal part.

Develop the diagnostic bond graph model of the process. If the aim is to estimate the leakage from the tank, discharge coefficient of the valve and heat loss to the environment, then construct the corresponding diagnostic sensitivity bond graph.

10.4. An LTI system is defined in Equation 10.34. The nominal parameter values are $\theta_1 = 1$, $\theta_2 = 100$, $\theta_3 = 0.2$ and $\theta_4 = 2$. The initial conditions are $x_0 = \left[1\ 0\ 0.2\right]^T$.

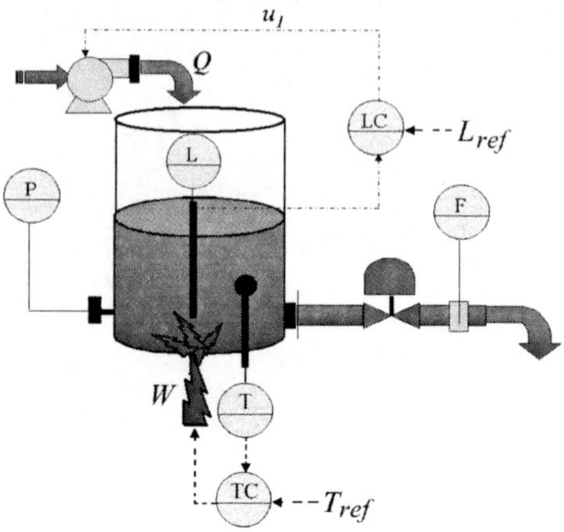

Fig. 10.45 A controlled thermo-fluid system

With unit step input, the data for the first 1 s duration is available in a sampling interval of 0.01 s.

$$\mathbf{A} = \begin{bmatrix} 0 & 0 & \theta_2 \\ 0 & 0 & -\theta_2 \\ \theta_1 & -\theta_4 & \theta_2\theta_3 \end{bmatrix}, \quad \mathbf{B} = \begin{bmatrix} 0 \\ 1 \\ 0 \end{bmatrix},$$

$$\mathbf{C} = \begin{bmatrix} 0 & \theta_1 & 0 \\ 0 & 0 & \theta_2 \end{bmatrix}, \quad \mathbf{D} = 0. \tag{10.34}$$

By assuming the guess parameter vector as $\begin{bmatrix} 0.5 & 120 & 0.5 & 1.8 \end{bmatrix}^T$ and guess initial conditions as zeroes, show that all seven unknowns (parameters and initial conditions) cannot be estimated. On the other hand, show that some parameters can be estimated by setting the target values of the residuals to zero. Show that one extra derivative will be required in one of the residuals. Average of the forward and the backward derivatives may be used to obtain better accuracy.

Chapter 11
Fault Tolerant Control

11.1 Introduction

Fault Tolerant Control (FTC) relates to recovery from fault such that the system is controlled under actual constraints without replacing part(s) of the faulty system. FTC approaches can be classified into two categories: passive approach (*e.g.* robust control) and active approach (*e.g.* adaptive control [112, 118]).

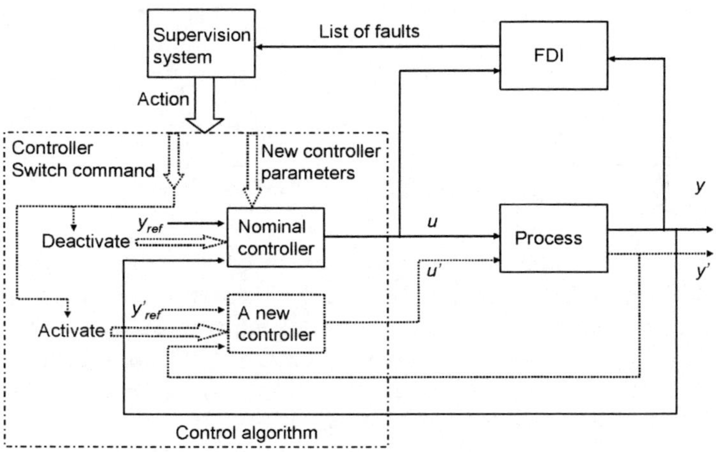

Fig. 11.1 Control reconfiguration for FTC

The general schematic representation of an active fault tolerant control system is shown in Figure 11.1. The FDI layer feeds the supervision system with the information of faulty components which in turn decides the consequent actions to be taken. This action may be to accommodate the fault through change of the controller parameters (the parameters have to be supplied by the supervision system) or to use a new redesigned controller (the controller must be generated on-the-fly, *i.e.* online and the controller algorithm and parameters have to be supplied by the supervision system). Whatever the action, it has to be done in a real-time environment.

Here we have used active FTC, where plant faults are diagnosed through FDI procedures, parameter estimation, feature extraction, *etc.* and subsequently the controller is redesigned (generally a feed-forward strategy) for fault accommodation.

A standard fault tolerant control problem is constituted by a set belonging to a control objective O, a class of control laws U, and a set of constraints C defined by a structure S and parameters Θ. An adaptive fault tolerant control problem can be defined as

$$\text{Solve } \langle O, S, \hat{\theta}, U \rangle, \tag{11.1}$$

where $\hat{\theta} \in \Theta$ is to be estimated as part of the prediction [8].

If a new controller (see Figure 11.1) is so redesigned that it takes the same reference (setpoints) and can produce the same output as the nominal process (see Figure 11.2), then it is referred to as a *model matching approach to control reconfiguration* [16, 268].

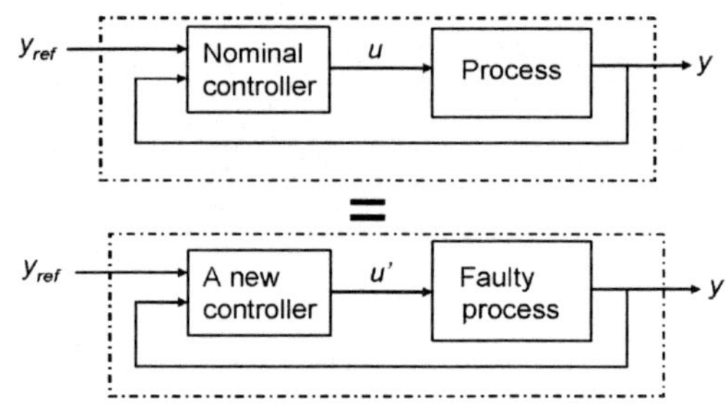

Fig. 11.2 Model matching approach to control reconfiguration

The active FTC handles multi-variable problems, takes care of actuator limitations and allows operation closer to constraints. Constraints are usually specified in terms of actuator ranges, actuator slew rates and output levels. Actuator ranges and slew rates are not only determined from the actuator limitations, but also from the type of power supply and power modulating device. These constraints are verified through system inversion [200], which provides the trajectory of effort and flow variables (consequently the power) needed to produce specified outputs. Actuators are selected (actuator sizing problem) from system inversion corresponding to desired output specifications. The process of finding an input sequence to satisfy a constraint given in the form of an output sequence is called system inversion. However, inverse systems are not physically realizable because the input to the inverse system is constructed in terms of the output and the state of the behavioral model of the system, which leads to the presence of differentiations at the input. So the inverse system models cannot be sufficiently represented by using conventional causality with bond graphs. However, the analytical models required for system inversion, system identification, actuator sizing, and other analysis can be conveniently derived by using the notion of bicausality [81, 83, 178–180].

This chapter is concerned with FTC problem, which can be implemented after the fault is localized and the corresponding faulty parameter is estimated.

11.2 Classical System Inversion Algorithms

The inversion algorithm given by Silverman [235] is related to performing successive differential and algebraic transformations of the output y so as to obtain a relationship which can be solved, if possible, for the input u in terms of the state x, the output y and its derivatives. This definition is applicable to both the linear and non-linear systems. For general linear systems this can be represented by using a state space model as follows [179, 180]:

$$\begin{aligned} \dot{x}(t) &= \mathbf{A}.x(t) + \mathbf{B}.u(t) \\ y(t) &= \mathbf{C}.x(t) + \mathbf{D}.u(t), \end{aligned} \tag{11.2}$$

where $u \in \mathbb{R}^m$ is the input vector, $y \in \mathbb{R}^m$ is the output vector, $x \in \mathbb{R}^n$ represents the state vector and **A**, **B**, **C**, **D** are the matrices of appropriate dimensions.

The inverse system (if it exists) can be written by using generalized state space representation as follows:

$$\begin{aligned} \dot{z}(t) &= \mathbf{F}.z(t) + \mathbf{G}(p).y(t) \\ u(t) &= \mathbf{H}.z(t) + \mathbf{J}(p).y(t), \end{aligned} \tag{11.3}$$

where **F** and **H** are constant matrices, while $\mathbf{G}(p)$ and $\mathbf{J}(p)$ are polynomial matrices in the differential operator $p \hat{=} d/dt$ and z is a generalized state vector of dimension $r, n \geq r$.

An alternative representation of the inverse system in the singular or descriptor form [255] can be written as

$$\begin{aligned} \mathbf{E}.\dot{X}(t) &= \mathbf{A}_s.X(t) + \mathbf{B}_s.y(t) \\ u(t) &= \mathbf{C}_s.X(t) + \mathbf{D}_s.y(t), \end{aligned} \tag{11.4}$$

where **E** is a singular matrix, X is a descriptor vector of dimension r_s ($r_s \geq n$) and $\mathbf{A}_s, \mathbf{B}_s, \mathbf{C}_s$ and $\mathbf{D}_s$ are constant matrices of appropriate dimensions.

In this case, we would specifically try to obtain the form given in Equation 11.3.

11.2.1 Linear Time-Invariant (LTI) System Inversion

The inversion of a general LTI system is shown in Figure 11.3, where $G(s)$ is the system model in frequency domain with $u(s)$ and $y(s)$ as the input and output, respectively; and $\bar{G}(s)$ is the inverse model with $\bar{u}(s) = y(s)$ as the input and $\bar{y}(s)$ as the output. An ideal inverse system should give $\bar{y}(s) = u(s)$.

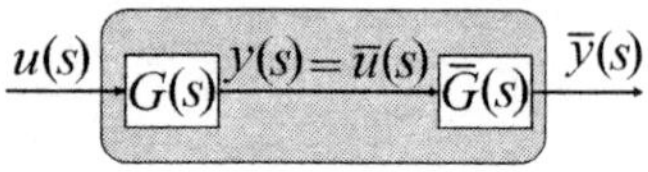

Fig. 11.3 Scheme of a general system inversion

In the single-input-single-output (SISO) case, the forward system is given by

$$G(s) = \frac{N(s)}{D(s)}, \tag{11.5}$$

where $N(s)$ and $D(s)$ are respectively the numerator and the denominator polynomials. The inverse system is then given by

$$\bar{G}(s) = \frac{D(s)}{N(s)}. \tag{11.6}$$

The inverse system is usually unrealizable, because the polynomial order of $D(s)$ is usually greater than that of $N(s)$. This calls for the use of a propering filter before the inverse system, *i.e.* $\bar{u}(s) = f(s).y(s)$, where $f(s)$ is the transfer function of the propering filter. The propering filter usually adds some high-frequency poles to the inverse system; but some sort of trade-off is needed in placement of the propering-poles because placing them too far left in s-plane makes the inverse system dynamics over-sensitive to input variations whereas lowly-damped propering poles add undesired time-constants. Therefore, the relative degrees of a system, described earlier in Chapter 3, play a vital role in selection of the propering filter and its order.

In state-space form, the forward LTI system model in multi-input-multi-output (MIMO) case is given by

$$\begin{aligned} \dot{x} &= \mathbf{A}x(t) + \mathbf{B}u(t), \\ y &= \mathbf{C}x(t) + \mathbf{D}u(t), \end{aligned} \tag{11.7}$$

where $x \in \mathbb{R}^n$, $u \in \mathbb{R}^m$ and $y \in \mathbb{R}^p$ are the state vector, the input vector and the output vector, respectively. Matrices **A**, **B**, **C** and **D** are of appropriate dimensions. The $p \times m$ feed-through matrix **D** is responsible for direct input injection to the output. If the feed-through matrix is regular, *i.e.* the number of inputs is equal to the number of outputs and matrix **D** is invertible, then

$$u(t) = \mathbf{D}^{-1}(y(t) - \mathbf{C}x(t)). \tag{11.8}$$

Note that in the SISO case, where $\mathbf{D}$ is 1×1 and $\mathbf{D} \neq 0$, the resulting transfer function has zero relative degree, *i.e.* polynomials $N(s)$ and $D(s)$ have same polynomial order.

Substituting Equation 11.8 in Equation 11.7

$$\begin{aligned}\dot{x} &= \mathbf{A}x(t) - \mathbf{B}\mathbf{D}^{-1}\mathbf{C}x(t) + \mathbf{B}\mathbf{D}^{-1}y(t), \\ &= (\mathbf{A} - \mathbf{B}\mathbf{D}^{-1}\mathbf{C})x(t) + \mathbf{B}\mathbf{D}^{-1}y(t).\end{aligned} \tag{11.9}$$

Then the inverse system can be represented by

$$\begin{aligned}\dot{\bar{x}} &= \bar{\mathbf{A}}\bar{x}(t) + \bar{\mathbf{B}}\bar{u}(t), \\ \bar{y} &= \bar{\mathbf{C}}\bar{x}(t) + \bar{\mathbf{D}}\bar{u}(t),\end{aligned} \tag{11.10}$$

where

$$\begin{aligned}\bar{\mathbf{A}} &= \mathbf{A} - \mathbf{B}\mathbf{D}^{-1}\mathbf{C}, \\ \bar{\mathbf{B}} &= \mathbf{B}\mathbf{D}^{-1}, \\ \bar{\mathbf{C}} &= -\mathbf{D}^{-1}\mathbf{C} \\ \bar{\mathbf{D}} &= \mathbf{D}^{-1}, \\ \bar{u}(t) &= y(t), \\ \bar{y}(t) &= u(t).\end{aligned}$$

For non-linear systems, inversion procedures are similar but invertibility is not global. Some singularities may appear [108] depending upon the constitutive relationships of elements. However, we assume that the singularity is avoidable by choosing appropriate form of causalities and constitutive relations.

11.2.2 Implicit Inversion of Strictly Proper Systems

Consider an LTI SISO system model in frequency domain in which the order of the polynomial $N(s)$ is m and the order of polynomial $D(s)$ is n. Then the system is called strictly proper, if $n > m$. Consider the block-diagram in Figure 11.4, where the original system to be inverted is placed in a conventional feedback loop.

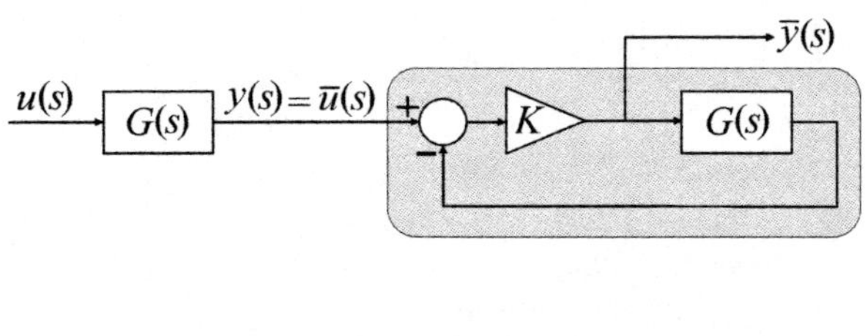

Fig. 11.4 Implicit proper inversion scheme

The transfer function between the input $u^*(t)$ and output $y^*(t)$ (see Figure 11.4) is given by

$$G^*(s) = \frac{y^*(s)}{u^*(s)} = \frac{K}{1+K.G(s)} = \frac{K.D(s)}{D(s)+K.N(s)}. \quad (11.11)$$

Considering large feedback gain K,

$$\lim_{K\to\infty} G^*(s) = \frac{D(s)}{N(s)} = \bar{G}(s). \quad (11.12)$$

For the MIMO case, the feedback should use a matrix gain. From Equation 11.12, it appears that the inverse system has m poles and n zeroes, and thus $\bar{G}(s)$ is improper. Note that in reality (see Equation 11.11), $\bar{G}(s)$ has as many poles as zeroes, but $n-m$ of them approach infinity as $K \to \infty$. This means $\bar{G}(s)$ is just proper and, in fact, the feedback serves as an implicit propering mechanism, which approaches the explicit inversion as $K \to \infty$. However, implicit inversion is easily applicable to large system models and some classes of weakly non-linear systems.

The stationary gain (at $s = 0$ or for zero-frequency steady-state input) of the approximate inverse system with finite feedback gain K is given by $G^*(0) = K.D(0)/(D(0)+K.N(0))$. Thus, a prefilter with gain $G_f = 1+1/K.G(0)$ has to be used before the inverse model. Note that $\lim_{K\to\infty} G_f = 1$, *i.e.* the prefilter is unnecessary for sufficiently high feedback gains. However, high feedback gains introduce large frequencies into the inverse system model, *i.e.* they increase its bandwidth and thus result in stiffer numerical integration.

11.2.3 Examples of System Inversion

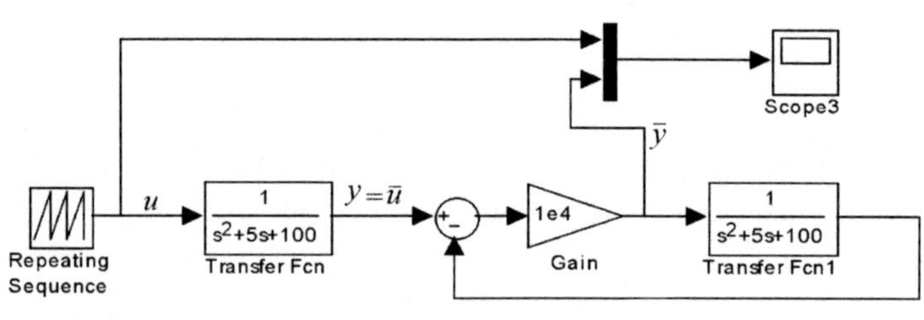

Fig. 11.5 Improper system inversion through high-gain feedback

Consider the inversion of a proper system, $G(s) = 1/(s^2+5s+100)$, through a high-gain feedback, as shown in Figure 11.5. The input is a repeating sequence of 20 s time-period and the output of the inverse system, shown in Figure 11.6, contains transients at sharp input-fluctuation points.

Note that the overshoots appear at points corresponding to discontinuities in successive derivatives of the input and these overshoots increase with increasing feedback gain. Lower feedback gain reduces the overshoots, but the inverse system output during smooth input period deviates from the desired result.

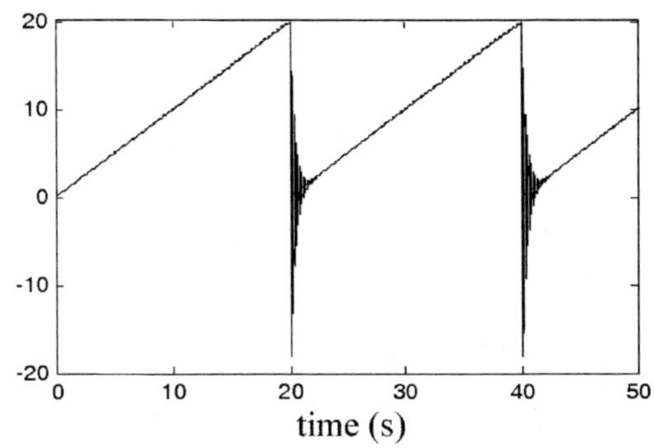

Fig. 11.6 Results of improper system inversion through high-gain feedback

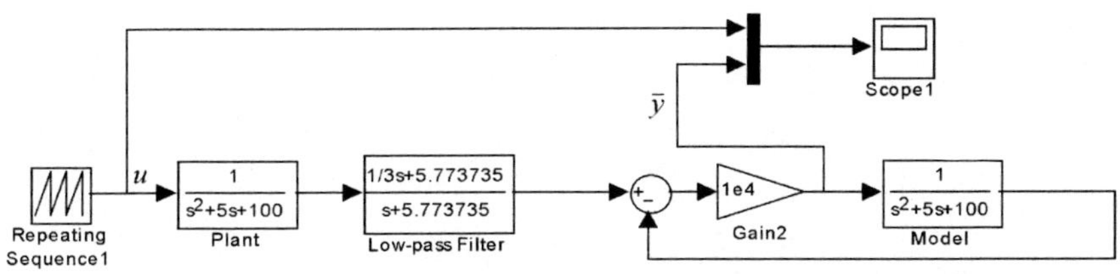

Fig. 11.7 Improper system inversion through high-gain feedback and a low-pass filter

To obtain a proper inverse model, a properly designed low-pass filter can be used as a prefilter. The objective of this prefilter is to sufficiently smoothen the outputs such that the effect of high-frequency poles, introduced due to high-gain feedback, are minimized. The inversion scheme with the prefilter and the output of the inverse model are shown in Figures 11.7 and 11.8, respectively.

11.2.4 Example of Input Reconstruction

Once an inverse model has been designed (*i.e.* the feedback gain and the required prefilter are selected), then it can be used to calculate the input sequence required to produce a given output sequence. Consider a just proper system given by $G(s) = (s^2+s+25)/(s^2+5s+100)$. This system may be realized in state-space form with the following coefficient matrices:

Fig. 11.8 Response of improper system inversion through high-gain feedback and a low-pass filter

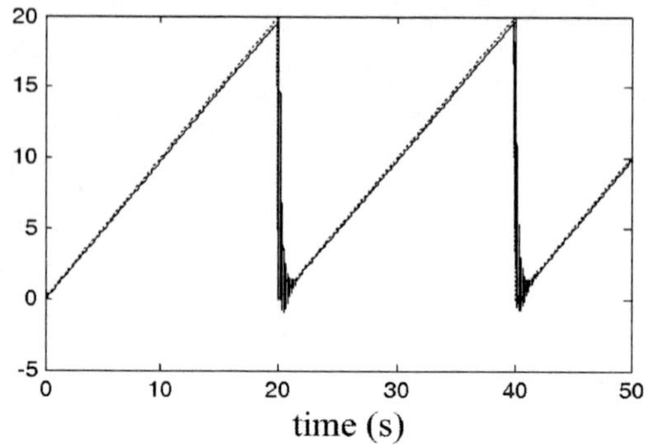

$$\mathbf{A} = \begin{bmatrix} -5 & -12.5 \\ 8 & 0 \end{bmatrix}, \quad \mathbf{B} = \begin{bmatrix} 4 \\ 0 \end{bmatrix}, \quad \mathbf{C} = \begin{bmatrix} -1 & -2.35 \end{bmatrix}, \text{ and } \mathbf{D} = 1.$$

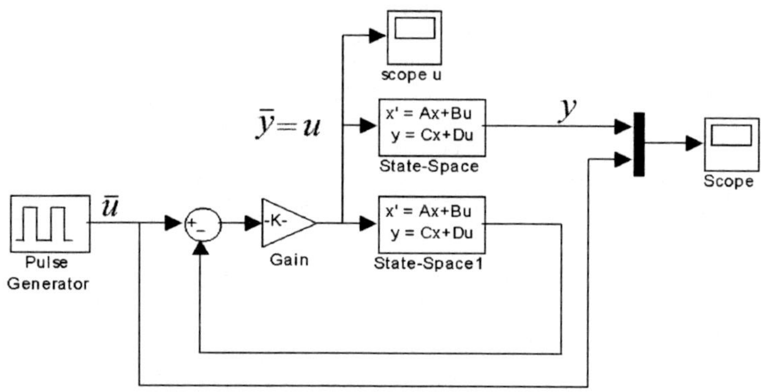

Fig. 11.9 Input reconstruction and testing for a proper system

Let the desired output be a unit pulse sequence with a time-period of 1 s. In the block-diagram shown in Figure 11.9, the desired output sequence is fed as the input to the inverse model and the output from the inverse model is fed to the forward model.

The simulation result showing the designed input, which is the output from the inverse model, is given in Figure 11.10a. The actual response from the forward system, when the designed input is applied to it, is given in Figure 11.10b. The inversion scheme in this case is reasonably perfect, because the process under consideration has zero relative-degree.

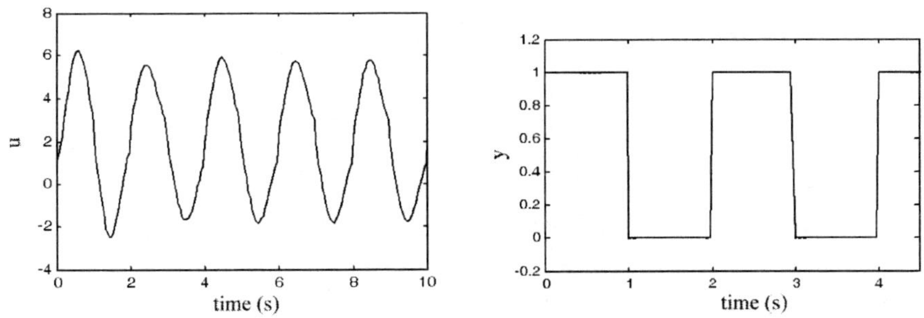

(a) (b)

Fig. 11.10 Results of input reconstruction and the test output

11.2.5 Example of Bond Graph Model Based Implicit System Inversion

Consider a spring-mass-damper system with an exciter on the mass and a velocity sensor, as shown in Figure 11.11a.

The cascaded forward and inverse bond graph models of the system are shown in Figure 11.11b, in which μ is the feedback gain. This cascaded model is represented in SYMBOLS software as shown in Figure 11.12a. The following parameter values were chosen: $m = 1$ kg, $K = 100$ N/m, $R = 1$ Ns/m, and $\mu = 1000$. The input ($F(t)$ or e_4) is taken as a repeating square-wave, which is compared with the output from the inverse model ($\bar{F}(t)$ or e_{11}) in Figure 11.12b. Note that the minor error in output from the inverse model disappears when the feedback gain is increased, *e.g.* $\mu = 10^4$.

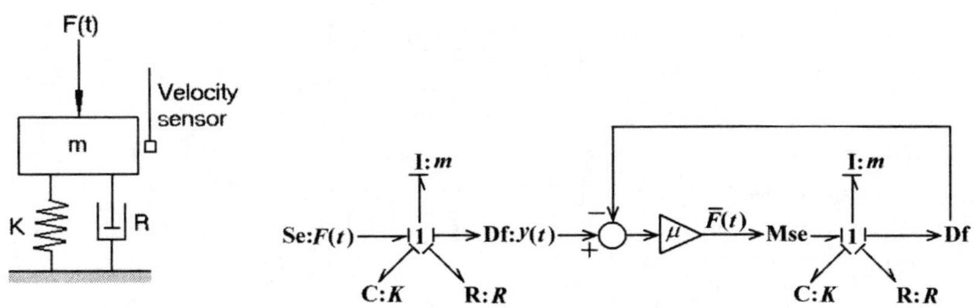

(a) (b)

Fig. 11.11 Bond graph based inversion of a spring-mass-damper system

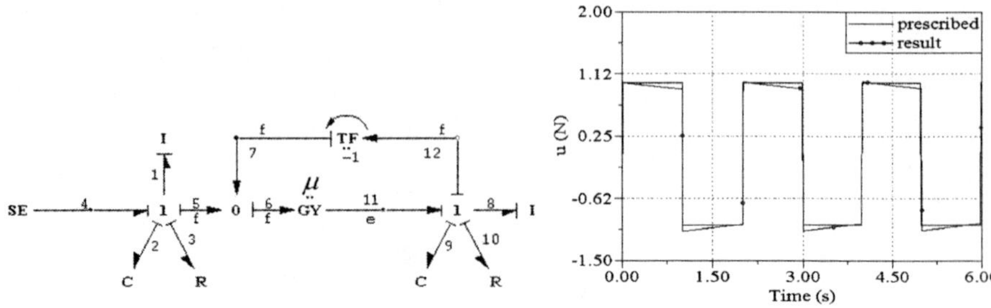

(a) (b)

Fig. 11.12 Inversion implementation in SYMBOLS software and the output

11.2.6 Bond Graph Model Based Explicit System Inversion

A minimal-order inverse model may be obtained by the so-called differential geometric approach. However, the involved transformations sometimes obscure the structural features and thus the physical interpretation of the inverse model dynamics becomes impossible. This motivated the authors of [179, 180] to develop the conditions for bond graph model based explicit system inversion. Those conditions may be stated as follows:

1. An m-input m-output BG model with no set of m disjoint I/O causal paths is not invertible.
2. An m-input m-output BG model with a unique set of m disjoint I/O causal paths is invertible.
3. An m-input m-output BG model with multiple sets of m disjoint I/O causal paths is invertible if the junction structure of the model in bicausality is solvable (*i.e.* no causal loops, non-linear algebraic loops and differential algebraic loops).

Simply stated, a bond graph model is invertible iff it can be bicausalled in such a way that all sources receive both effort and flow information whereas all sensors supply both effort and flow information, while ensuring at the same time that the resulting junction structure is solvable. Obviously, the solvability of the junction structure depends on the number of I/O pairs and their locations.

Note that the conditions in [179, 180] are stated in terms of power lines and shortest causal paths. A power line is an acausal concept which refers to the set of connections (bonds) through the junction structure starting from a source and ending at a sensor. Moreover, the output–input (O/I) propagation of bicausality must be carried out along a minimal-length I/O causal path, *i.e.* the causal path should contain minimum number of storage elements in integral causality. The latter ensures that the inverse model would require minimal order of output derivatives.

Consider the system shown in Figure 11.13, in which u_1 and u_2 are the two inputs and y_1 and y_2 are the two outputs.

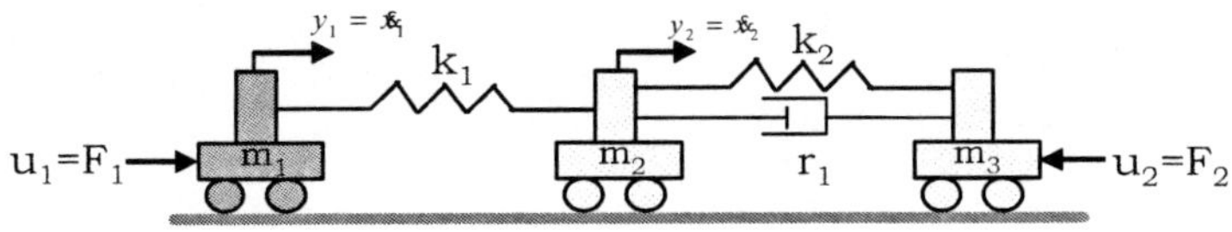

Fig. 11.13 A three DOF system taken from [179]

The integrally causalled bond graph model of the system is shown in Figure 11.14 where the two I/O power lines (bold lines) and two shortest I/O causal paths (dotted lines) are shown.

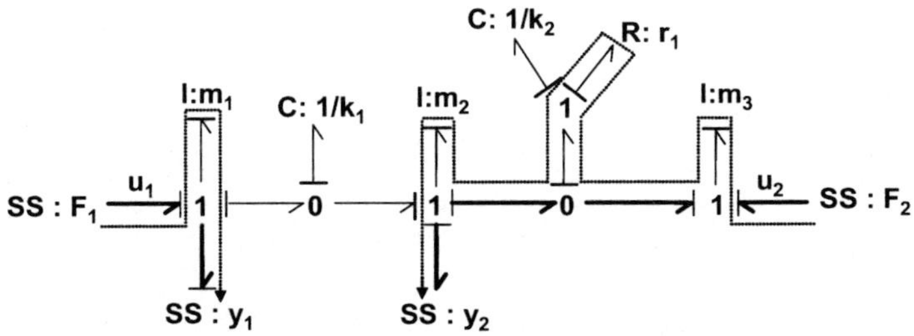

Fig. 11.14 Forward model showing I/O power lines and shortest I/O causal paths

The bond graph model of the system may be assigned bicausality as shown in Figure 11.15 such that sensors have become starting nodes and sources have become terminating nodes. During bicausality assignment, causalities of bonds appearing in the shortest causal path are changed. In Figure 11.15 there are two disjoint I/O causal paths (there is no common junction in causal paths). Moreover, these causal paths are unique, *i.e.* there is no alternative way of bicausalling the system. This means that the system is invertible for the given I/O locations.

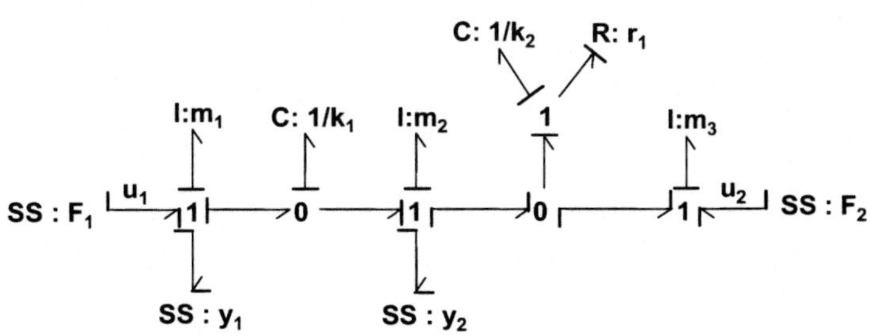

Fig. 11.15 Bicausality assignment with disjoint I/O paths

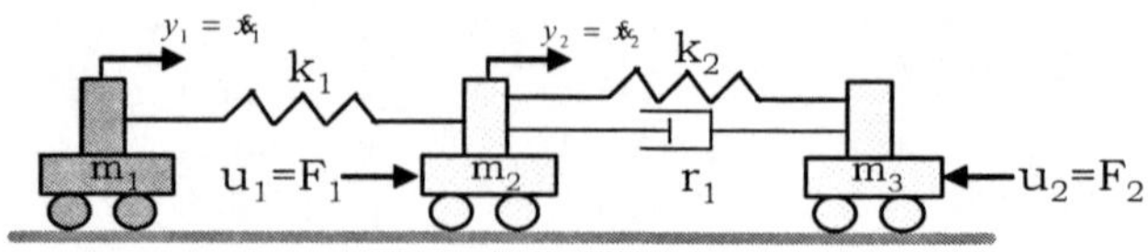

Fig. 11.16 A three DOF system taken from [179] with modified source location

On the other hand, consider the same system with a different location of input u_1, as shown in Figure 11.16. The corresponding bond graph model is shown in Figure 11.17.

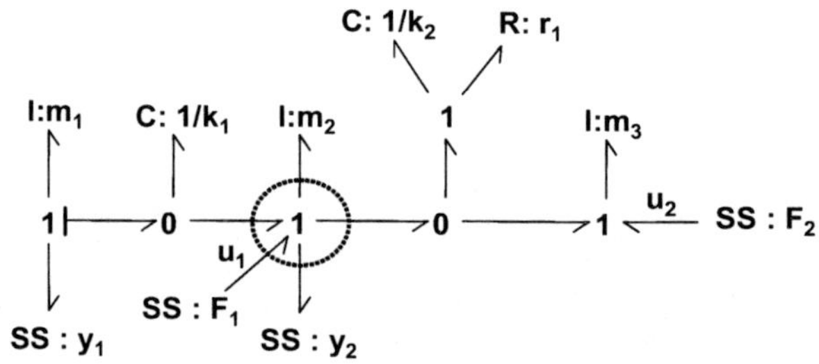

Fig. 11.17 Failure in bicausality assignment due to overlapping I/O paths

In the bond graph model shown in Figure 11.17, it is impossible to propagate bicausality from all sensors to all sources. This would violate the junction structure at the 1-junction shown within the dotted circle in Figure 11.17 because only two bonds can be bicausalled at a junction. Therefore, the system shown in Figure 11.16 is not invertible for the given I/O locations.

11.2.7 Example of Bond Graph Model Based Explicit System Inversion

Consider the simple example system shown Figure 11.18a.

Assume that the valve V_b was controlled by a complex controller to give a varying square-wave flow output between 0.08 kg/s and 0.14 kg/s at intervals of 50 s. Due to malfunctioning of the controller, we have removed it and kept the valve fully open. For fault tolerant control, we need to design an open-loop control by changing the pump flow, so that the desired flow through the valve is achieved.

The output response is prescribed as

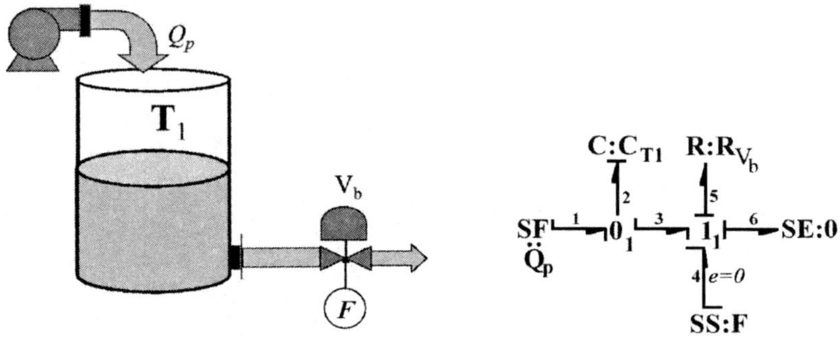

Fig. 11.18 A hydraulic system and its bond graph model

$$F = 0.11 + 0.03\psi(t),$$

$$\text{where } \psi(t) = \begin{cases} 1, \text{ for even values of } int(t/50), \\ 0, \text{ for odd values of } int(t/50)\,; \end{cases}$$

and *int* is the integer truncation operator. From the bi-causalled model for the system inversion (Figure 11.18b), the expression for pump flow is obtained as

$$f_1 = f_2 + f_3 = C_{T1}.\frac{de_2}{dt} + F = C_{T1}.\frac{de_3}{dt} + F = C_{T1}.\frac{de_5}{dt} + F,$$
$$Q_p = f_1 = C_{T1}.\frac{d\left(f_5^2/C_d^2\right)}{dt} + F = F + 2\frac{C_{T1}}{C_d^2}.F.\frac{dF}{dt}.$$

The desired output, F, is discontinuous and leads to very high flow requirements. To satisfy a constraint $Q_p \leq 0.2$ kg/s, we need smooth derivatives of F. For that, we applied a low pass filter with a cut-off frequency $= 2$ Hz, which results in a modified prescribed outflow profile as shown in Figure 11.19a. The synthesized command (flow Q_p) corresponding to the prescribed output is shown in Figure 11.19b. Figure 11.20 shows how closely the actual output tracks the prescribed out-flow.

11.3 Parameter Estimation

Parameter estimation requires that the so-called calculability property is satisfied. Let us assume a constraint relation $f(x_1, \cdots, x_n) = 0$, where $(x_1, \cdots, x_n) = X \in \mathbb{R}^n$ and we like to solve the relation explicitly for x_i in terms of x_j, $j = 1, \cdots, i-1, i+1, \cdots, n$; $j \neq i$. In analytical mathematical analysis the conditions for local solution is provided by the implicit function theorem [5]. The theorem states that for a given

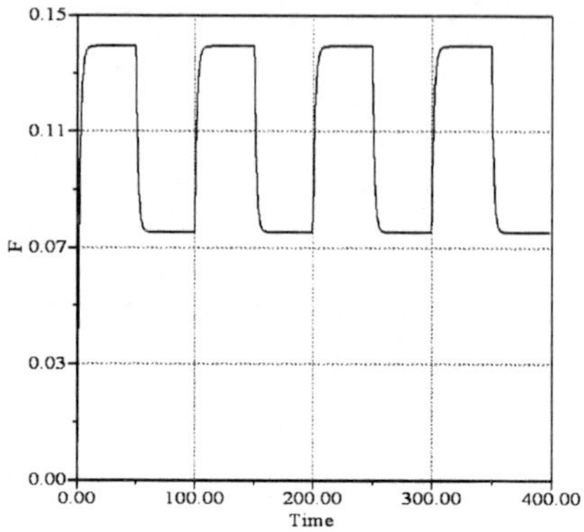

(a) (b)

Fig. 11.19 Prescribed flow profile and the generated command

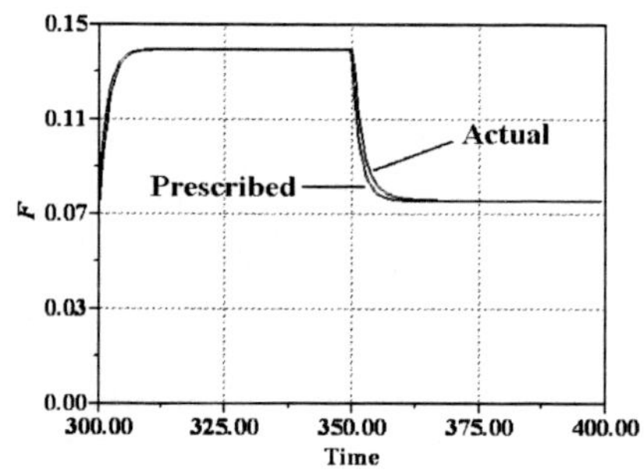

Fig. 11.20 Comparison of prescribed and actual responses

function $f : \mathbb{R}^n \rightarrow \mathbb{R}$ and a local point x_0 (a possible operating point), where $f(x_0) = 0$ and for which $\left.\frac{\partial f}{\partial x_i}\right|_{x_0} \neq 0,\ 1 \leq i \leq n$, then there exists a function g, defined on $\mathbb{R}$, such that $x_i = g(x_1, x_2, \cdots, x_{i-1}, x_{i+1}, \cdots, x_n)$, *i.e.* the variable x_i is calculable.

In the following, we present an algorithm for parameter estimation [223] by assuming that the calculability property is satisfied for the system under consideration:

1. After isolation of fault, the corresponding element must be assigned bicausality such that both effort and flow variables in its bond are known. If it is a field element, then effort and flow variables in all the connected bonds must be known.
2. Then the parameter associated with the element can be estimated by using its constitutive relation, which relates the power variables with the parameter.
3. From that element, bicausality is propagated to sensors. In the bicausality propagation, those storage elements which appear in causal paths linking the selected terminating node and the sensors are to be assigned preferred derivative causality (by bicausaling necessary sensors) to avoid dependence on unknown initial conditions. The derivatives of measurements are calculated by using the stored temporal information.

4. Those elements, whose constitutive relations are multi-valued functions or non-invertible functions, must be assigned appropriate causalities.

Naturally, the number of parameters which can be estimated simultaneously by using this approach (*i.e.* without using temporal information for optimization) is equal to the number of bicausalled bonds at the terminating nodes, with the assumption that each element's constitutive relation does not involve more unknown parameters than the independent equations available from the constitutive relation. For example, only n unknown parameters can be estimated from the constitutive relation of a $n \times n$ linear C-field involving n^2 number of parameters, when all n bonds connected to the C-field are bicausalled. Therefore, the maximum number of parameters which can be estimated is the number of non-redundant sensors in the plant. This type of parameter estimation, in which the model structure and most of its parameters are known, is termed partially known system identification.

Note that parametric or model uncertainty and process and measurement noise in the system have not been considered in the developed algorithm. When the uncertainties and noises are considered, the parameters can be estimated by processing ARRs or residual signals after appropriate filtering.

11.4 Benchmark Problem: Active FTC of a Two-tank System

The two-tank system considered here, as a case study, has appeared earlier in Figure 7.31 of Chapter 7. The nominal values for all parameters used in simulation are given in Table 7.9 of Chapter 7. Therein, a blockage fault in the valve V_1 was simulated by changing the discharge coefficient C_{d1} from nominal value of $1.8 \times 10^{-3}\sqrt{\text{kg.m}}$ to a faulty value of $1.2 \times 10^{-3}\sqrt{\text{kg.m}}$ at 500 s. The corresponding shift in residuals (see Figure 7.37 of Chapter 7) allowed us to detect and isolate that fault.

11.4.1 Fault Quantification with Single Fault Hypothesis

For partial blockage of valve V_1 (considered in the current test case), let the discharge coefficient of the valve after the fault be $\bar{C}_{d1}$, which has to be estimated. Many different ways of bicausality propagation are possible from the corresponding bond graph element R: R_{V1}; one of which is shown in Figure 11.21. Therefore, the discharge coefficient of valve V_1 can be estimated in different ways and considering two of them, it will be shown that they lead to the identical result. Note that the flow sensor $F : F_1$ was added in Chapter 7 to attain complete single fault isolability.

Theoretically, it needs bicausalling of a single suitable sensor to estimate a single parameter. However, we have tried to use the information available from other sensors (because they are available and useful) by converting some of the sensors to

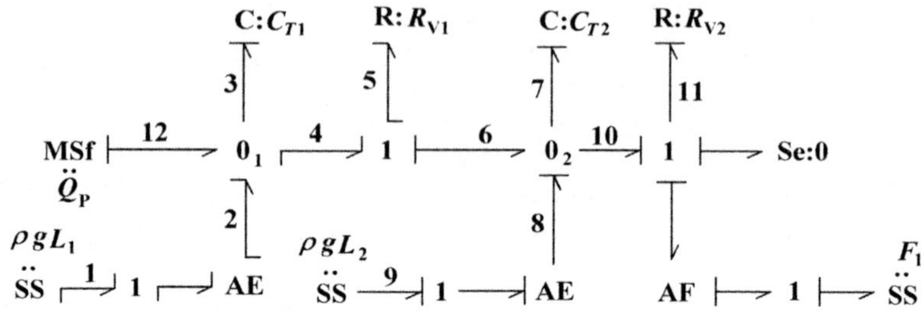

Fig. 11.21 Bicausality propagation to sensor L_1

sources, such that there are no algebraic loops in the solution and the order of derivatives of measurements needed for parameter estimation is kept to a minimum. Note that higher order derivatives of measurements lead to noisy parameter estimation. In Figure 11.21, bicausality is propagated to sensor L_1 (*i.e.* it becomes a flow and effort source at the same time), sensor L_2 is treated as an effort source and sensor F_1 is retained as a flow sensor, although it could as well have been converted into a flow source. This leads to the following relations:

$$\begin{aligned} e_2 &= \rho.g.L_1, e_8 = \rho.g.L_2, f_2 = 0, \\ e_5 &= e_4 - e_6 = e_2 - e_8 = \rho.g\left(L_1 - L_2\right), \\ f_5 &= f_4 = f_{12} - f_3 - f_2 = Q_p - C_{T1}\frac{d}{dt}\left(\rho.g.L_1\right) \\ &= Q_p - \frac{A_1}{g}.\frac{d}{dt}\left(\rho.g.L_1\right). \end{aligned} \tag{11.13}$$

Because the variables e_5 and f_5 are constrained by the constitutive law of the element R_{V1},

$$f_5 = \bar{C}_{d1}\sqrt{e_5},$$

from which one obtains

$$\bar{C}_{d1} = \frac{f_5}{\sqrt{e_5}} = \frac{Q_p - \frac{A_1}{g}.\frac{d}{dt}\left(\rho.g.L_1\right)}{\sqrt{\rho.g\left(L_1 - L_2\right)}}. \tag{11.14}$$

Note that because sensor F_1 was retained as a flow sensor, *i.e.* it was neither assigned bicausality nor made a source, it does not appear in Equation 11.14.

The recorded values of measurements for the current and the just previous sampling step are used in the estimation procedure. The simulation result obtained by using Equation 11.14 is given in Figure 11.22, which shows that the estimated value of $\bar{C}_{d1}$ tracks the actual value both before and after the fault. Note that the estimation gives the correct value immediately after the fault although the associated residuals are in the transient regime at that time (see Figure 7.37 of Chapter 7).

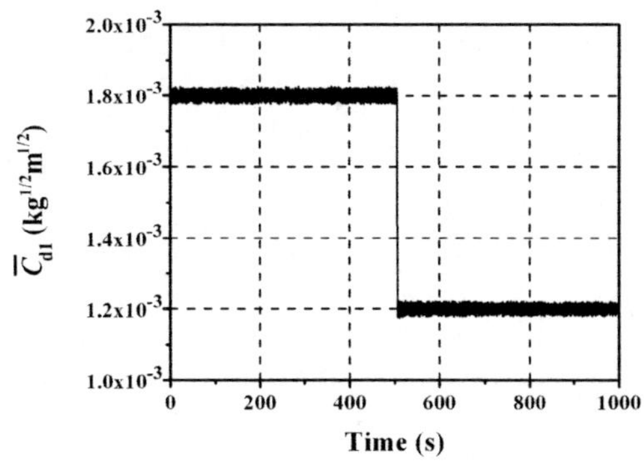

Fig. 11.22 Estimation of discharge coefficient of valve V_1

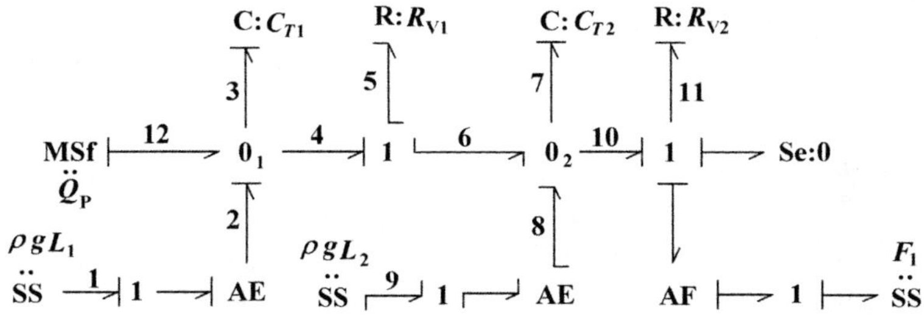

Fig. 11.23 Bicausal bond graph of two tank system: propagation towards sensor L_2

Alternatively, considering the computational causalities assigned to the model as given in Figure 11.23, where the bicausality is propagated towards the sensor L_2, one obtains

$$f_5 = f_6 = f_7 + f_8 + f_{10} = C_{T2}\frac{d}{dt}(\rho.g.L_2) + C_{d2}\sqrt{\rho.g.L_2} \tag{11.15}$$
$$= \frac{A_2}{g}.\frac{d}{dt}(\rho.g.L_2) + C_{d2}\sqrt{\rho.g.L_2},$$

and e_5 holds the same relation as before, *i.e.*

$$e_5 = \rho.g(L_1 - L_2).$$

Then, the value of $\bar{C}_{d1}$ may be estimated as

$$\bar{C}_{d1} = \frac{f_5}{\sqrt{e_5}} = \frac{\frac{A_2}{g}.\frac{d}{dt}(\rho.g.L_2) + C_{d2}\sqrt{\rho.g.L_2}}{\sqrt{\rho.g(L_1 - L_2)}}. \tag{11.16}$$

Note that addition of ARRs is also an ARR (although it does not become structurally independent as long as one of the basis ARRs is eliminated, *i.e.* orthogonality principle holds in the binary sense by replacing dot products by binary AND operators). By adding ARR_3 (Equation 7.24) and ARR_4 (Equation 7.25), one obtains

$$Q_p - \frac{A_1}{g} \cdot \frac{d}{dt}(\rho.g.L_1) - C_{d2}\sqrt{\rho.g.L_2} - \frac{A_2}{g} \cdot \frac{d}{dt}(\rho.g.L_2) = 0, \qquad (11.17)$$

or

$$Q_p - \frac{A_1}{g} \cdot \frac{d}{dt}(\rho.g.L_1) = C_{d2}\sqrt{\rho.g.L_2} + \frac{A_2}{g} \cdot \frac{d}{dt}(\rho.g.L_2). \qquad (11.18)$$

Both Equations 11.14 and 11.16 have the same denominator and Equation 11.18 proves that their numerators are equal. Therefore, Equations 11.14 and 11.16 lead to the identical parameter estimation; only the algebraic forms differ due to the chosen computational causalities. Likewise, bicausality may be propagated to sensor F_1 or any other combination of computational causalities (*i.e.* higher derivative orders) may be followed, and every such causal assignment leads to the same result.

11.4.2 Fault Quantification with Multiple Fault Hypotheses

Faults in components are isolatable even for multiple fault cases when the corresponding fault signatures are linearly independent, *i.e.* the set of residuals sensitive to those faults is structured and diagonal. For example, consider that residuals r_1, r_2 and r_5 are simultaneously abnormal, which leads to a coherence vector $C = [1,1,0,0,1]$. This coherence vector can only be obtained from a unique linear combination of fault signatures given in Table 7.8 and thus three components, *i.e.* pump, PI controller and V_2 can be directly isolated as the faulty components. This case is termed non-overlapping multiple fault scenario. On the other hand, if the obtained fault signature is not structured (overlapping multiple fault scenario) then it is not possible to structurally or directly isolate the faulty components. Consider a case where r_3 and r_4 are abnormal leading to a coherence vector $C = [0,0,1,1,0]$, from which a fault subspace or list of fault candidates, $\langle T_1, T_2, V_1 \rangle$, can be generated. In this case, fault isolation (called fault disambiguation, because there is some ambiguity involved in determining the exact component) is not possible by using the foregoing structural approach because the coherence vector may be obtained through different combinations of the following fault signatures: fault in V_1, fault in V_1 and T_1, fault in V_1 and T_2, fault in T_1 and T_2 or fault in V_1, T_1 and T_2. This situation is characterized by the inability to assign bicausality to all the elements corresponding to the fault candidates, with any scheme of bicausalling of the available sensors. For example, elements $C:C_{T1}$, $C:C_{T2}$ and $R:R_{V1}$cannot be simultaneously bicausalled in the bond graph model given in Figure 7.33; in fact such simultaneous bicausalling of them would require an additional flow sensor in V_1which in turn would result in a completely structured residual set and allow multiple fault isolation. Therefore, in this case, multiple parameters have to be simultaneously estimated and thus the simple procedure based on bicausality propagation cannot be applied. Indeed, one needs to use temporal information to determine a higher number of unknown variables than the known variables. The least-square optimization of a cost function defined by the cumulative sum of the square of the residuals, which has been discussed earlier in Chapter 10, is recommended for multiple parameter estimation. If parameters of a system change without changing the structure of the system then the estimation procedure, when applied over a moving time window, identifies the parameter values corresponding to the fault candidates.

In this section, we deal with the ideal non-overlapping multiple fault scenarios (*e.g.* coherence vector $C = [1,1,0,0,1]$ in Table 7.8) and some special cases of artificial non-overlapping multiple fault scenarios. Consider that residuals r_3 and r_5 (see Table 7.8) are simultaneously abnormal, which leads to a coherence vector $C=[0,0,1,0,1]$. This coherence vector can be broken into two parts: $C=C_s.\mathrm{OR}.C_u$, where $C_s = [0,0,0,0,1]$ is the structured part and $C_u = [0,0,1,0,0]$ is the unstructured part. From the structured part, V_2 can be directly isolated as one of the faulty components. The unstructured part C_u has a unique match with fault signature of T_1 in Table 7.8 and C_u cannot be synthesized by any other combination of component fault signatures in the unstructured part of the FSM. This means that both T_1 and V_2 are simultaneously faulty. Such a situation, *i.e.* when the coherence vector can be constructed from a unique combination of fault signatures from the structured and the unstructured parts of the FSM, is termed an artificial non-overlapping multiple fault scenario. Coherence vector $C = [0,0,0,1,1]$ is another example of an artificial non-overlapping multiple fault scenario. An interesting property of such a scenario can be observed by bicausalling the elements corresponding to the fault candidates (see Figure 11.24), which reveals that the non-overlapping faults correspond to the non-overlapping passive elements in causal paths. The parameters associated with the non-overlapping fault candidates can therefore be determined simultaneously from the bicausalled bond graph model.

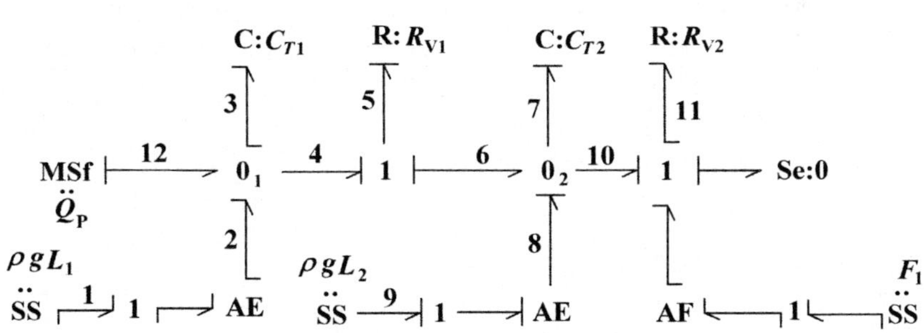

Fig. 11.24 Bicausality assignment for multiple non-overlapping fault parameter estimation

Note that parameters associated with storage elements cannot be estimated by using steady-state measurements. For example, in Figure 11.24, $f_3 = C_{T1} d(\rho g L_1)/dt$. In steady state, $f_3 = 0$, and hence C_{T1} cannot be estimated.

However, if an additive fault is considered, such as a flow source representing leakage fault from T_1 or a resistance through which leakage takes place (see Figure 10.29), then element C:C_{T1} can be assigned derivative causality and the additive fault represented by either SF or R element can be assigned bicausality to estimate its value both during the transient and the steady state operation. The fault tolerance problem discussed in the next section is concerned with recovery from additive faults.

11.4.3 Fault Accommodation Through Fault Tolerant Control

Once the fault magnitude is estimated, then the next step concerns accommodation of the identified fault by suitably changing the control laws, if possible. When valve V_1 of the two-tank system (see Figure 7.31) is partially blocked, the level in the tank T_1 reaches the setpoint and then the pump operates at a low load. Simultaneously, the level in the tank T_2 decreases because there is less input flow through the inlet valve and the output to consumer is open. Overall, the output flow rate to the consumer decreases. If the objective is to maintain a constant flow rate to the consumer, then the level in T_2 must be increased (to its steady state value during normal operation), which consequently requires more flow to be forced through the faulty valve, which in turn requires more level in the tank T_1. This qualitative analysis means that the level set-point in T_1 should be increased. The question now remains, by how much and at what rate? These two problems relate to the system inversion and actuator sizing problems [176–180], respectively.

The process of finding an input sequence to satisfy a constraint given in the form of an output sequence is called input reconstruction. System inversion and input reconstruction have been discussed earlier in this chapter. The general bond graph model-based inversion scheme for actuator sizing uses an SS element in the place of a prescribed output flow as shown in Figure 11.25, whereby the power variables (e_a, f_a) for the actuator and the power modulator (e_m, f_m) can be derived in terms of the state variables, the desired output and their derivatives.

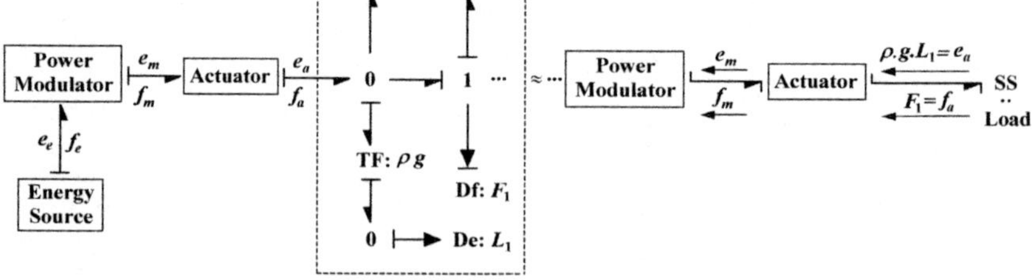

Fig. 11.25 Actuator sizing and system inversion for two-tank system

11.4.4 System Inversion

The value of the desired level setpoint for T_1, such that the steady state output of the system will match the output in the non-faulty case is determined from the bicausal bond graph model shown in Figure 11.21.

From the constitutive relation of the valve, one can write $f_5 = C_{d1}\sqrt{e_5} = C_{d1}\sqrt{\rho.g|L_1 - L_2|}$. The steady-state flow through V_1 in the non-faulty case, when the valve is fully open, is given by

$$f_n = C_{d1}\sqrt{\rho.g\,|L_{1s} - L_{2s}|}\,, \tag{11.19}$$

where f_n is the steady-state nominal flow rate, L_{1s} is the nominal level setpoint for tank T_1 and L_{2s} is the steady-state level in tank T_2 during non-faulty process operation. By using steady-state approximation, *i.e.* $C_{d1}\sqrt{\rho.g\,|L_{1s} - L_{2s}|} = C_{d2}\sqrt{\rho.g\,|L_{2s}|}$, one obtains

$$L_{2s} = \frac{C_{d1}^2}{C_{d1}^2 + C_{d2}^2} L_{1s}. \tag{11.20}$$

If the same amount of flow as the normal process output has to be forced through the defective valve by changing the level set-point of the tank T_1, then

$$\bar{C}_{d1}\sqrt{\rho.g\,|L_{1s}^* - L_{2s}|} = C_{d1}\sqrt{\rho.g\,|L_{1s} - L_{2s}|}, \tag{11.21}$$

where $\bar{C}_{d1}$ is the estimated discharge coefficient and L_{1s}^* is the new level setpoint for T_1.

Use of Equations 11.20 and 11.21 gives L_{1s}^* as

$$L_{1s}^* = \frac{C_{d1}^2}{C_{d1}^2 + C_{d2}^2} L_{1s} + \left(\frac{C_{d1}}{\bar{C}_{d1}}\right)^2 \left(1 - \frac{C_{d1}^2}{C_{d1}^2 + C_{d2}^2}\right) L_{1s}. \tag{11.22}$$

The fault can be accommodated if V_1 is not completely blocked ($\bar{C}_{d1} \neq 0$), tank T_1 is able to accommodate the prescribed level without overflowing (geometrical constraint) and the pump is able to give enough flow against the given pressure head. The last constraint relates to the actuator sizing problem.

11.4.5 Actuator Sizing

Consider that the actuating pump in this system, as is usual in any such other system, has been selected to deliver a rated flow against a given pressure difference, *i.e.* the pump's power rating is sufficient to achieve the normal level set-point as well as desired level as obtained through system inversion in Equation 11.22. Further consider the word bond graph of an actuating system, along with its power modulator, given in Figure 11.26 [177], in which different power variables are marked. Each of those power variables usually have a constraint as follows: $|e_m(t)| \leq E_m$, $|f_m(t)| \leq F_m$, $|e_m(t).f_m(t)| \leq W_m$, $|e_a(t)| \leq E_a$, $|f_a(t)| \leq F_a$, $|e_a(t).f_a(t)| \leq W_a$, $\forall t \in [0,T]$, where T is the time required to reach the steady state.

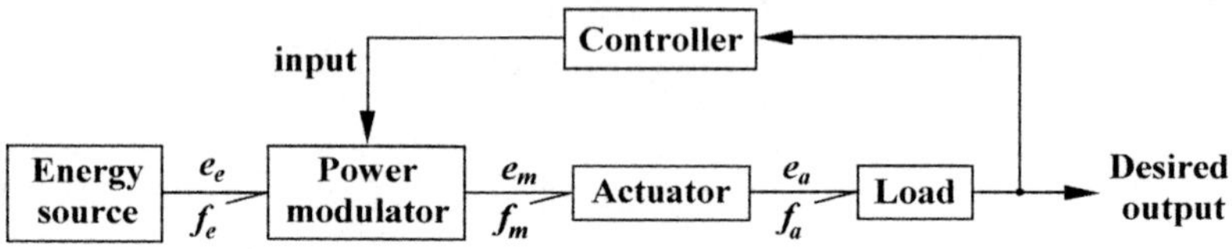

Fig. 11.26 Word bond graph of a feedback actuated system

A representative plot of the time evolution of different variables, in the constraint space, is shown in Figure 11.27. The hyperbolic curves represent constraints on power and the non-shaded area in the middle of each plot is the admissible operating regime.

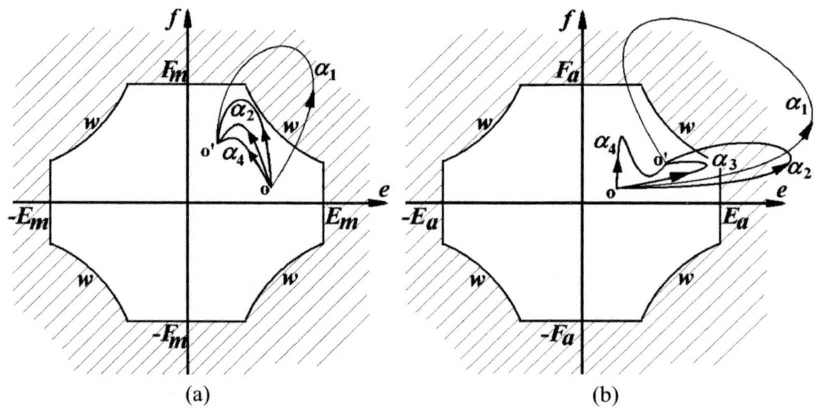

Fig. 11.27 Trajectory of power variables in the constraint space

The objective is to obtain an output profile (*e.g.* maximum allowable slew rate) and the corresponding input law, for which all the operating constraints are satisfied. The mass flow rate (f) through a pump is a non-linear function of pressure difference (ΔP) and it can be expressed as: $f = \Phi(\Delta P)$. We assume an impeller pump characteristic approximated by $f = \alpha.\Delta P^{-1}$, where α is a constant parameter corresponding to a fixed power rating of the pump (W). Normal operating mode for the plant is defined by the point, a, which is on the characteristic curve corresponding to 41.4% of maximum power output of the pump, W_{max} (see Figure 11.28). After the blockage fault in V_1, the operating point shifts to point b (a transient point; the steady state point will be exactly below point a, but we cannot wait till then and will accommodate the fault before that). Now the pump flow has to be increased to attain the pressure corresponding to the tank level L^*_{1s} $(= 0.79m)$, which is calculated by using Equation 11.22. Without using any optimal control theory, in which selection of the weighting matrices for the global and the transitional cost is subjective, we have chosen the path corresponding to maximum pump power output. Then the input pump flow is defined as: $f_{\max} = \alpha_{\max}.\Delta P^{-1}$, where $\alpha_{\max} = 10^3$. By using this input command and following the path c-d as shown in Figure 11.28, the pressure in tank T_1 is increased to a level sufficiently close to the desired level $(P_1 = \rho.g.L^*_{1s} = 7749.9\ \mathrm{N.m^{-2}})$. Thereafter, the FTC control law is automatically switched over to the original PI control law (with the new setpoint) to take care of small deviations and the path followed is shown as e–g in Figure 11.28, in which g is the final steady-state position.

The desired pressure level can as well be achieved by altering the PI controller setpoint after the fault diagnosis, but the actuator constraints would not be satisfied in that case. The corresponding transition should hypothetically follow a path shown as $b - c' - g$ (dotted lines) in Figure 11.28, which is not admissible under the given operating constraints.

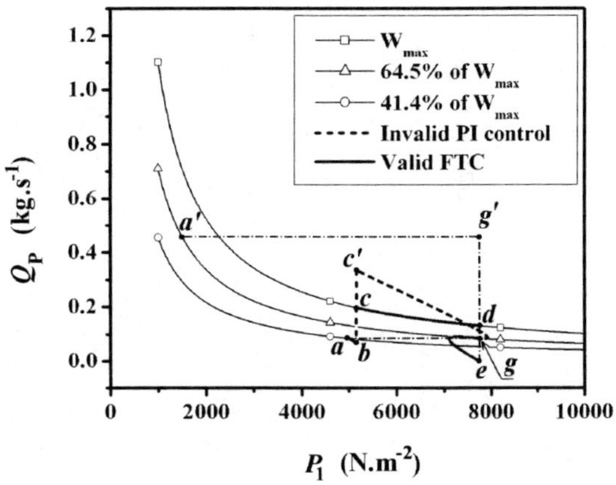

Fig. 11.28 Validation of input command for FTC implementation

Furthermore, consider a case where the normal operating point is a' instead of a and the desired operating point to reach is g' instead of g. In this case, in order to increase the pressure, the pump has to deliver more flow than what the pump can deliver while operating at its maximum power rating (see the operating curve corresponding to W_{max} in Figure 11.28) and hence, FTC is not possible in this case. The power modulator constraints can be determined similarly.

Note that one may solve an optimal control problem with a heavy cost on excessive input size, relaxed transitional cost on the outputs and relaxed global cost on the objective target, to determine an appropriate input sequence. However, determination of weighting matrix coefficients of the cost function is subjective and the solution does not guarantee that the pump will operate at its full efficiency. Furthermore, solution of the non-linear optimal control problem in real-time FTC applications may be time consuming when compared to the simpler method of deriving the control laws through system inversion.

The integrally causalled bond graph model of the two-tank system, given in Figure 7.32, was simulated with the parameter values given in Table 7.9 to study the system behavior, particularly with FTC implementation. The time responses of tank pressures are shown in Figure 11.29. Blockage fault in V_1 was introduced at 250 s (line L) and FTC was applied at 295 s (line M) after complete fault diagnosis including isolation, parameter estimation and required PI controller setpoint calculation was over. Thereafter, the fault tolerant control law was automatically switched over to the normal PI control (setpoint is changed) at 350 s (line N).

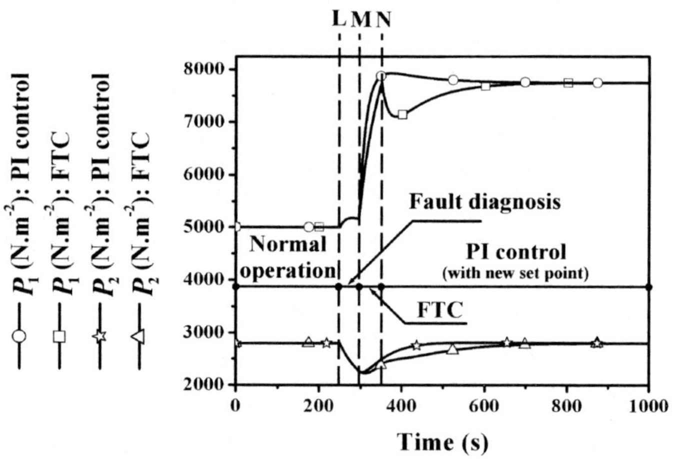

Fig. 11.29 Simulation results with FTC

The pressure responses resulting from direct modification of the setpoint of the PI controller while neglecting the actuator constraints, which is essentially physically impossible and thus a hypothetical case, are also plotted for comparison. It is observed that with the latter hypothetical strategy, the steady-state behavior could be quickly attained. However, note that the pump would be unable to operate according to the controller commands (saturate) and thus the latter approach has only academic value.

The FTC for the two-tank system is finally implemented in three steps:

1. The PI controller is suspended.
2. Pump is then operated at maximum power to bring the level in tank T_1 sufficiently close to a new set-point, L_{1s}.
3. Finally, the PI controller is reactivated with the new setpoint, which takes care of further small deviations.

Readers may refer to [279] for a bond graph model based scheme which covers a part of the problems associated with FTC.

One major problem with the active fault tolerance developed so far is that it is not sufficiently robust. It does not consider the uncertainties associated with parameter estimation and measurements. Moreover, the fault must be steady, *i.e.* parameter values should not be fluctuating continuously. In the next section, we develop a robust controller to address these problems.

11.5 Passive FTC: Robust Overwhelming Control

The first step in a commonly followed control system design scheme is to create a mathematical model of the system by using one of the established techniques. Thereafter, the model is usually linearized and represented in frequency domain for classical control or represented in a state space form for modern control, because there are well-established mathematical tools for controller design in both these domains. In [229], the authors have rightly criticized both these methods because these methods assume a given system and do not provide any guidance in designing that system in the first place. Furthermore, insight into the limitations of the physical system may be obscured when working with purely mathematical models. Therefore, they develop a bond graph model based method to design controllers in the physical domain. Inspired by the same arguments, an overwhelming control scheme was developed in [173] for robust trajectory control (tracking control). It has been successfully applied to various robotics problems such as trajectory control as well as force and impedance control in single and cooperative environments. This control strategy is also suitable for fault tolerant control, where the process must transit from one state to another according to some prescribed state-trajectories. Moreover, the robustness properties of the controller ensure that the system itself becomes fault tolerant, *i.e.* the output of the process can be maintained at desired levels even when there are gross parameter variations.

11.5.1 Overwhelming Controller Design

A robust overwhelming controller is easily derived for collocated control systems where the controller manifests itself as a driving point impedance, *i.e.* the output to be controlled is at the same point as the application of the input. However, the controller can be modified to control output variables at other places.

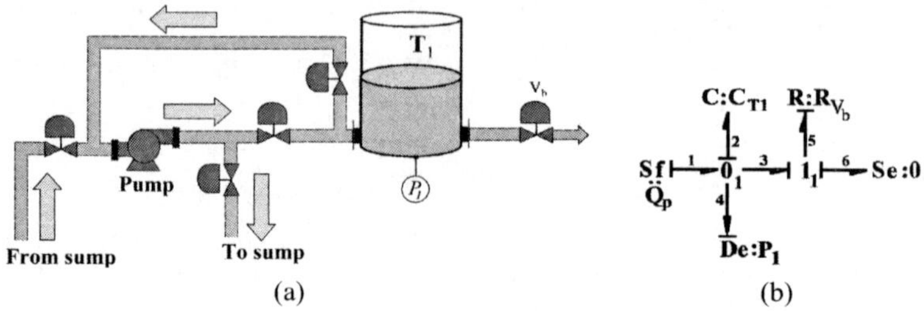

Fig. 11.30 An example taken for robust overwhelming control. **a** The system. **b** Its bond graph model

To illustrate overwhelming control strategy, let us consider Figure 11.30a which shows a tank connected to an output valve and a supply line. The objective is to change the water level in the tank according to a prescribed trajectory.

The pump, in the supply line, is assumed to give flow in a fixed direction. It is connected to the tank through a set of valves, which can be appropriately opened/closed so as to change the suction and discharge sides, *i.e.* the hydraulic network can either deliver to or discharge from the tank. The same configuration can be obtained by using a two-way spool valve. If the pump can deliver flow in both forward and reverse directions, *e.g.* gear pumps with internal or external teeth, then the hydraulic circuit is unnecessary.

The bond graph model of the system under consideration is shown in Figure 11.30b. An overwhelming controller uses a copy of the model, as shown in Figure 11.31. The output from the process is compared with the command (the trajectory) $P(t)$ to generate the input to the controller. As a result, some storage elements in the controller are forced to derivative causality. The driving input for the process is generated from the controller. The process model in the controller side acts as an inverse system. For more details see [95, 173, 193, 194].

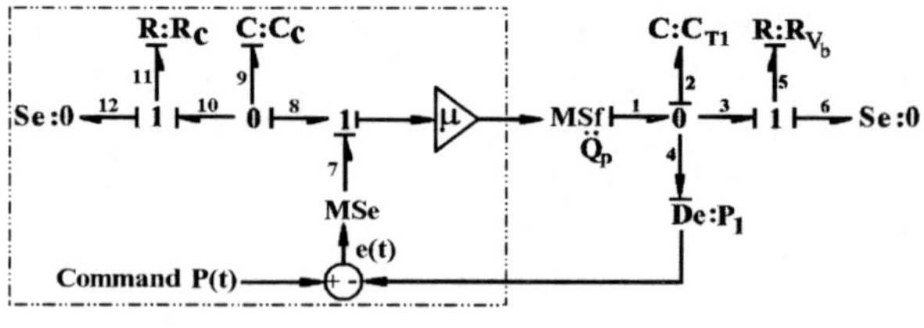

Fig. 11.31 Two-tank system model coupled with its overwhelming controller model

In the bond graph model in Figure 11.31, we have intentionally used different symbols for parameters. Let us initially assume that the system is linear. Then the corresponding signal flow graph of the integrated system may be drawn as shown in Figure 11.32a and it can be further reduced to the form shown in Figure 11.32b.

The transfer function between the command and the pressure in the tank may then be derived as

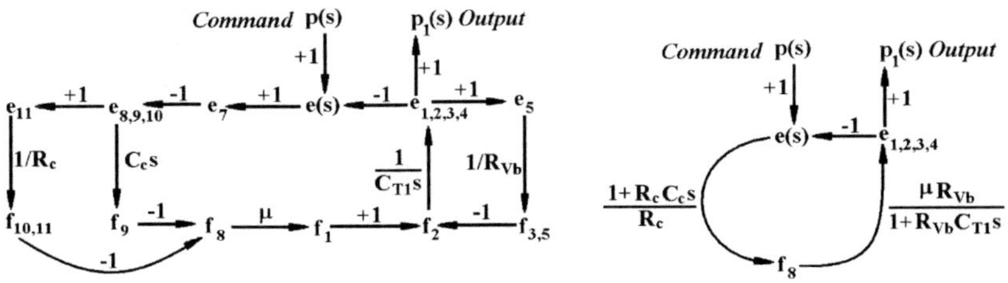

(a) (b)

Fig. 11.32 Full (**a**) and reduced (**b**) signal flow graphs of the integrated system and controller

$$G(s) = \frac{e_1(s)}{p(s)} = \frac{\mu\left(\frac{R_{vb}}{R_c}\right)\left(\frac{1+R_cC_cs}{1+R_{vb}C_{T1}s}\right)}{1+\mu\left(\frac{R_{vb}}{R_c}\right)\left(\frac{1+R_cC_cs}{1+R_{vb}C_{T1}s}\right)}. \tag{11.23}$$

If the process parameters values are known with absolute accuracy (which is rarely true) so that they are used in the controller, *i.e.* $R_c = R_{vb}$ and $C_c = C_{T1}$, then $G(s) = \frac{\mu}{1+\mu}$ and $Lim_{\mu\to\infty}G(s) = 1$. Note that if $R_c \neq R_{vb}$, $C_c \neq C_T$ and the numerator in Equation 11.23 is non-zero, even then $Lim_{\mu\to\infty}G(s) = 1$.

This proves the robustness of the controller, *i.e.* the pressure in the tank will follow its prescribed trajectory for values of $\mu >> 1$ even if the process parameters are not known with absolute accuracy. This also means that the controller becomes robust to parametric faults in the process.

This robustness property of the overwhelming controller is important because no real physical system perfectly behaves according to the mathematical model used to represent it. That is why one of the specifications in a robust controller design is that its properties should not change much if applied to a system slightly different from the mathematical one used for its synthesis. For control system synthesis, usually a simpler mathematical model is chosen (by model order reduction or system identification) so that the calculations are simplified. Even if a complete model is used to design the controller, all nominal parameters are never known with absolute precision; the control system should perform correctly even when it is connected to physical system with true parameter values different from the nominal values used in the controller design. The parameters of the model may be estimated online such that if there is sufficient variation in them then the controller can adjust itself to ensure the proper performance [44].

The controller acts as a virtual system (residing in the computer) appended to the actual system. The passivity of the interconnection may be proved by simple analogy with the physical model based control scheme developed in Chapter 3. Because of the virtual nature of the inverse system, which overwhelms plant dynamics and drags it according to the its own dynamics, the robust overwhelming controller developed here is also referred to as a Ghost controller in [173].

11.5.2 Example: A Robust Level Controller

Consider a two-tank system shown in Figure 11.33, which has also been considered in the last section.

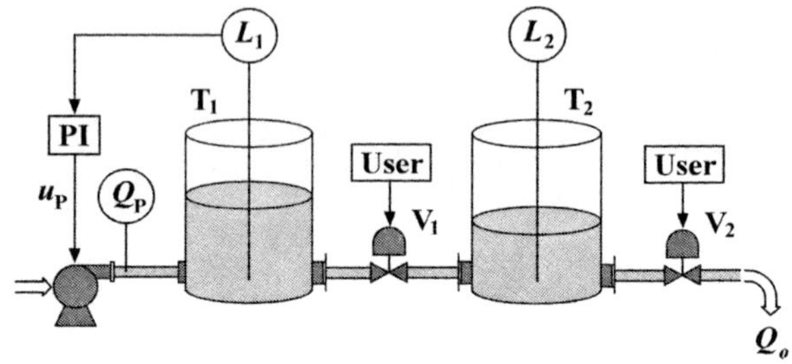

Fig. 11.33 A two-tank system in which the level in the first tank will be controlled

We will replace the PI controller in the system by a robust overwhelming controller. The process model and its overwhelming controller are shown in Figure 11.34. The controller is implemented in a computer.

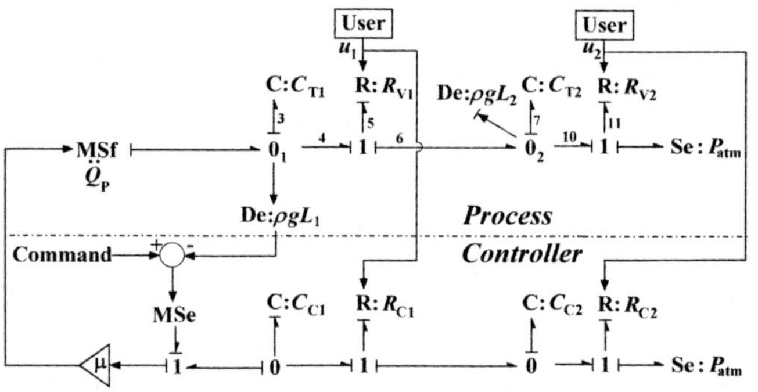

Fig. 11.34 Integrated model of the robust controller interfaced with the system

It is assumed that the valves follow non-linear characteristics defined earlier. For no specific reason, let us specify the following command:

$$P(t) = 2A + A\cos(\omega t), \tag{11.24}$$

which describes a oscillating liquid level in Tank_1 with a mean pressure of $2A$. Note that when equations are derived from the bond graph model in Figure 11.34, the time derivative of $P(t)$ is required due to the differentially causalled C-element. In this case, $\dot{P}(t) = -A\omega\sin(\omega t)$. If the derivative of the command cannot be computed symbolically, it may be numerically generated.

The model was simulated with the data given in Table.11.1. The lone state variable in the controller side was assigned an initial value of 0 whereas both the state variables in the process side were given an initial value 20. Note that the controller parameters C_{c1} and c_{dc2} are, respectively, different from plant parameters C_{T1} and c_{d2}.

Table 11.1 Parameters of the two-tank system and the ovwhelming controller

Symbol	Description	Value	Unit
c_{d1}	Discharge coefficient of valve V_1	1.8×10^{-3}	$kg^{\frac{1}{2}}m^{\frac{1}{2}}$
c_{d2}	Discharge coefficient of valve V_2	1.5964×10^{-3}	$kg^{\frac{1}{2}}m^{\frac{1}{2}}$
C_{T1}	Capacity of tank T_1	1.57×10^{-3}	$m.s^2$
C_{T2}	Capacity of tank T_2	1.57×10^{-3}	$m.s^2$
c_{dc1}	Discharge coefficient of valve V_1 in controller side	1.8×10^{-3}	$kg^{\frac{1}{2}}m^{\frac{1}{2}}$
c_{dc2}	Discharge coefficient of valve V_2 in controller side	1.0×10^{-3}	$kg^{\frac{1}{2}}m^{\frac{1}{2}}$
C_{c1}	Capacity of tank T_1 in controller	2.57×10^{-3}	$m.s^2$
C_{c2}	Capacity of tank T_2 in controller	1.57×10^{-3}	$m.s^2$
Q_{Pmax}	Maximum flow rate from pump	1	kg/s
A	Variable used in command	1×10^4	Pa
ω	Variable used in command	0.01	rad/s
μ	Controller gain	1×10^6	-

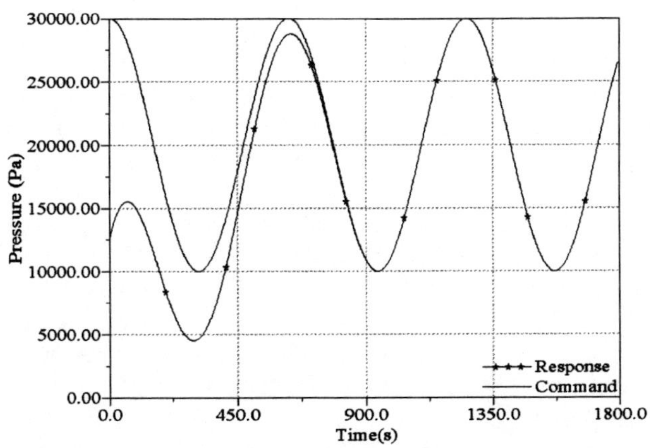

Fig. 11.35 Comparison between the prescribed and the followed level trajectories

The simulation results in Figure 11.35 show that the liquid level in $Tank_1$ follows the desired trajectory (Equation 11.24). The time taken for proper trajectory tracking does not change even when the value of the gain parameter μ is increased; because this is determined by intrinsic system characteristic (time constants) which can only be modified through pole-placement or other compensators. Note that we have a free measurement (L_2 or corresponding pressure P_2) which has not been used by the controller. This measurement can be used to compensate the effects of the part of the system after Valve1 (called additional passive degrees-of-freedom) and thereby achieve better trajectory tracking (see [173, 193, 194]).

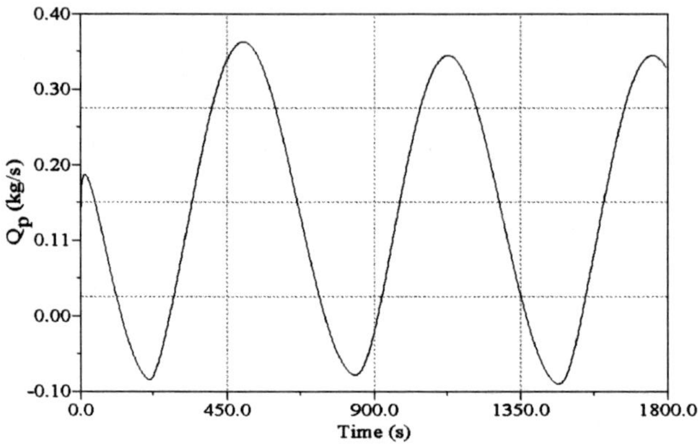

Fig. 11.36 The command (pump flow) generated by the overwhelming controller

The required input to the process (output of the pump) is plotted in Figure 11.36, which is the solution of the inverse problem. Actuator capacity and saturations should be tested on this output.

Note that even when process parameters are grossly different from the controller parameters, the process is non-linear, and the controller states are arbitrarily initialized; the control strategy works robustly. It can also be shown that the controller is not influenced by small measurement noises.

In complex processes it may not be possible to prescribe a trajectory beforehand. Moreover, even when a fixed trajectory is prescribed, the model structure or parameters may change during the course of command or trajectory tracking. Therefore, FTC algorithms may be implemented by following a receding horizon control strategy where the commands are computed for a very short time span at every sampling time. Then the first step of this calculated command is implemented and the output is sampled to be used again to recompute the next command step. This way, the target keeps on shifting forward.

Because the FTC algorithms are to be executed at every sampling step, *i.e.* in real-time environments, they should be extremely fast. This is why reduced order models are used in FTC. The small dynamics neglected during model order reduction may be considered as external disturbances and their influences on the controlled system's response are generally overwhelmed by the robust controller.

References

[1] Advanced decision support system for Chemical/Petrochemical manufacturing processes, CHEM Project,project No. G1RD-CT-2001-00466,http://www.chem-dss.org, 2000.

[2] G. Acosta, C. A. González, and B. Pulido. Basic tasks for knowledge-based supervision in process control. *Engineering Applications of Artificial Intelligence*, 14(4):441–455, 2001.

[3] J. S. Albus and F. G. Proctor. A reference model architecture for intelligent hybrid control systems. In *IFAC 13th Triennial World Congress*, pages 483–488, San Francisco (USA), 1996. IFAC.

[4] D. Angeli and E. D. Sontag. Monotone control systems. *IEEE Trans. Automat. Control*, 48:1684–1698, 2003.

[5] T. M. Apostol. *Mathematical Analysis II*. Addison-Wesley, second edition edition, 1974.

[6] J. Armengol, L. Travé-Massuyès, J. Vehí, and J. L. De la Rosa. A survey on interval model simulators and their properties related to fault detection. *Annual Reviews in Control*, 24:31–39, 2000.

[7] J. Armengol, J. Vehi, M. A. Sainz, and P. Herrero. Fault detection in a pilot plant using interval models and multiple sliding windows. In N. Eva Wu, editor, *SAFEPROCESS 2003*, pages 729–734, Washington DC, USA, June 2003. IFAC.

[8] K. Åström, P. Albertos, M. Blanke, and A. Isidori. *Control of Complex Systems*. Springer-Verlag, London, 2001.

[9] K. J. Åström and R. D. Bell. Drum-boiler dynamics. *Automatica*, 36:363–378, 2000.

[10] M. Bartys and S. De Las Heras. Benchmark simulation model in simulink. In *3rd DAMADICS workshop on robust methods in fault diagnosis*, The University of Hull, England., 2002.

[11] M. Bartys, R. Patton, M. Syfert, S. de las Heras, and J. Quevedo. Introduction to the DAMADICS actuator FDI benchmark study. *Control Engineering Practice*, 14(6):577–596, 2006.

[12] M. Basseville, G. Benveniste, and G. Moustakides. Detection and diagnosis of changes in the eigen structure of non stationary multivariable systems. *Automatica*, 23:479–489, 1987.

[13] M. Basseville and I. V. Nikiforov. *Detection of Abrupt Changes: Theory and Application*. Prentice Hall, ISBN 0-13-126780-9, 1993.

[14] J. S. Bay. *Fundamentals of linear state space systems*. McGraw-Hill, Boston, 1999.

[15] J. J. Beaman and P. C. Breedveld. Physical modeling with Eulerian frames and bond graphs. *Journal of Dynamic Systems, Measurement and Control*, 110:182–188, 1988.

[16] M. Blanke, M. Kinnaert, J. Lunze, and M. Staroswiecki. *Diagnosis and Fault Tolerant Control*. Springer-Verlag, 2003.

[17] G. Bloch, M. Ouladsine, and P. Thomas. On-line fault diagnosis of dynamic systems via robust parameter estimation. *Control Engineering Practice*, 12(3):1709–1717, 1995.

[18] L. S. Bonderson. Vector bond graphs applied to one-dimensional distributed systems. *Journal of Dynamic Systems, Measurement and Control*, 97(1):75–82, 1975.
[19] W. Borutzky. Supporting the generation of a state space model by adding tearing information to the bond graph. *Simulation Practice and Theory*, 7(5-6):419–438, 1999.
[20] W. Borutzky. *Bond Graphs - A Methodology for Modelling Multidisciplinary Dynamic Systems*. SCS Publishing House, Erlangen, San Diego, 2004.
[21] W. Borutzky. Residual bond graph sinks for numerical evaluation of analytical redundancy relations in model based single fault detection and isolation. In *Proc. 20th European Conference on Modelling and Simulation (ECMS'06)*, pages 166–172, Sankt Augustin, Germany, May 2006.
[22] W. Borutzky, B. Barnard, and J. Thoma. Describing bond graph models of hydraulic components in modelica. *Mathematics and Computers in Simulation*, 53:381–387, 2000.
[23] W. Borutzky, B. Barnard, and J. Thoma. An orifice flow model for laminar and turbulent conditions. *Simulation Modelling Practice and Theory*, 10:141–152, 2002.
[24] W. Borutzky and F. E. Cellier. Tearing algebraic loops in bond graphs. *SIMULATION:Transactions of the Society of Computer Simulation International*, 13(2):102–115, 1996.
[25] W. Borutzky and G. Dauphin-Tanguy. Incremental bond graph approach to the derivation of state equations for robustness study. *Simulation Modelling Practice and Theory*, 12(1):41–60, 2004.
[26] W. Borutzky, G. Dauphin-Tanguy, and C. Kam. Relations between two bond graph approaches to sensitivity analysis and study of robustness. *Mathematical and Computer Modelling of Dynamical Systems*, 12(2):141–157, 2006.
[27] A. M. Bos and J. L. Tiernego. Formula manipulation in the bond graph modelling and simulation of large mechanical systems. *Journal of Franklin Institute*, 319:51–65, 1985.
[28] G. E. Box and G. C. Tiao. *Bayesian inference in statistical analysis*. Addison-Wesley, Reading, Mass, 1973.
[29] P. C. Bredveld. Multibond graph elements in physical systems theory. *Journal of the Franklin Institute*, 319(1/2):1–36, 1985.
[30] P. C. Breedveld. *Physical Systems Theory in Terms of Bond Graphs*. PhD thesis, University of Twente, The Netherlands, 1984.
[31] P. C. Breedveld, R. C. Rosenberg, and T. Zhou. Bibliography of bond graph theory and application. *Journal of the Franklin Institute*, 328(5-6):1067–1109, 1991.
[32] J. W. Brewer. Progress in the bond graph representations of economics and population dynamics. *Journal of the Franklin Institute*, 328(5/6):675–696, 1991.
[33] J. W. Brewer and P. P. Craig. Bilinear dynamic single-ports and bond graphs of economic systems. *Journal of the Franklin Institute*, 313(4):185–196, 1982.
[34] J. W. Brewer, P. P. Craig., M. Hubbard, and K. Watt. The bond graph method for technological forecasting and ressource policy analysis. *Energy*, 7(6):505–537, 1982.
[35] J. F. Broenink. 20-sim software for hierarchical bond-graph/Block-diagram models. *Simulation Practice and Theory*, 7(5-6):481–492, 1999.
[36] J. F. Broenink and C. Kleijn. Computer-aided design of mechatronic systems using 20-SIM 3.0. In *Proc. WESIC 99*, Newport (S. Wales), UK, 1999.
[37] F. T. Brown. Direct application of loop rule to bond graphs. *Trans. ASME, Journal of Dynamic Systems, Measurement and Control*, 94(3):253–261, 1972.
[38] F. T. Brown. Convection bonds and bond graphs. *Journal of the Franklin Institute*, 325(5/6):871–886, 1991.
[39] F. T. Brown. *Engineering System Dynamics, a Unified Graph-Centered Approach*. Marcel Dekker, New York, 2001.
[40] F.T. Brown. Non-iterative evaluation of multiphase thermal compliances in bond graphs. *Proc. IMechE, Part I: Journal of Systems and Control Engineering*, 216(1):13–19, 2002.
[41] P. Bunus and P. Fritzson. Automated static analysis of equation-based components. *SIMULATION*, 80:321–345, 2004.

[42] F. Busson. *Les Bond Graphs Multiénergies Pour la Modélisation et la Surveillance En Génie Des Procédés*. PhD thesis, Université des Sciences et Technologies de Lille, 2002.

[43] R. Cacho, J. Felez, and C. Vera. Deriving simulation models from bond graphs with algebraic loops: The extension to multibond graph systems. *Journal of the Franklin Institute*, 337(5):579–600, 2000.

[44] P. Castaldi, W. Geri, M. Montanari, and A. Tilli. A new adaptive approach for on-line parameter and state estimation of induction motors. *Control Engineering Practice*, 13:81–94, 2005.

[45] S. Cauvin, M. O. Cordier, C. Dousson, P. Laborie, and F. Lévy. Monitoring and alarm interpretation in industrial environments. *AI Communications Journal: The European Journal on Artificial Intelligence*, 11(3-4):139–173, 1998.

[46] F. E. Cellier. *Continuous System Modelling*. Springer Verlag, New York, 1991.

[47] F. E. Cellier and J. Greifeneder. Object-oriented modeling of convective flows using the dymola thermo-bond graph library. In *International Conference on Bond Graph Modeling and Simulation (ICBGM'03)*, pages 89–103. SCS Publication, Vol.35, No.2, ISBN 1-56555-257-1, 2003.

[48] A. L. Chapman. *Heat Transfer*. Macmillan Publishing Company, 4th edition, 1984.

[49] J. Chen, R. Patton, and H. Zhang. Design of unknown input observers and robust fault detection filters. *International Journal of Control*, 63:85–105, 1996.

[50] J. Chen and R. J. Patton. *Robust Model-based fault diagnosis for dynamic systems*. Kluwer Academic Publisher:Dordrecht, 1999.

[51] L. Chittaro. Functional diagnosis and prescription of measurement using effort and flow variables. *IEE Control Theory and Applications*, 142:420–432, 1995.

[52] L. Chittaro, R. Ranon, and A. Soldati. Introducing deviations and multiple abstraction levels in the functional diagnosis of fluid transfer systems. *Artificial Intelligence in Engineering*, 12:355–373, 1998.

[53] E. Y. Chow and A. C. Willsky. Analytical redundancy and the design of robust failure detection system. *IEEE Transactions on Automatic Control*, 29(7):603–614, 1984.

[54] CODRA Engineering Informatique. *User's Manual of PANORAMA 7.0*. CODRA Engineering Informatique, 1997.

[55] C. Commault, J. M. Dion, O. Sename, and R. Motyeian. Observer-based fault detection and isolation for structured systems. *IEEE Transactions on Automatic Control*, 47(12):2074–2079, 2002.

[56] M. Darouach, M. Zasadzinski, and S. J. Xu. Full-order observers for linear systems with unknown inputs. *IEEE Transactions on Automatic Control*, 39(3):606–609, 1994.

[57] K. Dasgupta, J. Watton, and S. Pan. Open-loop dynamic performance of a servo-valve controlled motor transmission system with pump loading using steady-state characteristics. *Mechanism and Machine Theory*, 41:262–282, 2006.

[58] S. Dash and V. Venkatasubramanian. Challenges in the industrial applications of fault diagnostic systems. *Computers and Chemical Engineering*, 24(2-7):785–791, 2000.

[59] G. Dauphin-Tanguy. *Les Bond Graphs*. Hermès Science, 2000.

[60] G. Dauphin-Tanguy, P. Borne, and M. Lebrun. Order reduction of multi-time scale systems using bond graphs, the reciprocal system and the singular perturbation method. *Journal of the Franklin Institute*, 319:157–171, 1985.

[61] G. Dauphin-Tanguy, A. Rahmani, and C. Sueur. Bond graph aided design of controlled systems. *Simulation Practice and Theory*, 7(5-6):493–513, 1999.

[62] L. Dinca, T. Aldemir, and G. Rizzoni. Fault detection and identification in dynamic systems with noisy data and parameter/modeling uncertainties. *Reliability Engineering and System Safety*, 65(1):17–28, 1999.

[63] S. Diop. Elimination in control theory. *Mathematics of Control, Signals and Systems*, 4:17–32, 1991.

[64] M. A. Djeziri, R. Merzouki, B. Ould Bouamama, and G. Dauphin-Tanguy. Fault detection of backlash phenomenon in mechatronic system with parameter uncertainties using bond

graph approach. In *Proc. IEEE International Conference on Mechatronics and Automation*, pages 600–605, Luoyang, China, 2006.
[65] M. A. Djeziri, R. Merzouki, B. Ould Bouamama, and G. Dauphin-Tanguy. Robust fault diagnosis using bond graph approach. *IEEE/ASME Transactions on Mechatronics*, To appear, 2007.
[66] A. L. Dulmage and N. S. Mendelsohn. Coverings of bipartite graphs. *Canadian Journal of Mathematics*, 10:517–534, 1958.
[67] D. Dustegor, E. Frisk, V. Cocquempot, M. Krysander, and M. Staroswiecki. Structural analysis of fault isolability in the DAMADICS benchmark. *Control Engineering Practice*, 14(6):597–608, 2006.
[68] H. Elmqvist, D. Brûck, and M. Otter. *Dymola-User's Manual. Dynasim AB, see also http://www.Dynasim.se*. Lund, Sweeden, 1996.
[69] A. Emami-Naeini, M. M. Akhter, and S. M. Rock. Effect of model uncertainty on failure detection: the threshold selector. *IEEE Transactions on Automatic Control*, AC-33:1106–1115, 1988.
[70] M. J. Er, C. B. Low, K. H. Nah, M. H. Lim, and S. Y. Ng. Real-time implementation of a dynamic fuzzy neural networks controller for a SCARA. *Microprocessors and Microsystems*, 26(9-10):449–461, 2002.
[71] F. W. Fairman, S. S. Mahil, and L. Luk. Disturbance decoupled observer design via singular value decomposition. *IEEE Transactions on Automatic Control*, AC-29(1):84–86, 1984.
[72] J. Feenstra, P. J. Mosterman, G. Biswas, and P. C. Breedveld. Bond graph modeling procedures for fault detection and isolation of complex flow processes. In *International Conference on Bond Graph Modeling and Simulation (ICBGM'01)*, pages 77–82. Simulation Series, Vol.33, No.1, ISBN 1-56555-103-6, 2001.
[73] P. J. Feenstra, E. J. Manders, P. J. Mosterman, G. Biswas, and R. J. Barnett. Modeling and instrumentation for fault detection and isolation of a cooling system. In *Proc. IEEE South East Conference*, pages 365–372, Nashville, TN USA., 2000.
[74] P. Frank. Enhancement of robustness in observer-based fault detection. *International Journal of Control*, 59:955–981, 1994.
[75] P. M. Frank. *Introduction to System Sensitivity Theory*. Academic Press, New York, 1978.
[76] P. M. Frank. Fault diagnosis in dynamic systems using analytical and knowledge-based redundancy- a survey and some new results. *Automatica*, 26(3):459–474, 1990.
[77] P. M. Frank and X. Ding. Frequency domain approach to optimally robust residual generation and evaluation for model based fault diagnosis. *Automatica*, 30:789–904, 1994.
[78] P. M. Frank and X. Ding. Survey of robust residual generation and evaluation methods in observer-based fault detection systems. *Journal of Process Control*, 7(6):403–424, 1997.
[79] G. Gandanegara, X. Roboam, B. Sareni, and G. Dauphin-Tanguy. One model for one frequency range: Comparison of bond graph based simplification methods. In *International conference on bond graph modelling (ICBGM 03)*, volume 35. SCS publication, 2003. ISBN 1-56555-257-1.
[80] P. J. Gawthrop. Bicausal bond graphs. In *International Confernce on Bond Graph Modeling and Simulation (IBGM'95)*, pages 83–88, Las Vegas, USA, 1995.
[81] P. J. Gawthrop. Physical model-based control: A bond graph approach. *Journal of the Franklin Institute*, 332B(3):285–305, 1995.
[82] P. J. Gawthrop. Thermal modeling using mixed energy and pseudo bond graphs. *Proc. IMechE: Journal of Systems and Control engineering*, Vol. 213:201–216, 1999.
[83] P. J. Gawthrop. Physical interpretation of inverse dynamics using bicausal bond graphs. *Journal of the Franklin Institute*, 337(6):743–769, 2000.
[84] P. J. Gawthrop. Sensitivity bond graphs. *Journal of the Franklin Institute*, 337(7):907–922, 2000.
[85] P. J. Gawthrop. Bond graph based control using virtual actuators. *Proc. IMechE part-I: Journal of Systems and Control Engineering*, 218(4):251–268, 2004.

[86] P. J. Gawthrop, D. J. Balance, and G. Dauphin-Tanguy. Controllability indicators from bond graphs. In *Proc. International conference on bond graph modelling (ICBGM'99)*, volume 31 of *Simulation Series*. SCS Publication, 1999. ISBN 1-56555-155-9.
[87] P. J. Gawthrop, R. W. Jones, and S. A. Mackenzie. Identification of partially-known systems. *Automatica*, 28(4):831–836, 1992.
[88] P. J. Gawthrop and E. Ronco. Estimation and control of mechatronic systems using sensitivity bond graphs. *Control Engineering Practice*, 8(11):1237–1248, 2000.
[89] C. W. Gear. *Numerical Initial Value Problems in Ordinary Differential Equations*. Prentice-Hall, Englewood Cliffs, NJ, 1971.
[90] Gensym Corporation, Cambridge, MA. *G2 Reference Manual*, 2001.
[91] J. Gertler. Residual generation in model-based fault diagnosis. *Control-Theory and Advanced Technology*, 9(1):259–285, 1993.
[92] J. Gertler. Fault detection and isolation using parity relations. *Control Engineering Practice*, 5 No. 5:653–661, 1997.
[93] J. Gertler. *Fault Detection and Diagnosis in Engineering Systems*. Marcel Dekker, New York, 1998.
[94] J. Gertler and D. Singer. A new structural framework for parity equation-based failure detection and isolation. *Automatica*, 26(2):381–388, 1990.
[95] A. K. Ghosh, A. Mukherjee, and M. A. Faruqi. Computation of driving efforts for mechanisms and robots using bond graphs. *Trans. ASME, J. Dynamic Systems, Measurement and Control*, 113(4):744–748, 1991.
[96] S. K. Ghoshal. *Model-based Fault Diagnosis and Accommodation using Analytical Redundancy: A Bond Graph Approach*. PhD thesis, Indian Institute of Technology, Kharagpur, May 2006.
[97] S. K. Ghoshal and A. K. Samantaray. Multiple fault disambiguations through parameter estimation: A bond graph model-based approach. *International Journal of Intelligent Systems Technologies and Applications*, 4(3), 2008.
[98] C. A. González, G. Acosta, and C. De Prada. Knowledge based process control supervision and diagnosis: the AEROLID approach. *Expert Systems with applications*, 14:371–383, 1998.
[99] C. A. González, B. P. Junquera, G. A. Lazo, and C. L. Bello. On-line industrial supervision and diagnosis, knowledge level description and experimental results. *Expert Systems with Applications*, 20(2):117–132, 2001.
[100] J. J. Granda. The role of bond graph modeling and simulation in mechatronics systems: An integrated software tool: CAMP-G, MATLAB-SIMULINK. *Mechatronics*, 12:1271–1295, 2002.
[101] J. J. Granda. The Camp-G/Matlab-Simulink computer generated solution of bond graph derivative causality. In *Proc. ICBGM 03*, volume 35 of *Simulation Series*, pages 163–171, 2003. ISBN: 1-56555-257-1.
[102] H. Haffaf, B. Ould Bouamama, and G. Dauphin-Tanguy. Matroid algorithm for monitorability analysis of bond graphs. *Journal of the Franklin Institute*, 343(1):111–123, 2006.
[103] F. Hamelin and D. Sauter. Robust fault detection in uncertain dynamic systems. *Automatica*, 36(11):1747–1754, 2000.
[104] H. Hammouri, M. Kinnaert, and E. H. El Yaagoubi. *Lecture Notes in Computer Science*, volume 244, chapter Application of nonlinear observers to fault detection and isolation, pages 423–443. Springer:Berlin, 1999.
[105] B. Heim, S. Gentil, B. Celse, S. Cauvin, and L. Travé-Massuyès. FCC diagnosis using several causal and knowledge based models. In N. Eva Wu, editor, *SAFEPROCESS 2003*, pages 711–716, Washington DC, USA, June 2003. IFAC.
[106] C. Heny, D. Simanca, and M. Delgado. Pseudo-bond graph model and simulation of a continuous stirred tank reactor. *Journal of the Franklin Institute*, 337(1):21–42, 2000.
[107] D. M. Himmelblau. *Fault Detection and Diagnosis in Chemical and Petrochemical Processes*. Elsevier, Amsterdam, 1978.

[108] R. M. Hirschorn. Invertibility of multivariable nonlinear control systems. *IEEE Trans. Automatic Control*, 24(6):855–865, 1979.

[109] D. T. Horak. Failure detection in dynamic systems with modelling errors. *Journal of Guidance, Control, and Dynamics*, 11(6):508–516, 1998.

[110] S. Y. Huang and K. Youcef-Toumi. Zero dynamics of physical systems from bond graph models - part I: SISO systems. *Journal of Dynamic Systems, Measurement and Control, Transactions of the ASME*, 121(1):10–16, 1999.

[111] S. Y. Huang and K. Youcef-Toumi. Zero dynamics of physical systems from bond graph models - part II: MIMO systems. *Journal of Dynamic Systems, Measurement and Control, Transactions of the ASME*, 121(1):18–26, 1999.

[112] P. Ioannou and J. Sun. *Robust Adaptive Control*. PTR Prentice-Hall, Upper Saddle River, NJ, 1996. ISBN: 0134391004.

[113] R. Isermann. Process fault detection based on modelling and estimation methods : A survey. *Automatica*, 20:387–404, 1994.

[114] R. Isermann and B. Freyermuth. Process fault detection based on process model knowledge. Part I: Principles for fault diagnosis with parameter estimation. *ASME Transactions on Dynamic Systems, measurement and Control*, 113:620–626, 1991.

[115] R. Isermann and B. Freyermuth. Process fault detection based on process model knowledge. Part II: Case study experiments. *ASME Transactions on Dynamic Systems, measurement and Control*, 113:627–638, 1991.

[116] R. Issakovitch, V. Loguinov, and V. Popadko. *Chemical and gas process control*. Nedra, Moscow (in Russian), 1983.

[117] M. E. Janusz and V. Venkatasubramanian. Automatic generation of qualitative descriptions of process trends for fault detection and diagnosis. *Engineering Applications of Artificial Intelligence*, 4(5):329–339, 1991.

[118] M. Ji, Z. Zhang, G. Biswas, and N. Sarkar. Hybrid fault adaptive control of a wheeled mobile robot. *IEEE/ASME Transactions on Mechatronics*, 8(2):226–233, 2003.

[119] B. J. Joseph and H. R. Martens. The method of relaxed causality in the bond graph analysis of nonlinear systems. *Journal of Dynamic Systems, Measurement, and Control*, 1:95–99, 1974.

[120] W. S. Jung, S. H. Han, and J. Ha. A fast BDD algorithm for large coherent fault trees analysis. *Reliability Engineering and System Safety*, 83(3):369–374, 2004.

[121] D. Karnopp. Complementary formulations in vibrations: bond graph structures and modal transformations. *Journal of the Franklin Institute*, 310(6):303–316, 1980.

[122] D. C. Karnopp. Bond graphs in control: Physical state variables and observers. *Journal of the Franklin Institute*, 308(3):221–234, 1979.

[123] D. C. Karnopp. State variables and pseudo-bond graphs for compressible thermo-fluid systems. *Journal of Dynamic Systems, Measurement and Control*, 101(3):201–204, 1979.

[124] D. C. Karnopp. Computer simulation of stick-slip friction in mechanical dynamic systems. *Trans. ASME, Journal of Dynamic Systems, Measurement and Control*, 107:100–103, 1985.

[125] D. C. Karnopp. Bond graph models for electrochemical energy storage: Electrical, chemical and thermal effects. *Journal of the Franklin Institute*, 327:983–992, 1990.

[126] D. C. Karnopp and S. Azerbaijani. Pseudo bond graphs for generalised compartmental models in engineering and physiology. *Journal of the Franklin Institute*, 312(2):95–108, 1981.

[127] D. C. Karnopp, D. Margolis, and R. Rosenberg. *Systems Dynamics: A Unified Approach*. John Wiley, New York, second edition, 1990.

[128] D. C. Karnopp, D. L. Margolis, and R. C. Rosenberg. *System Dynamics: Modeling and Simulation of Mechatronic Systems*. Wiley, New York, 2000.

[129] M. Khemliche, B. Ould Bouamama, and H. Haffaf. Sensor placement for component diagnosability using bond-graph. *Sensors and Actuators A: Physical*, 132(2):547–556, 2006.

[130] M. Kinnaert. Design of redundancy relations for failure detection and isolation by constrained optimization. *International Journal of Control*, 63(3):609–622, 1996.

[131] M. Kinnaert. Robust fault detection based on observers for bilinear systems. *Automatica*, 35(11):1829–1842, 1999.

[132] M. Kinnaert and Y. Peng. Residual generator for sensor and actuator fault detection and isolation, A frequency domain approach. *International Journal of Control*, 61(6):1423–1435, 1995.

[133] A. Kobi, S. Nowakowski, and J. Ragot. Fault detection-isolation and control reconfiguration. *Mathematics and Computers in Simulation*, 37(2-3):111–117, 1994.

[134] A. Kobi, M. Ouladsine, and J. Ragot. Fault detection in ARX process. In *Proc. IEEE International Conference on Systems, Man and Cybernetics*, volume 3, pages 394–399, Le Touquet, France, 1993.

[135] T. Kohda, K. Inoue, and H. Asama. Computer aided failure analysis using system bond graphs. In *International Conference on Bond Graph Modeling and Simulation (ICBGM'01)*, pages 71–76. Simulation Series, Vol.33, No.1, ISBN 1-56555-103-6, 2001.

[136] T. Kohda, H. Katsubi, H. Fujihara, and K. Inoue. Identification of system failure causes using bond graph models. *IEEE Transactions on Systems, Man, and Cybernetics*, 5:269–274, 1993.

[137] J. Koj. The fault sources of pneumatic servo-motor control valve assembly. In *III Polish national conference on diagnostic of industrial processes*, pages 415–419, Jurata, Poland, 1998.

[138] P. V. Kokotovic, R. E. O Malley, and P. Sannuti. Singular perturbations and order reduction in control theory. An overview. *Automatica*, 12:123–132, 1976.

[139] J. Korbicz, J. M. Koscielny, Z. Kowalczuk, and W. Cholewa. *Fault Diagnosis: Models, Artificial Intelligence, Applications*. Springer-Verlag, Berlin, 2004. ISBN: 3540407677.

[140] M. Kothare, B. Mettler, M. Morari, P. Bendotti, and C. M. Falinower. Level control in the steam generator of a nuclear power plant. *IEEE Transactions on Control Systems Technology*, 8(1):55–69, 2000.

[141] M. Krysander. *Design and Analysis of Diagnostic Systems Utilizing Structural Methods*. PhD thesis, Linköpings universitet, 2003.

[142] J.D. Lamb, Asher G.M., and Woodall D.R. Causal loops and mason's rule for bond graphs. In *Proc. ICBGM' 93*, volume 25(2) of *Simulation series*, pages 67–72. SCS publication, 1993. ISBN:1-56555-019-6.

[143] I. D. Landau. *Identification et Commande Des Systèmes*. Hermes, Paris, 1988.

[144] W. Lee, C. Park, and G. Kelly. Fault detection in an air-handling unit using residual and recursive parameter identification methods. *ASHRAE Transactions (Part 1)*, 102(1):528–539, 1996.

[145] J. Lefévre and J. Barreto. A mixed block diagram bond graph approach for biochemical models with mass action rate law kinetics. *J. Franklin Institute*, 319(1/2):201–215, January/February 1985.

[146] W. Li, H. Raghavan, and S. Shah. Subspace identification of continuous time models for process fault detection and isolation. *Journal of Process Control*, 13(5):407–421, 2003.

[147] C. T. Lin. Structural controllability. *IEEE Trans. Automat. Control*, AC-19:201–208, 1974.

[148] S. F. Lin and A. P. Wang. Design of observers with unknown inputs using eigenstructure assignment. *International Journal of Systems Science*, 31(6):705–711, 2000.

[149] M. Lind. Modeling goals and functions of complex industrial plants. *Applied Artificial Intelligence*, 8:258–283, 1994.

[150] D. Linkens and H. Wang. Qualitative bond graph reasoning in control engineering: Fault diagnosis. In *International Conference on Bond Graph Modeling and Simulation (ICBGM'95)*, volume 27 of *Simulation Series*, pages 189–194, 1995. ISBN 1-56555-037-4.

[151] C.H. Lo, Y.K. wong, and A.B. Rad. Model-based fault diagnosis in continuous dynamic systems. *ISA Transactions*, 43:459–475, 2004.

[152] L. S. Louca. Bond graph based modal representations and model reduction of lumped parameter systems. *Simulation Modelling Practice and Theory, Special Issue: Bond Graph Modelling*, To appear, 2008.

[153] L. S. Louca and J. L. Stein. Energy-based model reduction of linear systems. In *Proc. International Conference on Bond Graph Modelling (ICBGM 99)*, volume 31. SCS publication, 1999. ISBN 1-56555-155-9.
[154] L. S. Louca and J. L. Stein. Ideal physical element representation from reduced bond graphs. *Proc. IMechE, Part I: Journal of Systems and Control Engineering*, 216:73–83, 2002.
[155] J. Lunze. Process supervision by means of qualitative models. *Annual Reviews in Control*, 24:41–54, 2000.
[156] J. Lunze, B. Nixdorf, and H. Richter. Hybrid system process supervision by means of a hybrid model. *Journal of Process Control*, 11(1):89–104, 2001.
[157] J. M. Maciejowski. Modelling and predictive control: Enabling technologies for reconfiguration. *Annual Reviews in Control*, 23:13–23, 1999.
[158] C. D. Maciel and C. Ritter. TCP/IP networking in process control plants. *Computers and Industrial Engineering*, 35(3-4):611–614, 1998.
[159] E. J. Manders, S. Narasimhan, G. Biswas, and P. J. Mosterman. A combined qualitative/quantitative approach for fault isolation in continuous dynamic systems. In *Proc. SAFEPROCESS'00*, pages 1074–1079, Budapest, Hungary., 2000.
[160] D. L. Margolis and G. E. Young. Reduction of models of large scale lumped structures using normal modes and bond graphs. *Journal of the Franklin Institute*, 304(1):65–79, 1977.
[161] K. Medjaher. *Contribution de l'Outil Bond Graph pour la Conception de Systémes de Supervision des Processus Industriels*. PhD thesis, Université des Sciences et Technologies de Lille, 2005.
[162] K. Medjaher, A. K. Samantaray, and B. Ould Bouamama. Diagnostic bond graphs for direct residual evaluation. In *International Conference on Bond Graph Modeling and Simulation (ICBGM'05)*, pages 307–312. Simulation Series, Vol.37, No.1, ISBN: 1-56555-287-3, 2005.
[163] K. Medjaher, A. K. Samantaray, B. Ould Bouamama, and M. Staroswiecki. Supervision of an industrial steam generator. Part II: Online implementation. *Control Engineering Practice*, 14(1):85–96, 2005.
[164] K. Medjaher, A.K. Samantaray, and B. Ould Bouamama. Bond graph model of a vertical u-tube steam condenser coupled with a heat exchanger. *Simulation Modelling Practice and Theory*, doi:10.1016/j.simpat.2007.10.002:In Press., 2008.
[165] R. Merzouki, K. Medjaher, M. A. Djeziri, and B. Ould Bouamama. Backlash fault detection in mechatronic system. *Mechatronics*, 17(6):299–310, 2007.
[166] R. Merzouki, B. Ould Bouamama, M. A. Djeziri, and M. Bouteldja. Modelling and estimation of tire-road longitudinal impact efforts using bond graph approach. *Mechatronics*, 17:93–108, 2007.
[167] C. A. Meyer, R. B. McClintock, G. J. Silvestri, and R. C. Spencer. *Steam Tables: Thermodynamic and Transport Properties of Steam, 6th Edition*. American Society of Mechanical Engineers (ASME), 6th edition, 1993.
[168] M. Modarres and S. W. Cheon. Function-centered modeling of engineering systems using the goal tree-sucess tree technique and functional primitives. *Reliability Engineering & System Safety*, 64(2):181–200, May 1999.
[169] P. O. Moksnes. *Modeling Two-Phase Thermo-Fluid Systems Using Bond Graph*. Dr. ing thesis, University of Sciences and Technology, Department of Marine Engineering, Norway, 1997.
[170] R. E. Moore. *Interval Analysis*. Prentice Hall, New Jersey, 1966.
[171] P. J. Mosterman and G. Biswas. Diagnosis of continuous valued systems in transient operating regions. *IEEE Trans. on Systems, Man and Cybernetics*, 29(6):554–565, 1999.
[172] A Mukherjee and R. Karmakar. *Modelling and Simulation of Engineering Systems through Bond Graphs*. Alpha Sciences International, Pangbourne, UK, 2000.
[173] A Mukherjee, R. Karmakar, and A. K. Samantaray. *Bond graph in modeling, simulation and fault identification*. CRC Press, ISBN: 978-8188237968, 1420058657, FL, USA, 2006.
[174] A. Mukherjee and A. K. Samantaray. System modelling through bond graph objects on SYMBOLS 2000. In *International Conference on Bond Graph Modeling and Simulation (ICBGM'01)*, volume 33, pages 164–170. Simulation Series, 2001. ISBN 1-56555-103-6.

[175] A. N. Nahavandi and R. F. Von Rollen. A space dependent dynamic analysis of a BWR. *Nuclear Science Engineering*, 20:392–413, 1964.

[176] R. F. Ngwompo and P. J. Gawthrop. Bond graph-based simulation of non-linear inverse systems using physical performance specifications. *Journal of the Franklin Institute*, 336(8):1225–1247, 1999.

[177] R. F. Ngwompo and S. Scavarda. Dimensioning problems in system design using bicausal bond graphs. *Simulation Practice and Theory*, 7:577–587, 1999.

[178] R. F. Ngwompo, S. Scavarda, and D. Thomasset. Inversion of linear time-invariant SISO systems modelled by bond graph. *Journal of the Franklin Institute*, 333(2):157–174, 1996.

[179] R. F. Ngwompo, S. Scavarda, and D. Thomasset. Physical model-based inversion in control systems design using bond graph representation - part 1: theory. *Proc. IMechE Part I Journal of Systems and Control Engineering*, 215:95–103, 2001.

[180] R. F. Ngwompo, S. Scavarda, and D. Thomasset. Physical model-based inversion in control systems design using bond graph representation - part 2: applications. *Proc. IMechE Part I Journal of Systems and Control Engineering*, 215:105–112, 2001.

[181] W. Nusselt. Die Oberflächenkondensation des Wasserdampfes. *VDI Zeitschrift*, 60(27):541–546 and 569–575, 1916.

[182] A. V. Oppenheim and R. W. Schafer. *Discrete-Time Signal Processing*. Prentice- Hall, 1989. 311 - 312.

[183] A. Y. Orbak, E. Eskinat, and O. S. Turkay. Physical parameter sensitivity of system eigenvalues and physical model reduction. *Journal of the Franklin Institute*, 341(7):631–655, 2004.

[184] A. Y. Orbak, O. S. Turkay, E. Eskinat, and K. Youcef-Toumi. Model reduction in the physical domain. *Proceedings of the Institution of Mechanical Engineers. Part I: Journal of Systems and Control Engineering*, 217(6):481–496, 2003.

[185] A. W. Ordys, A. W. Pike, M. A. Johnson, R. M. Katebi, and M. J. Grimble. *Modelling and Simulation of Power Generation Plants*. Springer-Verlag, 1994.

[186] J. OReilly. *Observers for Linear Systems*. Academic press,New York, 1983.

[187] B. Ould Bouamama. Model builder for thermo-fluid systems using a bond graph and functional modelling. In *13th European Simulation Symposium, Simulation in Industry, ESS'01*, pages 822–826, Marseille, France, 2001. ISBN-90-77039-02-3.

[188] B. Ould Bouamama. Bond graph approach as analysis tool in thermofluid model library conception. *Journal of the Franklin Institute*, 340(1):1–23, 2003.

[189] B. Ould Bouamama, K. Medjaher, M. Bayart, A. K. Samantaray, and B. Conrard. Fault detection and isolation of smart actuators using bond graphs and external models. *Control Engineering Practice*, 13(2):159–175, 2005.

[190] B. Ould Bouamama, K. Medjaher, A. K. Samantaray, and M. Staroswiecki. Supervision of an industrial steam generator. Part I: Bond graph modelling. *Control Engineering Practice*, 14(1):71–83, 2005.

[191] B. Ould Bouamama, A. K. Samantaray, K. Medjaher, M. Staroswiecki, and G. Dauphin-Tanguy. Model builder using functional and bond graph tools for FDI design. *Control Engineering Practice*, 13(7):875–891, 2005.

[192] B. Ould Bouamama, A. K. Samantaray, M. Staroswiecki, and G. Dauphin-Tanguy. Derivation of constraint relations from bond graph models for fault detection and isolation. In *International Conference on Bond Graph Modeling and Simulation (ICBGM'03)*, pages 104–109. Simulation Series Vol.35, No.2, ISBN 1-56555-257-1, 2003.

[193] P. M. Pathak, A. Mukherjee, and A. Dasgupta. Impedance control of space robots using passive degrees of freedom in controller domain. *Transactions of the ASME: Journal of Dynamic Systems, Measurement and Control*, 127(4):564–578, 2005.

[194] P. M. Pathak, A. Mukherjee, and A. Dasgupta. Impedance control of space robot. *International Journal of Modelling and Simulation*, 26(4):316–322, 2006.

[195] R. J. Patton and J. Chen. A review of parity space approaches to fault diagnosis. In *Proc. SAFEPROCESS 91*, volume 1, pages 239–255, 1991.

[196] R. J. Patton and J. Chen. Observer-based fault detection and isolation: Robustness and applications. *Control Engineering Practice*, 5(5):671–682, 1997.

[197] R. J. Patton, P. M. Frank, and R. N. Clark. *Fault Diagnosis in Dynamic Systems, Theory and Applications*. Prentice-Hall, Englewood Cliff, NJ, 1989.

[198] R. J. Patton, P. M. Frank, and R. N. Clark. *Issues of Fault Diagnosis for Dynamic Systems*. Springer-Verlag London Limited, 2000.

[199] H. M. Paynter. *Analysis and design of Engineering Systems*. M.I.T. Press, 1961.

[200] S. Péneau, J. P. Humeau, and Y. Jarny. Front motion and convective heat flux determination in a phase change process. *Inverse Problems in Engineering*, 4(1):53–91, 1996.

[201] I.R. Petersen and D.C. McFarlane. A methodology for robust fault detection in dynamic systems. *Control Engineering Practice*, 12:123–138, 2004.

[202] C. Pichardo-Almarza, A. Rahmani, G. Dauphin-Tanguy, and M. Delgado. High gain observers for non-linear systems modelled by bond graphs. *Proc. IMechE Part-I: J. Systems and Control Engineering*, 219:477–498, 2005.

[203] C. Pichardo-Almarza, A. Rahmani, G. Dauphin-Tanguy, and M. Delgado. Luenberger observers for linear time-invariant systems modelled by bond graphs. *Mathematical and Computer Modelling of Dynamical Systems*, 12(2-3):219–234, 2006.

[204] A. Pos, P. Borst, J. Top, and H. Akkermans. Reusability of simulation models. *Knowledge-Based Systems*, 9(2):119–125, 1996.

[205] F. Leon Pritchard. On implicit systems of differential equations. *Journal of Differential Equations*, 194(2):328–363, 2003.

[206] V. Puig and J. Quevedo. Passive robust fault detection using fuzzy parity equations. *Mathematics and Computers in Simulation*, 60:193–207, 2002.

[207] V. Puig, J. Quevedo, and S. Tornil. Robust fault detection: Active versus passive approaches. In *Proc. SAFEPROCESS 00, Budapest, Hungary*, 2000.

[208] V. Puig, A. Stancu, T. Escobet, F. Nejjari, J. Quevedo, and R.J. Patton. Passive robust fault detection using interval observers: Application to the DAMADICS benchmark problem. *Control Engineering Practice*, 14(6):621–633, 2006.

[209] J. Ragot and D. Maquin. An algorithm for obtaining redundancy equation of LTI systems. *Automatica*, 30(3):537–542, 1994.

[210] A. Rahmani, C. Sueur, and G. Dauphin-Tanguy. Pole assignment for systems modelled by bond graph. *Journal of the Franklin Institute*, 331(3):299–312, 1994.

[211] A. Rahmani, C. Sueur, and G. Dauphin-Tanguy. Approche des bond graphs pour l'analyse structurelle des systèmes linéaires,. *Linear Algebra and its Applications*, 259:101–131, 1997.

[212] P. Ralston, G. DePuy, and J. H. Graham. Computer-based monitoring and fault diagnosis: A chemical process case study. *ISA Transactions*, 40(1):85–98, 2001.

[213] F. Rambeaux, F. Hamelin, and D. Sauter. Optimal thresholding for robust fault detection of uncertain systems. *International Journal of Robust and Nonlinear Control*, 10:1155–1173, 2000.

[214] M. Robert. *Capteurs Intelligents et Méthodologie d'Evaluation*. Hermes, Paris, 1993.

[215] G. N Roberts and C. A. J. Tubb. Computer-aided design of mechatronic systems using 20-SIM 3.0. In G.N Roberts and C.A.J. Tubb, editors, *2nd Workshop on European Scientific and Industrial Collaboration (WESIC'99)*, Newport (S. Wales), UK, 1999.

[216] W. M. Rohsenow, J. P. Hartnett, and E. N. Ganic. *Handbook of Heat Transfer Fundamentals*. McGraw-Hill Book Company, 2nd edition, 1985.

[217] R. C. Rosenberg. State-space formulation for bond graph models of multiport systems. *Journal of Dynamic Systems, Measurement, and Control*, 93(1):35–40, 1971.

[218] R. C Rosenberg and D. C. Karnopp. A definition of the bond graph language. *Journal of Dynamic Systems, Measurement, and Control, Trans. ASME*, 3:179–182, 1972.

[219] B. Samanta and A. Mukherjee. Analysis of acoustoelastic systems using modal bond graphs. *Trans. ASME, J. Dynamic Systems, Measurement and Control*, 112(1):108–115, 1990.

[220] A. K. Samantaray. The home of bond graphs, web page: www.bondgraph.info, Jan 2007.

[221] A. K. Samantaray. Modeling and analysis of preloaded liquid spring/damper shock absorbers. *Simulation Modelling Practice and Theory (special issue on bond graph modelling)*, In Press, 2008.
[222] A. K. Samantaray and S. K. Ghoshal. Sensitivity bond graph approach to multiple fault isolation through parameter estimation. *Proceedings of IMechE, Part-I: Journal of Systems and Control Engineering*, 221(4):577–587, 2007.
[223] A. K. Samantaray and S. K. Ghoshal. Bicausal bond graphs for supervision: From fault detection and isolation to fault accommodation. *Journal of the Franklin Institute*, 345(1):1–28, 2008.
[224] A. K. Samantaray, S. K. Ghoshal, and S. Chakraborty. Bond graph model based design of supervision algorithm for distributed fault tolerant control systems. *Int. J. Automation and Control*, 1:28–47, 2007.
[225] A. K. Samantaray, S. K. Ghoshal, K. Medjaher, and B. Ould Bouamama. Reconfiguration of an industrial steam generator using bond graph modelling. *Int. J. Modelling, Identification and Control*, 2(2):154–168, 2007.
[226] A. K. Samantaray, K. Medjaher, B. Ould Bouamama, M. Staroswiecki, and G. Dauphin-Tanguy. Component based modelling of thermo-fluid systems for sensor placement and fault detection. *SIMULATION: Transactions of The Society for Modeling and Simulation International*, 80(7-8):381–398, 2004.
[227] A. K. Samantaray, K. Medjaher, B. Ould Bouamama, M. Staroswiecki, and G. Dauphin-Tanguy. Diagnostic bond graphs for online fault detection and isolation. *Simulation modelling practice and theory*, 14(3):237–262, 2005.
[228] A. Seidenberg. An elimination theory for differential algebra. *University of California, Publication Mathematics*, 3:31–65, 1956.
[229] A. Sharon, N. Hogan, and D. E. Hardt. Controller design in the physical domain. *Journal of the Franklin Institute*, 328:697–721, 1991.
[230] Z. Shi, F. Gu, B. Lennox, and A.D. Ball. The development of an adaptive threshold for model-based fault detection of a nonlinear electro-hydraulic system. *Control Engineering Practice*, 13:1357–1367, 2005.
[231] R. N. Shields and J. B. Pearson. Structural controllability of multi-input linear systems. *IEEE Trans. Automut. Control*, AC-21(2):203–212, 1976.
[232] R. Shoureshi and K. McLaughlin. Application of bond graph to thermofluid processes and systems. *Trans. ASME, J. Dynamic Systems, Measurement and Control*, 107:241–245, 1985.
[233] K. Sia and A. Naamane. Bond graph: A suitable tool for component faults diagnosis. In *International Conference on Bond Graph Modeling and Simulation (ICBGM'03)*, pages 89–103. Simulation Series Vol.35, No.2, ISBN 1-56555-257-1, 2003.
[234] C. Sié Kam and G. Dauphin-Tanguy. Bond graph models of structured parameter uncertainties. *Journal of the Franklin Institute*, 342(4):379–399, 2005.
[235] L. M. Silverman. Inversion of multivariable linear systems. *IEEE Trans. Autom. Control*, 14(3):270–276, 1969.
[236] S. Simani and C. Fantuzzi. Dynamic system identification and model-based fault diagnosis of an industrial gas turbine prototype. *Mechatronics*, 16(6):341Ű363, 2006.
[237] S. Simani, C. Fantuzzi, and R. J. Patton. *Model-based Fault Diagnosis in Dynamic Systems Using Identification Techniques*. Springer-Verlag, 2003.
[238] Simulink. *Dynamic System Simulation for Matlab*. Math Works, France, 1997.
[239] R. Slier and P. M. Frank. *Robust Observer-Based Fault Diagnosis in Non Linear Uncertain Systems*. Springer Verlag, advances in fault diagnosis for dynamic systems: model based approaches edition, 2000.
[240] M. Staroswiecki. Quantitative and qualitative models for fault detection and isolation. *Mechanical Systems and Signal Processing*, 14(3):301–325, 2000.
[241] M. Staroswiecki and M Bayart. Models and languages for the interoperability of smart instruments. *Automatica*, 32(6):859–873, 1996.
[242] M. Staroswiecki and G. Comtet-Varga. Analytical redundancy relations for fault detection and isolation in algebraic dynamic systems. *Automatica*, 37:687–699, 2001.

[243] J.L Stein and B.H. Wilson. An algorithm for obtaining proper models of distributed and discrete systems. *Transactions of the ASME: Journal of Dynamic Systems Measurement and Control*, 117(4):534–540, 1995.

[244] J. Stephan, M. Bodson, and J. Chiasson. Real-time estimation of the parameters and fluxes of induction motors. *IEEE Transaction on Industry Applications*, 30(3):746–759, 1994.

[245] C. Sueur and G. Dauphin-Tanguy. Structural controllability/observability of linear systems represented by bond graphs. *Journal of the Franklin Institute*, 326(6):869–883, 1989.

[246] C. Sueur and G. Dauphin-Tanguy. Bond graph approach for structural analysis of MIMO linear systems. *Journal of the Franklin Institute*, 328(1):55–70, 1991.

[247] C. Sueur and G. Dauphin-Tanguy. Controllability indices for structured systems. *Linear Algebra and its Applications*, 250:275–287, 1997.

[248] A. Sundarraman and R. Srinivasan. Monitoring transitions in chemical plants using enhanced trend analysis. *Computers and Chemical Engineering*, 27(10):1455–1472, 2003.

[249] P. Supavatanakul, J. Lunze, V. Puig, and J. Quevedo. Diagnosis of timed automata: Theory and application to the DAMADICS actuator benchmark problem. *Control Engineering Practice*, 14(6):609–619, 2006.

[250] M. Tagina, J. P. Cassar, G. Dauphin-Tanguy, and M. Staroswiecki. Bond graph models for direct generation of formal fault detection systems. *Systems Analysis, Modelling, Simulation.*, 23:1–17, 1996.

[251] M. Tagina, J. Ph. Cassar, G. Dauphin-Tangy, and M. Staroswiecki. Monitoring of systems modelled by bond graph. In *International Conference on Bond Graph Modeling and Simulation (ICBGM'95)*, pages 275–280. Simulation Series, Vol.27, No.1, ISBN 1-56555-037-4, 1995.

[252] P. Taillibert. Various improvement to diagnosing with temporal bands. In *9th Int. Workshop on Principles of Diagnosis, DX'98*, Massachussets, 1998.

[253] P. Taillibert. Residual evaluation without derivative computation. In A.M. Edelmayer and C. Banyasz, editors, *SAFEPROCESS 2000*. IFAC, Pergamon Press, ISBN:0-08043-250-6, 2000.

[254] P. Taillibert and E. Loiez. Polynomial temporal band sequences for analog diagnosis. In *International Joint. Conf. On Artificial Intelligence, IJCAI'97*, Nagoya, Japan, August 1997.

[255] S. Tan and J. Vandewalle. Inversion of singular systems. *IEEE Trans. Circuits and Systems*, 35(5):583–587, 1988.

[256] J. U. Thoma. *Introduction to Bond Graphs and their Applications*. Pergamon Press, Oxford; New York, 1975.

[257] J. U. Thoma. *Simulation by Bondgraphs. Introduction to a Graphical Method.* Springer-verlag, Berlin Heidelberg, 1990.

[258] J. U Thoma and B. Ould Bouamama. *Modelling and Simulation in Thermal and Chemical Engineering. Bond Graph Approach*. Springer Verlag, Telos, 2000.

[259] R. Tomovic and M. Vukobratovic. *General Sensitvity Theory, Modern Analytic and Computational Methods in Science and Mathematics*, volume 35. Elsevier, New York, 1972.

[260] Twentesim. *Users Manual of Twentesim (20sim)*. Controllab Products Inc., Enschede, Netherlands, 1996.

[261] C. Ündey, E. Tatara, and Ali Çinar. Real-time batch process supervision by integrated knowledge-based systems and multivariate statistical methods. *Engineering Applications of Artificial Intelligence*, 16(5-6):555–566, 2003.

[262] J. Van Amerongen and P. Breedveld. Modelling of physical systems for the design and control of mechatronic systems. *Annual Reviews in Control*, 27(1):87–117, 2003.

[263] J. Van Dijk and P. C. Breedveld. Simulation of system models containing zero-order causal paths-I. classification of zero-order paths. *Journal of the Franklin Institute*, 328:959–979, 1991.

[264] J. Van Dijk and P. C. Breedveld. Simulation of system models containing zero-order causal paths-II. numerical implications of class 1 zero-order paths. *Journal of the Franklin Institute*, 328:981–1004, 1991.

[265] V. Venkatasubramanian, R. Rengaswamy, K. Yin, and S. N. Kavuri. A review of process fault detection and diagnosis. Part I: Quantitative model-based methods. *Computers and Chemical Engineering*, 27:293–311, 2003.
[266] V. Venkatasubramanian, R. Rengaswamy, K. Yin, and S. N. Kavuri. A review of process fault detection and diagnosis. Part II: Qualitative models and search strategies. *Computers and Chemical Engineering*, 27:313–326, 2003.
[267] V. Venkatasubramanian, R. Rengaswamy, K. Yin, and S. N. Kavuri. A review of process fault detection and diagnosis. Part III: Process history based methods. *Computers and Chemical Engineering*, 27:327–346, 2003.
[268] D. Vink, D. Ballance, and P. Gawthrop. Bond graphs in model matching control. *Mathematical and Computer Modelling of Dynamical Systems*, 12(2-3):249–261, 2006.
[269] N. Viswanadham and R. Srichander. Fault detection using unknown-input observers. *Control-Theory and Advanced Technology*, 3(2):91–101, 1987.
[270] J. Wagner and R. Shoureshi. Nonlinear modeling and observer design for thermofluid system diagnostics. *International Journal of Modelling & Simulation*, 13(1):39 – 45, 1993.
[271] H. Wang and D. Linkens. Intelligent supervisory control, a qualitative bond graph reasoning approach. In *World scientific series in Robotics and intelligent Systems*, volume 14. World Scientific, Sinagapore, 1996.
[272] H. B. Wang, J. L. Wang, and J. Lam. Robust fault detection observer design: Iterative LMI approaches. *Journal of Dynamic Systems, Measurement, and Control*, 129:77–82, 2007.
[273] R. Wang. *Statistical theory*. Xian Jiaotong University Press:China, 2003.
[274] J. Watton. The dynamic performance of an electro-hydraulic servo-valve/motor system with transmission line effects. *ASME Journal of Dynamic Systems, Measurement, and Control*, 109:14–18, 1987.
[275] A. S. Willsky. A survey of design methods for failure detection systems. *Automatica*, 12:601–611, 1976.
[276] C. Wollhaf, K. Schultz, and S. Engell. BaSiP-batch process simulation with dynamically reconfigured process dynamics. *Computer and Chemical Engineering supplement*, 20:S1281–S1286, 1996.
[277] S. T. Wu and K. Youcef-Toumi. On relative degrees and zero dynamics from physical system modeling. *Journal of Dynamic Systems, Measurement and Control, Transactions of the ASME*, 117(2):205–217, 1995.
[278] E. Benjamin Wylie, Keith W. Bedford, and Victor L. Streeter. *Fluid Mechanics, 9th Edition*. WCB/McGraw-Hill, Boston, 1998. ISBN: 0071156003.
[279] H. Yang, Z.i Mao, and B. Jiang. Model-based fault tolerant control for hybrid dynamic systems with sensor faults. *ACTA Automatica Sinica*, 32(5):680–685, 2006.
[280] J. C. Yang and D. W. Clarke. The self-validating actuator. *Control Engineering Practice*, 7:249–260, 1999.
[281] S. H. Yang, X. Chen, and J. L. Alty. Design issues and implementation of internet-based process control systems. *Control Engineering Practice*, 11(6):709–720, 2003.
[282] Z. Yang and J. Stoustrup. Robust reconfigurable control for parametric and additive faults with FDI uncertainties. In *Proc. IEEE Conf. Decision and Control*, pages 4132–4137, 2000.
[283] P. Young. Parameter estimation for continuous time models - A survey. *Automatica*, 17(1):23 –39, 1981.
[284] A. Zaidi, N. Zanzouri, and M. Tagina. Graphical approaches for modelling and diagnosis of hybrid dynamic systems. *WSEAS Transactions on Systems*, 5(10):2322–2327, 2006.
[285] B. Zupancic. Extension software for real-time control system design and implementation with MATLAB-SIMULINK. *Simulation Practice and Theory*, 6(8):703–719, 1998.

Index

A

Abnormal event management, 2
Abrupt fault, 6, 178, 256, 399
Activated bond, 43
Active bond graph elements, 37
Actuator placement, 315, 329
Actuator sizing, 442, 443
Adaptability, 7
Adaptive control, 82, 84, 423
Adaptive threshold, 347, 374, 407
Additive fault, 441
Alarm, 196, 220, 259
  generation, 196
  interpretation, 8, 196
Algebraic loop, 101, 153, 239, 276
Amplifier of Effort, 298
Amplifier of Flow, 298
Analog differentiator, 284
Analytical redundancy relation, 180, 195
Augmented controller and observer, 104
Automated modeling, 59
Automaton, 336

B

Balanced reduction, 142, 143
Bank of observers, 181
Base sensors, 332
Basis ARRs, 195
Behavior model, 128
Behavioral equations, 32
Bicausality, 47, 130, 271, 297, 435, 436
Bipartite graph, 198, 201
Bisection technique, 411
Block diagram, 21, 93
Boiler, 236, 342
Bond
  activation, 43
  definition, 15
Bond graph
  bicausal, 47
  causal, 44
  definition, 15
  energy variables, 20
  for chemistry, 19
  for economic systems, 20
  for thermal processes, 53
  power variables, 19
  state variables, 48
  structure, 50
  thermofluid application, 68
  word bond graph, 18
Bond graph modeling, 10

C

C-element, 32
  economics, 35
Calculability property, 435
Caley-Hamilton theorem, 192
Causal loop, 101
Causal matching, 201, 202, 214
Causal path, 100, 274, 279, 331, 332
Causal stroke, 44
Causality, 43
  assignment, 44, 45, 55, 291
  compulsory causality, 56
  derivative causality, 45, 56
  integral causality, 45
Causality inversion approach, 214
Cavitation, 326
Characteristic polynomial, 92, 113
Choked-flow, 326

Coherence vector, 196, 220, 441
  structured part, 441
  unstructured part, 441
Collocated control system, 447
Compliance , 33
Condenser, 239
Conduction, 30
Constitutive equations, 32
Constraint space, 444
Control loop, 278
Control valve, 321, 326
Controllability matrix, 114
Convection, 241
Coupling capacitor, 153, 277
Coupling Element for Thermo-Fluids, 329, 395
Cumulative sum, 180, 257, 263, 400
Current estimator, 112

D

Decision procedure, 196, 256, 313
Decision rules, 180
Decision support system, 179, 261
Dedicated observer scheme, 181
Deduced redundancy, 202, 316, 332
Degraded operating mode, 268
Detection delay, 179
Diagnosis, 2
Diagnostic bond graph, 271, 391
  coupling, 282, 294
  definition, 275
Diagnostic sensitivity bond graphs, 415
Differential algebraic loop, 277, 292
Direct redundancy, 202, 300, 316, 332
Discrete-time residual evaluator, 392
Driving point impedance, 447

E

Eigenvalue decomposition, 142
Emissivity, 396
Entropy, 31
Entropy generation, 31
Equilibrium point, 170, 171
Equipment availability, 270, 336
Error dynamics, 105, 112, 113
Expert system, 258
Explicit system inversion, 432, 434
External model, 316

F

Factorization approach, 184
False alarm, 178, 196, 220, 224, 265
Fault
  classification, 6
  definition, 4
Fault accommodation, 2, 3, 315, 442
Fault candidates, 220
Fault detection, 3
Fault detection and isolation, 2
Fault disambiguation, 158
Fault events, 283
Fault hypothesis, 7
  generation, 154, 160
  validation, 154, 313
Fault isolation, 3, 220, 381
Fault quantification, 437
Fault scenario simulation, 222
Fault signature, 195, 440
Fault signature matrix, 196, 219, 279, 311
Fault signature validation, 221
Fault tolerant control, 249, 315, 423, 442
  active approach, 424, 437
  passive approach, 424, 447
Feature extraction, 8
Feedback linearization, 84
Film thickness, 240
Finite zeroes, 128, 131
Flow induced vibration, 95, 107
Forced convection, 395
Forward model, 128, 131
Functional model, 317, 320
Functional redundancy, 215, 309

G

Gain scheduling control, 84
Gauss-Newton optimization, 410, 411
Generalized displacement, 20
Generalized likelihood ratio, 180
Generalized momentum, 20
Generalized observer scheme, 182
Geometrical analysis, 178, 180
Geometrical constraint, 443
Ghost controller, 449
Gradient search optimization, 410
Gyrator
  definition, 42
  modulation, 42

H

Hagen-Poiseuille equation, 200
Hankel norm, 142
Hard failure, 3, 4
Hardware redundancy, 331, 332

Hardware-in-the-loop simulation, 224
Heat transfer, 239, 242
Hessian matrix, 411
Hybrid system, 121, 124
Hydraulic momentum, 66

I

I-element, 35
  economics, 37
Identification, 3
Implicit function theorem, 435
Implicit residual generation, 391
Implicit system inversion, 427, 431
Incidence matrix, 202, 295
Incipient fault, 1, 6, 178, 256
Incremental bond graph, 375, 415
Infinite zeroes, 128, 129
Information bond, 43
Input reconstruction, 429, 442
Intermittent fault, 6, 399
Internal feedback loop, 375
Interval model, 260
Invariant subspaces, 193
Inverse Laplace transform, 95, 99
Inverse model, 128, 130, 132
Inverse system, 425, 427, 448
Isolability, 7, 197, 259

J

Junction, 38
  one, 39, 40
  zero, 39
Just-constrained subsystem, 202

K

Kinetic energy, 37
Knowledgebase, 163, 258

L

Least squares optimization , 373
Linear fractional transformation, 375, 376
Linearization, 170
Lyapunov's stability criterion, 83

M

Material redundancy, 243, 257
Minimum-phase system, 133
Misdetection, 178, 196
Mission, 316, 318, 326
Modal bond graph, 147
Modal decomposition, 147
Modal truncation, 142, 143, 147
Model based FDI, 10
Model coupling, 282, 294, 305
Model matching technique, 315, 424
Model order reduction
  activity index, 100, 144
Model predictive control, 83
Model refinement, 58
ModelBuilder software, 218
Modeling uncertainties, 178, 181
Mollier chart, 237
Monitorability, 196, 219
Monostable characteristics, 347
Moving average, 180
Multiple fault hypotheses, 313, 374, 440
Multiplicative fault, 4
Multiport
  bond graph, 14
  definition, 13
Multivariable control, 83

N

Nelder-Mead's simplex algorithm, 386, 409
Network communication, 260
Non-minimum-phase system, 133
Non-overlapping multiple faults, 440
Null modes, 114
Numerical residual, 279
Nusselt's formulae, 240
Nyquist frequency, 111, 144

O

Objective function, 373
Observability matrix, 116, 193
Observable subspace, 290
Observer based FDI, 180
Observer based residuals, 183
Operating mode, 338
Operating mode management, 319
Optimal control, 446
Optimization of least squares of ARRs, 387
Over-constrained subsystem, 202, 214
Overlapping multiple faults, 440
Overwhelming controller, 447

P

Parameter estimation, 373, 381, 435, 438
Parametric faults, 449
Parity space residuals, 191

Partial system identification, 383
Passive approach to fault detection, 347
Passive elements, 26
Physical model based control, 93, 95
PID control, 82
Pole placement, 104, 105, 109
Port
  definition, 14
Potential energy, 33
Power bonds
  definition, 15
Power modulator, 443
Power variables, 14
Practical fault signature matrix, 197, 224, 253
Predictive maintenance, 178
Prefilter, 428
Principal component analysis, 261
Progressive fault, 6, 399
Propering filter, 426
Pseudo bond graph, 23
Pseudo-inverse, 411, 414
Pump, 235

Q

Qualitative reasoning, 154
  antecedents, 154, 155, 158
  conflict, 155, 156, 158
  consequences, 154, 155, 158
  critical parameter list, 159
  fault candidate elimination, 158
  fault candidates, 158
  fault tree, 156
  initial fault set, 156

R

R-element, 28
  economics, 29
  modulated, 28
R-multiport , 31
Real Time Work Shop, 281
Receding horizon control, 83, 452
Reconfiguration, 249, 261, 268, 315, 338
Refrigerator subsystem, 393
Relative degree, 128, 131, 133, 134, 426, 430
Relief valve, 308
Remote monitoring, 260
Remote supervision, 260
Residual, 177, 180, 184, 187, 219
  evaluation, 179, 180, 218, 305
  generation, 179, 180, 212
  generator, 184, 391
  normalization, 381
  sensitivity, 178, 221, 253, 409
  sensor, 274
  sink, 274, 294
  threshold, 178, 196, 224, 256
Robust control, 423
Robust overwhelming control, 447
Robustness, 7, 82
Robustness property, 449
Root cause analysis, 8
Root loci, 98, 128, 129
RS multiport, 31

S

Sensitivity, 410
Sensitivity bond graph, 414
Sensitivity layers, 418
Sensor and actuator loss, 335
Sensor failure, 335
Sensor placement, 221, 259, 279, 300, 315, 329
Sequential probability ratio, 180
Service, 317, 319
Servo-motor, 324
Servo-valve, 306
Shanon's sampling theorem, 111
  Shanon frequency, 111
Shortest causal path, 134
Signal flow graph, 164, 448
  branches and gains, 85
  conversion, 84
  example, 87
  forward path, 90, 102
  loop gain, 90, 103
  Mason's gain rule, 89, 103
  nodes, 84
  receptors, 85
Signal sensor, 271
Silverman's algorithm, 425
Simulink®, 105, 281, 285
Single fault hypothesis, 157, 220, 312, 347, 437
Singular perturbation, 142
Singular value decomposition, 142
Smart actuator, 316, 320
Smart sensor, 316
Software, 59
Source
  effort, 38
  flow, 38
  modulated, 38
Source loading, 214
Source-Sensor, 130, 298
Sources of failure, 2

Specific enthalpy, 237
Specific volume, 237
Standard interconnection model, 375
State equation, 48
State space, 48
State variable, 57
State-transition matrix, 111
Steam expansion system, 238
Steam generator, 229, 339
Steam quality, 237, 242
Steam table, 237, 245
Stefan-Boltzmann Law, 395
Structural analysis, 113, 132, 197, 201
Structural controllability, 114, 331
Structural equations, 38
Structural independence, 195
Structural observability, 116, 331
Structural rank, 113, 118
Structurally independent residual, 197
Structured residuals, 197, 311
Supervision, 257
  definition, 2
System inversion, 442, 443

T

Technological specifications, 218
Temporal causal graph, 163
  fault hypothesis
    generation, 164
    validation, 166, 168
  hybrid diagnosis, 169
  propagation
    backward, 164
    forward, 164, 166, 168
  qualitative trend, 166, 168
Terminating node, 300, 305, 436, 437
TF-element, 41
  modulated , 41
Thermal equilibrium, 236
Thermodynamic equilibrium, 229
Threshold generator, 377
Trajectory tracking, 452
Transfer function, 128, 448
  from bond graph model, 101
  from signal flow graph, 89
  from state space model, 91
Transformer
  definition, 41
Tree graph, 159
  algebra node, 160
  characteristic functions, 161
  fault propagation, 162
  function node, 160
  qualitative equations, 160
  starting nodes, 161
  terminating nodes, 161
True bond graph, 19

U

Uncertainty in parameter estimation, 382
Under-constrained subsystem, 202, 250
Unicausality, 130, 298
Unit circle, 112, 113
Unknown input observer, 178, 185
User operating mode, 316, 318

V

Valve characteristics, 199, 238
Valve positioner, 322
Vector bond graph, 418
Vector bonds, 416
Vena contracta, 324, 326
Versions of service, 319
Virtual sensor, 272
Von Mises asymptote, 200

W

Word bond graph, 22, 53
  definition, 22

Z

z-Transform, 111
Zero dynamics, 133
Zero-order-causal-path, 277
Zero-order-hold , 111
Zero-phase digital filtering, 400

Other titles published in this series (continued):

Soft Sensors for Monitoring and Control of Industrial Processes
Luigi Fortuna, Salvatore Graziani, Alessandro Rizzo and Maria G. Xibilia

Adaptive Voltage Control in Power Systems
Giuseppe Fusco and Mario Russo

Advanced Control of Industrial Processes
Piotr Tatjewski

Process Control Performance Assessment
Andrzej W. Ordys, Damien Uduehi and Michael A. Johnson (Eds.)

Modelling and Analysis of Hybrid Supervisory Systems
Emilia Villani, Paulo E. Miyagi and Robert Valette

Process Control
Jie Bao and Peter L. Lee

Distributed Embedded Control Systems
Matjaž Colnarič, Domen Verber and Wolfgang A. Halang

Precision Motion Control (2nd Ed.)
Tan Kok Kiong, Lee Tong Heng and Huang Sunan

Optimal Control of Wind Energy Systems
Iulian Munteanu, Antoneta Iuliana Bratcu, Nicolaos-Antonio Cutululis and Emil Ceangă

Identification of Continuous-time Models from Sampled Data
Hugues Garnier and Liuping Wang (Eds.)

Dry Clutch Control for Automated Manual Transmission Vehicles
Pietro J. Dolcini, Carlos Canudas-de-Wit and Hubert Béchart
Publication due May 2008

Real-time Iterative Learning Control
Xu Jian-Xin, Sanjib K. Panda and Lee Tong Heng
Publication due July 2008

Model Predictive Control Design and Implementation Using MATLAB®
Liuping Wang
Publication due July 2008

Magnetic Control of Tokamak Plasmas
Marco Ariola and Alfredo Pironti
Publication due July 2008

Design of Fault-tolerant Control Systems
Hassan Noura, Didier Theilliol, Jean-Christophe Ponsart and Abbas Chamseddine
Publication due October 2008